Flexible Supercapacitors

Flexible Supercapacitors

Materials and Applications

Edited by

Guozhen Shen
State Key Laboratory for Superlattices and Microstructures
Beijing, China

Zheng Lou
State Key Laboratory for Superlattices and Microstructures
Beijing, China

Di Chen
University of Science and Technology
Beijing, China

WILEY

This edition first published 2022

Registered Office
John Wiley & Sons, Inc., 111 River Street, Hoboken, NJ 07030, USA

Editorial Office
111 River Street, Hoboken, NJ 07030, USA

For details of our global editorial offices, customer services, and more information about Wiley products visit us at www.wiley.com.

Wiley also publishes its books in a variety of electronic formats and by print-on-demand. Some content that appears in standard print versions of this book may not be available in other formats.

Library of Congress Cataloging-in-Publication Data
Names: Shen, Guozhen (Electrical engineer), editor. | Lou, Zheng, editor. | Chen, Di, editor.
Title: Flexible supercapacitors : materials and applications / edited by Guozhen Shen, State Key Laboratory for Superlattices and Microstructures, Beijing, China, Zheng Lou, State Key Laboratory for Superlattices and Microstructures, Beijing, China, Di Chen, University of Science and Technology, Beijing, China.
Description: First edition. | Hoboken, NJ : John Wiley & Sons, Inc., 2022. | Includes bibliographical references and index.
Identifiers: LCCN 2021052431 (print) | LCCN 2021052432 (ebook) | ISBN 9781119506164 (hardback) | ISBN 9781119506188 (obook) | ISBN 9781119506171 (epdf) | ISBN 9781119506157 (epub)
Subjects: LCSH: Supercapacitors. | Flexible electronics.
Classification: LCC TK7872.C65 F555 2022 (print) | LCC TK7872.C65 (ebook) | DDC 621.31/5–dc23/eng/20211207
LC record available at https://lccn.loc.gov/2021052431
LC ebook record available at https://lccn.loc.gov/2021052432

Cover Design: Wiley
Cover Image: © draganab/Getty Images

Set in 9.5/12.5pt STIXTwoText by Straive, Pondicherry, India

10 9 8 7 6 5 4 3 2 1

MIX
Paper from responsible sources
FSC® C013604

Contents

List of Contributors

Songshan Bi
Key Laboratory of Advanced Energy Materials Chemistry (Ministry of Education)
College of Chemistry, Nankai University
Tianjin, 300071, P. R. China

Hongmei Cao
Key Laboratory of Advanced Energy Materials Chemistry (Ministry of Education)
College of Chemistry, Nankai University
Tianjin, 300071, P. R. China

Cao Guan
Frontiers Science Center for Flexible Electronics, Institute of Flexible Electronics
Northwestern Polytechnical University
Xi'an, 710072, P.R. China

Mengmeng Hu
State Key Laboratory of Advanced Welding and Joining
Harbin Institute of Technology (Shenzhen)
Shenzhen 518055, Guangzhou, China; and
Flexible Printed Electronics Technology Centre
Harbin Institute of Technology (Shenzhen)
Shenzhen 518055, Guangzhou, China

Yan Huang
State Key Laboratory of Advanced Welding and Joining
Harbin Institute of Technology (Shenzhen)
Shenzhen 518055, Guangzhou, China; and
Flexible Printed Electronics Technology Centre
Harbin Institute of Technology (Shenzhen)
Shenzhen 518055, Guangzhou, China; and
School of Materials Science and Engineering
Harbin Institute of Technology (Shenzhen)
Shenzhen 518055, Guangzhou, China

Muhammad Imran
Department of Chemistry, Faculty of Science
King Khalid University
Abha 61413, Saudi Arabia

Muhammad S. Javed
School of Physical Science and Technology
Lanzhou University
Lanzhou 730000, PR China; and
Siyuan Laboratory, Guangdong Provincial Engineering Technology Research Center of Vacuum Coating Technologies and New Energy Materials, Department of Physics
Jinan University
Guangzhou 510632, PR China

Wen-Yong Lai
State Key Laboratory of Organic Electronics and Information Displays (SKLOEI), Institute of Advanced Materials (IAM)
Nanjing University of Posts & Telecommunications
9 Wenyuan Road, Nanjing 210023, Jiangsu, China; and
Frontiers Science Center for Flexible Electronics (FSCFE), MIIT Key Laboratory of Flexible Electronics (KLoFE)
Northwestern Polytechnical University
127 West Youyi Road, Xi'an 710072, Shaanxi, China

La Li
State Key Laboratory for Superlattices and Microstructures
Institute of Semiconductors, Chinese Academy of Science
Beijing 100083, China

Dun Lin
MOE of the Key Laboratory of Bioinorganic and Synthetic Chemistry, The Key Lab of Low-Carbon Chem and Energy Conservation of Guangdong Province, School of Chemistry
Sun Yat-Sen University, Guangzhou
Guangzhou, 510275, People's Republic of China

Jie Liu
State Key Laboratory of Advanced Welding and Joining
Harbin Institute of Technology (Shenzhen)
Shenzhen, 518055, Guangzhou, China; and
Flexible Printed Electronics Technology Centre
Harbin Institute of Technology (Shenzhen)
Shenzhen, 518055, Guangzhou, China

Qingjiang Liu
State Key Laboratory of Advanced Welding and Joining
Harbin Institute of Technology (Shenzhen)
Shenzhen, 518055, Guangzhou, China; and
Flexible Printed Electronics Technology Centre
Harbin Institute of Technology (Shenzhen)
Shenzhen, 518055, Guangzhou, China

Yao Liu
State Key Laboratory of Advanced Welding and Joining
Harbin Institute of Technology (Shenzhen)
Shenzhen, 518055, Guangzhou, China; and
Flexible Printed Electronics Technology Centre
Harbin Institute of Technology (Shenzhen)
Shenzhen, 518055, Guangzhou, China

Xihong Lu
MOE of the Key Laboratory of Bioinorganic and Synthetic Chemistry, The Key Lab of Low-Carbon Chem and Energy Conservation of Guangdong Province, School of Chemistry
Sun Yat-Sen University, Guangzhou
Guangdong, 510275, People's Republic of China

Wenjie Mai
Siyuan Laboratory, Guangdong Provincial Engineering Technology Research Center of Vacuum Coating Technologies and New Energy Materials, Department of Physics
Jinan University
Guangzhou, 510632, PR China

Ting Meng
Frontiers Science Center for Flexible Electronics, Institute of Flexible Electronics
Northwestern Polytechnical University
Xi'an, 710072, P. R. China

Zhiqiang Niu
Key Laboratory of Advanced Energy Materials Chemistry (Ministry of Education), College of Chemistry
Nankai University
Tianjin, 300071, P. R. China

Guozhen Shen
State Key Laboratory for Superlattices and Microstructures
Institute of Semiconductors, Chinese Academy of Science
Beijing, 100083, China

Peng Sun
Siyuan Laboratory, Guangdong Provincial Engineering Technology Research Center of Vacuum Coating Technologies and New Energy Materials, Department of Physics
Jinan University
Guangzhou, 510632, PR China

Hua Wang
State Key Laboratory of Advanced Welding and Joining
Harbin Institute of Technology (Shenzhen)
Shenzhen, 518055, Guangzhou, China; and
Flexible Printed Electronics Technology Centre
Harbin Institute of Technology (Shenzhen)
Shenzhen, 518055, Guangzhou, China

Jiaqi Wang
State Key Laboratory of Advanced Welding and Joining
Harbin Institute of Technology (Shenzhen)
Shenzhen, 518055, Guangzhou, China; and
Flexible Printed Electronics Technology Centre
Harbin Institute of Technology (Shenzhen)
Shenzhen, 518055, Guangzhou, China

Panpan Wang
State Key Laboratory of Advanced Welding and Joining
Harbin Institute of Technology (Shenzhen)
Shenzhen, 518055, Guangzhou, China; and
Flexible Printed Electronics Technology Centre
Harbin Institute of Technology (Shenzhen)
Shenzhen, 518055, Guangzhou, China

Qiufan Wang
Key laboratory of Catalysis and Energy Materials Chemistry of Ministry of Education & Hubei Key Laboratory of Catalysis and Materials Science, Hubei R&D Center of Hyperbranched Polymers Synthesis and Applications
South-Central University for Nationalities
182 Minzu Road, Wuhan, 430074, China

Rui Wang
Key Laboratory of Advanced Energy Materials Chemistry (Ministry of Education)
College of Chemistry, Nankai University
Tianjin, 300071, P. R. China

Xiang Wu
School of Materials Science and Engineering
Shenyang University of Technology
China

Jing Xu
School of Materials Science and Engineering
Nanjing University of Science and Technology
Xuanwu District, Nanjing 210094, China; and
Herbert Gleiter Institute of Nanoscience
Nanjing University of Science and Technology
Xuanwu District, Nanjing 210094, China

Daohong Zhang
Key laboratory of Catalysis and Energy Materials Chemistry of Ministry of Education & Hubei Key Laboratory of Catalysis and Materials Science, Hubei R&D Center of Hyperbranched Polymers Synthesis and Applications
South-Central University for Nationalities
182 Minzu Road, Wuhan 430074, China

Xiyue Zhang
MOE of the Key Laboratory of Bioinorganic and Synthetic Chemistry, The Key Lab of Low-Carbon Chem and Energy Conservation of Guangdong Province, School of Chemistry
Sun Yat-Sen University
Guangzhou, Guangdong, 510275, People's Republic of China

Yizhou Zhang
State Key Laboratory of Organic Electronics and Information Displays (SKLOEI), Institute of Advanced Materials (IAM)
Nanjing University of Posts & Telecommunications
9 Wenyuan Road, Nanjing, 210023, Jiangsu, China; and
Institute of Advanced Materials and Flexible Electronics (IAMFE), School of Chemistry and Materials Science
Nanjing University of Information Science & Technology
219 Ningliu Road, 210044, Nanjing, Jiangsu, China

Preface

As an emerging and exciting research field, flexible electronics have attracted tremendous interests from both the academic and industrial communities. Till now, many kinds of flexible electronic devices and systems have been developed, such as flexible displays, electronic skins, health monitoring bioelectronics, chemical and biosensors, wearable smart textile, and intelligent soft robots, etc. This area develops very fast and some flexible products are already commercially available. For example, flexible organic light-emitting diode displays have been widely used in smart phones, smart watches, and tablet personal computers.

The booming development of flexible electronics has driven the demand for compatible flexible energy storage devices, ideally to make the whole electronic system flexible. Although conventional energy storage devices, such as lithium-ion batteries, lead acid batteries, supercapacitors, have been widely used in our modern society and affected our daily life, their rigid shape, heavy weight, and thickness make them not suitable for flexible electronics. Among different energy storage devices, supercapacitors have the advantages of simple device structure, high power density, short charge and discharge time, long cycle life and wide operating temperature range. When making supercapacitor flexible, it will also possesses the required features of excellent flexibility, portability, stretchability, miniaturized size, ultrathin thickness for flexible electronic devices. During the past several years, researches on flexible supercapacitors are very active and this field expanded very fast. Thus, it is considered timely to provide a survey of a number of important developments in this filed.

This book provides an up-to-date survey of the state of flexible supercapacitors. It contains a selection of 11 chapters contributed by a number of research teams. All the contributors are active researchers in the field of flexible supercapacitors. The most important topics related to flexible supercapacitors are included in this book, ranging from the selection and design of different active electrode materials, the design of different device structures, suitable fabrication techniques, and different functions. I hope this book will be a source of inspiration for graduate students, researchers, and industrial engineers, and will stimulate new developments in this challenging but exciting field.

Guozhen Shen, Professor
Beijing, China

1

Flexible Asymmetric Supercapacitors

Design, Progress, and Challenges

Dun Lin, Xiyue Zhang and Xihong Lu

MOE of the Key Laboratory of Bioinorganic and Synthetic Chemistry, The Key Lab of Low-Carbon Chem and Energy Conservation of Guangdong Province, School of Chemistry, Sun Yat-Sen University, Guangzhou, Guangdong, 510275, People's Republic of China

1.1 Introduction

Recently, flexible electronic products, such as flexible microphones [1], elastic circuits [2–4], pressure and strain sensors [5–7], artificial skin sensors [8–10], intelligent garments [11], and wearable health monitoring devices have boomed as a new and important field of modern electronics (Figure 1.1). Therefore, the development of suitable energy storage devices, which can serve as an excellent power supply while sustaining high mechanical flexibility, are becoming increasingly necessary to power these electronics [13–21]. Supercapacitors (SCs), also known as electrochemical capacitors or ultracapacitors, have emerged as the bridge between batteries and traditional capacitors due to their promising merits of high power density (about $10\,\mathrm{kW\,kg^{-1}}$), good reversibility, excellent cyclic stability (over 10^6 cycles), and safety [22, 23]. Meanwhile, accompanied with the advanced development of lightweight, foldable, and stretchable materials, substantial effort has been invested in the fabrication of flexible supercapacitors (FSCs) [24–28].

In order to satisfy the further demand for practical usage, the configuration of the two electrodes as well as the geometry of the devices are of vital importance and worth careful considerations [29]. The major obstacle of early designed FSCs is their relatively low energy density (E) to mismatch basic requirements of future applications. Thus, tremendous efforts have been denoted to optimize the overall performance of FSCs according to the Eq. (1.1), without sacrificing their power density and service life.

$$E = \frac{1}{2}CV^2 \tag{1.1}$$

In general, either enhanced capacitance (C) or enlarged operating voltage (V) of the device should make sense. Of which, the C of a FSC device can be equivalent to the negative electrode capacitance (C_n) and positive electrode capacitance (C_p) connected in series (Figure 1.2a), which can be calculated using Eq. (1.2)

$$\frac{1}{C} = \frac{1}{C_n} + \frac{1}{C_p} \tag{1.2}$$

Flexible Supercapacitors: Materials and Applications, First Edition. Edited by Guozhen Shen, Zheng Lou and Di Chen.

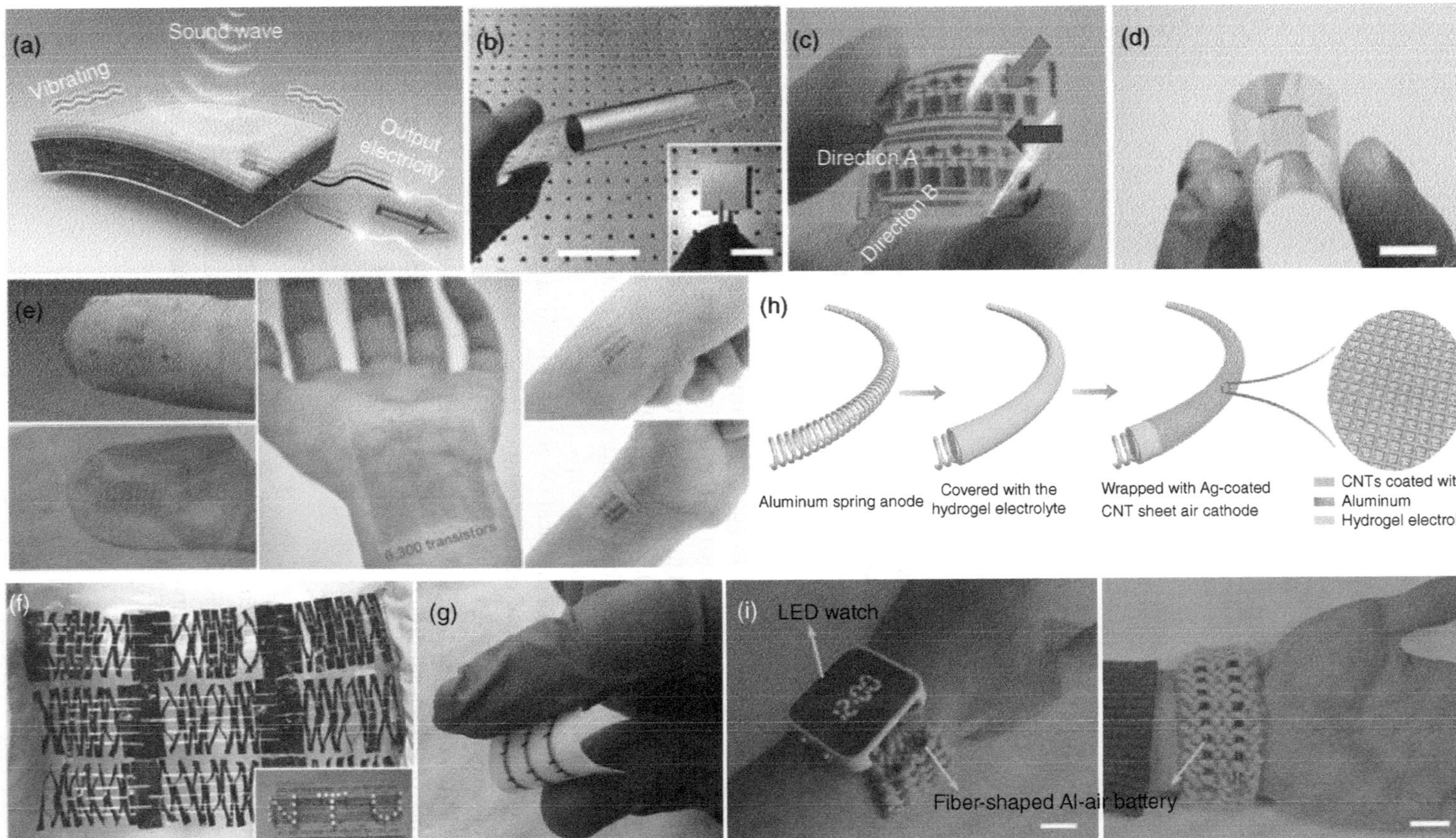

Figure 1.1 (a, b) Scheme and optical image of a flexible acoustic device. *Source:* Reproduced with permission from Ref. [121], © 2017, Springer Nature. Optical image of (c) a flexible circuit *Source:* Reproduced with permission from Ref. [2], © 2018, NPG, (d) multiplexed fingerprint sensor. Scale bar, 1 cm. *Source:* Reproduced with permission from Ref. [5], © 2018, NPG and (e) artificial skin electronics *Source:* Reproduced with permission from Ref. [8], © 2018, NPG. (f) 3 × 3 honeycomb-like supercapacitor array powering LED panel. *Source:* Reproduced with permission from Ref. [13], © 2017, Wiley. (g) Image of an array of field-effect heterojunctions on textile. *Source:* Reproduced with permission from Ref. [14], © 2017, NPG. (h, i) Fabrication and optical image of the fiber-shaped Al-air battery. *Source:* Reproduced with permission from Ref. [15], © 2016, Wiley.

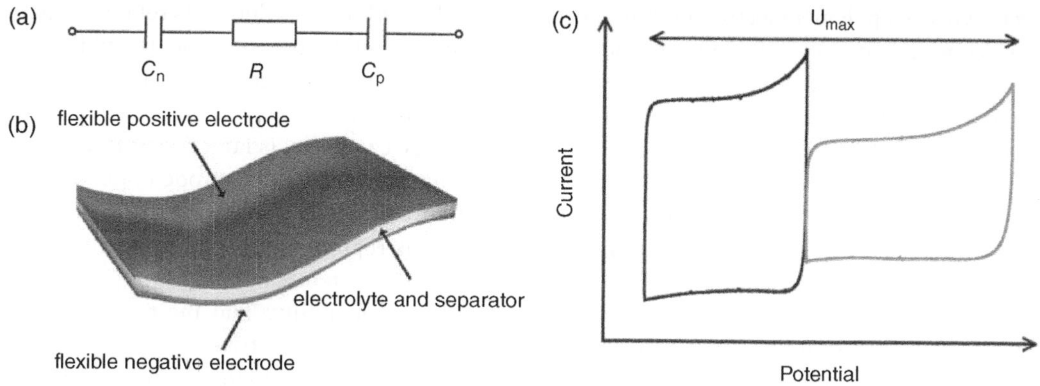

Figure 1.2 (a) The equivalent circuit of an AFSC. (b) Schematic illustration of the typical configuration of AFSCs and (c) Cyclic voltammograms (CV) curves as schematic illustrations of typical AFSCs. *Source:* Reproduced with permission [30]. © 2016, Royal Society of Chemistry.

The maximum C value of the FSCs can be reached when C_n is equal to C_p. Thus, early investigations focused on the symmetric flexible supercapacitors (SFSCs) with cathodes and anodes being identical for achieving higher device capacitance [31–34]. However, due to their limited potential voltage (<1 V in aqueous electrolyte), the energy density of SFSCs is still unsatisfactory. Notably, the V of a FSC device related to the capacitive potential range of electrodes. Thus, asymmetric flexible supercapacitors (AFSCs), also called hybrid SCs or battery-capacitor SCs, are designed with different electrodes configured together (Figure 1.2b) [24–28]. By making use of the distinct capacitive potential range, AFSCs have been widely proven to effectively achieve high operating voltages (even >2 V in aqueous electrolyte) as well as optimized capacitance after balancing the charge between the specific positive and negative electrodes (Figure 1.2c) [30, 35]. In addition, they have several important advantages including small size, low weight, ease of handling, excellent reliability, and a wider range of operating temperatures. Therefore, AFSCs have become one of the most promising energy storage devices for flexible and wearable electronics.

In this context, to achieve high electrochemical performance while maintaining good mechanical stability, FSCs with asymmetric structure could realize further gains, and thus arouse global efforts in relative research. This chapter enumerates some typical newly developed AFSCs in terms of structure design of electrode materials and device's configuration engineering. We first focus on the guidelines on the material design and charge balance of a typical AFSC device. Furthermore, different types of various newly developed AFSCs, including sandwich-type, fiber-type, and the other type of AFSCs devices, are illustrated based on various electrode materials. Finally, the future developing trends and challenges are discussed to provide certain reference to readers on how to contrive this device.

1.2 Configurations of AFSCs Device

Specifically, AFSCs device can be fabricated by constructing two flexible dissimilar electrodes (a Faradaic positive electrode and a capacitor-type negative electrode), a separator and, in most cases, quasi-solid-state electrolyte in a soft package. Among various types of quasi-solid-state electrolytes, gel polymer electrolytes have been extensively used in FSCs due to its relatively high ionic conductivity [36–40]. Soft and bendable plastics including polyethylene terephthalate (PET) [41–44], polydimethylsiloxane (PDMS) [45], and ethylene/vinyl acetate copolymer (EVA) film [46] are typically used as packaging materials for FSC devices.

Considering that the fundamental limit of energy storage capability is largely determined by the electrode material, either the material choice or structure design of electrode materials are of vital importance. Apart from directly fabricated freestanding films like carbon nanotube (CNT) films [47, 48] and graphene films [49, 50], previous reports for FSCs indicate that the flexible electrodes can also rely on a flexible substrate such as thin metal foils [51, 52], polymer substrates [53], textiles [54], and papers [55], to provide flexibility. The main differences between the AFSCs and SFSCs are that the AFSCs require that the positive and negative electrodes are not the same, but they need to be matched well. Electrode materials that are dominated by Faradaic reactions such as metal oxides (RuO_2 [56, 57], MnO_2 [58–64], CoO [60, 65–67] NiO [68–70], V_2O_5 [71, 72], etc.), metal sulfides ($NiCo_2S_4$ [73–75], MoS_2 [76], CoS_2 [77, 78], NiS [58, 63, 79, 80], etc.) and conductive polymers (polyaniline (PANI) [32, 81], polypyrrole (PPy) [82, 83], poly (3,4-ethylenedioxythiophene) (PEDOT) [84, 85] etc.) are normally applied as positive electrodes in AFSCs due to their high specific capacitance and relatively higher potential window. Notably, carbon-based materials (activated carbon [60, 66, 67], graphene [59, 86], CNTs [87], carbon fibers [88–90] etc.), metal nitrides (TiN [20], VN [91], MoN [92], etc.), and some metal oxides (Fe*Ox* [93], Mo*Ox* [94] etc.) are usually employed as negative electrodes because of their fast charging/discharging rate and suitable working window at negative potential.

However, before they are assembled in an AFSC, the matching problems of the two electrodes with different theoretical capacitance need to be solved [91]. As for an AFSC, the charge balance will follow the relationship $q+ = q-$. The charge stored by each electrode depends on the specific capacitance (C), the potential range for the charge/discharge process (E) and the mass of the active electrode material (m), following the Eq. (1.3):

$$q = C \times E \times m \tag{1.3}$$

In order to get $q+ = q-$ at the typical current density, the mass balancing will follow the Eq. (1.4):

$$\frac{m_+}{m_-} = \frac{C_- \times E_-}{C_+ \times E_+} \tag{1.4}$$

In this way, the suitable mass ratio between the positive electrode (m+) and negative electrode (m−) is defined, which is much closed to the mass loading of the active materials of positive and negative electrode in typical AFSCs.

1.3 Progress of Flexible AFSCs

1.3.1 Sandwich-Type AFSCs

To date, the most widely applied configuration of AFSCs is sandwich-type AFSCs, which stacks two flexible flat electrodes face-to-face with an ionic conductive separator and liquid/gel electrolyte in the middle of the two electrodes to isolate direct contact. AFSCs with such shape holds great potential in future applications in flexible planar electronic devices, such as flexible display, wristbands, membrane-type sensors, etc. [24]

1.3.1.1 Carbon-Based Anodes

The most reliable anode materials for sandwich-type AFSCs are carbon-based materials with significant excellence in conductivity and mechanical stability, such as graphene, CNTs, carbon fibers, etc. [24, 25, 27, 56, 95–97] For example, Zhai et al. [98] successfully synthesized hydrogenated MnO_2 nanorods (H-MnO_2) on carbon cloth (CC) via electrodeposition followed by annealing in hydrogen atmosphere (Figure 1.3a), and loaded reduced graphene oxide (RGO) on CC using vacuum process. The obtained H-MnO_2 cathode and RGO anode were assembled as flexible solid-state AFSC with LiCl/PVA gel electrolyte and a separator sandwiched in between. The as-fabricated sandwich-type AFSC (denoted as H-MnO_2//RGO) exhibited a reliable operating voltage window as wide as 1.8 V and extraordinary mechanical tolerance to bending (Figure 1.3b). Owing to the significantly wide potential window, the device achieved a high energy density of 0.25 mWh cm^{-3} at power density of 1.01 W cm^{-3}, which has surpassed many SFSCs and some AFSCs previously reported. To verify the feasibility of the AFSCs device as energy storage device for wearable electronics, two H-MnO_2//RGO devices were tailored on a laboratory coat in series and able to power an electronic watch (Figure 1.3c). Recently, Yu and his co-workers [25] reported a sandwich-type AFSC with CNT-textile anode and MnO_2/graphene-textile cathode, which achieved an operating potential window of 1.5 V and a maximum energy density of 12.5 Wh kg^{-1}. Choi et al. [56] developed a solid-state AFSC based on an ionic liquid functionalized chemically modified graphene (IL-CMG) film as anode and a hydrous RuO_2-ILCMG composite film as cathode, which reached a high output voltage of 1.8 V and thus delivered a maximum energy density of 19.7 Wh kg^{-1} and

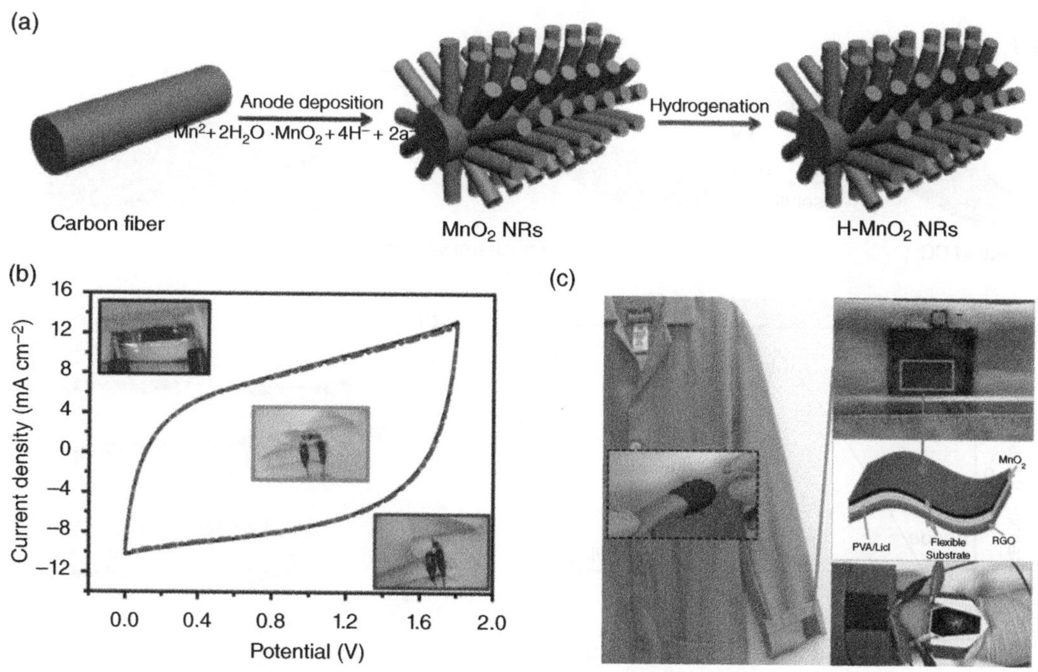

Figure 1.3 (a) Schematic diagram illustrates the growth process for preparing H-MnO_2 NRs on carbon cloth substrate. (b) CV curves obtained at different bent conditions at 200 mV s^{-1}. Insets are the photos of ASCs device on finger. (c) Schematic diagram and photo images of wearable ACS in real applications (sewing on the clothes model and powering electronic watch). *Source:* Reproduced with permission [98].

maximum power density of 6.8 kW kg^{-1}. Moreover, the as-fabricated device exhibited superior cyclic stability even when bent or twisted.

Unfortunately, most carbon-based anodes are relatively low in capacitance due to the electrochemical double layer energy storage mechanism. To this end, effective strategies of achieving high-energy-dense AFSCs has been extensively developed by employing pseudocapacitive anodes such as functionalized carbon, transition metal oxides, transition metal nitrides, conductive polymers, etc. [90] Recently, Wang et al. [90] creatively applied electrochemical activation to CC (Figure 1.4a), which were rarely employed as SC electrode materials because of its intrinsic low capacitance as a result of the small surface area and poor electrochemical activity [32]. The obtained electrochemically activated carbon cloth (EACC) anode was coupled with MnO_2@TiN loaded on CC as cathode to fabricate a novel sandwich-type AFSC (denoted as MnO_2@TiN//EACC) with an extended operation voltage window of 2 V (Figure 1.4b). Besides the broadened voltage window, the impressively boosted capacitance of EACC due to the roughened surface and the introduction of oxygen-containing groups on the surface for redox reactions also contribute to an excellent energy density as high as 1.5 mWh cm^{-3}, which enables its successful application in powering light emitting diode (LED) indicator even under bent condition (Figure 1.4c).

1.3.1.2 Transition Metal Oxide Anodes

Transition metal oxides can generate reversible redox reactions on the surface or even in the bulk during charging/discharging, which results in much higher capacitances compared to carbon-based anodes. By integrating transition metal oxides with flexible current collectors as

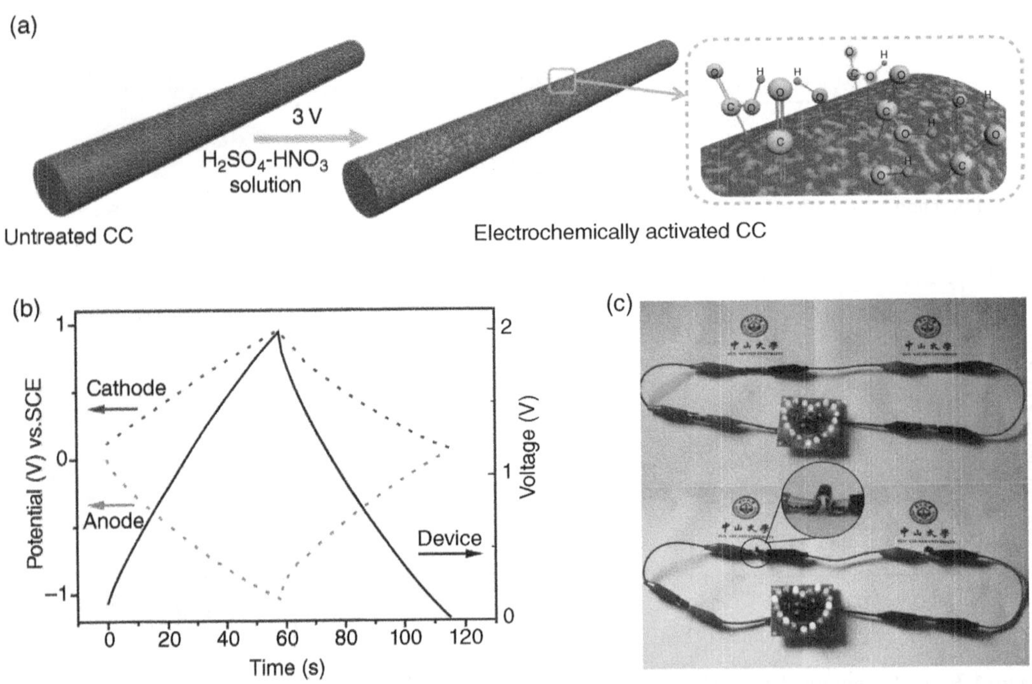

Figure 1.4 (a) Schematic diagram of the CC activation process. (b) Galvanostatic charge/discharge curve and potential distribution curve for the MnO_2@TiN//EACC device at 6 mA cm^{-2}. (c) A LED indicator powered by the tandem straight and bended MnO_2@TiN//EACC devices. *Source:* Reproduced with permission [90]. © 2015, Wiley-VCH.

pseudocapacitive anodes, the energy density of AFSC can be drastically enhanced. For example, Yang et al. [54] grew α-MnO_2 nanowires (NWs) and amorphous Fe_2O_3 nanotubes (NTs) on flexible carbon textile as the pseudocapacitive cathode and anode respectively (Figure 1.5a) to fabricate a sandwich-type flexible asymmetric pseudocapacitor (Figure 1.5b). The as-fabricated sandwich-type AFSC operates at a maximum cell voltage of 1.6 V (Figure 1.5c) and delivers high energy density of 0.55 mWh cm^{-3} (Figure 1.5d). Two devices connected in series can readily operate a blue LED after charging (Figure 1.5d inset), indicating the potential of the AFSC in future applications. Similarly, a novel flexible all-solid-state asymmetric SC fabricated with a carbon-fabric-loaded WO_{3-x}/MoO_{3-x} core/shell nanowires anode and a polyaniline cathode was reported by Xiao and his co-workers. The device showed satisfactory energy density (1.9 mWh cm^{-3}), impressive cyclic stability, as well as good mechanical flexibility.

1.3.1.3 Transition Metal Nitride Anodes

Owing to the high conductivity and transition metal sites with multiple valence states, transition metal nitrides are emerging as promising pseudocapacitive anode materials with fast and reversible redox reactions. Many transition metal nitrides have been exploited for AFSCs, such as

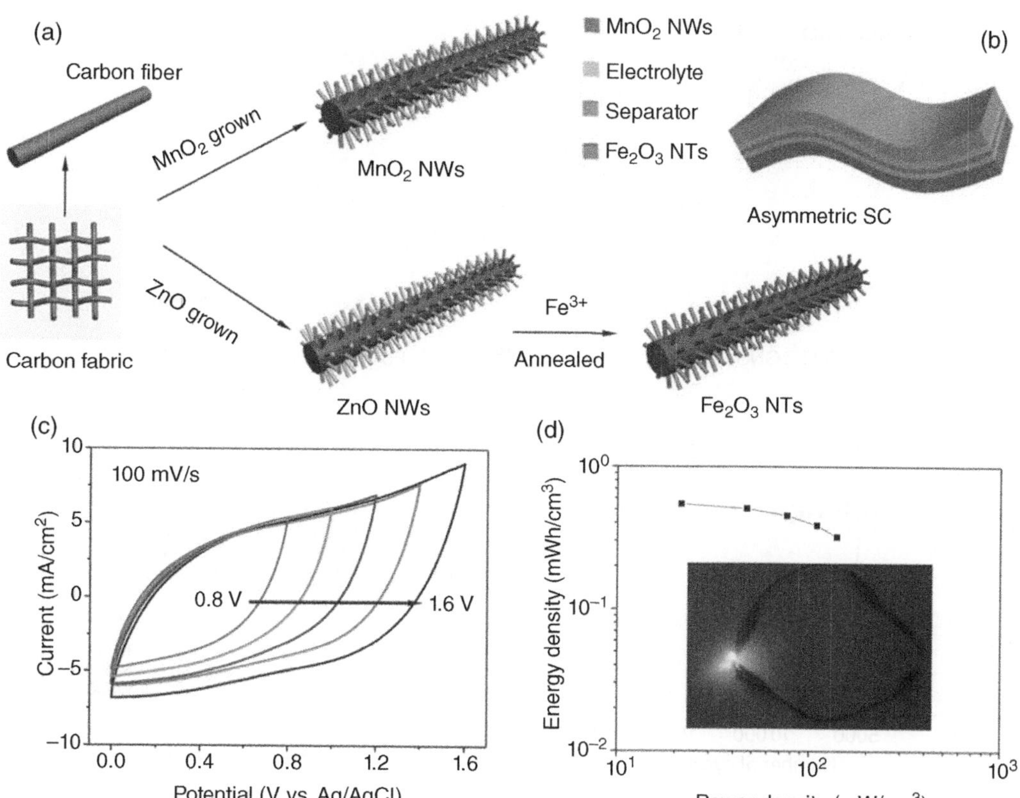

Figure 1.5 (a) Schematic diagram illustrating the synthesis procedure of MnO_2 NWs and Fe_2O_3 NTs on carbon cloth. (b) Schematic sketch illustrating the designed asymmetric supercapacitor device. (c) CV curves of the assembled solid-state AFSC device collected in different scan voltage windows. (d) Ragone plots of the solid-state AFSC device. Inset shows a blue LED powered by the tandem AFSC devices. *Source:* Reproduced with permission [54]. © 2014, American Chemical Society.

titanium nitride, vanadium nitride, tungsten oxynitride, iron nitride, etc., with high performances comparable to transition metal oxide anodes [91, 99–101]. For instance, Fan's group successfully fabricated an all-metal nitrides solid-state asymmetric SC, where the titanium nitride (TiN) cathode and iron nitride (Fe_2N) anode were grown on CC-loaded graphene nanosheets (GNS) using atomic layered deposition followed by calcination under ammonia atmosphere (Figure 1.6a).

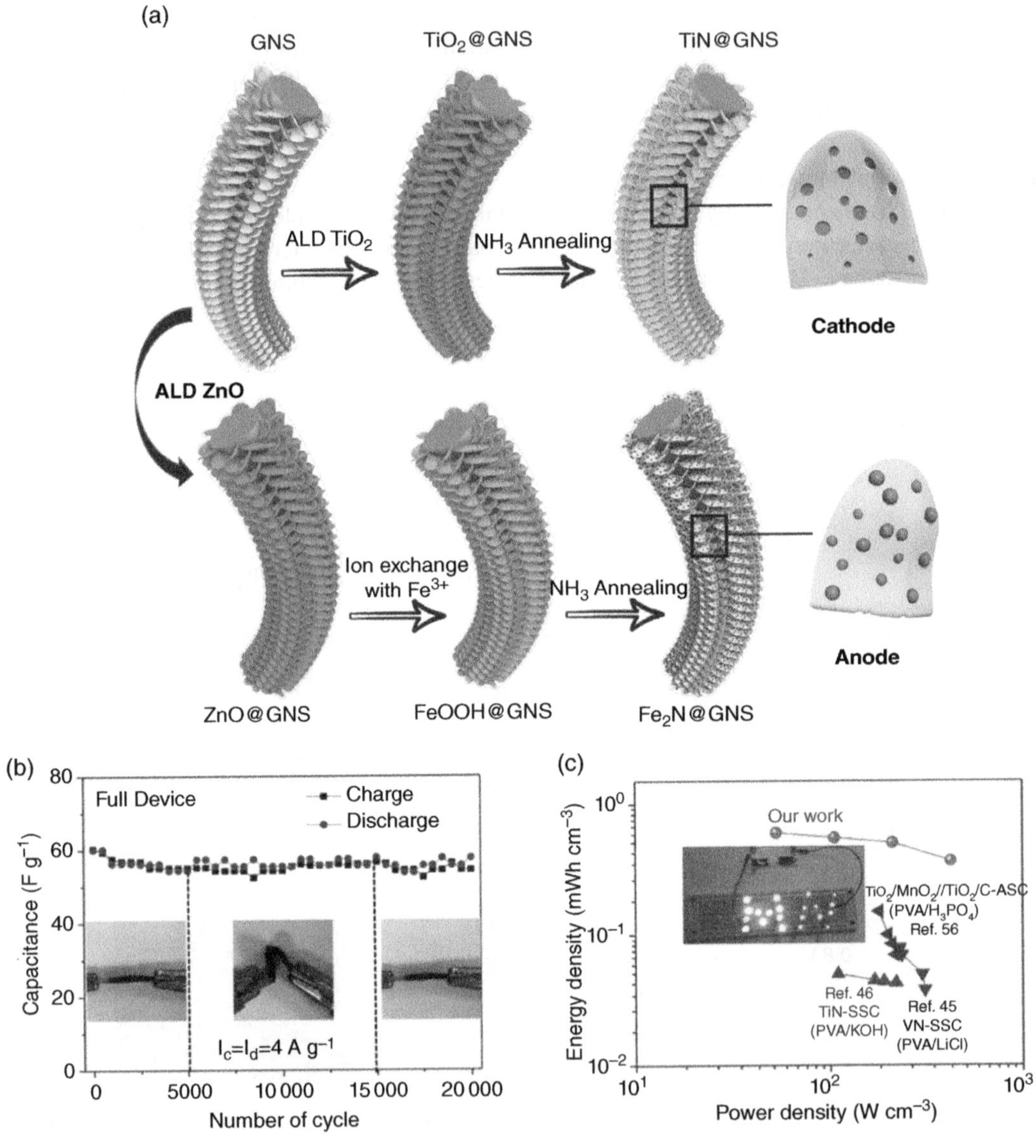

Figure 1.6 (a) Schematics of the fabrication processes of metal nitride cathode and anode materials. (b) Cycling performance of full device at $4\,A\,g^{-1}$ in 20 000 cycles with different bending situations. (c) Ragone plots of quasi-solid-state TiN-Fe_2N AFSC in comparison with other PVA-based solid electrolyte SFSCs and AFSCs. Inset: pink and white LEDs in parallel are lit up by two full devices in tandem. *Source:* Reproduced with permission [101]. © 2015, Wiley-VCH.

The porous configuration of TiN and homogeneous distribution of Fe_2N nanoparticles contribute to the extraordinary cycling durability (≈98% capacity retention after 20 000 cycles) of the fabricated quasi-solid-state AFSC device using PVA/LiCl polymer gel as neutral electrolyte (Figure 1.6b). The AFSC device achieved a maximum energy density of 0.61 mWh cm^{-3} and a maximum power density of 422.7 mW cm^{-3}, which were substantially higher than those of transition-metal-nitride-based SCs and PVA-based solid-state SCs (Figure 1.6c). Lu's group has reported various AFSCs using CC-loaded transition metal nitride anodes in recent years. For example, they used neutral PVA/LiCl polymer gel electrolyte to effectively stabilize porous VN NWs anode, and paired it with VO_x NWs cathode to assemble a stable and high-performance quasi-solid-state AFSC device with a high output voltage of 1.8 V [91]. Furthermore, the VO_x//VN-AFSC device was able to deliver an impressive volumetric capacitance of 1.35 F cm^{-3}, a highest energy density of 0.61mWh cm^{-3} and extraordinary cycling stability with 12.5% loss of capacitance after 10 000 cycles. They also prepared holey tungsten oxynitride (WON) nanowires on CC through the annealing of WO_3 precursor nanowires in ammonia atmosphere [100]. The as-fabricated AFSC device with WON NWs anode and MnO_2 cathode could deliver a high working voltage of 1.8 V and volumetric capacitance of 2.73 F cm^{-3}. The maximum energy density of MnO_2//WON AFSC device was 1.27 mWh cm^{-3} at a power density of 0.62 W cm^{-3}, which has transcended many reported AFSC devices.

1.3.1.4 Conductive Polymer Anodes

Conductive polymers are promising candidates as pseudocapacitive materials owing to their good conductivity and reversible redox reactions during charging/discharging, but they are mostly applied as cathode materials while rarely studied as anode materials for AFSCs. Recently, Wang et al. synthesized 150 WO_3@PPy nanowires on carbon fibers as the anode and grew $Co(OH)_2$ nanowires on carbon fabric as the cathode for AFSC device. The as-fabricated AFSC device exhibited apparent pseudocapacitive behavior within a stable potential range of 0–1.6 V. The maximum volumetric capacitance of 2.8 F cm^{-3} was achieved at a scan rate of 20 mV s^{-1}. Moreover, the asymmetric supercapacitor (ASC) device delivered an energy density as high as 1.03 mWh cm^{-3}.

1.3.2 Fiber-Type ASCs

Despite distinct advances, the planar-shaped SCs are still insufficient in deformability for weaving into textiles or integrating into linear-shaped electronics. In this regard, researchers have creatively assembled electrodes with one-dimensional geometry to fabricate fiber-type AFSCs. Fiber-shaped AFSCs have been developed into multiple configurations including parallel type, wrap type, coaxial-helix type and two-ply yarn type, in order to effectively meet the demands of different wearable energy textiles, including sensing [102–104], communication [105], and storage [106].

1.3.2.1 Parallel-Type Fiber AFSCs

For a parallel-type fiber AFSC, two fiber-shaped electrodes are assembled side-by-side, separated by gel/polymer electrolyte, and finally supported on a flat substrate [60, 107–109]. For instance, Yu et al. [109] reported a parallel type all-solid-state asymmetric micro-SC using MnO_2-deposited rGO/SWCNT fiber as the cathode (denoted as GCF/MnO_2-10) and an N-doped rGO/SWCNT fiber as the anode (denoted as GCF/N2) (Figure 1.7a). By fully utilizing the potential window of both cathode (0 ~ 0.9 V) and anode (−0.9 ~ 0 V), the device showed a high output voltage of 1.8 V (Figure 1.7b). Excellent electrochemical performances such as good cycling stability

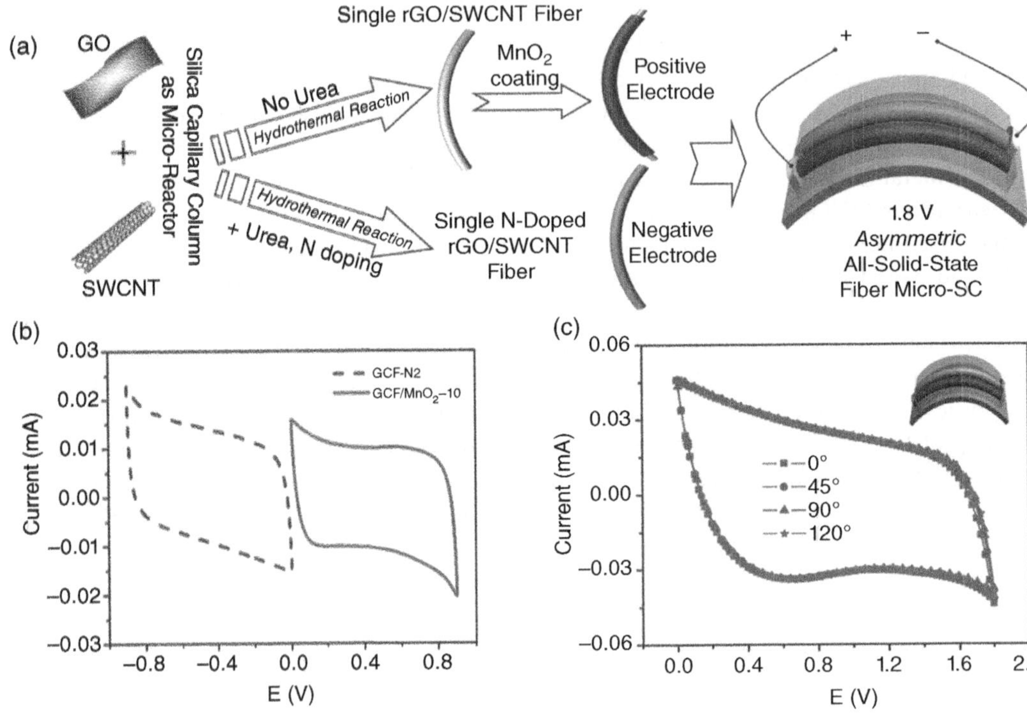

Figure 1.7 (a) Schematic illustration of the design and fabrication of the asymmetric fiber-based micro-SC. (b) Comparative CV curves obtained for the GCF-N2 and GCF/MnO_2-10 fibers at the scan rate of $10\,mV\,s^{-1}$. (c) CV curves of one asymmetric micro-SC, which are bended at different angles. *Source:* Reproduced with permission [109]. © 2014, Wiley-VCH.

(87% capacitance retention after 10 000 cycles), high energy density ($5\,mW\,h\,cm^{-3}$) and power density ($929\,mW\,cm^{-3}$) were also achieved. Furthermore, such device geometry exhibited promising mechanic stability under different bending states (Figure 1.7c). This asymmetric micro-SC device was testified as a reliable power source for a ZnO film-based UV photodetector, suggesting its promising potential in future applications.

1.3.2.2 Wrap-Type Fiber AFSCs

The design of a wrap-type AFSC is very similar to that of a parallel-type fiber AFSC, which encapsulates two electrodes into a protective flexible tube instead of placing them on a flexible substrate [53, 59, 62, 110–112]. Recently, Lu and his co-workers [112] successfully synthesized N and low valence-state Mo dual-doped MoO_3 nanowires on carbon fibers, which was coupled with MnO_2@TiN-loaded carbon fiber cathode and sealed with heat-shrinkable tube to fabricate a wrap type solid-state ASC (denoted as MnO_2@TiN//N-MoO_{3-x}) (Figure 1.8a). The galvanostatic charge/discharge (GCD) curves of MnO_2@TiN//N-MoO_{3-x} with different current densities in Figure 1.8b indicate that the stable operating voltage of the device reaches a significantly high value of 2.0 V. The ASC device also shows superior rate capability when current density increased by 15 folds (Figure 1.8c). More importantly, the excellent flexibility and mechanic robustness enabled the fiber AFSC device to perfectly maintain its electrochemical performances in bent and even knotted conditions (Figure 1.8d). Benefiting from the ultrahigh output voltage and Faradaic

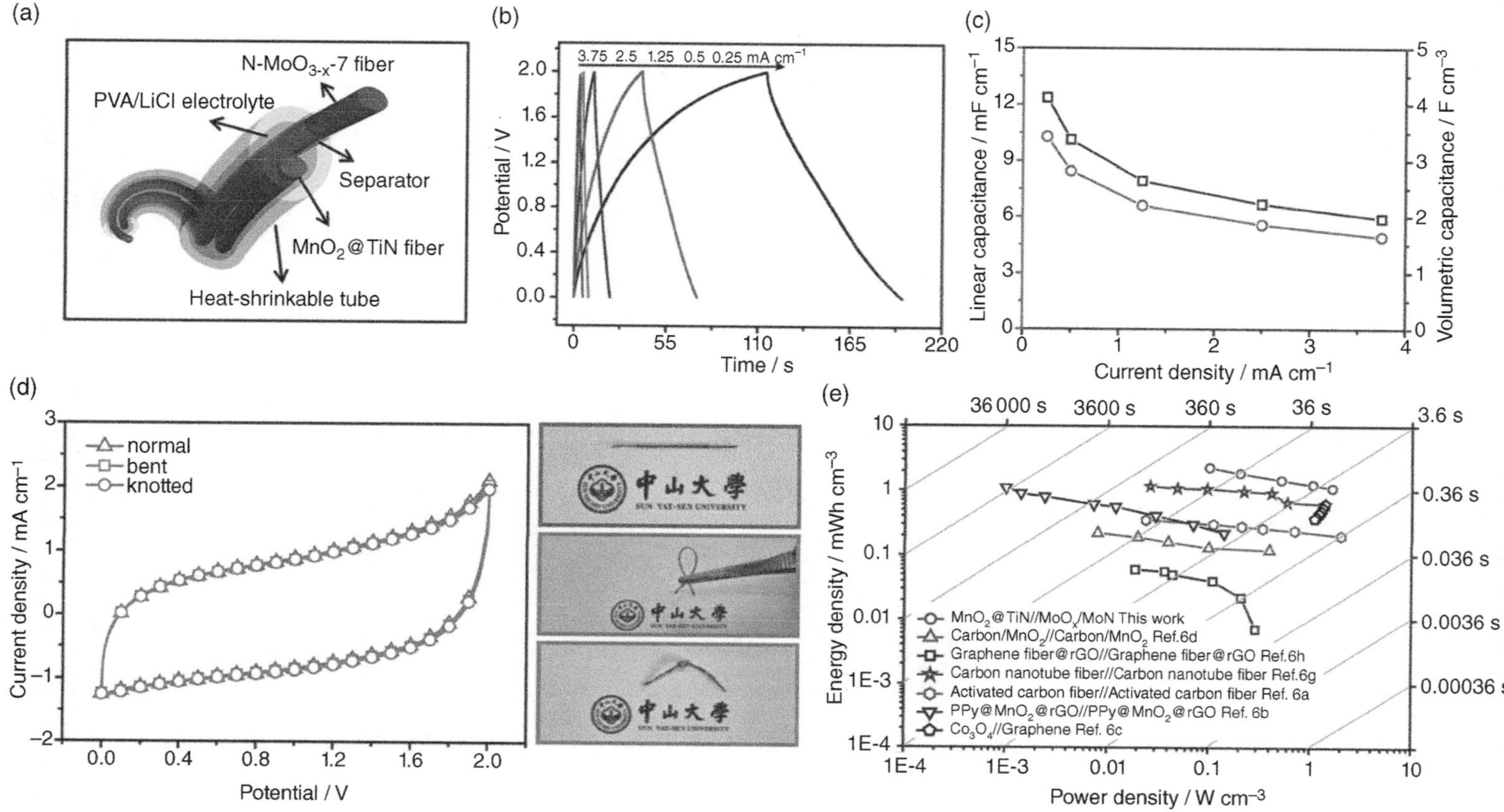

Figure 1.8 (a) Schematic illustration of the as-assembled fiber-shaped MnO_2@TiN//N-MoO_{3-x}-ASC device. (b) GCD curves of our fiber-shaped AFSC device. (c) Linear capacitances and volumetric capacitances of the fiber-shaped AFSC device as a function of current density. (d) CV curves collected at 100 mV s^{-1} for the fiber-shaped AFSC device under different conditions (left) and corresponding device pictures (right). (e) Ragone plots for the fiber-shaped AFSC device and other recently reported fiber-shaped FSCs. *Source:* Reproduced with permission [112]. © 2016, Wiley-VCH.

electrodes with improved conductivity, the MnO_2@TiN//N-MoO_{3-x} device exhibited a maximum energy and power density of 2.29 mWh cm^{-3} and 1.64 W cm^{-3} respectively, outperforming many other fiber-shaped SC devices reported (Figure 1.8e).

1.3.2.3 Coaxial-Helix-Type Fiber AFSCs

By helically wrapping a wire shape axial electrode with another wire electrode, coaxial-helix-type ASCs with core–shell cable-like structures have been creatively explored [86, 113–117]. For example, the Thomas group [117] fabricated a novel cable-like coaxial-helix-type AFSC as illustrated in

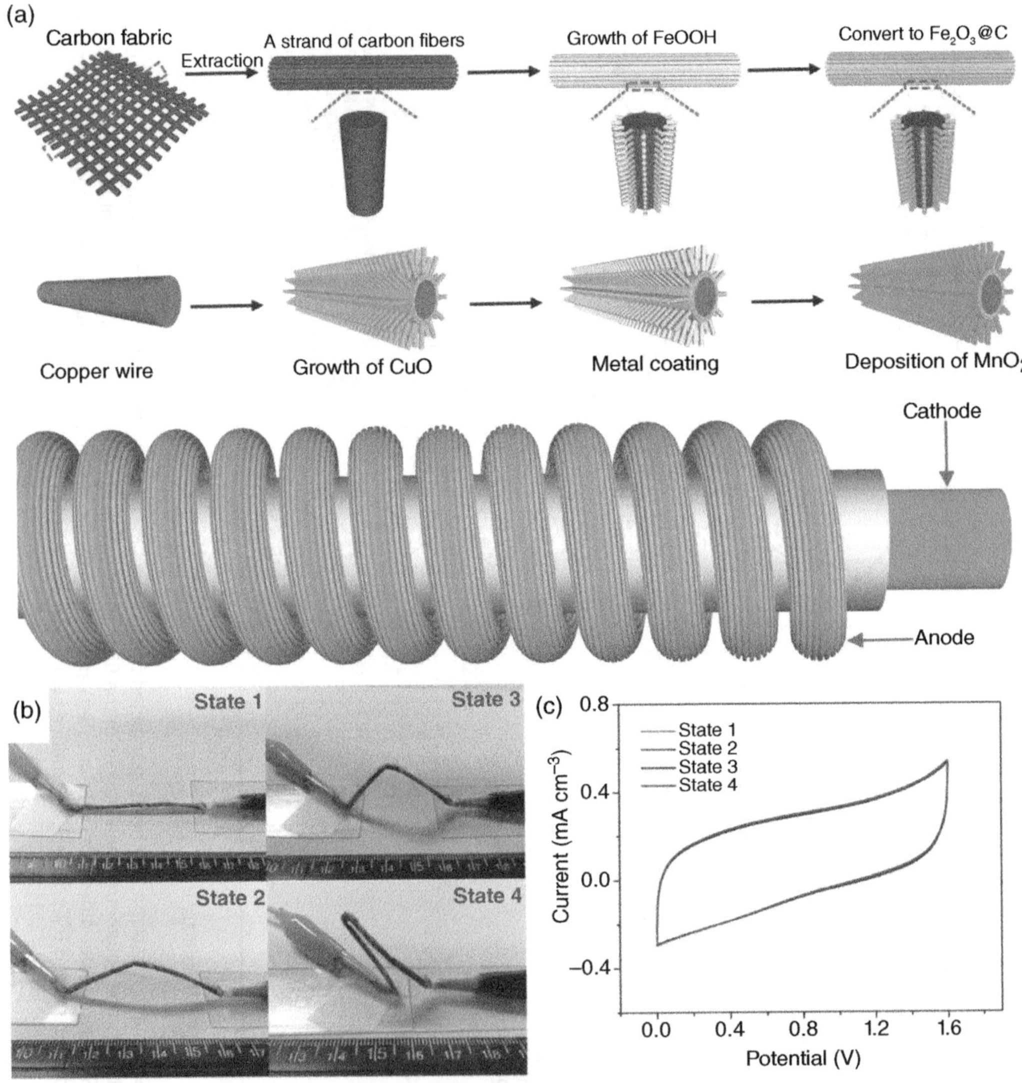

Figure 1.9 (a) Schematics illustration shows the fabrication process of an anode and a cathode, respectively, and the structure of a coil-type asymmetric supercapacitor electrical cable. (b) Optical images of a coil-type asymmetric supercapacitor electrical cable at different bending states. (c) CV curves obtained at different bending states at 200 mV s^{-1}. *Source:* Reproduced with permission [117]. © 2015, Wiley-VCH.

Figure 1.9a. The AFSC device was coaxially assembled using wire-shaped electrodes, where the MnO_2@AuPd@CuO nanowiskers@copper wire core-shell-structured cathode was fabricated via a three-step synthesis, and the Fe_2O_3@carbon fiber core-shell-structured anode was prepared through hydrothermal growing followed by conversion. The AFSC device could tolerate different bending states from 0° to nearly 180° (Figure 1.9b), while its promising electrochemical performances such as quasi-rectangular-shaped CV curves, maximum output voltage of 1.8 V and volumetric capacitance of $1.6\,F\,cm^{-3}$ did not sacrifice (Figure 1.9c). Furthermore, the AFSC device could serve as a bi-functional integrated electrical cable for both electricity storage and electricity transmission.

1.3.2.4 Two-Ply-Yarn-Type AFSCs

Fiber AFSCs with stand-alone two-ply-yarn type configuration has been proposed for efficient weaving into textiles, which are constructed by coating two fiber electrodes with polymer electrolyte and then twisting them together without outer packaging [118–120]. As shown in Figure 1.10a, Jin et al. [118] recently employed carbon-fiber-thread@polyaniline as cathode and functionalized carbon-fiber thread as anode to fabricate a two-ply-yarn-type AFSC device with PVA-H_3PO_4 gel electrolyte. The device successfully reached a high operating voltage of 1.6 V with energy density up to $2\,mW\,h\,cm^{-3}$. In order to demonstrate its potential in practical applications, the device was woven into a glove (Figure 1.10b) and exhibited excellent mechanical tolerance against bending (Figure 1.10c). The flexibility of the device was also examined via stretching to even 100% (Figure 1.10d), while the capacitance of the stretched device merely changed (Figure 1.10e).

1.4 Summary

To conclude, AFSCs have been universally accepted as one of the most promising energy storage devices, which effectively utilize the different potential windows of the pseudocapacitive cathodes and electric double-layer capacitive anodes to increase the operating voltage of the device, thus contributing to the significant boosting of their energy density. Furthermore, researchers have developed AFSCs with impressive lightweight, small size, and high flexibility to satisfy the growing demand of portable/wearable electronics. Many novel and efficient configurations have been designed for easy integration with textiles and miniature electronics.

Still, enormous effort should be paid to the future improvement and popularization of AFSCs. Firstly, further optimizing the overall electrochemical performances of the flexible electrodes and the devices remains as the biggest obstacle for AFSCs. Although the employment of pseudocapacitive anodes could well improve the limited energy density of AFSCs due to the relatively low capacitance of carbon-based anodes, their poor conductivity results in unsatisfactory power density, which is worth more effective solutions. Secondly, the fabrication of a high-mass-loading electrode with good electrochemical performances and simultaneously good mechanical properties is rather challenging. Thirdly, most of the AFSCs are arduously handmade, which is difficult for industrial scale-up with machines. Finally, the smart integration of AFSCs as power sources with energy harvesting units (solar cells, fuel cells, nanogenerators, etc.) or energy consuming units (sensors, LEDs, displays) would be a promising strategy for rational arrangement and utilization of energy, which is accordant with the future goals of sustainable development.

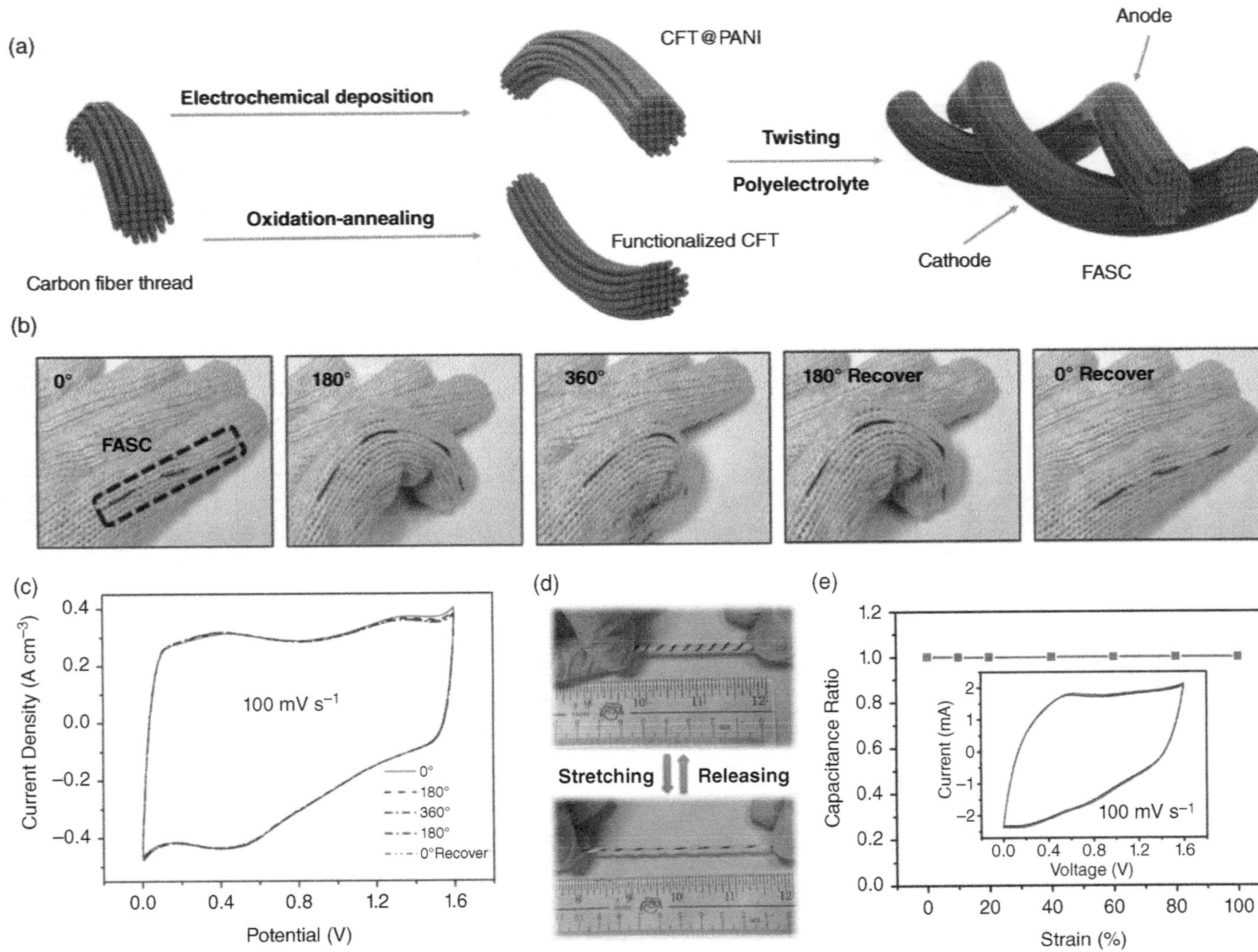

Figure 1.10 (a) Schematic diagram of the fabrication procedure of an AFSC. (b) Digital images of an AFSC woven into a glove at different bending states. (c) CV curves of solid-state AFSC at different bending states with a scan rate of $100\,mV\,s^{-1}$. (d) Digital images of the stretchable AFSC before and after stretching. (e) Capacitance ratio of the AFSC under different applied strains (0%, 10%, 20%, 40%, 60%, 80%, and 100%) at a scan rate of $100\,mV\,s^{-1}$. Inset shows the CV curves of the AFSC measured at different applied tensile strains. *Source:* Reproduced with permission [118]. © 2015, Elsevier.

References

1 Li, Z., Le, T., Wu, Z. et al. (2015). *Adv. Funct. Mater.* 25: 464.
2 Ji, D., Li, T., Zou, Y. et al. (2018). *Nat. Commun.* 9: 2339.
3 Zhou, L., Mao, J., Ren, Y. et al. (2018). *Small* 14: 1703126.
4 Li, D., Lai, W.Y., Zhang, Y.Z., and Huang, W. (2018). *Adv. Mater.* 30: 1704738.
5 An, B.W., Heo, S., Ji, S. et al. (2018). *Nat. Commun.* 9: 2458.
6 Amjadi, M., Turan, M., Clementson, C.P., and Sitti, M. (2016). *ACS Appl. Mater. Interfaces* 8: 5618.
7 Yan, C., Wang, J., Kang, W. et al. (2014). *Adv. Mater.* 26: 2022.
8 Wang, S., Xu, J., Wang, W. et al. (2018). *Nature* 555: 83.
9 Zhao, X., Hua, Q., Yu, R. et al. (2015). *Adv. Electron. Mater.* 1: 1500142.
10 Ge, J., Sun, L., Zhang, F.R. et al. (2016). *Adv. Mater.* 28: 722.
11 He, X., Zi, Y., Guo, H. et al. (2017). *Adv. Funct. Mater.* 27: 1604378.
12 Shadike, Z., Zhou, Y.N., Chen, L.L. et al. (2017). *Nat. Commun.* 8: 566.
13 Lv, Z., Luo, Y., Tang, Y. et al. (2017). *Adv. Mater.* 30: 1704531.
14 Carey, T., Cacovich, S., Divitini, G. et al. (2017). *Nat. Commun.* 8: 1202.
15 Xu, Y., Zhao, Y., Ren, J. et al. (2016). *Angew. Chem. Int. Ed.* 55: 7979.
16 Gates, B.D. (2009). *Science* 323: 1566.
17 Park, S., Vosguerichian, M., and Bao, Z. (2013). *Nanoscale* 5: 1727.
18 Kim, S., Kwon, H.J., Lee, S. et al. (2011). *Adv. Mater.* 23: 3511.
19 Koo, M., Park, K.I., Lee, S.H. et al. (2012). *Nano Lett.* 12: 4810.
20 Lu, X., Liu, T., Zhai, T. et al. (2014). *Adv. Energy Mater.* 4: 1300994.
21 Zhang, X., Zhang, H., Lin, Z. et al. (2016). *Sci. China Mater.* 59: 475.
22 Senthilkumar, B., Khan, Z., Park, S. et al. (2015). *J. Mater. Chem. A* 3: 21553.
23 Luo, H.M., Yang, Y.F., Mu, B. et al. (2016). *Carbon* 100: 214.
24 Wu, L., Lu, H., Xiao, L. et al. (2014). *J. Mater. Chem. A* 2: 16424.
25 Yu, G., Hu, L., Vosgueritchian, M. et al. (2011). *Nano Lett.* 11: 2905.
26 Algharaibeh, Z. and Pickup, P.G. (2011). *Electrochem. Commun.* 13: 147.
27 Chen, P.C., Shen, G., Shi, Y. et al. (2010). *ACS Nano* 4: 4403.
28 Ghosh, D., Mandal, M., and Das, C.K. (2015). *Langmuir* 31: 7835.
29 Choudhary, N., Li, C., Moore, J. et al. (2017). *Adv. Mater.* 29: 1605336.
30 Yu, M., Wang, Z., Han, Y. et al. (2016). *J. Mater. Chem. A* 4: 4634.
31 Liao, Q., Li, N., Jin, S. et al. (2015). *ACS Nano* 9: 5310.
32 Wang, G., Wang, H., Lu, X. et al. (2014). *Adv. Mater.* 26: 2676.
33 Wang, L., Feng, X., Ren, L. et al. (2015). *J. Am. Chem. Soc.* 137: 4920.
34 Wang, Q., Wang, X., Liu, B. et al. (2013). *J. Mater. Chem. A* 1: 2468.
35 Akinwolemiwa, B., Peng, C., and Chen, G.Z. (2015). *J. Electrochem. Soc.* 162: A5054.
36 Wang, G., Lu, X., Ling, Y. et al. (2012). *ACS Nano* 6: 10296.
37 Hu, X.L., Hou, G.M., Zhang, M.Q. et al. (2012). *J. Mater. Chem.* 22: 18961.
38 Kang, Y.J., Chun, S.J., Lee, S.S. et al. (2012). *ACS Nano* 6: 6400.
39 Yu, H., Wu, J., Fan, L. et al. (2012). *Sources* 198: 402.
40 Huang, C.W., Wu, C.A., Hou, S.S. et al. (2012). *Adv. Funct. Mater.* 22: 4677.
41 Huang, C., Jiang, J., Lu, M. et al. (2009). *Nano Lett.* 9: 4297.
42 Ge, J., Cheng, G., and Chen, L. (2011). *Nanoscale* 3: 3084.
43 Niu, Z., Du, J., Cao, X. et al. (2012). *Small* 8: 3201.
44 Han, J., Dou, Y., Zhao, J. et al. (2012). *Small* 9: 98.

45 Li, X., Wang, G., Wang, X. et al. (2013). *J. Mater. Chem. A* 1: 10103.
46 Zhang, Z., Zhai, T., Lu, X. et al. (2013). *J. Mater. Chem. A* 1: 505.
47 Li, R., Wang, Y., Zhou, C. et al. (2015). *Adv. Funct. Mater.* 25: 5384.
48 Zhao, J., Chen, J., Xu, S. et al. (2014). *Adv. Funct. Mater.* 24: 2938.
49 He, Y., Chen, W., Li, X. et al. (2013). *ACS Nano* 7: 174.
50 Gwon, H., Kim, H.-S., Lee, K.U. et al. (2011). *Energy Environ. Sci.* 4: 1277.
51 Lu, X., Zheng, D., Zhai, T. et al. (2011). *Energy Environ. Sci.* 4: 2915.
52 Duffy, N.W., Baldsing, W., and Pandolfo, A.G. (2008). *Electrochim. Acta* 54: 535.
53 Zhang, Y., Zhang, X., Yang, K. et al. (2018). *J. Mater. Chem. A* 6: 12250.
54 Yang, P., Ding, Y., Lin, Z. et al. (2014). *Nano Lett.* 14: 731.
55 Zhang, L., Zhu, P., Zhou, F. et al. (2016). *ACS Nano* 10: 1273.
56 Choi, B.G., Chang, S.J., Kang, H.W. et al. (2012). *Nanoscale* 4: 4983.
57 Kong, L., Zhang, C., Wang, J. et al. (2015). *ACS Nano* 9: 11200.
58 Wang, Y., Jin, Y., Zhao, C. et al. (2017). *Mater. Lett.* 191: 169.
59 Cai, W., Lai, T., Lai, J. et al. (2016). *Sci. Rep.* 6: 26890.
60 Feng, C., Zhang, J., He, Y. et al. (2015). *ACS Nano* 9: 1730.
61 Xia, H., Shirley Meng, Y., Yuan, G. et al. (2012). *Electrochem. Solid-State Lett.* 15: A60.
62 Liu, W.J., Liu, N.S., Shi, Y.L. et al. (2015). *J. Mater. Chem. A* 3: 13461.
63 Li, Q., Liang, C.L., Lu, X.F. et al. (2015). *J. Mater. Chem. A* 3: 6432.
64 Jabeen, N., Xia, Q., Savilov, S.V. et al. (2016). *ACS Appl. Mater. Interfaces* 8: 33732.
65 Xie, Z., Lu, W., Yang, L. et al. (2017). *Talanta* 170: 358.
66 Xu, W., Chen, J., Yu, M. et al. (2016). *J. Mater. Chem. A* 4: 10779.
67 Xie, Y., Fan, M., Shen, T. et al. (2016). *Mater. Technol.* 31: 646.
68 Luan, F., Wang, G., Ling, Y. et al. (2013). *Nanoscale* 5: 7984.
69 Jiao, Y., Liu, Y., Yin, B. et al. (2014). *Nano Energy* 10: 90.
70 Yu, M., Wang, W., Li, C. et al. (2014). *Npg Asia Mater.* 6: e129.
71 Tang, K., Li, Y., Li, Y. et al. (2016). *Electrochim. Acta* 209: 709.
72 Van Lam, D., Shim, H.C., Kim, J.H. et al. (2017). *Small* 13: 1702702.
73 Yang, J., Yu, C., Fan, X. et al. (2016). *Energy Environ. Sci.* 9: 1299.
74 Xiao, J., Wan, L., Yang, S. et al. (2014). *Nano Lett.* 14: 831.
75 Shen, L., Wang, J., Xu, G. et al. (2015). *Adv. Energy Mater.* 5: 1400977.
76 Yang, Y., Fei, H., Ruan, G. et al. (2014). *Adv. Mater.* 26: 8163.
77 Su, C., Xiang, J., Wen, F. et al. (2016). *Electrochim. Acta* 212: 941.
78 Ranaweera, C.K., Wang, Z., Alqurashi, E. et al. (2016). *J. Mater. Chem. A* 4: 9014.
79 Das, H.T., Mahendraprabhu, K., Maiyalagan, T., and Elumalai, P. (2017). *Sci. Rep.* 7: 15342.
80 Xia, H., Hong, C., Shi, X. et al. (2015). *J. Mater. Chem. A* 3: 1216.
81 Wang, D.W., Li, F., Zhao, J. et al. (2009). *ACS Nano* 3: 1745.
82 Shi, Y., Pan, L., Liu, B. et al. (2014). *J. Mater. Chem. A* 2: 6086.
83 Davies, A., Audette, P., Farrow, B. et al. (2011). *J. Phys. Chem. C* 115: 17612.
84 Laforgue, A. and Power, J. (2011). *Sources* 196: 559.
85 Tang, P., Han, L., and Zhang, L. (2014). *ACS Appl. Mater. Interfaces* 6: 10506.
86 Xu, P., Wei, B.Q., Cao, Z.Y. et al. (2015). *ACS Nano* 9: 6088.
87 Hong, X., Kim, J., Shi, S.F. et al. (2014). *Nat. Nanotechnol.* 9: 682.
88 Liu, Y., Lu, Q., Huang, Z. et al. (2018). *J. Alloys Compd.* 762: 301.
89 Yu, D., Goh, K., Wang, H. et al. (2014). *Nat. Nanotechnol.* 9: 555.
90 Wang, W., Liu, W.Y., Zeng, Y.X. et al. (2015). *Adv. Mater.* 27: 3572.
91 Lu, X., Yu, M., Zhai, T. et al. (2013). *Nano Lett.* 13: 2628.

92 Ruan, D., Lin, R., Jiang, K. et al. (2017). *ACS Appl. Mater. Interfaces* 9: 29699.
93 Samuel, E., Joshi, B., Jo, H.S. et al. (2017). *Chem. Eng. J.* 328: 776.
94 Zheng, L., Xu, Y., Jin, D., and Xie, Y. (2011). *Chem. Asian J.* 6: 1505.
95 Singh, A. and Chandra, A. (2015). *Sci. Rep.* 5: 15551.
96 Tang, Q.Q., Chen, M.M., Wang, L. et al. (2015). *Sources* 273: 654.
97 Amitha, F.E., Reddy, A.L.M., and Ramaprabhu, S. (2009). *J. Nanopart. Res.* 11: 725.
98 Zhai, T., Xie, S.L., Yu, M.H. et al. (2014). *Nano Energy* 8: 255.
99 Liu, Y., Xiao, R., Qiu, Y. et al. (2016). *Electrochim. Acta* 213: 393.
100 Yu, M., Han, Y., Cheng, X. et al. (2015). *Adv. Mater.* 27: 3085.
101 Zhu, C., Yang, P., Chao, D. et al. (2015). *Adv. Mater.* 27: 4566.
102 Shim, B.S., Chen, W., Doty, C. et al. (2008). *Nano Lett.* 8: 4151.
103 Cherenack, K., Zysset, C., Kinkeldei, T. et al. (2010). *Adv. Mater.* 22: 5178.
104 Laxminarayana, K. and Jalili, N. (2005). *Text. Res. J.* 75: 670.
105 Marculescu, D., Marculescu, R., Zamora, N.H. et al. (2003). *Proc. IEEE* 91: 1995.
106 Chen, Z., Ren, W., Gao, L. et al. (2011). *Nat. Mater.* 10: 424.
107 Zheng, B.N., Huang, T.Q., Kou, L. et al. (2014). *J. Mater. Chem. A* 2: 9736.
108 Wang, J., Luo, C., Mao, J. et al. (2015). *ACS Appl. Mater. Interfaces* 7: 11476.
109 Yu, D., Goh, K., Zhang, Q. et al. (2014). *Adv. Mater.* 26: 6790.
110 Senthilkumar, S.T. and Selvan, R.K. (2014). *Phys. Chem. Chem. Phys.* 16: 15692.
111 Yu, N., Yin, H., Zhang, W. et al. (2016). *Adv. Energy Mater.* 6: 1501458.
112 Yu, M., Cheng, X., Zeng, Y. et al. (2016). *Angew. Chem. Int. Edit.* 55: 6762.
113 Zhang, Z., Xiao, F., and Wang, S. (2015). *J. Mater. Chem. A* 3: 11215.
114 Xu, H.H., Hu, X.L., Sun, Y.M. et al. (2015). *Nano Res.* 8: 1148.
115 Dong, X., Guo, Z., Song, Y. et al. (2014). *Adv. Funct. Mater.* 24: 3405.
116 Wang, X., Liu, B., Liu, R. et al. (2014). *Angew. Chem. Int. Edit.* 53: 1849.
117 Yu, Z., Moore, J., Calderon, J. et al. (2015). *Small* 11: 5289.
118 Jin, H.Y., Zhou, L.M., Mak, C.L. et al. (2015). *Nano Energy* 11: 662.
119 Su, F. and Miao, M. (2014). *Nanotechnology* 25: 135401.
120 Cheng, X.L., Zhang, J., Ren, J. et al. (2016). *J. Phys. Chem. C* 120: 9685.
121 Li, W., Torres, D., Díaz, R. et al. (2017). Nanogenerator-based dual-functional and self-powered thin patch loudspeaker or microphone for flexible electronics. *Nat Commun* 8: 15310.

2

Stretchable Supercapacitors

La Li and Guozhen Shen

State Key Laboratory for Superlattices and Microstructures, Institute of Semiconductors, Chinese Academy of Science, Beijing, 100083, China

2.1 Overview of Stretchable Supercapacitors

Stretchable supercapacitors (SCs) that possess both flexibility and stretchability in terms of mechanical property and easy integration in terms of whole electro circuit design have attracted plenty of interest because they fulfill the demands of wearable or skin-attachable electronic devices on energy storage [1–5]. By couple with stretchable SCs, portable/ wearable devices could easily realize the special health monitor and chemical, physical, biological, etc. signal detection without impacting on the size, volume, mass of the wearable electronic [6–8]. Current stretchable SCs are composed of deformable substrate, electrode materials, and all-solid-state electrolyte, which are much simplified in comparison with traditional SCs that contain two other components: current collector and separator.

The energy storage mechanism of stretchable SCs is similar to traditional SCs, which can be divided into pseudocapacitors and electrical double-layer capacitors (EDLCs) according to the used electrode materials [9–12]. These two kinds of SCs have their merit and shortcoming, pseudocapacitors presented by the metal oxides, conductive polymers and nanocomposites possess the advantages of much higher capacitance owing to the reversible faradaic reactions at the electrode/ electrolyte interface, for example, the theoretical-capacitance value of two-dimensional layered double hydroxide electrode materials is as high as $3000\,F\,g^{-1}$, but suffer from poor cycling life, inferior rate capability and relatively low conductivity [13–17]. Porous carbon materials based EDLCs with benefits of ultra-long cycling life, high power density, good chemical stability, non-toxic and environmental friendliness are of importance for SCs, but their specific capacitance is limited to fewer than about $300\,F\,g^{-1}$ [18–22]. The electrochemical performance like specific capacitance, stability, lifespan etc. under different deformation is a basal parameter in the process of development of stretchable SCs, must be considered.

Stretchable substrate, which could assist the energy storage to be fabricated or integrated on clothes and the human body, is highly desirable for wearable electronic devices because human skin and human body activity often go along with a certain extent of stretchability [23, 24]. Polydimethylsiloxane (PDMS) is the most frequent use substrate in the stretchable devices owing to its high strain (>100%), which is much stronger than that of stainless steel mesh (~20%) [25–27]. There are many other typical stretchable substrates, such as polyurethane (PU) [28], elastic yarns, thermoplastic polyurethanes (TPU), thermoplastic elastomer (TPE) [29] with properties

Flexible Supercapacitors: Materials and Applications, First Edition. Edited by Guozhen Shen, Zheng Lou and Di Chen.

of large strain and free shape, which make stretchable devices work excellent under various deformation. Self-healability as an attractive high elasticity has been included into stretchable substrate, which could get the wearable devices back to life while the devices inevitably encounter malfunction during their lifetime. Several kinds of self-healable substrates like polyvinyl alcohol (PVA) [30], polyacrylic acid (PAA) [31], polyacrylamide (PAM) [32] have been synthesized in recent years.

All-solid-state gel electrolytes are one of the essential components for stretchable SCs in design of fully wearable energy storage devices, which could protect SCs from the risks of liquid leakage and simplify the device configurations of the SCs by removing the extra separator and substrates of the devices. On the basis of solvent type, gel electrolytes are generally considered to fall into the below two categories: hydrogel electrolyte and organic ions/solvent-based gel electrolyte. The ionic conductivity of hydrogel electrolyte can reach $10^{-4}\,S\,cm^{-1}$, up to $10^{-1}\,S\,cm^{-1}$ [2]. Organic gel electrolyte exhibits excellent stability in air and enhanced electrochemical performance by adding organic ions to electrolyte, which is a unique way to improve the electrochemical properties of SCs devices.

To date, many achievements have been made to fabricate stretchable SCs that include both 1D fiber SCs, 2D planar SCs and 3D structured SCs [33, 34]. The fiber shaped energy devices have been woven into flexible energy textiles while and the planar devices have been proved that they could be attached to the human body. Figure 2.1 presents the typical 1D, 2D, and 3D stretchable SCs and their application in wearable electronics, as well as the multifunctional stretchable SCs contain self-healable SCs, compressible SCs, and their integration with other wearable devices.

In this chapter, we focus on the recent progress in stretchable SCs and their potential application in wearable electronics. First, the main approaches to assemble stretchable SCs consist of both 1D fiber and 2D planar devices are presented. Then, we describe the main electrochemical and mechanical performances in the field of the stretchable SCs. The multifunctional SCs such as self-healable SCs, compressible SCs and the self-powered integrated system are also highlighted. Finally, we discuss the existing challenges and future development trends for stretchable SCs.

2.2 Fabrication of Stretchable Supercapacitor

As previously mentioned, viewed from device dimension, SCs have three main categories, 1 D fiber SCs, 2D planar, and 3D SCs. Here, the devices, structure design and strategies for making SCs stretchable in three dimensions are summarized. These approaches can be used to fabricate stretchable devices, which have potential applications beyond stretchable energy storage, including all-in-one stretchable integrated system, where all unites with the same substrate, high stretchability and general applicability are desirable.

2.2.1 Structures of Stretchable Fiber-Shaped SCs

Recently, 1D SCs, also known as fiber, wire, or yarn-shaped SCs with advantages of light weight and easy integration have received significant attention as one ideal form for wearable electronics because they can be weaved into textiles or smart clothes [43–46]. 1D fiber electrode held together in different configurations can form several kinds of fiber shaped SCs, including parallel, twisted, and coaxial structure. Parallel SCs refer to two fiber electrodes constructed side by side, these two fiber electrodes can be attached on the same two fiber substrates or both sides of one substrate in a parallel state [47]. Twisted configuration is a structure of two intertwining fiber electrodes,

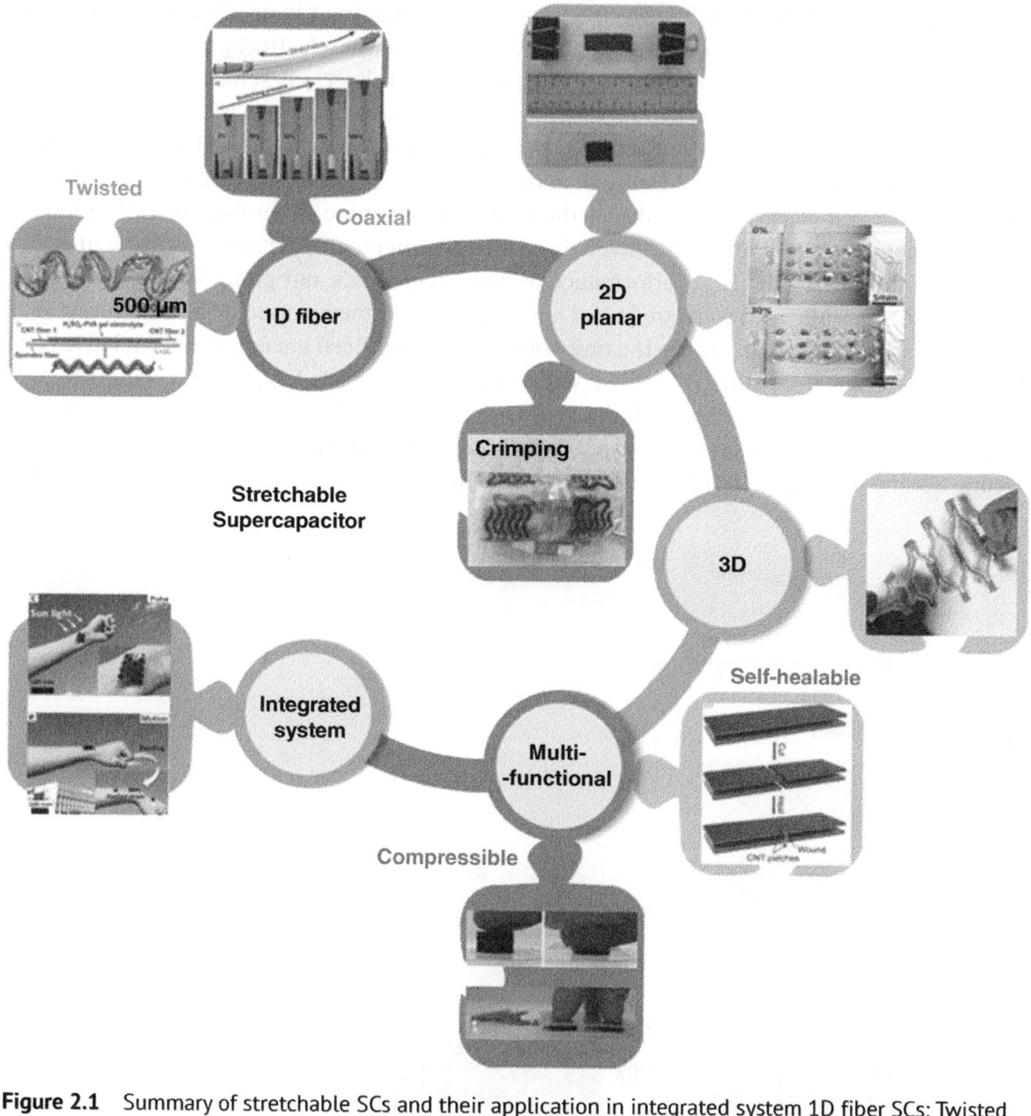

Figure 2.1 Summary of stretchable SCs and their application in integrated system 1D fiber SCs: Twisted SCs, *Source:* Reproduced with permission [35]. © 2014, Wiley-VCH. Coaxial SCs, *Source:* Reproduced with permission [36]. © 2013, Wiley-VCH. 2D planar SCs, the individual elements are: *Source:* Reproduced with permission [37]. © 2013, Wiley-VCH. Reproduced with permission [38]. © 2013, American Chemical Society. Reproduced with permission [39]. © 2017, Wiley-VCH. 3D SCs, *Source:* Reproduced with permission [40]. © 2016, American Chemical Society. Multifunctional: self-healable SCs, *Source:* Reproduced with permission [31]. © 2015, Nature Publishing Group. Compressible SCs, *Source:* Reproduced with permission [41]. © 2015, WILEY-VCH. Integrated system, *Source:* Reproduced with permission [101]. © 2018, Elsevier Ltd.

separated by all-solid-state gel electrolyte in the middle, which is the most popular structures in fabrication of fiber SCs [48]. The layout of two electrodes of coaxial SCs can be geometrically viewed as a column situated within a cylinder sharing the same center axis, the inter electrode and outer electrode are separated by gel electrolyte or separator, showing a core–sheath configuration [44, 49, 50].

By design of different structure, the electrochemical performance of fiber SCs can be improved. For example, Peng's group [51] made a comparative study to analyze electrochemical performance of coaxial and twisted SCs, in detail, carbon nanotube (CNT) sheet was scrolled on the another CNT sheets with a gel electrolyte coat to fabricate coaxial devices and two CNT sheets fiber electrodes were intertwined to obtain twisted SCs, the electrochemical tests showed that coaxial SCs exhibit larger specific capacitance of 59 F/g than that of twisted SCs (4.5 F g^{-1}) because the distance between positive and negative electrodes of the coaxial SCs is more closer and the utilization of materials is more effective compared to the twisted one. In order to understand how the configuration effect on the electrochemical performance of fiber shaped SCs, our group built a theoretical model using ANSYS Maxwell software [52]. The simulated parameters used in simulation system were particularly organized to ensure the unity of volume, superficial area, materials, relative permittivity, and positive/negative voltage between coaxial and twisted geometries. At first, two electrode models were carried out with the same electrode material and electrolyte parameters, the voltage of electrode was set from −0.4 V to 0.4 V. The analysis results on electric field under the same conductions were displayed, as shown in Figure 2.2. Figure 2.2a and b depicted the

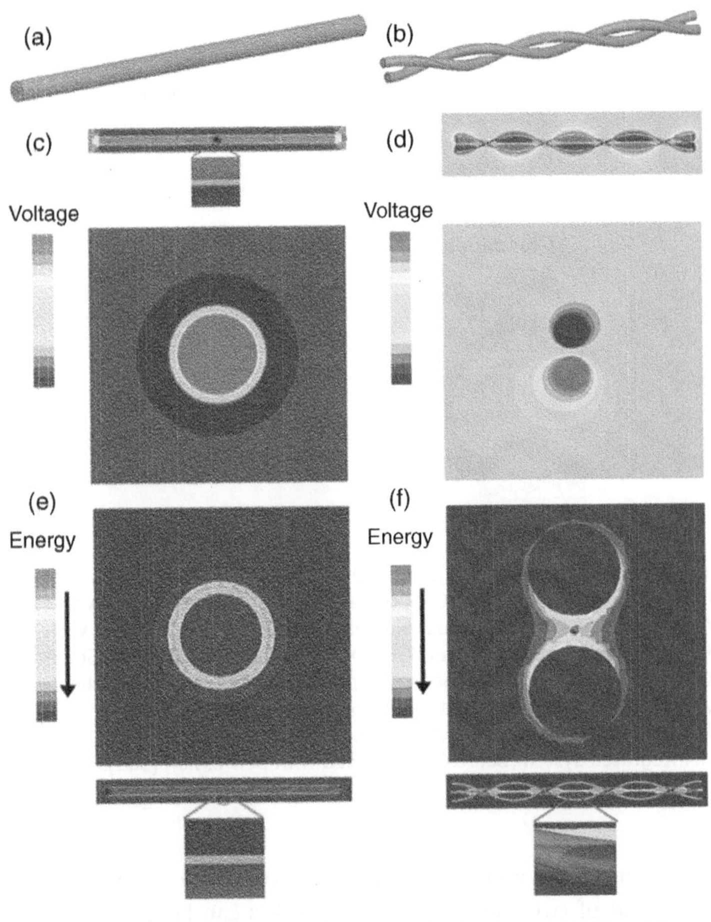

Figure 2.2 Structure and voltage, energy distribution of 1D fiber SCs: twisted and coaxial. (a, b) Schematic diagram of the coaxial and twisted fiber SCs. (c–f) The voltage distribution and energy distribution of the two structures simulated by ANSYS Maxwell software. *Source:* Reproduced with permission [52]. © 2018, Wiley-VCH.

schematic diagram of the coaxial and twisted fiber SCs used in the theoretical model. Figure 2.2c and d displayed the corresponding electrostatic potential distribution around the two fiber SCs. Compared to the scattered, asymmetrical, and sigmoid potential distribution of the twisted devices, the potential distribution of coaxial SCs is more homogeneous, leading to the better electric charge-transfer effect and lower contact resistance. Figure 2.2e and f showed the energy distribution simulation results of the twisted and coaxial devices.

According to the following equation, the energy densities (E) and capacitance (C) can be calculated:

$$\mathrm{C} = \frac{Q}{V} dE = dQ * V \tag{2.1}$$

The Eq. (2.2) was obtained as:

$$\begin{aligned} Q_1 &= C_{10}V_1 + C_{12}\left(V_1 - V_2\right) \\ Q_2 &= C_{20}V_2 + C_{21}\left(V_2 - V_1\right) \end{aligned} \tag{2.2}$$

Then capacitance matrix is derived from Eq. (2.2) as:

$$\begin{pmatrix} Q_1 \\ Q2 \end{pmatrix} = \begin{bmatrix} C_{10} + C_{12} & -C_{12} \\ C_{20} + C_{21} & -C_{21} \end{bmatrix} \begin{pmatrix} V_1 \\ V_2 \end{pmatrix} \tag{2.3}$$

Where Q represents the quantity of electric charge, V stands for voltage.

A capacitance ratio of 5.14 : 1 was obtained between the two structures, confirming and explaining the better electrochemical performances of the coaxial SCs coming from the shorter distance between positive and negative electrode, more uniform potential distribution and better conductivity. But the fabrication process of the coaxial SCs is much more complicated in comparison with twisted devices. These simulation results provide a theoretical support to the design of the fiber SCs from the views of the device configuration, which can be applied and extended to other filed that refers to the devices structures.

2.2.1.1 Fabrication of Stretchable Parallel SCs

The fabrication of stretchable, fiber-shaped SCs with parallel structure is relatively easy. There are two typical ways to realize stretchability: one is by packaging the devices into stretchable substrate, another is by employing stretchable substrate or designing special structure, like screw spring, serpentine, etc. to implement it [53–55]. Figure 2.3a concretely displayed the first group of stretchable SCs by embedding the whole devices into pre-strained stretchable substrate [56]. The CNT fibers, which were prepared by drawing and twisting CNT strips out of a vertically aligned CNTs array, were used as the wire-shaped current collector. The highly aligned inner structures can provide both excellent mechanical properties and high electrical conductivity. Afterwards,($KMnO_4$) was used. Then, the two MnO_2 coated CNT fibers were placed parallelly into pre-strained stretchable substrate, after releasing, the stretchable fiber SCs were obtained with serpentine structures. Owing to the excellent electrochemical performance of MnO_2 nanomaterials, the outstanding electrical conductivity of CNT fibers as well as the pre-straining method, the fabricated fiber SCs showed a specific volumetric capacitance of ≈409.4 F cm^{-3} at 0.75 A cm^{-3}at 40% of tensile strength. Figure 2.3b exhibited the cyclic voltammetry (CV) curves of the stretchable devices under stretching and folding, nearly invariable CV curves under different deformation state demonstrated the successful assembly of the stretchable fiber SCs. Although this method achieved the stretchability,

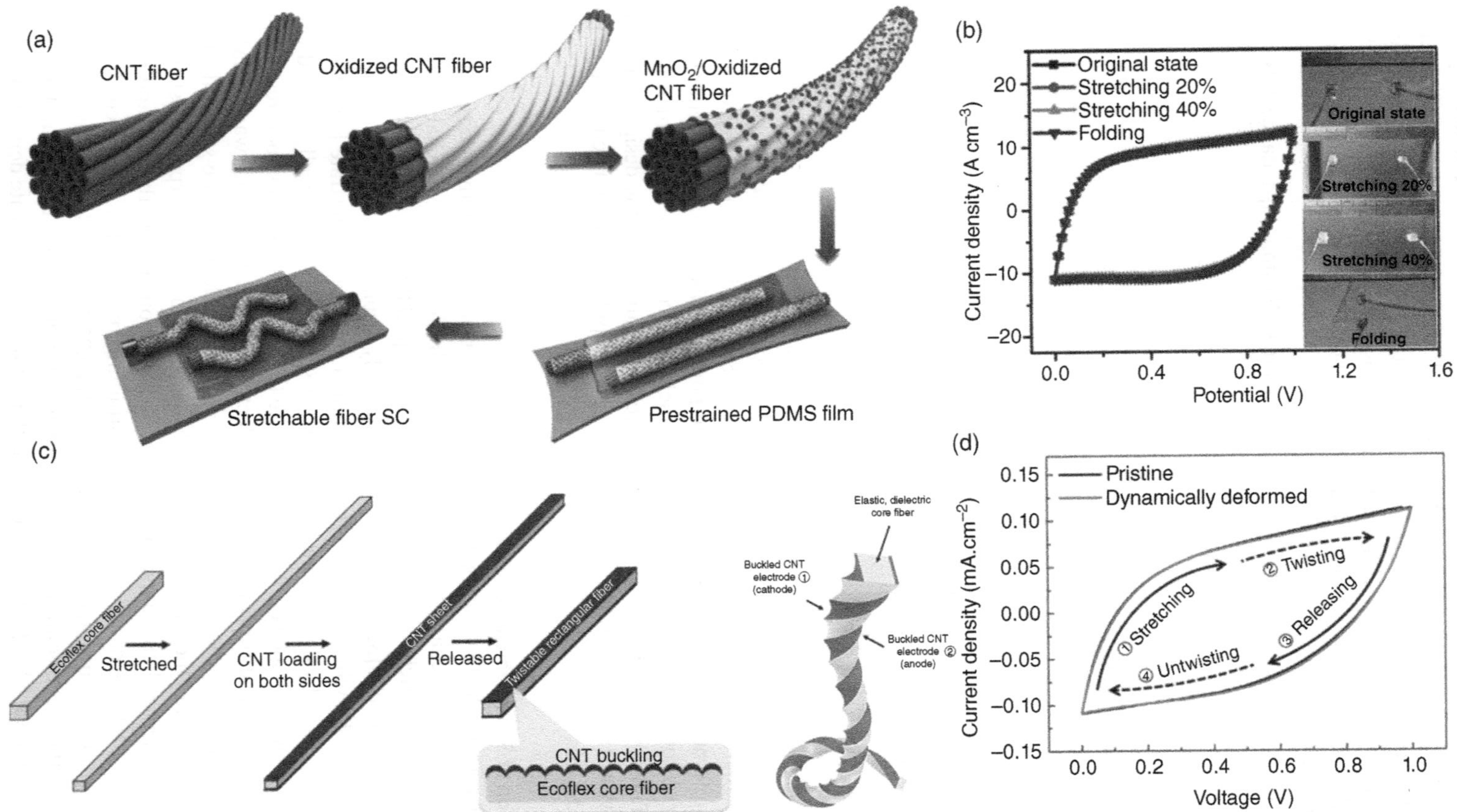

Figure 2.3 Typical fabrication methods of 1D parallel fiber SCs: (a) Schematic illustration of the fabrication process of the fiber SCs by embedding in pre-strained stretchable substrate, (b) CV curves of the stretchable devices under stretching and folding. *Source:* Reproduced with permission [56]. © 2017, Wiley-VCH. (c) Schematic illustration of the fabrication process of the fiber SCs by twisting sandwich fiber with double side electrode, (d) CV curves measured during successive stretch, twist, release and untwist process. *Source:* Reproduced with permission [57]. © 2016, American Chemical Society.

the SC device itself did not live up to the conception of stretchability. Therefore, they always suffer from low tensile strength and poor stretching cycle stability.

Different from the first embedded method, the second parallel SCs always employ the elastic materials as the stretchable substrate. In 2016, Choi et al. introduced stretchable SCs with sandwich fiber, the schematic illustration of the fabrication process was presented in Figure 2.3c [57]. The assembled method can be summarized as four steps: (i) Preparation of fiber shaped substrate. The dielectric silicone rubber (Ecoflex 0030) was processed as a sheet with a length of 30 mm by the conventional reaction of commercially available resins in a rectangular and then cut into ~2 mm thick. (ii) Pre-stretching the fiber shaped silicone rubber. The as-prepared core fiber was stretched along the axial direction by 300% of its length. (iii) Covering electrode material on the surface of the core fiber. CNT aerogel sheets that extract from a multiwalled carbon nanotube (MWNT) forest served as electrode materials, which is attached on both sides of the silicone rubber fiber. The direction normal to the CNT electrode was coved with a PU film to prevent electrical shortage by residual CNT bundles during CNT wrapping. The strong adhesion between the CNT electrode and the silicone rubber fiber was guaranteed by continual dropping ethanol during CNT wrapping, which also protects the CNT surface from delamination. Then, the PU mask was mechanically removed from the silicone rubber fiber with CNT coat. 4: Release the fiber. The CNT electrode materials after release archived a uniform waved electrode. The electrical resistance of the CNT electrode materials is about 179 $\Omega\,cm^{-1}$ at the CNT loading of 2.31 $mg\,cm^{-2}$ for each side, which is benefit to the electrochemical performances of the stretchable fiber SCs. Figure 2.3d showed the CV curves of the fiber SCs measured during successive stretch, twist, release and untwist process. Importantly, when applied a successive stretch, twist, release and untwist deformation during charge and discharge, the electrochemical performance of the fiber devices didn't change. The stretch test also showed an imperceptible nuance under various strains ranging from 0% to 200%.

2.2.1.2 Fabrication of Stretchable Twisted SCs

The preparation technology of the stretchable twisted SCs has been well developed in the recent years. Several kinds of strategies toward stretchable twisted SCs were summarized in Figure 2.4. Figure 2.4a showed the schematic illustration of a fabrication method through twisting two pre-stretched elastic wires wrapped with aligned CNT sheet as electrode materials [58]. First, the elastic wire was stretched to 100%. Then, the aligned CNT sheet was transformed to the pre-stretched wire. Next, two poly (vinyl alcohol) (PVA)/H_3PO_4 coated elastic wires with CNT electrode were twisted. After releasing the pre-strain wire to its nature state, the stretchable twisted SCs were prepared, which showed extremely high strain (>300%) without detrimental effects to the conductivity of the electrode. Figures 2.4b exhibited the top-view and cross-section SEM (scanning electron microscope) images of the twisted SCs. From the SEM images, we can see that the thickness of PVA/H_3PO_4 and CNT sheets were 50 and 10 mm, respectively. The electrochemical performances of the fabricated twisted SCs were also carried out, as displayed in Figure 2.4c. The inset is the photos of the twisted devices under various strains ranging from 0% to 370%. The normalized capacitance (a ratio of the capacitance under a specific strain to that at 0 strain) as a function of the tensile strain showed a 13% of increase for the device based on CNT wrapped elastic wire electrodes as the tensile strain increased from 0% to 250%. This can be explained by the strain-induced enhancement in the contact between the two twisted electrodes upon stretching. When the tensile strength was above 250%, a slight decrease in the capacitance was appeared caused by the imperfect contact between CNT and/or PVA/H_3PO_4, but it was still 10% higher than the initial capacitance. Moreover, in this paper, the CNT/PEDOT-PSS (poly (3,4-ethylenedioxythiophene)-poly

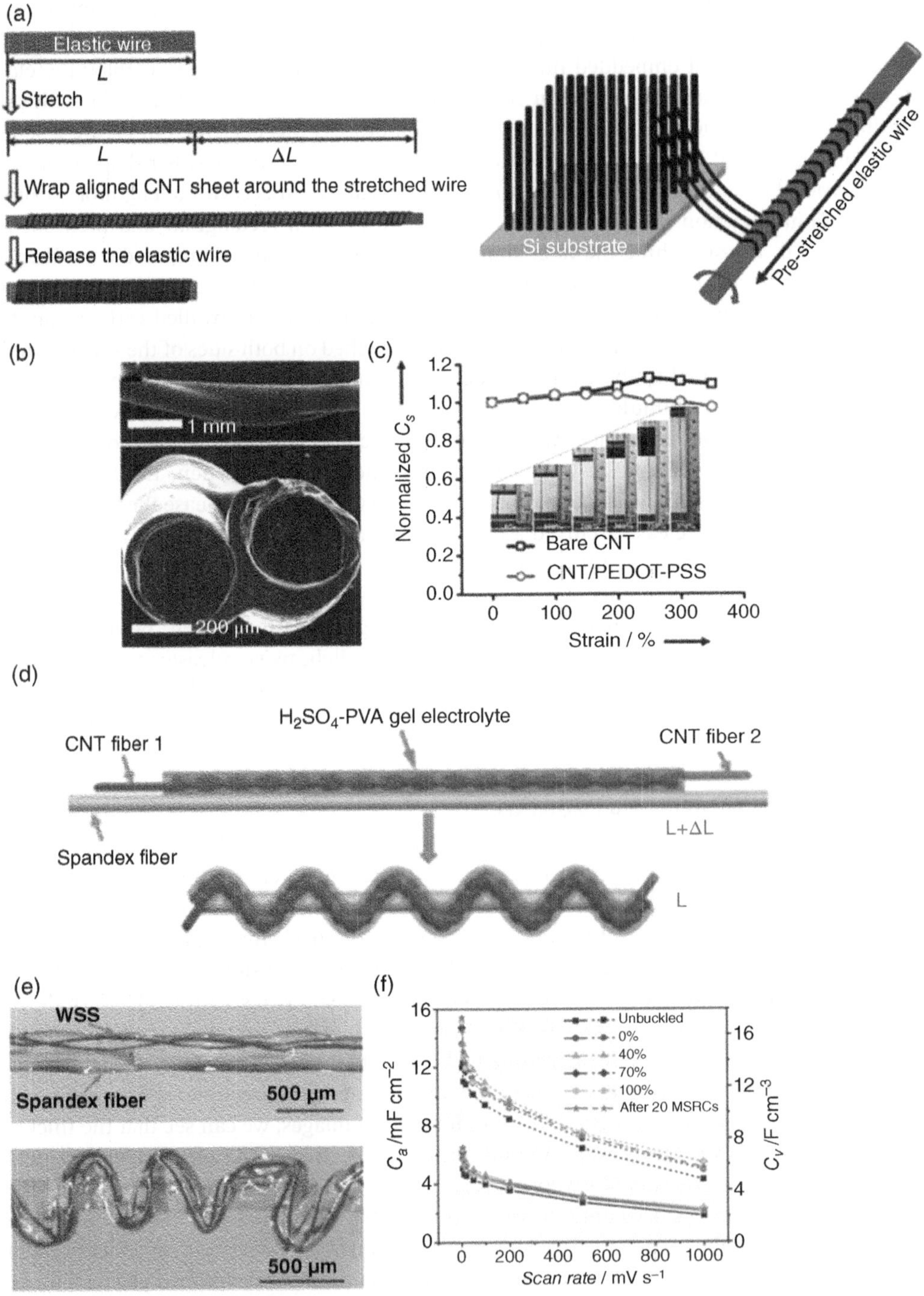

Figure 2.4 (a) Schematic illustration of fabricating twisted SCs by wrapping aligned CNT sheet around a pre-stretched elastic wire. (b) Top-view and cross section SEM images of the fiber SCs. (c) Changes in the normalized capacitance under stretching. *Source:* Reproduced with permission [58]. © 2014, Wiley-VCH. (d) Schematics of the fabrication procedures for stretchable fiber SCs by twisting the devices. (e) Optical microscopy images the fiber devices. (f) Specific capacitance of the fiber SCs under stretching. *Source:* Reproduced with permission [35]. © 2014, Wiley-VCH.

(styrene sulfonate)) based stretchable twisted SCs were also fabricated using the same technology, which depicted an excellent capacitance retention under different stretching state, indicating that it's a feasible method to large-scale produce different kinds of stretchable twisted SCs.

Another group of stretchable twisted SC have no ability of stretchability itself, with the whole devices twined around the elastic substrate. For example, Xu et al. proposed a fabrication method by using spandex fiber as the substrate (Figure 2.4d) [35]. Specifically, a fiber-shaped SC was fabricated by twisting two CNT based fiber electrodes. Then, next, spandex fiber was pre-stretched to 100% with a PDMS coat to prevent influence from acidic H_2SO_4-PVA gel electrolyte. And then the fiber-shaped SCs were "glued" onto the spandex fiber. When the substrate was released, the SCs with stretchability over 100% were accomplished. Figure 2.4e showed the photography of the serpentiform stretchable SCs. A slightly enlarged of the CV loop areas suggested the enhanced capacitance when the SC was stretched to a large strain of up to 100%. Moreover, the specific capacitance of the fiber SCs under stretching (Figure 2.4f) also surpassed those of the initial unbuckled SCs after 20 cycles of deep stretching. The outstanding stretchability can contribute to the buckled structures formed after releasing the strain. This fabrication method could make most of the twisted SCs stretchable, but the devices and the stretchable substrate are two separate parts, which is unfavorable to keeping the electrochemical performances under repeated deformation. Nevertheless, more advanced technologies for solving the above-mentioned problems are demanded imminently to realize its practical application in wearable electronics.

2.2.1.3 Fabrication of Stretchable Coaxial SCs

Among the three structures of fiber SCs, the fabrication of stretchable coaxial SCs is the most complicated, but this configuration makes the device more integrated than the parallel and twisted structures [59]. The biggest difficulty lies in the fabrication of sheath/shell-like electrode (outer electrode). To date, many efforts have been devoted to the development of novel coaxial SCs. For example, Lee et.al proposed an assembly method by wrapping separator and carbon nanofiber (CNF) film on the surface of the core fiber electrode, but it makes the contact resistance large and the electrode materials waste [44]. In contrast to this structure, Peng's group reported a stretchable coaxial SCs that fabricate layer by layer, from inside to outside, in sequence of inner electrode, gel electrolyte and outer electrolyte without separator, as shown in Figure 2.5a and b [36]. Figure 2.5a showed the schematic illustration of the corresponding fabrication process. An elastic fiber was used as stretchable core fiber substrate. To start with, a coat of PVA-H_3PO_4 gel electrolyte was made on the surface of the elastic fiber. Next, the aligned CNT sheets were wrapped around the gel electrolyte coted elastic fiber as the inner electrode. Followed by coating the second layer of electrolyte as separator and wrapping with another CNT sheet as the outer electrode. Finally, the third layer of gel electrolyte was spread around the outer electrode. The fabricated SCs exhibited a specific capacitance of 20 $F\,g^{-1}$, mass energy of 0.515 $W\,h\,kg^{-1}$and power densities of 421 $W\,kg^{-1}$. Figure 2.5b presented the CV profiles of the stretchable coaxial SCs with increasing strains. The overall stretchability could reach as high as 100% with a good specific capacitance retention. The specific capacitance was remained above 90% after 1000 charge-discharge cycles when applied a strain of 75%, showing a promising stability under deformation. Although they achieved the goal of produce stretchable coaxial SCs, unfortunately, reports so far on high performance have been very limited. Great efforts should also be focused on promoting the electrochemical properties of stretchable coaxial devices.

Recently, Yu et al. fabricated a stretchable helical coaxial asymmetrical SCs with high energy and power density by employing two CNT@MnO_2 and CNT@PPy as positive and negative electrode respectively without a stretchable polymer core substrate [60]. Figure 2.5c depicted the schematics

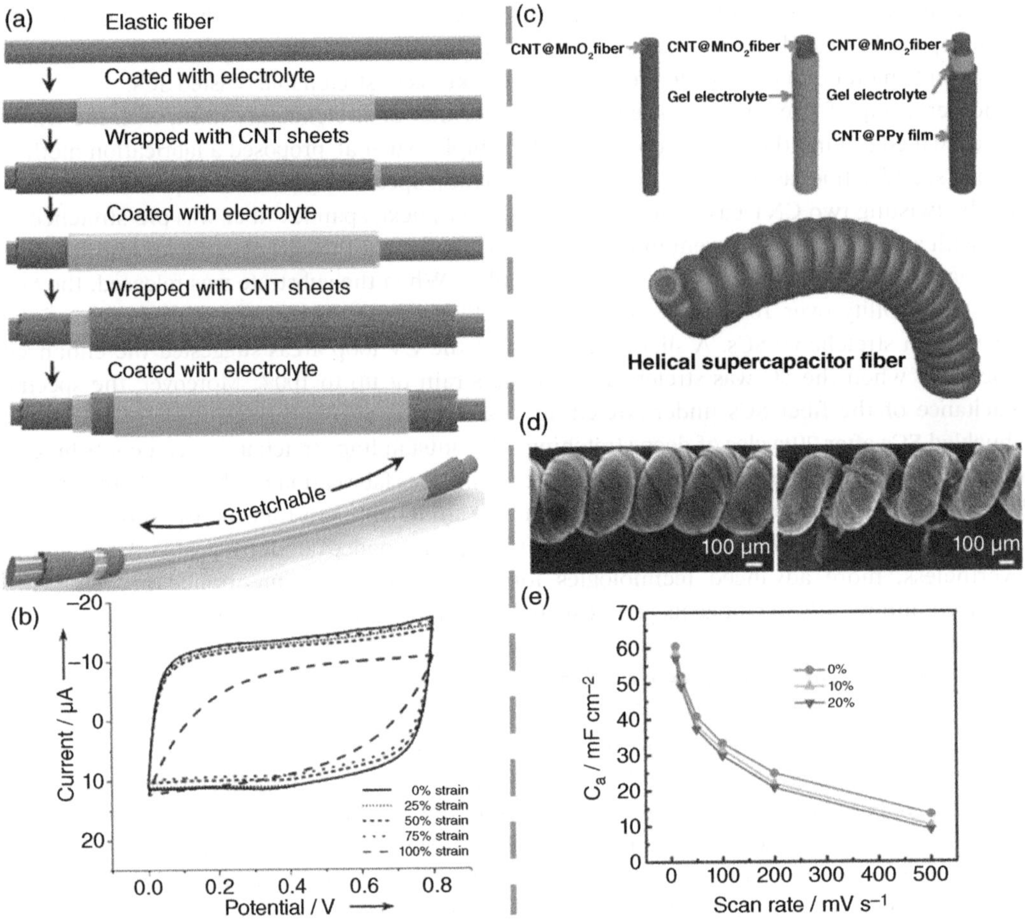

Figure 2.5 Schematics of the fabrication procedures for coaxial SCs and corresponding electrochemical performances at different tensile strains. (a, b) Pre-stretching substrate and electrode. *Source:* Reproduced with permission [36]. © 2013, Wiley-VCH. (c, d) Over-twisting the SCs device into a helical structure. (e) Areal specific capacitance variations with scan rates. *Source:* Reproduced with permission [60]. © 2016, Wiley-VCH.

of the fabrication procedures. To start with, a CNT fiber was decorated with MnO_2 nanomaterials, which served as inter electrode. Next, the gel electrolyte was coated around the CNT@MnO_2 fiber. Then the CNT@PPy film was wrapped on the gel electrolyte as the outer electrode. The cross-section SEM images of asymmetric coaxial SCs at 0% and 50% strain were displayed in Figure 2.5d. From the SEM images, we can clearly see the two concentric circular areas, indicating the core-sheath structures of the fabricated SCs. And the SCs possess a helical structure formed by over-twisting the devices which make the SCs stretchable. Owing to the utilization of two different electrode materials, the fabricated SCs exhibited a specific capacitance of 60.435 $mF\,cm^{-2}$ at the scan rate of 10 $mV\,s^{-1}$ and a high energy density of 18.88 $\mu W\,h\,cm^{-2}$ with an extended potential window of 1.5 V. The specific areal capacitance curves in Figure 2.5e showed a slightly decrease when the SC was stretched to 20%. Despite the limited stretchability of the coaxial SCs, the electrochemical performance was much improved, paving a great support to the real application of the stretchable SCs.

2.2.2 Planar Stretchable SCs

Stretchable 2D planar SCs with excellent properties of small size, low weight, excellent lifespan, high security and easy integration have become a preferred choice as energy storage to power the wearable electronics [37, 61–63]. There are two main categories including layer by layer sandwich structure and micro supercapacitors (MSCs). The fabrication method of the 2D planar stretchable SCs is similar to 1D fiber shaped SCs. One is via embedding rigid independent devices to the stretchable substrate or establishing serpentine interconnects between rigid devices to realize stretchability. Another one is replacing the rigid unit by stretchable component. In this section, we will introduce the typical fabrication method reported during the last few years.

2.2.2.1 Fabrication of the Stretchable Planar SCs with Sandwich Structure

Sandwiched planar SCs is the most developed and widely applied structure in the field of the SCs. The main issues that impede the successful fabrication of stretchable SCs is how to make a tight connection between electrode materials and elastic substrate. As early as 2009, Yu et al. proposed a strategy via directly transferring the single-walled carbon nanotube (SWCNT) film to pre-strained PDMS substrate [64]. It's worth mentioning that the PDMS substrate was treated by UV light to forming a strong bong with SWCNT film. The fabrication process was presented in Figure 2.6a. Figure 2.6b showed the SEM images of the SWCNT film with buckled microstructures, which directly contribute to the stretchability of the fabricated SCs. Figure 2.6c displayed the CV profiles of the stretchable SCs measured at 30% strain. From the CV curves, we can see that the electrochemical performance of the stretchable SCs remains unchanged even under 30% applied tensile strain. But the utilization of a polymeric separator restricted its stretchability. To overcome this problem, Niu et al. reported stretchable SCs based on periodically sinusoidal oriented SWNT film without separator and liquid electrolyte using the same pre-strained-buckled-release method, which showed increased stretchability of over 120%.

Most of the deformable substrate used in the field of stretchable SCs is PDMS. In 2014, Xie et al. reported a flat Ni foam based stretchable all-solid-state SC with wavy shaped polyaniline (PANI)/graphene electrode [65]. Figure 2.6d showed the schematics of the fabrication process for fabricating the PANI/graphene electrodes based stretchable SCs. First, a flat Ni foam with a thickness of 200 mm was manually made into a wavy shape, next, the porous graphene was synthesized on the buckled Ni foam via atmospheric pressure chemical vapor deposition (CVD). Then the graphene coated Ni foam was put in a solution of 3 M HCl to etch nickel foam to obtain wavy-shaped graphene film. In order to improve the electrochemical performance of the SCs, the PANI was deposited on the wavy shaped graphene film. Finally, two PANI covered graphene films with PVA/H_3PO_4 wall were encapsulated into Elastic substrate (Ecoflex). Figure 2.6e depicted the CV curves of the stretchable SCs at different tensile strains. The initial specific capacitances calculated from CV curves were 261 $F\,g^{-1}$. It can be seen that no obvious change appeared when the SC was stretched to 30%. Moreover, the stretching cycle tests revealed that the SC maintained high mechanical strength and stability over 100 cycles.

2.2.2.2 Omnidirectionally Stretchable Planar SCs

From the aforementioned stretchable actions, it can be concluded that stretchability can be realized without affecting on the electrochemical performances of the SCs. There are even some operations with enhanced electrochemical performance when the SCs devices was stretched due to the more contact between electrode and gel electrolyte under stretching. Unfortunately, these actions can only be stretched along one direction, hence, if it is possible to make SCs isotropic stretchable,

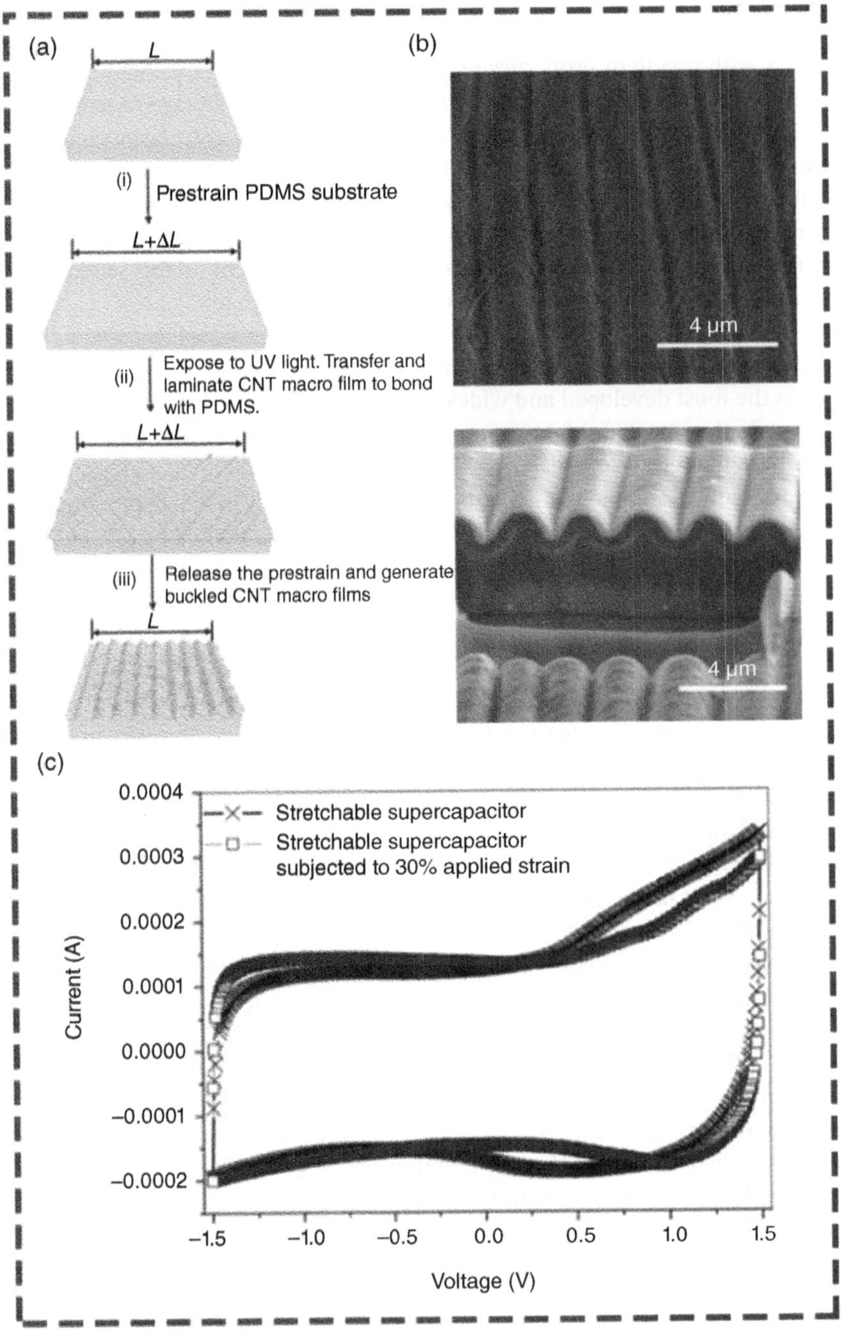

Figure 2.6 (a) Fabrication process of the stretchable SCs by buckling electrode materials on an elastomeric PDMS substrate. (b) SEM image of a buckled CNT macro film. (c) CV profiles of the stretchable SCs measured at 30% strain. *Source:* Reproduced with permission [64]. © 2009, Wiley-VCH. (d) Schematics of the stretchable SCs fabrication. (e) CV curves of the stretchable SCs at different tensile strains. *Source:* Reproduced with permission [65]. © 2014, The Royal Society of Chemistry.

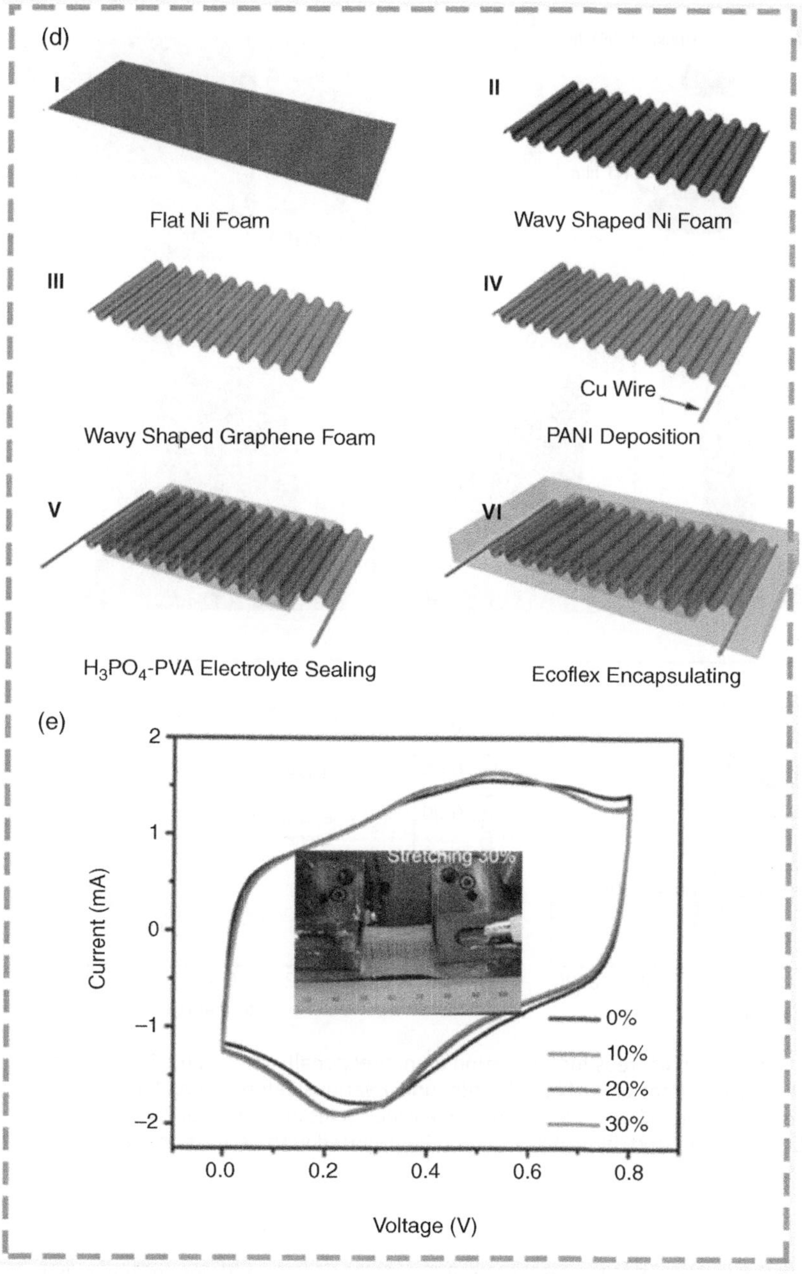

Figure 2.6 (Continued)

the electrochemical performance, such as specific capacitance, cycle stability etc. also could be improved in a certain degree.

In 2016, Yu et al. designed a novel isotropic wavy shaped CNT film electrode based omnidirectionally stretchable SCs, as shown in Figure 2.7 [66]. Figure 2.7a showed the fabrication process. In detail, the PDMS elastic film was uniformly pre-strained in all direction, and then CNT film was transformed to the PDMS substrate, finally, after relaxation, the omnidirectionally stretchable SCs

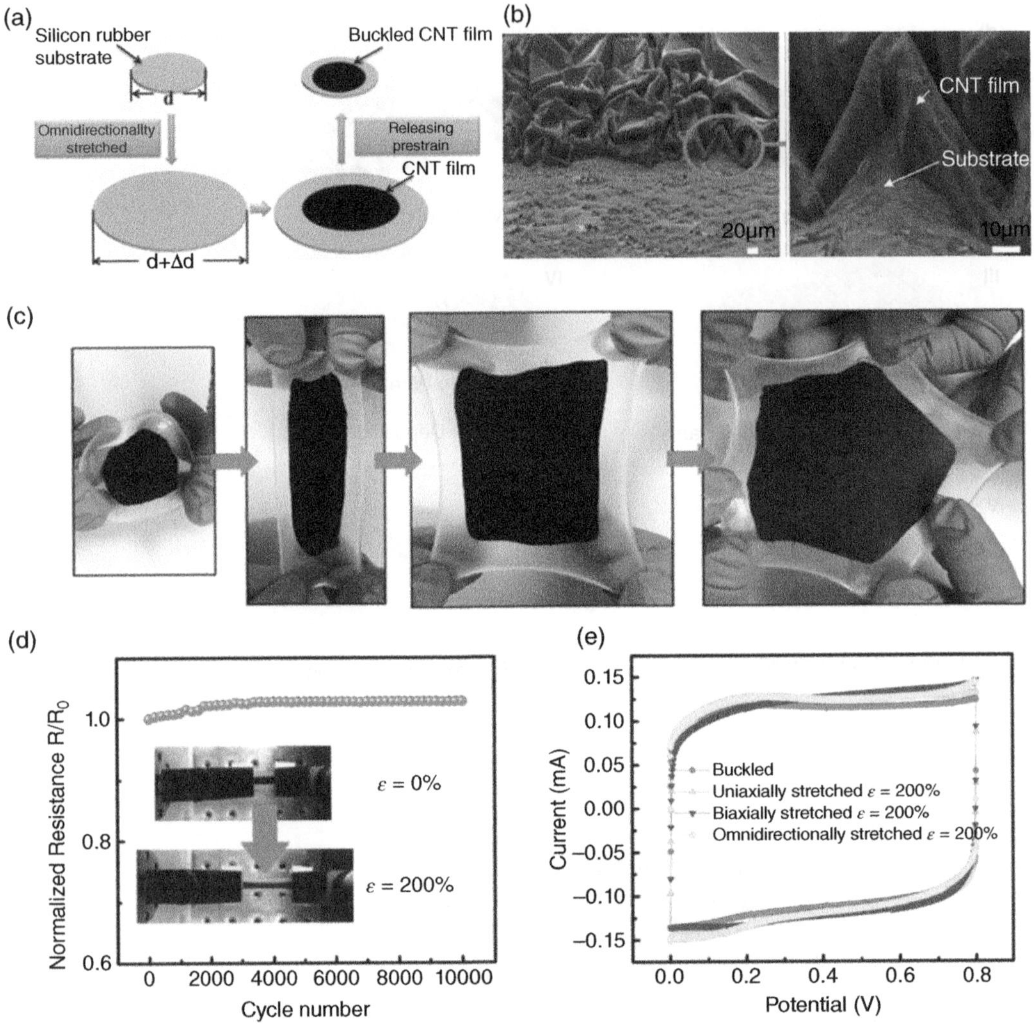

Figure 2.7 (a) Schematic illustration of steps for fabricating omnidirectionally stretchable SC. (b) SEM image of the buckled CNT film. (c) Photos of buckled CNT film under various deformations. (d) Normalized electrical resistance of the stretchable SC under stretching–releasing cycles at a strain of 200%. (e) CV curves of the fabricated SC at various stretching states. *Source:* Reproduced with permission [66]. © 2016, American Chemical Society.

were assembled. The SEM images of side view of buckled CNT film on silicon rubber substrate were presented in Figure 2.7b. The buckled CNT film and tight connection will benefit the stretchability. Figure 2.7c demonstrated the excellent isotropic stretchability of the buckled CNT film. The resistance variation after 10 000 uniaxially stretching–releasing cycles was also measured (Figure 2.7d), which showed a little increase (< 3%) at tensile strains of 200%. Figure 2.7e displayed the CV curves of the fabricated SCs with various deformable states. No significant change was observed in the CVs with 200% applied omnidirectional strains. Moreover, the specific area capacitance was improved from 1160.43 to 1230.61 mF cm^{-2} during uniaxial, biaxial, and omnidirectional elongations, which showed wide potential application in stretchable electronics.

2.2.2.3 Stretchable On-Chip Micro Supercapacitors (MSCs)

In the past decades, great progress has been achieved in the development of on chip stretchable 2D devices due to their advantages of ultra-thin thickness, low weight, easy handing in appearance, feasibility of integration into miniaturized different kinds of wearable electronics like sensors, detector, nanorobot, etc. on the same elastic substrate and excellent mechanical performances under various deformation [16]. To overcome the low operation voltage windows of single on-chip MSC, MSC arrays with series or parallel connection often directly designed [38, 67–71]. To date, several papers refer to stretchable on-chip MSC have been published. For example, Ha's group fabricated a multi-walled carbon nanotubes (MWCNTs) @ Mn_3O_4 electrode based stretchable and patchable MSC array by dry-transformation method [2]. Figure 2.8a showed the fabrication process of the planar MSC array on a stretchable substrate. At first, typical photolithography technology was employed to fabricate single MSC devices, which were then embedded into Ecoflex substrate with microchannels to build interconnections between MSC arrays. The cross-section scheme of the single MSC was displayed in Figure 2.8b. It can be observed that single MSC itself had no stretch ability, but all-solid-state organic solvent-based gel electrolyte made it easy to be transformed into Ecoflex substrate. Figure 2.8c depicted the digital photography of the stretchable and patchable MSC arrays. No observable change in CV curves under repeated bending, twisting, and stretching state presented in Figure 2.8d, suggested the stable electrochemical performance of the MSC array. Importantly, even in water, a micro-light-emitting diode (μ-LED) can be easily lit by the fabricated MSC arrays. Moreover, the long-term stability with 85% of capacitance retention for two weeks in ambient air without encapsulation opens up the possibility of promising commercial potential.

Serpentine interconnects have been deemed to an effective method to prepare stretchable MSCs. A related work was done by Ha and co-workers in 2013, as shown in Figure 2.9 [39]. To achieve the stable electrochemical performance over deformation, the stretchable MSC arrays were fabricated as follows (Figure 2.9a): 400 nm thick polyimide (PI) film was spread on the SiO_2/Si substrate via spin-coating, then a Ti/Au (5/50 nm) film was sputtered on the PI film to form current collector with serpentine interconnections. The second PI film was then spread to form a neutral mechanical plane for Ti/Au electrode. After that, SWCNT as active materials was spread on the metal current collector. The MSC precursor was the transfer from rigid SiO_2/Si substrate to elastic PDMS substrate though tape. After drop-casting of ion-gel electrolyte and encapsulation, the stretchable MSC arrays were finally fabricated.

The obtained SWCNT electrodes based MSC array exhibited a capacitance of 100 μF at the scan rate of $0.5\,V\,s^{-1}$, power density of $70.5\,kW\,kg^{-1}$ at energy density of $11.5\,Wh\,kg^{-1}$. It also showed a stable electrochemical performance at the tensile strain of 30% due to the smooth deformation of serpentine interconnections. The MSC array with voltage window of 3 V can easily light a μ-LED, as shown in Figure 2.9b. Even under bending and stretching, no noticeable degradation can be observed, demonstrating its good mechanical stability and wide application in wearable and portable electronics.

Design of the wave shaped electrode is also an efficient way to obtain a stretchable MSC arrays. Our group proposed a 3D print assisted technology to assembled MWCNT/PANI electrode based stretchable MSC array, as illustrated in Figure 2.10 [72]. Figure 2.10a displayed the fabrication procedures. In details, we used 3D printer to prepare a photosensitive resin-based mold with a convex wavy electrode. Next, this convex wavy channel was transferred to PDMS film. Then 50 nm of Au film PDMS was sputtered on the surface of PDMS as the current collector. MWCNT/PANI electrode and polyvinylidene fluoride (PVDF) was mixed together to form an adhesive slurry, which was then injected to the concave wavy electrode. After coating the PMMA-PC-$LiClO_4$ gel

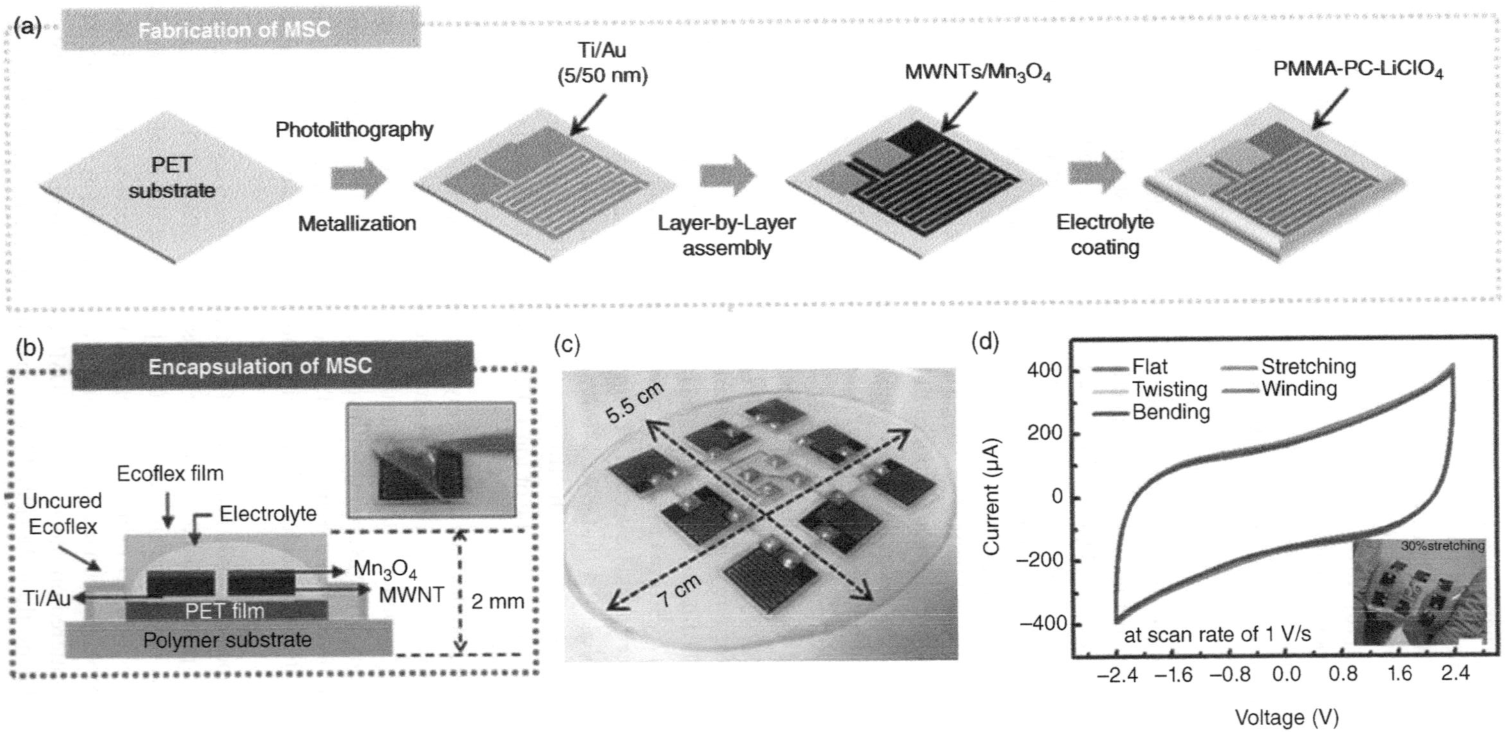

Figure 2.8 (a) Schematics of the fabrication procedures for a MWNT/Mn_3O_4 based planar stretchable MSC. (b) The encapsulation of the MSC. (c) The stretchable MSCs array with embedded liquid metal interconnections. (d) CV curves of the MSC array measured under different types of deformations. *Source:* Reproduced with permission [2]. © 2015, The Royal Society of Chemistry.

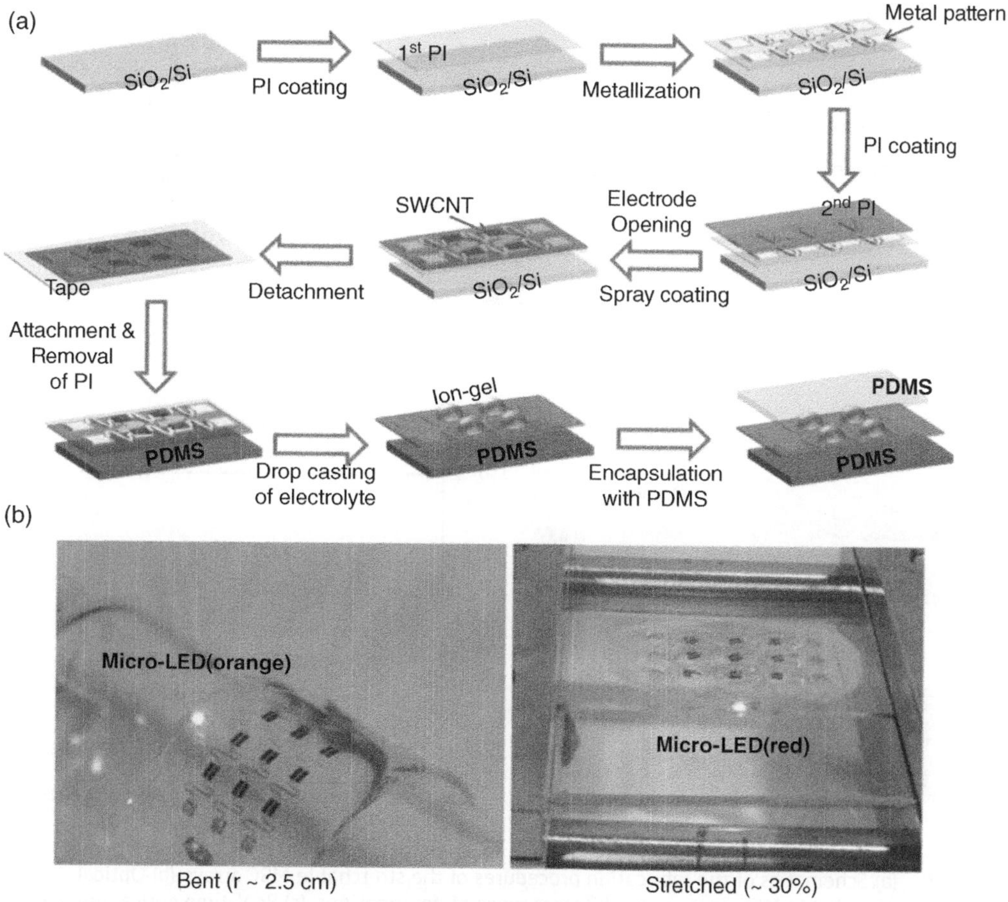

Figure 2.9 (a) Schematics of fabricating a stretchable MSC array on a PDMS substrate with serpentine interconnects. (b) Photographs of the μ-LEDs lighting test under bent and 30% stretched state. *Source:* Reproduced with permission [39]. © 2013, American Chemical Society.

electrolytes, the MSC array was prepared. The initial specific area capacitance calculated from CV curves was 44.13 mF cm^{-2} for the single MSC without applied strain, which showed negligible difference under various strain ranging from 0% to 40%. The maximum energy density was 0.004 mWh cm^{-2} at power density of 0.07 mW cm^{-2}. Moreover, the fabricated MSC device showed an excellent cycling stability with capacitance retention of 87% even after 20 000 charge-discharge cycles. Figure 2.10b showed the optical images of the stretchable MSC array and rolled up devices. The MSC arrays were connected as 4S (series) + 2P (parallel), so it can be charged to 3.2 V. During the test, we found after charging, the MSC arrays can maintain at 1.8 V for more than 20 000 seconds, which could provide a stable power to light a LED (Figure 2.10c). Even after 20 minutes, the LED could work normally. Significantly, the brightness of the LED didn't change after three days under air ambient conditions because of the air-stable gel electrolyte and encapsulation of the PDMS. Figure 2.10d depicted the normalized capacitance (C/C_0) of the device under different deformation, respectively. The retention of the capacitance was kept nearly 100% under stretching, twisting, crimping as well as winding, confirming the electrochemical stability of the MSC arrays, which is a promising candidate for powering other stretchable electronics.

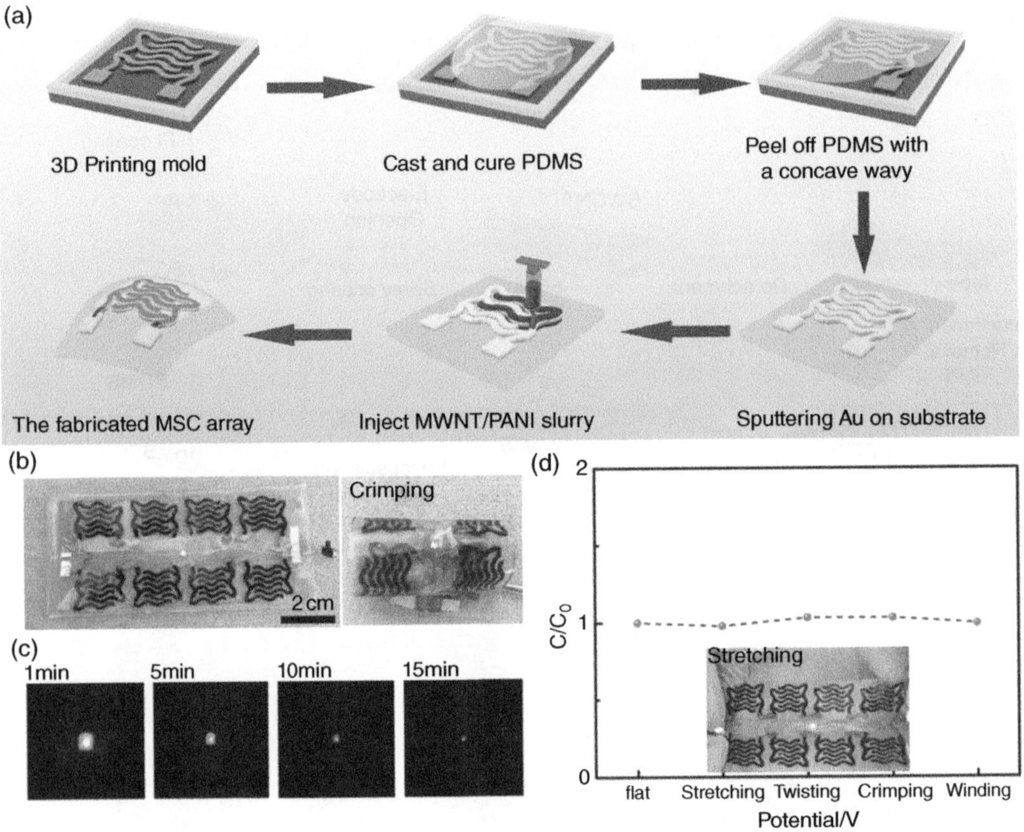

Figure 2.10 (a) Schematics of the fabrication procedures of the stretchable MSC arrays. (b) Optical images of the stretchable MSC array under different types of deformations. (c) Real-time optical images of LED powered by MSC array. (d) normalized capacitance (C/C_0) measured before and after deformation, respectively. *Source:* Reproduced with permission [72]. © 2017, Wiley-VCH.

2.2.3 3D Stretchable SCs

The stretchability of 3D stretchable SCs are typically achieved by the configuration design, such as 3D cellular and pyramid structure that omits the utilization of elastic substrate, which is quite different with the strategy toward 1D fiber SCs and 2D planar SCs. Kirigami or patterning-based editable technique is always employed to assemble the 3D stretchable SCs. Recently, many efforts have been devoted to design the 3D stretchable SCs, which can be divided into two types: cellular structure and editable SCs. The first one is to realize stretchability through cellular electrodes or embedding MSC arrays into a cellular structured elastic substrate. The later one represents the SCs devices arbitrary shape, which can be adjusted according to the demand of wearable electronics.

2.2.3.1 Cellular Structure

In nature, many animals like the North American elk or bird have a cellular bone that provides large deformations under attacking or during flying. Inspired by the biological materials, cellular structure is introduced to stretchable SCs because it can resist a broad spectrum of deformations including bending and stretching. A typical work has been reported by Peng's group, as shown in Figure 2.11a–c [40]. Figure 2.11a displayed the optical images of the stretchable cellular CNT film

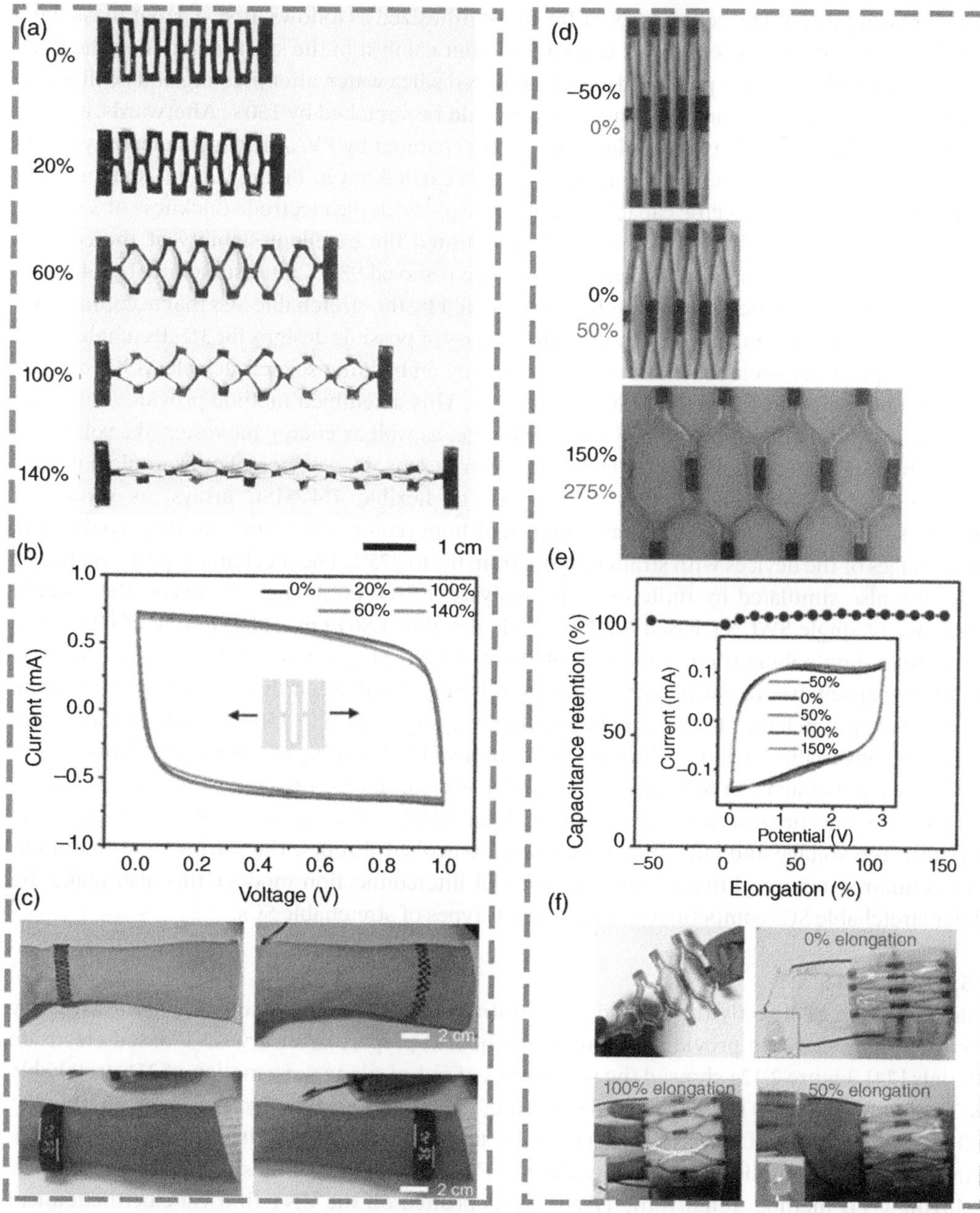

Figure 2.11 (a) Optical images of the stretchable cellular CNT film under increasing strain. (b) CV curves under stretching. (c) Photographs of the "watch strap" powered by cellular MSC array. *Source:* Reproduced with permission [40]. © 2016, The Royal Society of Chemistry. (d) Optical images of honeycomb 4 × 4 MSC arrays under different stretching state. (e) Capacitance retention versus elongation (the inset figure is the CV curves for 0–150% elongation, respectively). (f) LED powered by a honeycomb MSC device under stretching. *Source:* Reproduced with permission [73]. © 2016, American Chemical Society.

under increasing strain. The cellular CNT film was synthesized as follows: first, a paper mask with exposed cellular pattern was employed to coat a cellular catalyst in the silica wafer, then the CNT film was prepared by CVD on patterned catalyst covered silica wafer, after pressing and peeling off, the cellular CNT film was finally obtained, which could be stretched by 150%. Afterwards, a sandwiched stretchable SCs with two cellular CNT films separated by PVA/H_3PO_4 gel electrolyte was assembled. The electrochemical performance was also carried out in Figure 2.11b. The fabricated stretchable SCs showed a specific capacitance of 42.4 F g^{-1} with the electrode thickness of 38.3 μm. The identical CV curves for 0–140% strain demonstrated the excellent stability of the cellular stretchable SCs. The high specific capacitances can be reserved 98.3% after stretching by 140% for 3000 cycles. Figure 2.11c depicted a watch strap powered by the stretchable SCs that accommodate the deformation of the arm size, suggesting a novel class of possible designs for 3D stretchable SCs.

Another type of 3D stretchable SCs are fabricated by embedding several flexible MSCs devices into a cellular form thus make the MSCs stretchable. This assembled method provides a general integration way, not only in the area of energy storage, as well as energy harvester like solar cell, wireless charging units and wearable electronics such as sensors, detectors. For example, Pu et al. introduced a stretchable cellular PDMS support for flexible 4*4 MSC arrays, as shown in Figure 2.11d–f [73]. The mechanical performance of honeycomb MSC array was displayed by the optical images of the devices with strain ranging from 0% to 275%. The mechanical performance of the devices also simulated by finite element analysis (FEA). From the CV curves, the specific capacitance of single SWCNT based MSC was calculated to 1.86 F cm^{-3} at scan rate of 0.05 V S^{-1}, the corresponding volumetric capacitance of the 4*4 MSC arrays was 0.15 F cm^{-3}. Figure 2.11e showed the capacitance retention versus strains, it is very clearly seen that the capacitance of the MSC arrays kept unchanged when applied strains varying from 0 to 150%, which was also suggested by the invariable CV curves (Inset). A commercial LED lighting test driven by a honeycomb MSC devices that attach to a Nike wrist band under stretching was provided in Figure 2.11f, demonstrating the mechanical stability of the cellular MSC arrays at different stretching state. Noticeably, the voltage and current can be easily controlled through the intrinsic configuration of the cellular structure (different series or parallel interconnection modes), this also make the cellular stretchable SCs competitive among various types of stretchable SCs.

2.2.3.2 Editable SCs

Editable SCs can be directly transferred into arbitrary shapes with stretchability along arbitrary direction. Figure 2.12a–c provided a mature method to prepare arbitrary CNT based electrode materials [74]. Figure 2.12a showed the schematics of out-of-plane deformation of 3D stretchable SCs. To obtain the SWCNT with pyramid structure, the pyramid catalyst containing 5 nm thick of Al_2O_3 and 1.2 nm thick of Fe was sputtered on the silicon wafer by photolithography and electron beam evaporation deposition. Then the aligned CNT array was synthesized on the silicon by employing CVD method. Finally, the PANI was deposited on the CNT film via electrochemical deposition to improve the specific capacitance of the electrode materials. Figure 2.12b showed the optical images of the CNT film with increasing strain along the z axis. From the images, we can see the structure and integrity of the CNT film was not damaged during the stretching process. The stretchable SCs were assembled by two pieces of PANI@CNT film and a PVA/H_3PO_4 gel electrolyte in the middle. The specific capacitances of the fabricated stretchable SCs calculated from the charge–discharge curves reached an areal capacitance of 61.4 mF cm^{-2}, which was maintained at 93.3% after 3000 strain-release cycles along the y and z axes. Figure 2.12c displayed the gavanostatic charge–discharge profiles of the fabricated stretchable SCs with increasing strains from 0% to 16% along the x and y axis. It can be seen that there is no degradation in both

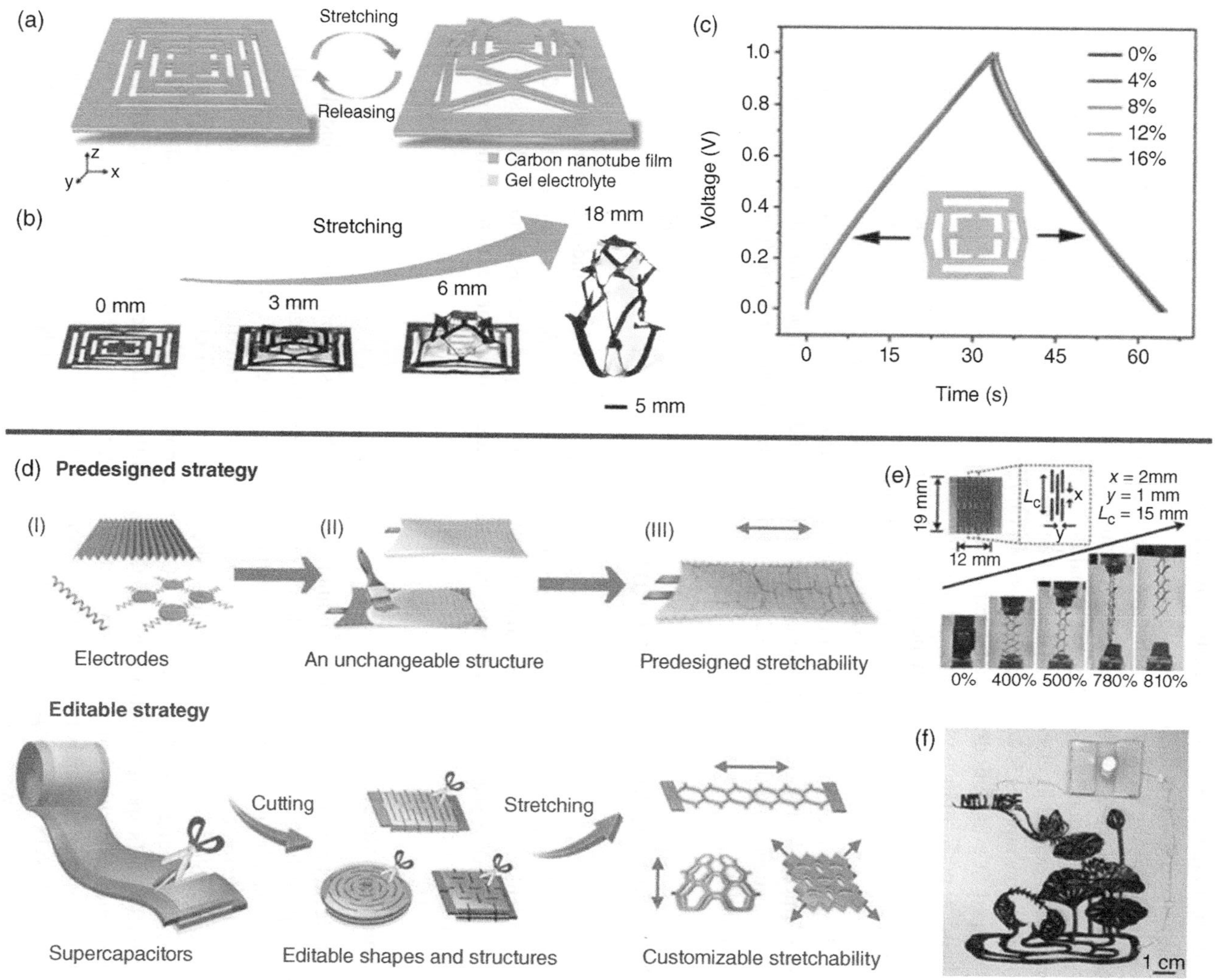

Figure 2.12 (a) Schematics of a 3D stretchable SC. (b) Optical images of the CNT film being gradually stretched along the z axis. (c) GCD curves with increasing strains from 0% to 16% along the x or y axis. *Source:* Reproduced with permission [74]. © 2016, The Royal Society of Chemistry. (d) Schematics for the assembling process of stretchable SCs. (e) Digital images of 3D SCs under different strain tests. (f) Lighting test driven by tailored paper-cutting SC. *Source:* Reproduced with permission [41]. © 2017, Wiley-VCH.

charge–discharge time and operation voltage window, showing an outstanding stability under stretching.

By the kirigami technique, the SCs can be designed with stretchability, as shown in Figure 2.12d [41]. In comparison with prestrain-release based stretchable SCs, the form and shape of editable are more free. Cellular structure, pyramid structure, living-hinge structure etc. can be easily obtained and the stretchable direction are not restricted. Traditional sandwiched SCs always use gel electrolyte as the separator, which caused the weak intermolecular interaction thus cannot effectively serve as a separator to prevent short circuit of SCs during editing process. Therefore, the authors employed nanocellulose fibers as separator and PVA/H_3PO_4 as gel electrolyte. MnO_2 nanowires@CNT served as electrode materials. Figure 2.12e displayed the photography of fabricated honeycomb shaped SCs with increasing strain varying from 0% to 810%. The fabricated honeycomb shaped SCs exhibited a specific capacitance of 227.2 mF cm^{-2} and no degradation of electrochemical performance when applied a strain of 500%, providing an excellent mechanical stability of the editable SCs. In addition, nearly 98% of initial capacitance of the honeycomb shaped editable SCs interlocked by PU fibers was remained even after 10 000 stretch-and-release cycles under reversible 400% tensile strain. As mentioned above, the editable SCs endow advantages of tunable voltage by different interconnection in parallel and/or series. As a proof of concept, four SCs were connected in series and tailored into delicate and artistic patterns to power a LED (Figure 2.12f), suggesting that editable SCs can realize more complicated and free patterns with high stretchability.

2.3 Multifunctional Supercapacitor

With the development of the highly stretchable SC, its application was expanded to many areas, including portable, wearable energy storage, electronic skins, sensors, detectors, implantable medical devices, which also raise new demands to the energy storage, such as compressible, self-healable etc. In recent years, many attempts have been made to incorporate such functions into stretchable energy storage.

2.3.1 Compressible SCs

Compressible SCs have emerged as a new branch of stretchable SCs. Compared to stretchable electronics, compressible SCs endow the large levels of strain without sacrificing basic electrochemical performance. Traditional compressible SCs usually use graphene, SWCNTs or their composite foam as electrode materials, which can't be compressed into one unit due to the liquid electrolyte and incompressible encapsulation. To address these problems, Niu et.al proposed a nanostructured composite sponge based all-solid-state SCs with a remarkable compression tolerance [75]. The PANI@SWCNTs sponge electrode materials were synthesized as follows (Figure 2.13a): the sponge was cut into required shapes and then repeatedly dipped with SWCNTs solution. After drying, the electrochemical deposition process was performed to coat a PANI nanomaterials on the surface of the SWCNTs covered sponge. The mass loading of SWCNT and PANI was 2.2 and 4.1 mg cm^{-2}, respectively.

Figure 2.13b showed the schematic diagram of the cross-sectional scheme of the fabricated sandwiched compressible SCs. Two patterned Au film on poly(ethyleneterephthalate) (PET) served as the current collector attaching to the outer-top and outer-bottom side of sponge electrodes,

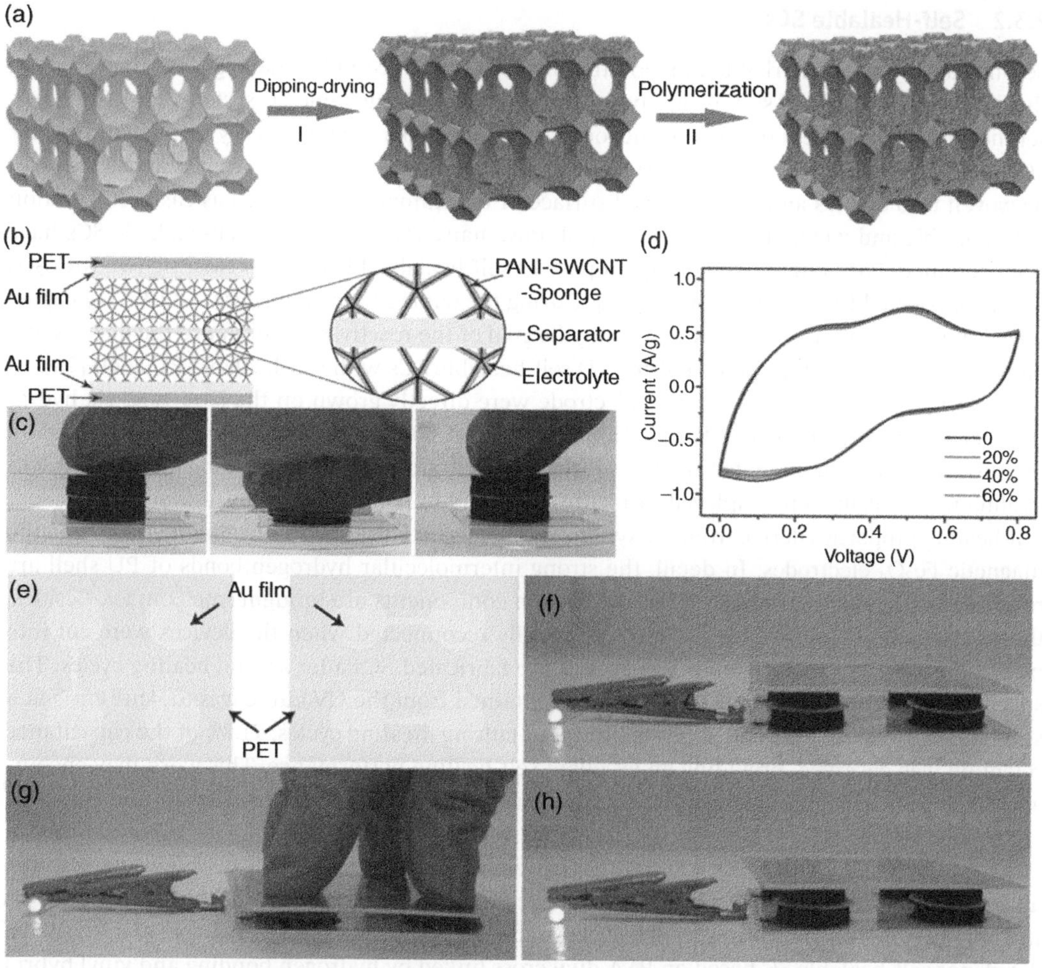

Figure 2.13 (a) Schematic diagram of synthesizing PANI@SWCNTs sponge composite electrodes. (b) Cross-sectional scheme of the fabricated PANI@SWCNTs sponge based SCs. (c) Real-time optical images of the SCs under compressing. (d) CV curves of the compressible SCs with increasing strains from 0% to 60%. (e) Photography of the patterned d Au current collector on PET plate. (f–h) Lighting test driven by four SCs showing the compressing and recovering process. *Source:* Reproduced with permission [75]. © 2015, Wiley-VCH.

separated by filter paper and PVA/H_2SO_4 gel electrolyte. The single device during stress-release cycle was illustrated in Figure 2.13c. Figure 2.13d displayed the CV curves under different compressible strain. The specific capacitance of PANI@SWCNT sponge electrodes based all-solid-state SC calculated from CV curves was $216\,F\,g^{-1}$. Almost 97% of capacitance retention was maintained even under 60% compressible strain. Furthermore, four SCs (Figure 2.13e and f) were connected in series to improve the overall output potential, which was powerful enough to light up an LED. We can see the brightness of the LED under compressible strain (Figure 2.13g and h) did not show any noticeable change with a compressing releasing cycle, demonstrating the excellent stability of the compressible all-solid-state SCs, which will pave the way for practical applications of SCs in the field of compressible energy storage devices to fit the demands of the compression-tolerant electronics.

2.3.2 Self-Healable SCs

Self-healability as a remarkable function that helps SCs from complete damage and then prolong the lifespan of the energy storage has been introduced into stretchable SC systems [76]. Similarly, self-healable SCs possess two main categories, including fiber shaped energy storage and planar SCs. 1D self-healable SCs devices have received tremendous attention in recent years as they can be woven into textiles and fit the curved surface of the human body as well as easily integration with portable and wearable electronics. Until now, many 1D fiber shaped self-healable SCs have been designed. For example, our group proposed a self-healable fiber SCs by twisting two $NiCo_2O_4$ electrode covered PVA/KOH hydrogel [77]. During the damaging-healing cycles, 82.19% of capacity was remained after fourcycles, achieving the goal of the reactivated work once the devices was damaged. Zhi's group reported a twisted 1D self-healable SCs with self-healable PU shell [28]. In their work, magnetic Fe_3O_4 materials electrode were directly grown on the bare yarn via hydrothermal and annealed process, then a 2 μm thickness of polypyrrole (PPy) film was deposited on the surface of Fe_3O_4 nanomaterials. The schematic illustration of the self-healing process and self-healing mechanism were displayed in Figure 2.14a and b. It can be concluded that the excellent self-healing properties derive from the synergistic effects between the self-healing PU shell and the magnetic Fe_3O_4 electrodes. In detail, the strong intermolecular hydrogen bonds of PU shell are reversible and could reestablish when the broken components are brought into contact. Besides, the magnetic Fe_3O_4 electrodes can also electrically reconnected when the devices were cut into pieces. Figure 2.14c showed the CV curves of the fabricated SCs after several healing cycles. The specific capacitance of the self-healing device obtained from the CV curve was 61.4 mF cm^{-2} at a scan rate of 10 mV s^{-1}. We can see even after four cutting–healing cycles, 71.8% of the capacitance retention was observed, providing an excellent self-healing performance. This work may encourage the design and fabrication of the 1D fiber shaped self-healable energy storage and wearable electronics.

Self-healable 2D planar SCs are another attractive energy storage because they possess the advantage of 2D devices like small size, low weight, ease of handing in appearance as well as the ultralong lifespan of the stretchable electronics. As a typical example, Huang et.al fabricated a 2D planar self-healable SC based on PAA dual cross linked by hydrogen bonding and vinyl hybrid silica nanoparticles (VSNPs) [31]. Figure 2.14d illustrated the schematics of the fabrication process of the highly stretchable and self-healable SCs. As for electrode materials, CNT papers were synthesized by CVD, and then deposited with the PPy, which were attached on the both side of the self-healable VSNPs-PAA gel electrolyte-based film to prepare a self-healable planar SCs. Figure 2.14e showed the demonstration of and ionic conductivity of the self-healed substrate. The wound in the VSNPs-PAA film could be autonomously repaired via the intermolecular hydrogen bonds among the cross-linked polymer chains on the VSNPs in 10 mins under the ambient condition, which has no effect on the ionic conductivity and mechanical properties of the VSNPs-PAA film. The CV curves with different cut-healing times were depicted in the Figure 2.14f. The fabricated self-healable SCs exhibited a specific capacitance of 61.4 mF cm^{-2} at a scan rate of 10 mV s^{-1}, which was kept unchanged even after four healing cycles.

2.3.3 Stretchable Integrated System

Stretchable integrated systems have demonstrated a great real-word application in wearable electronics, such as electronic skin that monitor heart rate, wrist pulse, body temperature, voice etc., robots that move as instructed, bio-medicine carrier that reflect signal, while leaving people's

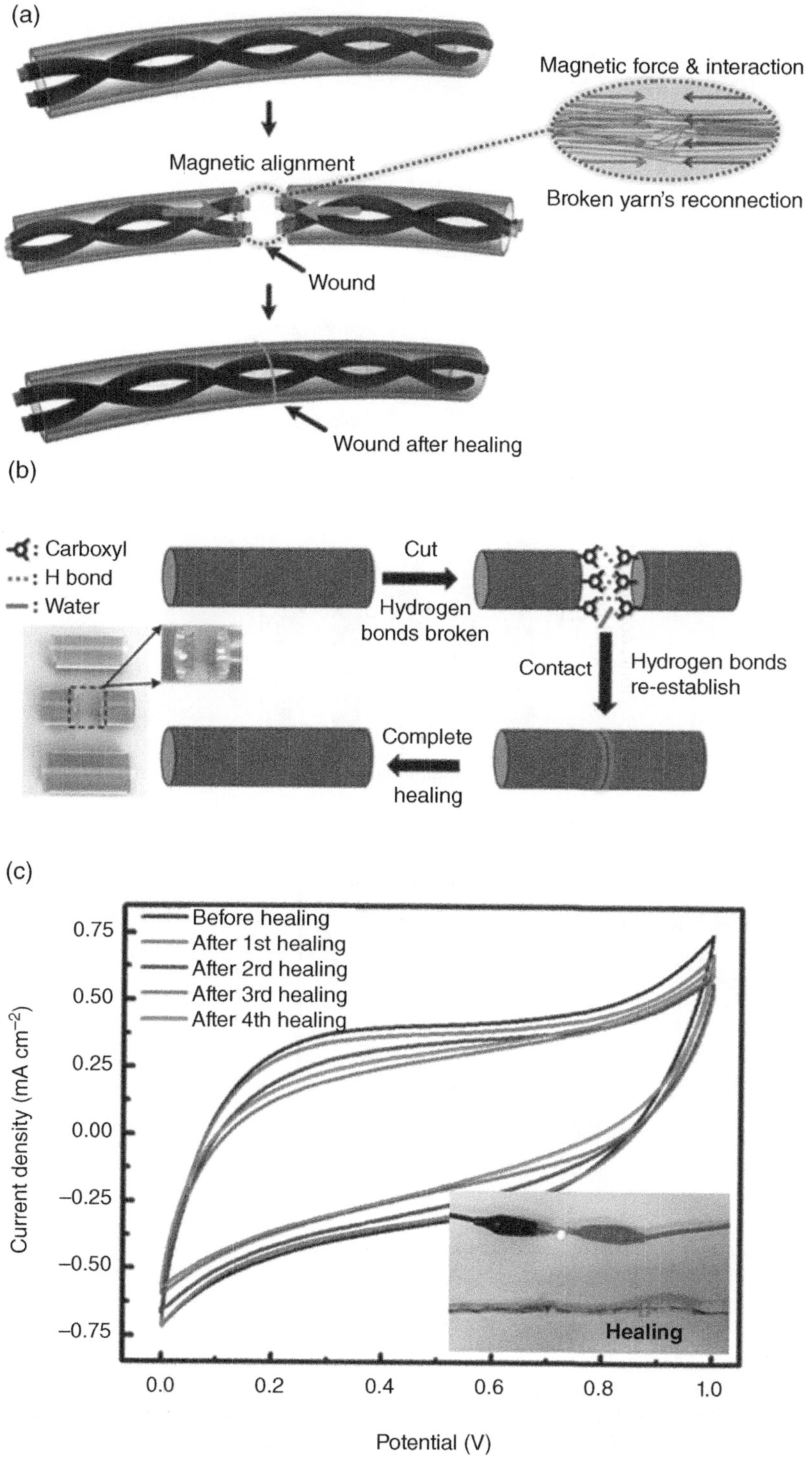

Figure 2.14 (a) Schematic illustration of the self-healing process. (b) Schematic diagram of the self-healing mechanism. (c) CV curves after several healing cycles. *Source:* Reproduced with permission [28]. © 2015, American Chemical Society.

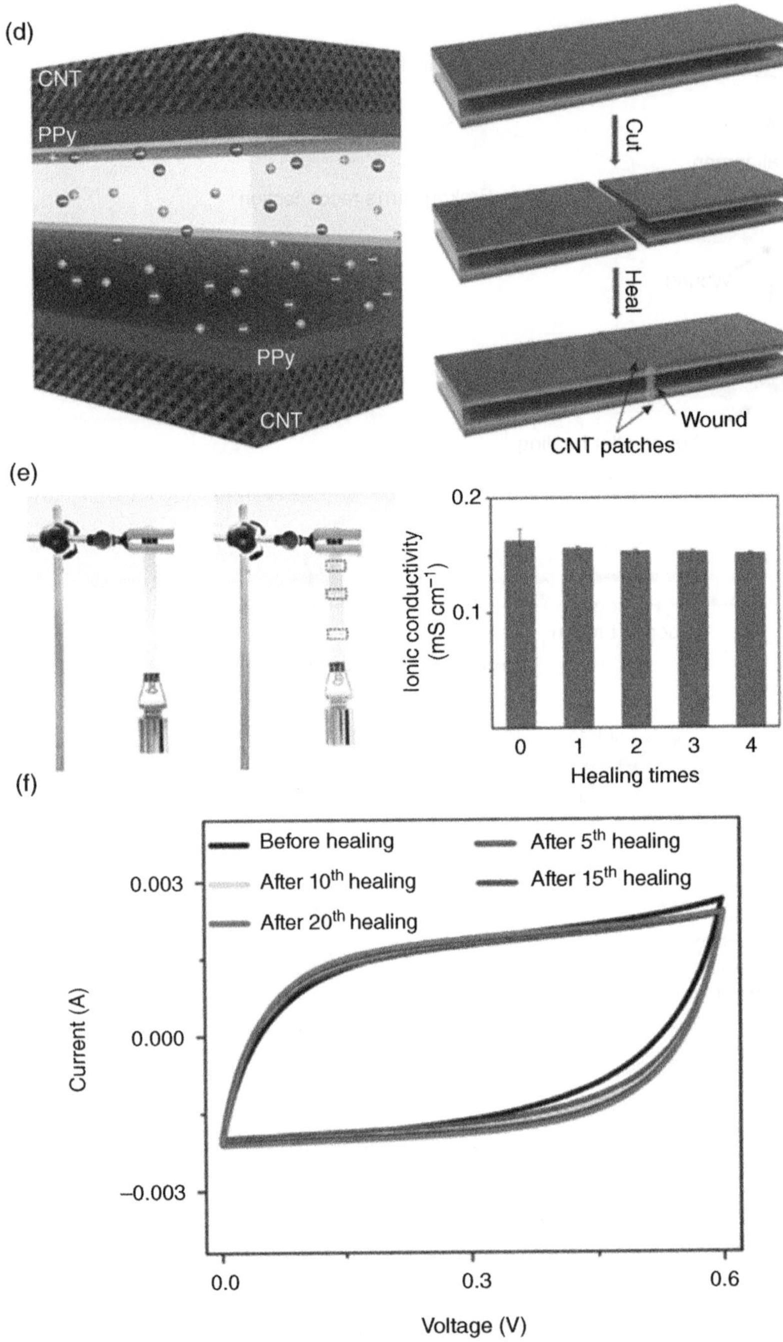

Figure 2.14 (Continued) (d) Schematics of fabrication strategies for highly stretchable and healable SCs. (e) Demonstration and ionic conductivity of the self-healed substrate. (f) CV curves with different cut-healing times. *Source:* Reproduced with permission [31]. © 2015, Nature Publishing Group.

normal activities of daily life unaffected [78–82]. These integrated systems that compose of the energy harvester, energy storage, and the functional sensing/detecting units are usually connected on the same deformable substrate [83–87]. In comparison with Li-ion battery, fuel cells, and other types of battery, SCs have gained extensive attention out of security consideration. To date, several types of integrated devices toward the direction of portable and wearable electronics, like energy storage-sensor unit and self-powering system have been successfully fabricated. In harmony with the configuration of the stretchable SCs, the fiber substrate-based 1D integrated device, PDMS based 2D integrated system and elastic materials-based 3D devices have shown a rapid development momentum in the field of deformable integrated system. For integration of different functional devices with stretchability, there are three vital factors need to be considered: (i) coordinate work, each unit or functional partition could realize their unique function after connection, the parameters like size, work current etc. should be matched to each other; (ii) a low cost and facile fabrication procedure, where every component could be stably and reliably connected in an easy way; and (iii) integral encapsulation. A favorable package could make the whole integrated system work longer and even remain their function in extreme environment.

The integrated power pack comprising either wireless power transmission or internal power generator is highly desired for wearable electronics. Nanogenerator is one of the most popular power generators used in an integrated system, which could collect the energy produced by human activities [88–95]. Researches on nanogenerators was first reported in 2006 by Prof. Zhonglin Wang, and then a series of integrated systems with nanogenerator were designed [96–99]. For example, Guo et al. provided a stretchable all-in-one integrated system that contains triboelectric nanogenerator, SCs, and an electric watch, which could harvest all kinds of mechanical energy from human motions (bending, stretching) and transfer to SCs for powering the wearable watch [98]. Solar cells that convert sunlight into electricity are considered as the most promising energy conversion devices, which are also introduced to integrated system [42, 100]. Most recently, Yun et.al reported on the fabrication of stretchable integrated system including solar cells, all-solid-state MSC, and a strain sensor [101]. In this integrated system, the PPy@CNT electrode based MSC arrays were connected on the PI substrate, resembling a serpent in form. The graphene foam base strain sensor was directly prepared on the deformable PDMS substrate. MSC array and solar cells were separately embedded onto the PDMS substrate. When the MSC array was placed into the deformable Ecoflex substrate, the PI film was removed. The obtained integrated system was attached on human's wrist to detect externally applied strains and the arterial pulse using the energy stored in MSCs, charged with SCs. Noticeably, the charge/discharge behavior maintain their value even after 1000 stretching/releasing cycles, demonstrating the outstanding cycle stability and stretchability of the fabricated devices.

Despite the successful integration with SCs and sensors, solar cells suffer from the inherent instability caused by light intensity, which constrain the practical application of solar cells in wearable electronics, especially in some integrated systems that need continuous monitoring [102, 103]. Alternatively, the wireless charging device holds a significant place in stable and reliable energy generation of the integrated system [104, 105]. Most recently, Ha's group reported a stretchable multifunctional sensing system integrated with a wireless charging unit, as shown in Figure 2.15 [106]. This stretchable integrated device could realize not only the monitoring of bio signals such as the human pulse, motion, and voice, but also the detection to environmental signal like gas, ultraviolet (UV) light. Figure 2.15a and b displayed the schematic illustration and circuit diagram of the stretchable 2D multifunctional integrated system, which operated a RF power receiver, MSC array, graphene foam-based strain sensor and UV/NO_2 gas sensor on the same elastic PDMS substrate. The digital photographs of integrated system attached to human body were

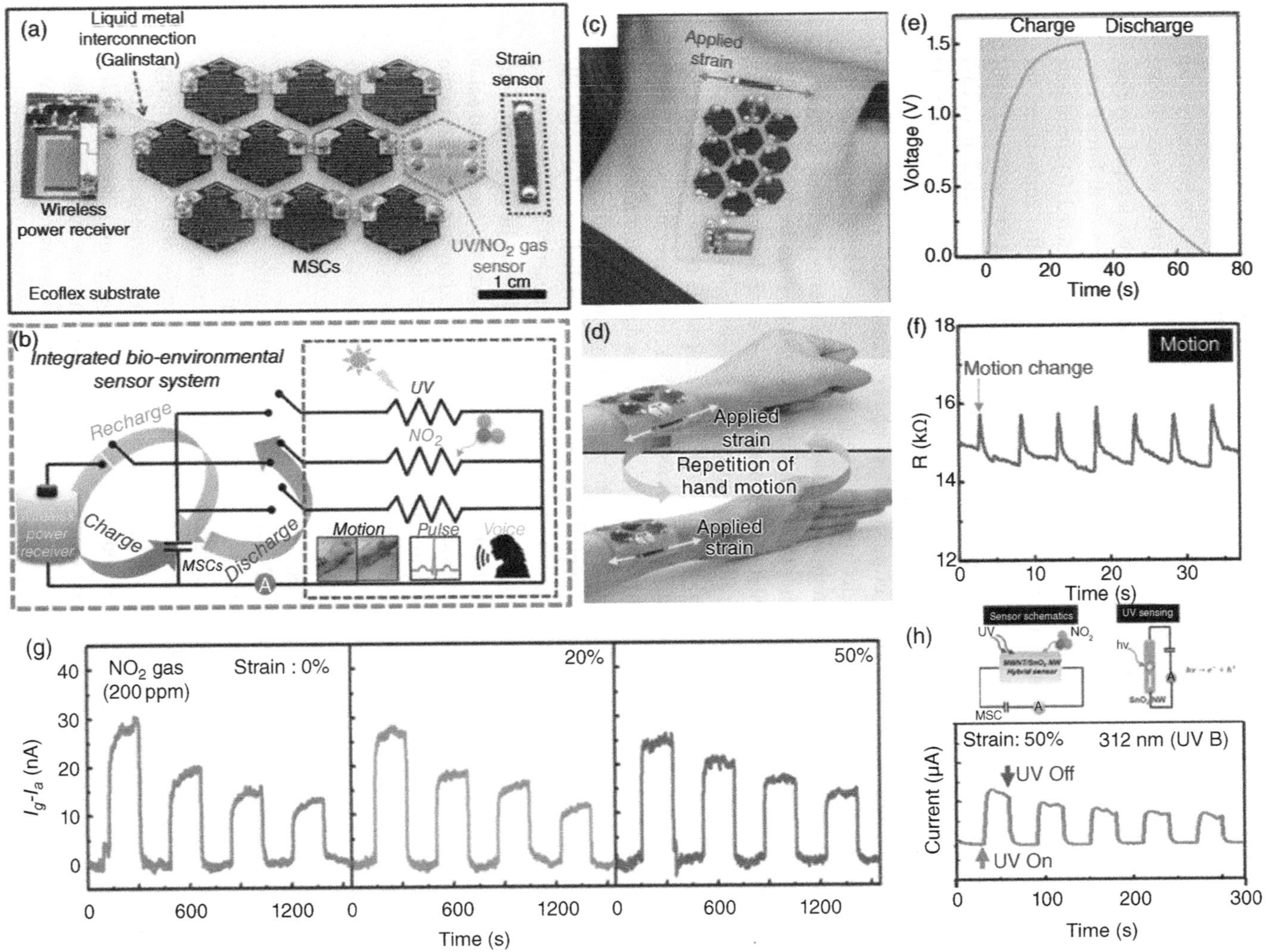

Figure 2.15 (a, b) Schematic illustration and circuit diagram of 2D multifunctional integrated system, containing a RF power receiver, a MSC array, strain sensor, and UV/NO_2 gas sensor. (c, d) Photograph of integrated system attached to human body. (e) Charge/discharge curve of integrated system powered by RF power source. (f) Carotid pulse curve of the strain sensor. (g) NO_2 gas response at strain of 0%, 20%, 50%. (h) UV detection under 0% and 50% strain with exposure of 312 nm. *Source:* Reproduced with permission [106]. © 2015, Wiley-VCH.

presented in Figure 2.15c and d. The stable and repeatable wireless charging of MSCs with the integrated RF power receiver was depicted in Figure 2.15e. The carotid pulse curve (Figure 2.15f) together with the resistance versus hand motion and swallowing saliva detection curve of the fragmentized graphene foam-based strain sensor demonstrated the strain sensor has a stable response to the bio-signals like body motion, voice, swallowing of saliva, and the carotid artery pulse. Figure 2.15g showed the I-t curves of the MWCNT/SnO_2 nanowire-based gas sensor using the energy supplied by integrated MSC array under stretching varying from 0% to 50%, indicating the excellent performance and stretchability of both gas sensor and MSC array. Figure 2.15h presented the mechanism in response to NO_2 gas and UV light of MWCNT/SnO_2 hybrid sensor. The UV sensitivity was unchanged under a strain between 0% and 50%. All the results obtained in this paper suggested the fabricated stretchable multifunctional integrated sensing system can be used as next-generation body-attached healthcare and environmental sensor devices for continuous nondestructive monitoring.

2.3.4 Perspective

As a new member of the SC family, stretchable devices have been greatly developed in the past few years, buoyed by the portable and wearable electronics, which need a stretchable energy storage to form a complete and safe system to monitor electrical and biomedical signal generated by human activities, thus achieving the practical application of wearable electronics in the field of biomimetic E-skins, interactive human-machine interface, "big health" and "big data." In this review, we systematically summarize the recent progress in stretchable SCs from the perspective of the three dimension and corresponding configuration of the stretchable device, as well as the fabrication process and strategies toward the stretchable SCs. The stretchable integrated system is also concluded in this chapter, all of them realized stable response to the physical and bio-signal under different stretching or deformations when attached to the human body, showing potential for wearable electronics.

Despite the considerable achievements in stretchable SCs which have been achieved, there are great challenges still remaining for future practical applications. These challenges and the direction for future development can be summarized as follows: (i) Electrochemical performance of SCs under deformation still requires to be improved. Electrode material is the decisive factor of the electrochemical performance. But most of the stretchable SCs depend on the CNTs, which lead to the low specific capacitance of the devices. Discovering novel kinds of electrode materials and design of various configuration for stretchable SCs are one of the promising directions. (ii) The elastic substrate with high stretchability and skin-touched feature is desired. The most popular elastic substrate used in stretchable devices is PDMS. However, it possesses hydrophobic properties. Electrode materials with binder like PVDF covered on the PDMS usually need to face the detaching problem, which is not of benefit to the mechanical stability of the devices. A stretchable substrate that ensures electrode materials direct growth on it may be a great way to develop stretchable devices. (iii) More attention should be focused on the development of air stable gel electrolyte. Most of the existing all-solid-state gel electrolyte is water-based polymer gel electrolyte, which have a short lifespan under air ambient conditions. How to improve the lifespan of the water-based gel electrolyte in the further development of stretchable SCs must be a significant topic. (iv) The integration and encapsulation method is important for wearable integrated system. For the stretchable multifunctional integrated systems powered by SCs containing several kinds of stretchable devices, the integration and encapsulation method that makes every component work tuneful must be considered. The embedding method is a mature way but can't be the only one for packaging the integrated system. (v) More low-cost fabrication technology for large scale production of stretchable SCs should be introduced to meet the requirement of the practical application. From the angle of actual application, the large-scale production of stretchable SCs with low cost also should be considered. Overall, stretchable SCs have been proved as a promising energy storage to power the portable and wearable electronics and occupy an indispensable position in the future development and applications of wearable electronics.

References

1 Li, Lou, Z., Chen, D. et al. (2018). Recent advances in flexible/stretchable supercapacitors for wearable electronics. *Small* 14 (43): e1702829.

2 Lee, G., Kim, D., Kim, D. et al. (2015). Fabrication of a stretchable and patchable array of high performance micro-supercapacitors using a non-aqueous solvent based gel electrolyte. *Energy Environ. Sci.* 8 (6): 1764–1774.

3 Hu, H., Pei, Z., and Ye, C. (2015). Recent advances in designing and fabrication of planar micro-supercapacitors for on-chip energy storage. *Energy Storage Mater.* 1: 82–102.

4 Li, L., Wu, Z., Yuan, S. et al. (2014). Advances and challenges for flexible energy storage and conversion devices and systems. *Energy Environ. Sci.* 7 (7): 2101–2122.

5 Chu, X., Zhang, H., Su, H. et al. (2018). A novel stretchable supercapacitor electrode with high linear capacitance. *Chem. Eng. J.* 349: 168–175.

6 Xu, J., Wu, H., Lu, L. et al. (2014). Integrated photo-supercapacitor based on Bi-polar TiO_2 nanotube arrays with selective one-side plasma-assisted hydrogenation. *Adv. Funct. Mater.* 24 (13): 1840–1846.

7 Chen, X., Lin, H., Chen, P. et al. (2014). Smart, stretchable supercapacitors. *Adv. Mater.* 26 (26): 4444–4449.

8 Huang, Y., Liang, J., and Chen, Y. (2012). An overview of the applications of graphene-based materials in supercapacitors. *Small* 8 (12): 1805–1834.

9 Zheng, Y., Yang, Y., Chen, S. et al. (2016). Smart, stretchable and wearable supercapacitors: prospects and challenges. *CrystEngComm* 18 (23): 4218–4235.

10 Senthilkumar, B., Vijaya, S.K., Sanjeeviraja, C. et al. (2013). Synthesis and physico-chemical property evaluation of PANI–$NiFe_2O_4$ nanocomposite as electrodes for supercapacitors. *J. Alloys Compd.* 553: 350–357.

11 Mahmood, Q., Park, S.K., Kwon, K.D. et al. (2016). Transition from diffusion-controlled intercalation into extrinsically pseudocapacitive charge storage of MoS_2 by nanoscale heterostructuring. *Adv. Energy Mater.* 6 (1): n/a-n/a.

12 Hsia, B., Marschewski, J., Wang, S. et al. (2014). Highly flexible, all `solid-state micro-supercapacitors from vertically aligned carbon nanotubes. *Nanotechnology* 25 (5): 055401.

13 An, C.H., Wang, Y.J., Huang, Y.N. et al. (2014). Porous $NiCo_2O_4$ nanostructures for high performance supercapacitors via a microemulsion technique. *Nano Energy* 10: 125–134.

14 Wu, H., Jiang, K., Gu, S. et al. (2015). Two-dimensional $Ni(OH)_2$ nanoplates for flexible on-chip microsupercapacitors. *Nano Res.* 8 (11): 3544–3552.

15 Yu, Z.-Y., Chen, L.-F., and Yu, S.-H. (2014). Growth of $NiFe_2O_4$ nanoparticles on carbon cloth for high performance flexible supercapacitors. *J. Mater. Chem. A* 2 (28): 10889.

16 Li, L., Lou, Z., Han, W. et al. (2016). Flexible in-plane microsupercapacitors with electrospun NiFe2O4 nanofibers for portable sensing applications. *Nanoscale* 8 (32): 14986–14991.

17 Ai, Y., Lou, Z., Li, L. et al. (2016). Meters-long flexible $CoNiO_2$-nanowires@carbon-fibers based wire-supercapacitors for wearable electronics. *Adv. Mater. Technol.* 1 (8): 1600142.

18 Zhao, X., Zheng, B., Huang, T. et al. (2015). Graphene-based single fiber supercapacitor with a coaxial structure. *Nanoscale* 7 (21): 9399–9404.
19 Wu, Z.S., Parvez, K., Feng, X. et al. (2013). Graphene-based in-plane micro-supercapacitors with high power and energy densities. *Nat. Commun.* 4: 2487.
20 Wu, Z.S., Feng, X., and Cheng, H.M. (2013). Recent advances in graphene-based planar micro-supercapacitors for on-chip energy storage. *Natl. Sci. Rev.* 1 (2): 277–292.
21 Liu, T., Zhang, F., Song, Y. et al. (2017). Revitalizing carbon supercapacitor electrodes with hierarchical porous structures. *J. Mater. Chem. A* 5 (34): 17705–17733.
22 Beidaghi, M. and Wang, C. (2012). Micro-supercapacitors based on interdigital electrodes of reduced graphene oxide and carbon nanotube composites with ultrahigh power handling performance. *Adv. Funct. Mater.* 22 (21): 4501–4510.
23 Yan, C. and Lee, P.S. (2014). Stretchable energy storage and conversion devices. *Small* 10 (17): 3443–3460.
24 Qi, D., Liu, Z., Liu, Y. et al. (2015). Suspended wavy graphene microribbons for highly stretchable microsupercapacitors. *Adv. Mater.* 27 (37): 5559–5566.
25 Yun, J., Lim, Y., Jang, G.N. et al. (2016). Stretchable patterned graphene gas sensor driven by integrated micro-supercapacitor array. *Nano Energy* 19: 401–414.
26 Núñez, C.G., Navaraj, W.T., Polat, E.O. et al. (2017). Energy-autonomous, flexible, and transparent tactile skin. *Adv. Funct. Mater.* 27 (18): 1606287.
27 Liu, Z., Qi, D., Guo, P. et al. (2015). Thickness-gradient films for high gauge factor stretchable strain sensors. *Adv. Mater.* 27 (40): 6230–6237.
28 Huang, Y., Huang, Y., Zhu, M. et al. (2015). Magnetic-assisted, self-healable, yarn-based supercapacitor. *ACS Nano* 9 (6): 6242–6251.
29 Chen, S., Lou, Z., Chen, D. et al. (2016). Polymer-enhanced highly stretchable conductive fiber strain sensor used for electronic data gloves. *Adv. Mater. Technol.* 1 (7): 1600136.
30 Wang, K., Zhang, X., Li, C. et al. (2015). Chemically crosslinked hydrogel film leads to integrated flexible supercapacitors with superior performance. *Adv. Mater.* 27 (45): 7451–7457.
31 Huang, Y., Zhong, M., Huang, Y. et al. (2015). A self-healable and highly stretchable supercapacitor based on a dual crosslinked polyelectrolyte. *Nat. Commun.* 6: 10310.
32 Huang, Y., Zhong, M., Shi, F. et al. (2017). An intrinsically stretchable and compressible supercapacitor containing a polyacrylamide hydrogel electrolyte. *Angew. Chem. Int. Ed. Engl.* 56 (31): 9141–9145.
33 Zhang, X., Zhang, H., Lin, Z. et al. (2016). Recent advances and challenges of stretchable supercapacitors based on carbon materials. *Sci. China Mater.* 59 (6): 475–494.
34 An, T. and Cheng, W. (2018). Recent progress in stretchable supercapacitors. *J. Mater. Chem. A* 6 (32): 15478–15494.
35 Xu, P., Gu, T., Cao, Z. et al. (2014). Carbon nanotube fiber based stretchable wire-shaped supercapacitors. *Adv. Energy Mater.* 4 (3): 1300759.
36 Yang, Z., Deng, J., Chen, X. et al. (2013). A highly stretchable, fiber-shaped supercapacitor. *Angew. Chem. Int. Ed. Engl.* 52 (50): 13453–13457.
37 Niu, Z., Dong, H., Zhu, B. et al. (2013). Highly stretchable, integrated supercapacitors based on single-walled carbon nanotube films with continuous reticulate architecture. *Adv. Mater.* 25 (7): 1058–1064.
38 Kim, H., Yoon, J., Lee, G. et al. (2016). Encapsulated, high-performance, stretchable array of stacked planar micro-supercapacitors as waterproof wearable energy storage devices. *ACS Appl. Mater. Interfaces* 8 (25): 16016–16025.
39 Kim, D., Shin, G., Kang, Y.J. et al. (2013). Fabrication of a stretchable solid-state micro-supercapacitor array. *ACS Nano* 7 (9): 7975–7982.

40 He, S., Cao, J., Xie, S. et al. (2016). Stretchable supercapacitor based on a cellular structure. *J. Mater. Chem. A* 4 (26): 10124–10129.
41 Lv, Z., Luo, Y., Tang, Y. et al. (2018). Editable supercapacitors with customizable stretchability based on mechanically strengthened Ultralong MnO_2 nanowire composite. *Adv. Mater.* 30 (2).
42 Xu, J., Ku, Z., Zhang, Y. et al. (2016). Integrated photo-supercapacitor based on PEDOT modified printable perovskite solar cell. *Adv. Mater. Technol.* 1 (5): 1600074.
43 Sun, H., Zhang, Y., Zhang, J. et al. (2017). Energy harvesting and storage in 1D devices. *Nat. Rev. Mater.* 2 (6): 17023.
44 Le, V.T., Kim, H., Ghosh, A. et al. (2013). Coaxial fiber supercapacitor using all-carbon material electrodes. *ACS Nano* 7 (7): 5940–5947.
45 Xiao, X., Li, T., Yang, P. et al. (2012). Fiber-based all-solid-state flexible supercapacitors for self-powered systems. *ACS Nano* 6 (10): 9200–9206.
46 Li, P., Jin, Z., Peng, L. et al. (2018). Stretchable all-gel-state fiber-shaped supercapacitors enabled by macromolecularly interconnected 3D graphene/nanostructured conductive polymer hydrogels. *Adv. Mater.* 30 (18): e1800124.
47 Liu, B., Liu, B., Wang, X. et al. (2014). Constructing optimized wire electrodes for fiber supercapacitors. *Nano Energy* 10: 99–107.
48 Ren, J., Li, L., Chen, C. et al. (2013). Twisting carbon nanotube fibers for both wire-shaped micro-supercapacitor and micro-battery. *Adv. Mater.* 25 (8): 1155–1159, 1224.
49 Kou, L., Huang, T., Zheng, B. et al. (2014). Coaxial wet-spun yarn supercapacitors for high-energy density and safe wearable electronics. *Nat. Commun.* 5: 3754.
50 Wang, Q., Wang, X., Xu, J. et al. (2014). Flexible coaxial-type fiber supercapacitor based on NiCo2O4 nanosheets electrodes. *Nano Energy* 8: 44–51.
51 Chen, X., Qiu, L., Ren, J. et al. (2013). Novel electric double-layer capacitor with a coaxial fiber structure. *Adv. Mater.* 25 (44): 6436–6441.
52 Li, L., Lou, Z., Chen, D. et al. (2018). Hollow polypyrrole sleeve based coaxial fiber supercapacitors for wearable integrated Photosensing system. *Adv. Mater. Technol.* 3 (8): 1800115.
53 Wang, H., Wang, C., Jian, M. et al. (2018). Superelastic wire-shaped supercapacitor sustaining 850% tensile strain based on carbon nanotube@graphene fiber. *Nano Res.* 11 (5): 2347–2356.
54 Ren, J., Xu, Q., and Li, Y.-G. (2018). Flexible fiber-shaped energy storage devices: principles, progress, applications and challenges. *Flex. Print. Electron.* 3 (1): 013001.
55 Choi, C., Kim, S.H., Sim, H.J. et al. (2015). Stretchable, weavable coiled carbon nanotube/MnO2/ polymer fiber solid-state supercapacitors. *Sci. Rep.* 5: 9387.
56 Li, M., Zu, M., Yu, J. et al. (2017). Stretchable fiber supercapacitors with high volumetric performance based on buckled MnO_2/oxidized carbon nanotube fiber electrodes. *Small* 13 (12).
57 Choi, C., Lee, J.M., Kim, S.H. et al. (2016). Twistable and stretchable sandwich structured fiber for wearable sensors and supercapacitors. *Nano Lett.* 16 (12): 7677–7684.
58 Chen, T., Hao, R., Peng, H. et al. (2015). High-performance, stretchable, wire-shaped supercapacitors. *Angew. Chem. Int. Ed. Engl.* 54 (2): 618–622.
59 Zhang, Q., Sun, J., Pan, Z. et al. (2017). Stretchable fiber-shaped asymmetric supercapacitors with ultrahigh energy density. *Nano Energy* 39: 219–228.
60 Yu, J., Lu, W., Smith, J.P. et al. (2017). A high performance stretchable asymmetric fiber-shaped supercapacitor with a core-sheath helical structure. *Adv. Energy Mater.* 7 (3): 1600976.
61 Yun, J., Kim, D., Lee, G. et al. (2014). All-solid-state flexible micro-supercapacitor arrays with patterned graphene/MWNT electrodes. *Carbon* 79: 156–164.
62 Gu, S., Lou, Z., Li, L. et al. (2015). Fabrication of flexible reduced graphene oxide/Fe_2O_3 hollow nanospheres based on-chip micro-supercapacitors for integrated photodetecting applications. *Nano Res.* 9 (2): 424–434.

63 Lv, T., Yao, Y., Li, N. et al. (2016). Highly stretchable supercapacitors based on aligned carbon nanotube/molybdenum disulfide composites. *Angew. Chem. Int. Ed. Engl.* 55 (32): 9191–9195.

64 Yu, C., Masarapu, C., Rong, J. et al. (2009). Stretchable supercapacitors based on buckled single-walled carbon-nanotube macrofilms. *Adv. Mater.* 21 (47): 4793–4797.

65 Xie, Y., Liu, Y., Zhao, Y. et al. (2014). Stretchable all-solid-state supercapacitor with wavy shaped polyaniline/graphene electrode. *J. Mater. Chem. A* 2 (24): 9142–9149.

66 Yu, J., Lu, W., Pei, S. et al. (2016). Omnidirectionally stretchable high-performance supercapacitor based on isotropic buckled carbon nanotube films. *ACS Nano* 10 (5): 5204–5211.

67 El-Kady, M.F. and Kaner, R.B. (2013). Scalable fabrication of high-power graphene micro-supercapacitors for flexible and on-chip energy storage. *Nat. Commun.* 4: 1475.

68 Lim, Y., Yoon, J., Yun, J. et al. (2014). Biaxially stretchable, integrated array of high performance microsupercapacitors. *ACS Nano* 8 (11): 11639–11650.

69 Wang, J., Li, F., Zhu, F. et al. (2018). Recent progress in micro-supercapacitor design, integration, and functionalization. *Small Methods* 3 (8): 1800367.

70 Moon, Y.S., Kim, D., Lee, G. et al. (2015). Fabrication of flexible micro-supercapacitor array with patterned graphene foam/MWNT-COOH/MnOx electrodes and its application. *Carbon* 81: 29–37.

71 Qi, D., Liu, Z., Yu, M. et al. (2015). Highly stretchable gold nanobelts with sinusoidal structures for recording electrocorticograms. *Adv. Mater.* 27 (20): 3145–3151.

72 Li, L., Lou, Z., Han, W. et al. (2017). Highly stretchable micro-supercapacitor arrays with hybrid MWCNT/PANI electrodes. *Adv. Mater. Technol.* 2 (3): 1600282.

73 Pu, J., Wang, X., Xu, R. et al. (2016). Highly stretchable microsupercapacitor arrays with honeycomb structures for integrated wearable electronic systems. *ACS Nano* 10 (10): 9306–9315.

74 He, S., Qiu, L., Wang, L. et al. (2016). A three-dimensionally stretchable high performance supercapacitor. *J. Mater. Chem. A* 4 (39): 14968–14973.

75 Niu, Z., Zhou, W., Chen, X. et al. (2015). Highly compressible and all-solid-state supercapacitors based on nanostructured composite sponge. *Adv. Mater.* 27 (39): 6002–6008.

76 Guo, Y., Zheng, K., and Wan, P. (2018). A flexible stretchable hydrogel electrolyte for healable all-in-one configured supercapacitors. *Small* 14 (14): e1704497.

77 Jia, R., Li, L., Ai, Y. et al. (2018). Self-healable wire-shaped supercapacitors with two twisted NiCo2O4 coated polyvinyl alcohol hydrogel fibers. *Sci. China Mater.* 61 (2): 254–262.

78 Chen, C., Cao, J., Wang, X. et al. (2017). Highly stretchable integrated system for micro-supercapacitor with AC line filtering and UV detector. *Nano Energy* 42: 187–194.

79 Li, L., Fu, C., Lou, Z. et al. (2017). Flexible planar concentric circular micro-supercapacitor arrays for wearable gas sensing application. *Nano Energy* 41: 261–268.

80 Lou, Z., Li, Wang, L. et al. (2017). Recent progress of self-powered sensing systems for Wearable Electronics. *Small* 13 (45): 1701791.

81 Chen, D., Lou, Z., Jiang, K. et al. (2018). Device configurations and future prospects of flexible/stretchable lithium-ion batteries. *Adv. Funct. Mater.* 28 (51): 1805596.

82 Hagleitner, C., Hierlemann, A., Lange, D. et al. (2001). Smart single-chip gas sensor microsystem. *Nature* 414 (6861): 293–296.

83 Chen, C., Cao, J., Lu, Q. et al. (2017). Foldable all-solid-state supercapacitors integrated with photodetectors. *Adv. Funct. Mater.* 27 (3): 1604639.

84 Hou, X., Liu, B., Wang, X. et al. (2013). SnO_2-microtube-assembled cloth for fully flexible self-powered photodetector nanosystems. *Nanoscale* 5 (17): 7831.

85 Hu, Y., Cheng, H., Zhao, F. et al. (2014). All-in-one graphene fiber supercapacitor. *Nanoscale* 6 (12): 6448–6451.

86 Xu, J. and Shen, G. (2015). A flexible integrated photodetector system driven by on-chip microsupercapacitors. *Nano Energy* 13: 131–139.

87 Kim, D., Yun, J., Lee, G. et al. (2014). Fabrication of high performance flexible micro-supercapacitor arrays with hybrid electrodes of MWNT/V_2O_5 nanowires integrated with a SnO2 nanowire UV sensor. *Nanoscale* 6 (20): 12034–12041.

88 Cao, X., Jie, Y., Wang, N. et al. (2016). Triboelectric nanogenerators driven self-powered electrochemical processes for energy and environmental science. *Adv. Energy Mater.* 6 (23): 1600665.

89 Cha, S., Kim, S.M., Kim, H. et al. (2011). Porous PVDF as effective sonic wave driven nanogenerators. *Nano Lett.* 11 (12): 5142–5147.

90 Choi, D., Choi, M.Y., Choi, W.M. et al. (2010). Fully rollable transparent nanogenerators based on graphene electrodes. *Adv. Mater.* 22 (19): 2187.

91 Chun, J., Ye, B.U., Lee, J.W. et al. (2016). Boosted output performance of triboelectric nanogenerator via electric double layer effect. *Nat. Commun.* 7: 12985.

92 Fan, F.R., Tang, W., and Wang, Z.L. (2016). Flexible nanogenerators for energy harvesting and self-powered electronics. *Adv. Mater.* 28 (22): 4283–4305.

93 Kwon, J., Seung, W., Sharma, B.K. et al. (2012). A high performance PZT ribbon-based nanogenerator using graphene transparent electrodes. *Energy Environ. Sci.* 5 (10): 8970.

94 Lee, K.Y., Gupta, M.K., and Kim, S.W. (2015). Transparent flexible stretchable piezoelectric and triboelectric nanogenerators for powering portable electronics. *Nano Energy* 14: 139–160.

95 Lee, K.Y., Kim, D., Lee, J.-H. et al. (2014). Unidirectional high-power generation via stress-induced dipole alignment from ZnSnO3Nanocubes/polymer hybrid piezoelectric nanogenerator. *Adv. Funct. Mater.* 24 (1): 37–43.

96 Wang, Z.L. (2012). Self-powered nanosensors and nanosystems. *Adv. Mater.* 24 (2): 280–285.

97 Fu, Y., Wu, H., Ye, S. et al. (2013). Integrated power fiber for energy conversion and storage. *Energy Environ. Sci.* 6 (3): 805.

98 Guo, H., Yeh, M.H., Lai, Y.C. et al. (2016). All-in-one shape-adaptive self-charging power package for wearable electronics. *ACS Nano* 10 (11): 10580–10588.

99 Zi, Y., Lin, L., Wang, J. et al. (2015). Triboelectric-pyroelectric-piezoelectric hybrid cell for high-efficiency energy-harvesting and self-powered sensing. *Adv. Mater.* 27 (14): 2340–2347.

100 Zhang, Z., Chen, X., Chen, P. et al. (2014). Integrated polymer solar cell and electrochemical supercapacitor in a flexible and stable fiber format. *Adv. Mater.* 26 (3): 466–470.

101 Yun, J., Song, C., Lee, H. et al. (2018). Stretchable array of high-performance micro-supercapacitors charged with solar cells for wireless powering of an integrated strain sensor. *Nano Energy* 49: 644–654.

102 Bi, D., Xu, B., Gao, P. et al. (2016). Facile synthesized organic hole transporting material for perovskite solar cell with efficiency of 19.8%. *Nano Energy* 23: 138–144.

103 Freitag, M., Teuscher, J., Saygili, Y. et al. (2017). Dye-sensitized solar cells for efficient power generation under ambient lighting. *Nat. Photonics* 11 (6): 372–378.

104 Akhtar, F. and Rehmani, M.H. (2015). Energy replenishment using renewable and traditional energy resources for sustainable wireless sensor networks: a review. *Renew. Sust. Energ. Rev.* 45: 769–784.

105 Yue, Y., Yang, Z., Liu, N. et al. (2016). A flexible integrated system containing a microsupercapacitor, a photodetector, and a wireless charging coil. *ACS Nano* 10 (12): 11249–11257.

106 Kim, D., Kim, D., Lee, H. et al. (2016). Body-attachable and stretchable multisensors integrated with wirelessly rechargeable energy storage devices. *Adv. Mater.* 28 (4): 748–756.

3

Fiber-shaped Supercapacitors

Mengmeng Hu[1,2], Qingjiang Liu[1,2], Yao Liu[1,2], Jiaqi Wang[1,2], Jie Liu[1,2], Panpan Wang[1,2], Hua Wang[1,2] and Yan Huang[1,2,3,]*

[1] *State Key Laboratory of Advanced Welding and Joining, Harbin Institute of Technology (Shenzhen), Shenzhen, 518055, Guangdong, China*
[2] *Flexible Printed Electronics Technology Centre, Harbin Institute of Technology (Shenzhen), Shenzhen, 518055, Guangdong, China*
[3] *School of Materials Science and Engineering, Harbin Institute of Technology (Shenzhen), Shenzhen, 518055, Guangdong, China*

3.1 Introduction

Flexible and wearable electronics have attracted much attention in both academia and industry for their wide application possibilities in various fields such as medical monitoring devices, portable military equipment, and smart textiles, etc. For the realization of fully flexible and wearable electronics, lightweight, energy supplying devices with flexibility are highly demanded [1–5]. Supercapacitors, also called electrochemical capacitors, have attracted much attention because of fast charge/discharge rate, high power density and long lifetime [6]. Flexible solid-state supercapacitors are considered as the promising candidate to power flexible and wearable electronics. Compared to the traditional planar structured supercapacitors, fiber-shaped supercapacitors (FSSCs) have the merits of tiny volume, high flexibility and deformability, as well as great electrochemical performance, which is considered to be the great candidate in wearable electronics [7, 8].

In recent years, FSSCs have been substantially studied to meet the requirements of flexible and wearable electronics. The research focused on electrolytes, electrode materials from traditional (such as carbon-based materials) to emerging materials (such as Mxene), electrode substrates including metal fibers, carbon-based fibers and cotton wires, and novel structures. Many flexible and high energy density FSSCs have been fabricated and demonstrate their potential application in wearable electronics. Apart from the continuous devotion to improve the energy storage performance and fabricate miniaturized configurations, considerable efforts have also been made to make the FSSCs multifunctional. Besides the basic use of storing electric energy, the integration of other functions such as self-healability, stretchability, and electrochromic ability into the supercapacitors can ensure their use in various circumstances.

Here, we discuss the topic of FSSCs by summarizing the important development and progress of recent years. We first give an introduction on the device structure and the electrolyte. According to the device configuration, FSSCs can be classified to parallel-like, twisted, and core-shell type. Electrolyte is a key component in supercapacitors which plays an important role in electrochemical properties. For FSSCs, gel polymer electrolytes (GEs) are extensively used because of high ionic

*e-mail: yanhuanglib@hit.edu.cn

Flexible Supercapacitors: Materials and Applications, First Edition. Edited by Guozhen Shen, Zheng Lou and Di Chen.

conductivity and good mechanical strength. The GEs are introduced from four subsections: aqueous GEs (AGEs), organic GEs (OGEs), ionic liquid GEs (IGEs), and redox-active GEs (RGEs). We then summarize the capacitive performance of various electrode materials. In this section, traditional electrode materials including carbon-based macro/nano materials, conducting polymers (CPs) and metal oxides are discussed in detail. Additionally, many new electrode materials are introduced such as metal nitrides, Mxenes, metal–organic frameworks (MOFs), polyoxometalates (POMs) and BP. Third, we focus on the electrode substrates and the electrochemical performance of the configured FSSCs. The most commonly used fiber-shaped substrates include metal fiber, carbon-based fiber, cotton wires, and composite fibers. Fourth, FSSCs with unique functionalities such as self-healability, stretchability, electrochromic ability, shape-memory, and photodetection are summarized. Last, a short conclusion combining the future perspectives in this field is given. This chapter aims to give a comprehensive illustration of FSSCs.

3.2 Structure of FSSCs

FSSCs integrate the current collector, two active electrodes, electrolyte and/or separator of the devices in 1D system with tiny volume, high flexibility, and excellent deformability. According to the device configuration, FSSCs can be classified to three types of parallel-like, twisted and core-shell. The parallel-like FSSC is fabricated by two parallel fiber-shaped electrodes with the gel electrolyte as the separator (Figure 3.1a). The twisted FSSC is composed by twisting two fiber-shaped electrodes with the gel electrolyte as the separator (Figure 3.1b) while the core-shell FSSC wraps the electrodes and electrolyte in one fiber which can provide larger efficient areas (Figure 3.1c).

(a)
(b)
(c)

Figure 3.1 (a) Schematic diagram of parallel-like fiber supercapacitor. Source: Reproduced with permission [9]. © 2015, Royal Society of Chemistry. (b) Schematic diagram of twisted fiber supercapacitor. *Source:* Reproduced with permission [10]. © 2016, American Chemical Society. (c) Schematic diagram of core-shell fiber supercapacitor. *Source:* Reproduced with permission [11]. © 2014, Wiley-VCH.

FSSCs with light weight and great flexibility can be either directly woven into electric fabrics or stitched into textiles. However, two things need to be noticed. First, the short circuit between positive and negative electrodes should be avoided. Second, the internal resistance of a FSSC increases with the increase of device length, which generates much heat and reduces electrochemical performance when it is performed as the power source [12].

3.3 Electrolyte

As a key component of a supercapacitor, electrolyte plays an important role in electrochemical properties such as energy density and cycle stability. For FSSCs, the key to achieving high-performance is to fabricate a suitable electrolyte with good mechanical strength, high ionic conductivity, and superior stability [13]. GEs are extensively used in FSSCs because of these afore-mentioned advantages (e.g. high ionic conductivity of 10^{-4} to 10^{-3} S cm^{-1} under ambient condition) [14]. GE is composed of a polymer framework host, a solvent filling in the framework and a conducting electrolytic material dissolved in the solvent, in which the polymer network holds a large amount of solvent and as a result GE possesses the traits of both liquid and solid [15]. Many polymer matrices are commonly used such as polyvinyl alcohol (PVA), polyacrylate (PAA), poly(methyl methacrylate) (PMMA), poly(acrylonitrile) (PAN), and poly(ethylene oxide) (PEO). LiCl, $LiClO_4$, H_2SO_4, H_3PO_4, and KOH (potassium hydroxide) are the most used conducting electrolytic material which provides free/mobile ions that take part in the conduction process. In this section, we discuss GEs from the following four subsections: (i) aqueous GEs (AGEs), (ii) organic GEs (OGEs), (iii) ionic liquid GEs (IGEs), (iv) redox-active GEs (RGEs).

AGEs are the most extensively studied GEs in FSSCs with high ionic conductivity, low-cost and nontoxicity. AGE is composed of a polymer framework host (PVA, PEO, PAA et al.), water as the solvent and a conducting electrolytic salt which can be acid (H_2SO_4, H_3PO_4), alkali (KOH) or neutral salt (LiCl, $LiClO_4$). PVA has been widely used as a polymer host in AGEs due to its easy preparation, nontoxic characteristic and good film-forming property. Figure 3.2a shows the field emission scanning electron microscopy (FE-SEM) image of cross-linked gPVAP(20) which is obtained by freeze drying after removing polyvinylpyrrolidone (PVP) in the PVA/PVP composite membrane in which the weight ratio of PVP and PVA is 20 : 80 [16]. The ionic conductivity of gPVAP(20)-H_2SO_4 gel electrolyte membrane at 20 °C is 24.9 mS cm^{-1}. A supercapacitor using this gel as the electrolyte and graphene/carbon nanocomposite as the electrode can operate at 1 V voltage and extends to 3 V for a cell connected in series (Figure 3.2b).

With the recent development in alkaline rechargeable batteries and SCs, the research of alkaline AGEs is of great importance. Lewandowski et al. have fabricated a totally solid state electric double layer capacitor using an alkaline polymer electrolyte (polyethyleneoxide–KOH–H_2O) and an activated carbon (AC) powder as electrode material [20]. The capacity of ca. 1.7–3.0 F for a 1 V voltage range has been measured [20]. Many efforts have been done to increase the capacity of alkaline supercapaitors. A quasi-solid-state supercapacitor was fabricated by using alkaline PVA and potassium iodide (PVA–KOH–KI) as gel electrolyte and activated carbons as electrodes [21]. KI as redox additive in the AGEs is used to increase the ionic conductivity of electrolyte, and to improve the pseudocapacitance of the electrode. The supercapacitor had a high specific capacitance of 236.90 F g^{-1}, increasing by 74.28% compared to PVA–KOH system at the same current density, leading to an energy density of 7.80 Wh kg^{-1} and a power density of 15.34 kW kg^{-1}. In addition, the supercapacitor exhibited excellent cycle-life stability.

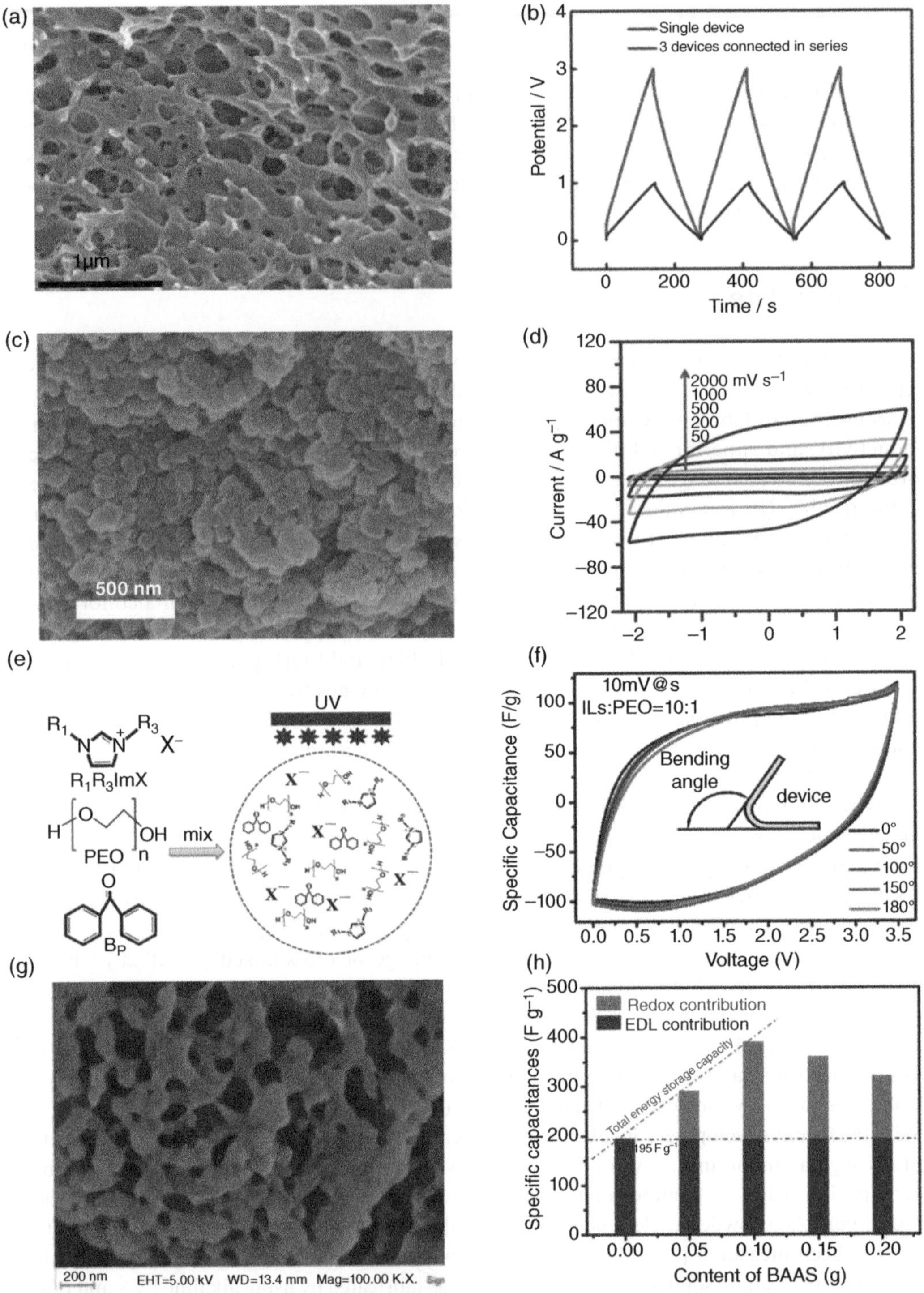

Figure 3.2 (a) Cross-sectional FE-SEM image of gPVAP(20) hydrogel polymer membrane; (b) Charge-discharge curves at $0.5\,A\,g^{-1}$ of single GCB//gPVAP(20)-H_2SO_4//GCB supercapacitor and three supercapacitors in series. *Source:* Reproduced with permission [16]. © 2014, Elsevier. (c) SEM image of the PAN-b-PEG-b-PAN copolymer; (d) Cyclic voltammograms of symmetric two-electrode capacitors. *Source:* Reproduced with permission [17]. © 2012, Wiley-VCH. (e) Schematic representation of the preparation procedure and structure of the gel polymer electrolyte formed via the cross-linking of ionic liquid and PEO chains; (f) CV curves at a scan rate of $10\,mV\,s^{-1}$ of the flexible EDLC made of the 10EMImTFSI-PEO-FPC electrolyte film under different bending angles. *Source:* Reproduced with permission [18]. © 2017, Elsevier. (g) SEM image of the gel sample; (h) The C_s of device with different BAAS content at $0.8\,A\,g^{-1}$. *Source:* Reproduced with permission [19]. © 2016, Royal Society of Chemistry.

The proton and hydroxide ion-conducting AGEs suffer from a narrow voltage window of ≤1 V, limited by water decomposition reaction, while neutral AGEs have been reported to have stability window ≥1.5 V with carbon electrodes due to the high overpotential on both hydrogen and oxygen evolution reactions [22–25]. The energy density of the device will be improved as the energy density correlates to the voltage window squared ($E = 1/2CV^2$). Therefore, a neutral polymer electrolyte comprised of lithium sulfate (Li_2SO_4) and polyacrylamide (PAM) was developed. [22] The Li_2SO_4-PAM electrolyte film showed an ionic conductivity up to $10\,mS\,cm^{-1}$ in 45% RH conditions. The voltage window of the double layer capacitor using carbon nanotube (CNT)-graphite electrode and Li_2SO_4-PAM solid electrolyte was about 2 V, which was significantly larger than that reported for proton- or hydroxyl-conducting AGEs. The capacitor exhibited excellent stability with over 95% capacitance retention after 10 000 charge/discharge cycles. The improvement in voltage window with neutral pH AGEs marks a significant step toward high energy density storage devices.

Although AGEs are the most extensively studied in FSSCs, their narrow operating voltage window greatly limits their wide application. OGEs have attracted much interest for their higher operating voltage window. The difference between OGEs and AGEs is that OGEs use an organic solvent as the plasticizer while AGEs use water. Organic solvents such as propylene carbonate (PC), ethylene carbonate (EC) and dimethyl formamide (DMF) have been used as common plasticizers to improve the working voltage range [14]. Huang et al. have fabricated an OGE using a PAN-b-PEG-b-PAN copolymer composed by PAN blending poly(ethylene glycol) (PEG) swollen with DMF dissolving $LiClO_4$ [17]. SEM image shows the external morphology of PAN-b-PEG-b-PAN (Figure 3.2c) [17]. The small granules with diameter size of <200 nm ensured copolymer granules highly accessible to the lithium salt solution, which eventually incorporated into polymer matrix to form the OGE with ionic conductivity of $6.9\,mS\,cm^{-1}$ [17]. A capacitor fabricated by the OGE film and two carbon electrodes could operate in a 2.1 V voltage window with high-rate voltammograms (Figure 3.2d), which greatly improved the operating voltage than AGEs supercapacitor.

While AGEs have a narrow electrochemical window (~1.2 V) and OGEs suffer from serious health and safety problems, IGEs have attracted much attention because of the merits of excellent thermal stability, high ionic conductivity, nonvolatility, non-flammability, low melting point and wide potential windows (up to 3.5 V) [14, 18, 26, 27]. Similar to AGE and OGE, IGE is composed of polymer host (such as PVA, PEO, and PVDF–HFP) polyvinylidene fluoride - hydrofluoric acid) and ionic liquid salt (IL) which is liquid at room temperature and composed of dissociated ions with no intervening solvent [28]. An IGE film was prepared by cross-linking ionic liquid (EMImTFSI) with poly (ethylene oxide) (PEO) and benzophenone (Bp) followed by ultraviolet (UV) irradiation (Figure 3.2e) [18]. A flexible supercapacitor assembled by the IGE film, and two activated carbon electrodes showed a very broad electrochemical window (3.5V) with a high energy density of $30.13\,Wh\,kg^{-1}$ and a high-power density of $874.8\,W\,kg^{-1}$ at a current density of $1\,A\,g^{-1}$. This flexible supercapacitor still retained its electrochemical performance at different bending angles, indicating that the IGEs are suitable for application in wearable and flexible electronics (Figure 3.2f).

R GE which is to modify/add redox-active species to the electrolyte rather than as a solid form at the electrode is used to improve the electrochemical performance of a supercapacitor [14, 19]. Redox additives can add extra pseudocapacitance by oxidizing at the positive electrode and reducing at the negative electrode reversible faradaic reactions and facilitate fast electron transfer at the electrode–electrolyte interface [14, 19]. The following properties are significant for redox-active additives: involvement in redox reaction, better solubility, excellent electrochemical stability, and easy preparation [19]. Here are some redox couples: iodides (KI), [14, 29, 30], $K_3Fe(CN)_6$, [14, 30] organic redox mediators such as hydroquinone (HQ) [14, 31], P-phenylenediamine (PPD) [14, 32], and methylene blue (MB) [14, 33]. Feng et al. have fabricated a PVA–H_2SO_4–BAAS (bromamine acid sodium) gel polymer [19]. Figure 3.2g shows a FE-SEM image of the polymer, which indicates that porous microstructures occur ubiquitously in the polymer [19]. This continuous and interconnected porous microstructure network is beneficial for ion transport and diffusion in gel systems. The polymer exhibited a high ionic conductivity of 21.4 mS cm^{-1}. The supercapacitor was prepared with activated carbon as electrode and the PVA–H_2SO_4–BAAS gel polymer as both electrolyte and separator. The polymer provided a reversible redox reaction for enhanced supercapacitor performance. As shown in Figure 3.2h, whatever the content of BAAS in the gel polymer, the BAAS enhanced the supercapacitor performance remarkably. The supercapacitor exhibited a high energy density of 30.5 Wh kg^{-1} and a power density of 600 W kg^{-1}. The device also displayed stable capacitive performance even under 100% tensile strain or under 2000 kPa pressure, indicating that the gel polymer endows the device with good flexibility and electrochemical stability.

In this section, we describe basic composition and their benefits and disadvantages of four extensively used GEs, named aqueous GEs (AGEs), organic GEs (OGEs), ionic liquid GEs (IGEs), and redox-active GEs (RGEs). They are different in the nature of electrolyte, ionic conductivity and mechanical stability, which affect the performance of supercapacitors. When choosing the electrolyte to assemble a high-performance supercapacitor, the above aspects should be considered seriously.

3.4 Electrode

Electrode materials are the most critical component of supercapacitors which play a decisive role in the electrochemical performance of the entire device. Traditional electrode materials include carbon-based micro/nano materials, conducting polymers, and metal oxides. Additionally, many emerging electrode materials have been researched, such as metal nitrides, MXenes, MOFs, POMs, and black phosphorusare (BP). This section will briefly introduce the traditional and emerging electrode materials.

3.4.1 Carbon-Based Materials

Carbon-based materials with high specific surface area, high conductivity, high stability and nontoxicity, which include activated carbon, CNTs and graphene, are commonly used in supercapacitors. They are considered as electric double-layer (EDL) electrode materials, where the capacitance results from the pure electrostatic charge accumulated at the electrode/electrolyte interface, which is strongly dependent on the surface area of the electrode materials [34]. Activated carbon (AC) is widely used resulting from its large surface area, relatively good electrical conductivity and low cost. It comes from a wide range of raw materials, mainly divided into plant (e.g. wood, coconut shell, bamboo) and mineral raw materials (e.g. coal, petroleum coke, asphalt). It is prepared via two steps of carbonization and activation. Activation (physical or chemical) can effectively regulate the specific surface area and pore structure. Designing ACs with narrow pore size distribution (accessible to the electrolyte ions) and interconnected pore structure would be beneficial for enhancing the energy density of supercapacitors, without deteriorating their high-power density and cycle life [34].

CNTs with their unique pore distribution, high electrical conductivity, and excellent mechanical and thermal stability, have attracted much attention as the electrode material for supercapacitors [34, 35]. CNTs can be classified as single-wall (SWNTs) and multi-walled carbon nanotubes (MWNTs), both of which have been applied as energy storage electrode materials [34, 36]. Owing to their unique mesoporous microstructure, these materials have high charge accumulation ability in the electrode/electrolyte interface [36]. One thing that should be noticed is that aligned CNTs have faster ionic transportation than entangled CNTs because aligned porous structure provides a large electrolyte-accessible surface for charge delivery [34, 37]. In order to enhance the electrochemical performances of the CNTs-based devices, CNTs are usually modified by transition metal oxides and/or conducting polymers to introduce pseudocapacitance. For example, an all-solid-state supercapacitor with CNT/polyaniline (PANI) nanocomposite thin films as electrodes and H_2SO_4-PVA gel as electrolyte was fabricated [38]. The device performed outstanding electrochemical performances with a high specific capacitance of 350 $F\,g^{-1}$ for the electrode materials and a specific capacitance of 31.4 $F\,g^{-1}$ for the entire device. It also showed good stable cycle life with only 8.1% decay in specific capacitance after 1000 charge–discharge cycles.

Graphene, as a well-known 2D monolayer composed of sp^2-boned carbon with good electronic and mechanical properties, large surface area and chemical stability, has been investigated for use as electrode materials in the supercapacitors. There are various approaches for fabricating graphene, such as mechanical stripping, chemical stripping, chemical vapor deposition and redox method. In 2008, Ruoff and co-workers first reported the chemically modified graphene as electrode material of the supercapacitor [39]. They obtained graphene with a specific surface area of 705 $m^2\,g^{-1}$ by using hydrazine hydrate to reduce graphene oxide. The specific capacitance of the supercapacitor composed of the graphene in aqueous and organic electrolyte is 135 and 99 $F\,g^{-1}$, respectively. Inspired by this work, various works on graphene as electrode materials have been explored. The capacitance of most solid-state supercapacitors based on freestanding graphene materials ranges from 80 to 118 $F\,g^{-1}$, which are far below the corresponding theoretical capacitance (550 $F\,g^{-1}$) [39, 40]. The reason is that restacked graphene sheets reduce active surface area of graphene material and slow the ion transport/diffusion within the active material. El-Kady et al. have developed a new strategy to fabricate graphene-based flexible SCs whereby they could effectively avoid the restacking of graphene [40]. They used a standard LightScribe DVD optical drive to reduce graphite oxide films to graphene. Figure 3.3a shows the SEM image of the laser-scribed graphene (LSG), which has the open network which prevents the agglomeration of graphene sheets. The network structure of LSG has open pores, which helps facilitate the electrolyte accessibility to the electrode surfaces and improves the ionic diffusion of the electrolyte in LSG electrodes. The produced films showed specific surface area (1520 $m^2\,g^{-1}$) and high electrical conductivity (1738 $S\,m^{-1}$), allowing them to be used directly as SC electrodes without binders or current collectors. The device made with graphene as electrodes and PVA–H_3PO_4 as a gel electrolyte exhibited ultrahigh energy density while maintained the high-power density and excellent cycle stability (>97% over 10 000 cycles). Moreover, the as-fabricated device exhibited only 5% change of the initial capacitance when it was tested in a bent state for more than 1000 cycles.

In summary, carbon-based materials (AC, CNTs and graphene) are commonly used electrode materials in supercapacitors. Their high specific surface area and high conductivity are factors for achieving high power density of supercapacitors. However, their performance still needs to be improved to meet the need for high energy density.

3.4.2 Conducting Polymers

Conducting polymers (CPs) show great potential as pseudocapacitive electrode materials owing to the merits of good flexibility, high capacitance, lightweight and low cost. The mostly used CPs include polypyrrole (PPy), PANI, poly (3,4-ethylenedioxythiophene) (PEDOT) and their derivatives. Yuan et al. have assembled all-solid-state SCs with PANI networks/Au/paper electrodes using a 40 mm-thick H_3PO_4/PVA membrane electrolyte. [46] The all-solid-state SC showed a power density of around 3 W cm^{-3} and an energy density of around 0.01 Wh cm^{-3}, with long-term stability after 10 000 charge–discharge cycles at 1 mA cm^{-2}. PPy is another widely used electrode material in supercapacitors. PPy and its composites with various morphology and/or structure are fabricated by different synthesis strategies which are generally divided into chemical polymerization and electrochemical polymerization. Figure 3.3b displays the SEM image of PPy nanowires (NWs) via electrochemical polymerization [42]. However, the poor cycling stability in long-term charge-discharge processes due to structural pulverization and counter-ion stability greatly limits its further application. Several effective strategies are developed to solve these troublesome problems: (i) fabrication of PPy with a designed morphology or structure; (ii) synthesis of PPy composites with carbon-based or metal oxides. Through these ways, PPy-based supercapacitors can enhance its cycling stability significantly, meanwhile achieve high energy density and power density.

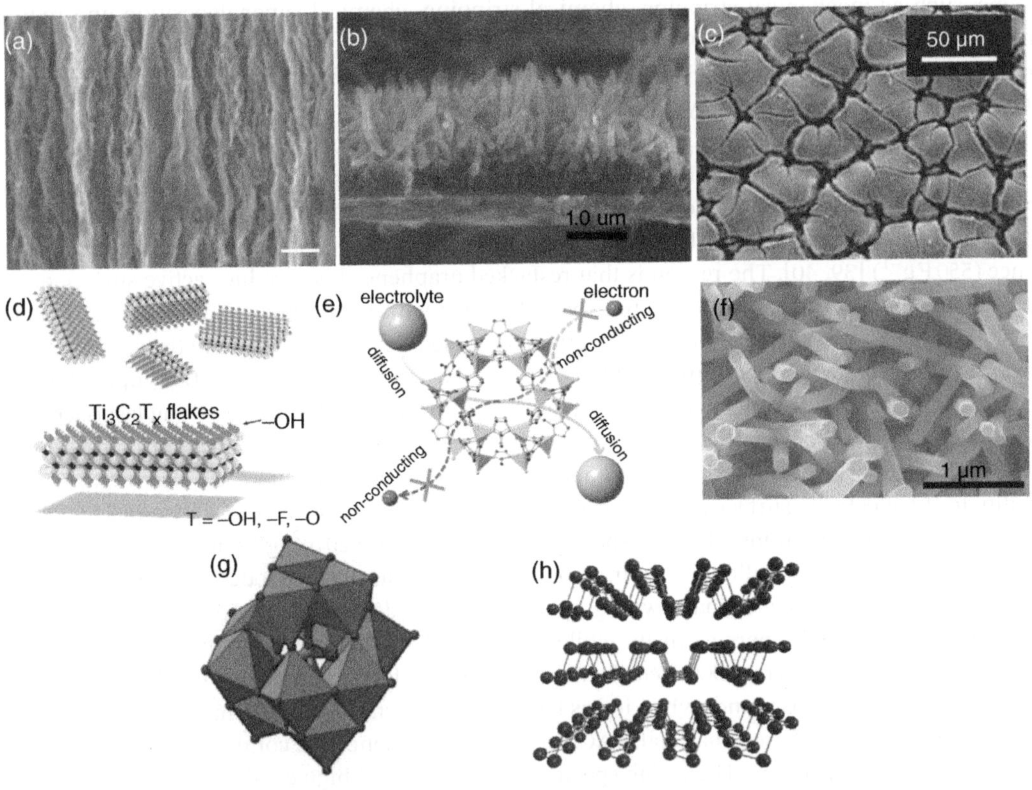

Figure 3.3 (a) The cross-sectional SEM images of LSG film. *Source:* Reproduced with permission [92]. © 2012, Wiley. (b) Side view of PPy nanowires with length of 1.5 μm. *Source:* Reproduced with permission [42]. © 2010, Royal Society of Chemistry. (c) SEM image (top-view) of RuO_2 electrode after 300 electrodeposition cycles. *Source:* Reproduced with permission [14]. © 2015, Wiley-VCH. (d) Schematic of $Ti_3C_2T_x$ flakes. *Source:* Reproduced with permission [43]. © 2016, Wiley-VCH. (e) Illustration that electron cannot migrate along or access the skeleton of MOF crystal, while electrolyte can diffuse in and out of MOF pores. *Source:* Reproduced with permission [44]. © 2015, American Chemical Society. (f) SEM image of the Cu–CAT NWAs grown on carbon fiber paper. *Source:* Reproduced with permission [45]. © 2017, Wiley-VCH. (g) Schematic of phosphomolybdate (PMo_{12}). *Source:* Reproduced with permission [14]. © 2015, Royal Society of Chemistry. (h) Structure of black phosphorus. *Source:* Reproduced with permission [30]. © 2014, Elsevier.

3.4.3 Metal-Based Materials

Tremendous efforts have been made in investigating metal oxide materials (e.g. RuO_2, MnO_2, V_2O_5, Fe_2O_3, MoO_3, $NiCo_2O_4$) because of their much higher specific capacitance than carbon-based and conducting polymer materials. The metal oxides provide pseudocapacitance in nature which store charge through fast, reversible surface redox reactions and exhibit large capacitance. Among these metal oxides, RuO_2 is the first studied material because of high proton conductivity, remarkably high specific capacitance, wide potential window and a long cycle life. [47] Figure 3.3c displays the microscopic appearance of RuO_2 [48]. Interestingly, the RuO_2/PEDOT:poly(styrene sulfonate) (PSS)-based conductive ultra-thin film synthesized by an aerosol spray strategy could be used to make transparent and flexible supercapacitors [47]. However, the high cost and environment unfriendliness of Ru-based electrode materials greatly limit their use in supercapacitors. Thus, cheaper and environmental benign metal oxides as alternatives are urgently needed to be developed. Manganese-based oxides with low cost, environmental friendliness, and high theoretical specific capacitance (~1400 $F\,g^{-1}$) are considered as promising alternative electrode materials for RuO_2. Chodankar et al. have fabricated flexible solid-state supercapacitors based on MnO_2 thin films deposited on a flexible stainless-steel substrate with PVP-$LiClO_4$ gel electrolyte [49]. The cell achieved a high operating voltage of 1.6 V with high energy density of 23 $Wh\,kg^{-1}$ and a power density of 1.9 $kW\,kg^{-1}$. Tremendous efforts have been made on MnO_2-based materials for supercapacitors, which include synthetic methods, MnO_2 polymorph (from amorphous to highly crystalline) and nanostructures (film, powder composite, advanced 3D architecture etc.) to improve electrochemical performance. Vanadium-based oxides are also considered to be promising alternative materials for RuO_2 in supercapacitors owing to their high specific capacitance. For example, V_2O_5 can be used as both positive and negative electrode for asymmetric supercapacitors. Various methods are investigated to prepare V_2O_5 electrodes, such as sol-gel process, quenching, hydrothermal method, and electrodeposition. Many other metal oxide materials such as Fe_2O_3, MoO_3, WO_3, SnO_2, and InO_2 also have gained significant attention as electrode materials for supercapacitors. However, metal oxide materials always suffer problems of poor conductivity, instability, or sluggish kinetics, which limit their electrochemical properties to some extent.

Metal nitrides are an emerging class of electrode materials for high-performance supercapacitors owing to their excellent electric conductivity (4000–55 500 $S\,cm^{-1}$) and high capacitance, such as VN (vanadium nitride) and TiN [50, 51]. Xiao et al. fabricated flexible freestanding mesoporous VN nanowires (MVNNs)/CNTs hybrid electrodes by vacuum-filtering method [52]. The synergistic effects from the high electrochemical performance of MVNNs and the high conductivity and mechanical consolidation of CNTs were well achieved. The supercapacitor was constructed based on these hybrid electrodes with an PVA/H_3PO_4 electrolyte. The device showed high volumetric capacitance, energy, and power of 7.9 $F\,cm^{-3}$, 0.54 $mWh\,cm^{-3}$ and 0.4 $W\,cm^{-3}$, respectively. In addition, the device displayed 82% capacity retention over 10 000 cycles and excellent flexibility with almost no change in the CV (cyclic voltammetry) profile under different bending conditions. Lu et al. fabricated a flexible supercapacitor with TiN nanowires (NWs) grown on a carbon cloth and PVA/KOH gel as the electrolyte [53]. The gel electrolyte suppressed the oxidation reaction on electrode surface and stabilized the electrode. As a result, the supercapacitor exhibited extraordinary stability with 83% of its initial capacitance retained after 15 000 cycles and a high energy density of 0.05 $mWh\,cm^{-3}$. These results open up new opportunities of high-performance metal nitride-based supercapacitors. Similarly, many other metal sulphides have been explored for their potential as electrode materials for supercapacitors [54].

3.4.4 Mxenes

MXenes are a large family of two-dimensional (2D) transition metal carbides and/or nitrides with the formula $M_{n+1}X_nT_x$, where M stands for a transition metal (e.g. Ti, Zr, V, or Mo), X is carbon and/or nitrogen, and T_x is the surface termination (–OH, =O, or –F) [43, 55–58]. High conductivity, good mechanical property, and hydrophilic surface make MXenes promising electrode materials for energy storage application since the first supercapacitor based on 2D $Ti_3C_2T_x$ was reported by Gogotsi and co-workers in 2011 [57–60]. MXenes are produced by etching of element 'A' (group IIIA and IVA elements, mostly Al) from layered ternary metal carbides/nitrides (MAX phases) via HF-contained [57, 58] or acidic solutions of fluoride salts (LiF in HCl) [61]. Their properties such as surface chemistry and conductivity, are significantly affected by the synthesis method. For example, HF etching primarily produces fluorine-containing functional groups, while LiF-HCl treatment yields predominantly oxygen-containing surface groups [62, 63].

Among many MXenes, $Ti_3C_2T_x$ is one of the most commonly studied materials for supercapacitor electrode application. Figure 3.3d and e display the schematic and SEM image of nacre-like layered stacking of the $Ti_3C_2T_x$ respectively. The exclusive layered structure of MXenes is thought to effectively prevent dense stacking which always happens in graphene and CPs, which leads to relatively porous structures that are beneficial for ion transport. An all-solid-state supercapacitor has been fabricated based on $Ti_3C_2T_x$ nanosheet and graphene as the electrode and PVA/H_3PO_4 gel electrolyte as the electrolyte [60]. In the electrode, the small MXene between the graphene layers served as active material and ideal "buffer" for enhanced electrolyte shuttling, as well as a conducting spacer which prevented the irreversible π–π stacking between the graphene sheets [64–66]. The device exhibited outstanding volumetric capacitances of up to $216\,F\,cm^{-3}$. Moreover, it delivered a high energy density of $3.4\,mWh\,cm^{-3}$ and a high power density of $1600\,mW\,cm^{-3}$.

3.4.5 Metal Organic Frameworks (MOFs)

MOFs are a new type of porous materials constructed by connecting metal ions/ion clusters and organic ligands, which are commonly prepared by using solvothermal, hydrothermal or microwave-assisted method [44, 67, 68]. MOFs have the advantage of high surface, adjustable pore size and an abundance of active sites. Benefited by their high surface area, ions in the electrolyte can diffuse into the micropores and/or mesopores in MOFs, which generate large electric double-layer (EDL) capacitance (Figure 3.3f) [44]. However, the application of pristine MOFs as electrode materials in supercapacitors is limited by their poor electrical conductivity. An effective strategy to improve conductivity is to prepare nanocomposites with other conducting materials, such as carbon-based species (e.g. CNT and graphene) and CPs (e.g. PANI and PPy) [44, 69]. For example, Co-based MOF crystals (ZIF-67) were coated on carbon cloth with electrochemically polymerized PANI [44]. PANI acted as bridges for electron transportation between the external circuit and the internal surface of MOFs to overcome the problem of insulated MOFs. The hybrid-structured electrode integrated the advantages of high EDLC (electric double-layer capacitor) capacitance from the large surface area of MOFs and effective pseudocapacitance generated by PANI. The supercapacitor yielded a remarkable areal capacitance of $35\,mF\,cm^{-2}$, a power density of $0.833\,W\,cm^{-3}$ and more than 80% capacitance retention after 2000 cycles. Recently, conducting MOFs have been designed and prepared. Li et al. have fabricated conductive MOF nanowire arrays (Cu-CAT) and demonstrated their application as the sole electrode for solid-state supercapacitors [45]. Cu-CAT was constructed by coordinating Cu ions with 2,3,6,7,10,11-hexahydroxytriphenylene (HHTP) ligands in the ab plane to create a 2D hexagonal lattice. Then the structure packed along the c-axis with a slipped-parallel AB stacking model to form a honeycomb-like porous structure. Cu-CAT displayed 1D channels along the c-axis with an open-window size of ~1.8 nm. Cu-CAT also exhibited good charge transportation because of the effective orbital overlaps between the Cu ions and organic ligands. The high-magnification SEM image in Figure 3.3g shows the morphology of Cu-CAT nanowires which have uniform hexagonal prism shape and hexagonal top facet [45]. The electrode was consisted by the conductive MOF (Cu–CAT) grown on carbon fiber (CF) paper with high EDL capacitance of $120\,F\,g^{-1}$. The resulted solid-state supercapacitor achieved a high areal capacitance of ~$22\,\mu F\,cm^{-2}$ and good cycling stability of more than 85% capacitance retention after 5000 cycles. This work provides a great prospect for developing MOFs as a good electrode material for supercapacitors.

3.4.6 Polyoxometalates (POMs)

POMs comprise a 3D framework redox-active molecular clusters that combine oxygen and early transition metals (i.e. M = V, Nb, Ta, Mo, W) at their highest oxidation states [70]. POMs are molecular oxides that contain tens to hundreds of metal atoms that reach nuclearity as high as 368 metal atoms in one single cluster molecule [14, 70]. The smallest and simplest POM presents the Keggin structure (corresponding to the general composition $[AM_{12}O_{40}]^{n-}$) which is stable in acidic aqueous electrolytes. Notably, phosphomolybdate (PMo_{12}) and phosphotungstate (PW_{12}) represent the ultimate degree of dispersion for an oxide nanocluster because all 12 MO_6 moieties (M: Mo or W) are at the surface of the cluster (Figure 3.3h). Moreover, they are well suited to achieve a high capacity for energy storage application because of their fast and reversible multi-charge transfer reactions. Due to the high solubility of these POMs in aqueous solutions and polar organic solvents, when they are used as electrode materials, most of them are anchored onto conducting substrates such as conducting polymers (CPs) and carbon-based materials to prevent the dissolution of the POMs into the electrolyte. The anchoring of anionic POMs on the surface of nanocarbons and/or within the network of CPs not only maintains the integrity and activity of inorganic clusters, but also benefits the conducting properties of the hybrid composite structure. Various strategies have been used to fabricate POM-modified electrodes, including chemisorption, electrodeposition, sol–gel techniques, layer-by-layer self-assembly, and entrapment in conductive polymers [71, 72].

Dubal et al. prepared a high-performance hybrid electrode based on phosphomolybdic acid ($H_3PMo_{12}O_{40}$) (PMo_{12} in short) nanodots anchored onto rGO (rGO-PMo_{12}) for an all-solid-state symmetric supercapacitor [70]. It provided an extended potential window of 0.3 V which allows to realize a broad voltage of 1.6 V for the rGO-PMo_{12} supercapacitor, compared to the potential window of 1.3 V for rGO cell. The 3D hierarchical porous architecture and ultrathin graphene nanosheets decorated with tiny yet numerous PMo_{12} clusters endowed rGO-PMo_{12} excellent volumetric capacitance of over $3.18\,F\,cm^{-3}$ and high energy density of $1.07\,mWh\,cm^{-3}$. Yang and co-workers developed flexible POM-doped PPy nanopillar array (POM-PPy/nanopillar) films through a soft-lithography and post-electrodeposition process [73]. The anchoring of POMs in PPy was achieved by electrodepositing POMs during the electro-polymerization of pyrrole monomers. The POM-PPy/nanopillar film electrode exhibited a superior areal capacitance as high as $77.0\,mF\,cm^{-2}$ and excellent long-term stability with ~100% retention of its initial capacitance after 2000 cycles. Further work on the fabrication of POM-based hybrid electrodes needs to be done to meet more critical requirements of energy storage devices.

3.4.7 Black Phosphorus (BP)

BP as a new member of 2D nanomaterials has attracted enormous attention because of its distinct structure with corrugated planes of P atoms, which are connected by strong intralayer P–P bonding and weak interlayer van der Waals forces (Figure 3.3i) [74]. BP crystals are produced by using mechanical exfoliation (high-energy ball-milling of red phosphorus), heating toxic white or red phosphorus under high pressure or transforming white phosphorus in liquid metal [30]. Single- and few-layered BP nanoflakes are commonly obtained by liquid exfoliation approach in organic solvents (such as acetone and N-methyl-2-pyrrolidone) [30]. Few-layer BP has a direct band gap which is tunable between 0.30 and 2.2 eV by controlling the number of layers. Moreover, BP has a large spacing of 5.3 Å between adjacent puckered layers, which is larger than the 3.6 Å of graphite and comparable to the 6.15 Å of the 1T MoS_2 phase [75]. Similar to other 2D materials, BP nanosheet is a promising candidate for energy storage application. The exploration of BP in supercapacitors is still in start stage [75]. Recently, a liquid-exfoliated BP nanoflake-based all-solid-state supercapacitor on a PET (polyethylene terephthalate) substrate was developed with a PVA/H_3PO_4 gel electrolyte [75]. It displayed volumetric capacitance of 17.78 F cm^{-3} (59.3 F g^{-1}), energy density ranging from 0.123 to 2.47 mWh cm^{-3}, a maximum power density of 8.83 W cm^{-3} and capacitance retention of 71.8% after 30 000 cycles.

3.5 Electrode Design of FSSCs

FSSCs with the merits of small volume, great flexibility and compatibility with fabrics have gained much attention in flexible and wearable electronics. Fiber-shaped electrodes are fabricated by activated materials coated on the fiber-shaped substrate through electrodeposition, hydrothermal reaction, etc. The commonly used fiber-shaped substrates mainly include metal fiber, carbon-based fiber, and cotton fiber.

3.5.1 Metal-Fiber Supported Electrode

Intrinsically conductive metallic fibers, such as Ni, Ti, Au and stainless-steel fibers exhibit capabilities of high conductivity, good mechanical strength, and high flexibility. Thus, the metal fiber can serve as physical support for active materials and excellent current collectors. Gu and coworkers presented a coaxial-type symmetric FSSC, which was fabricated by $CuCo_2O_4$ on Ni wires as electrodes together with a PVA/KOH electrolyte gel as electrolyte and separator [76]. Figure 3.4a shows the schematic diagram of fabrication process of the solid-state FSSC. The device showed an excellent rate capability of 11.09 F g^{-1} at a current of 2 mA and great cycling stability with 93.5% retention after 4000 charge/discharge cycles. In addition, the FSSC showed high flexibility and electrochemical stability with no apparent degradation in the bending test. Besides the Ni fiber, Ti, Au and stainless-steel fiber have also been used for FSSCs. Vellacheri et al. demonstrated a symmetric solid-state cable-type supercapacitor with working voltage of 1.1 V, which was made of PEDOT deposited on D-TiO_2/Ti as electrodes and PVA-LiCl as electrolyte [77]. Figure 3.4b–d displays SEM images of D-TiO_2/Ti, which was made by two-step process. In the first step, the Ti wire (purity of 99.9%, diameter of 0.5 mm) was subjected in anodization environment to grow TiO_2 nanotubes (TiO_2/Ti). In the second step, TiO_2 of TiO_2/Ti was converted to D-TiO_2 by a cathodic polarization process. Through the process of anodization and cathodic polarization, D-TiO_2/Ti electrode obtained advantageous structural features (such as high surface area and short ion

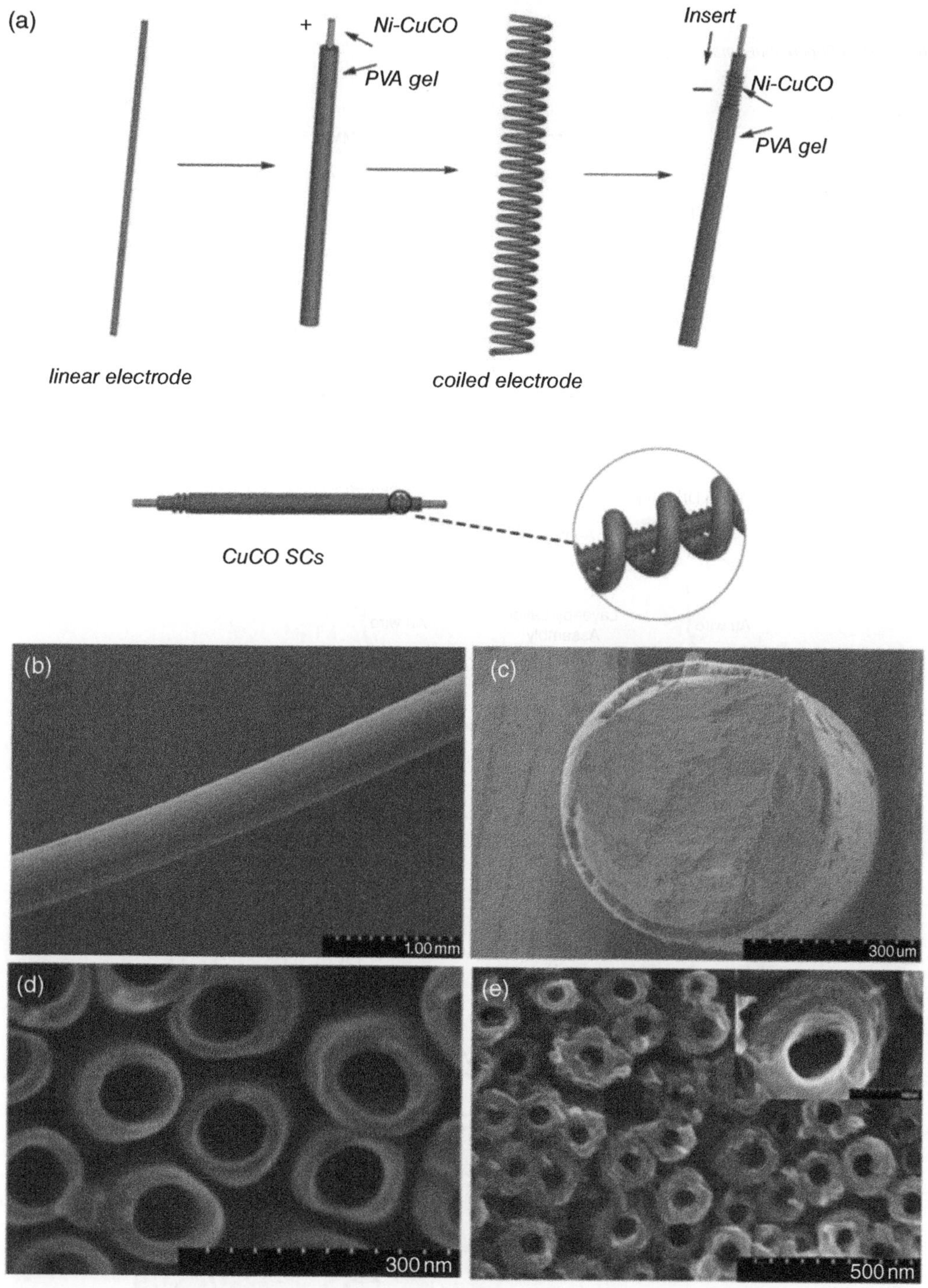

Figure 3.4 (a) Schematic diagram showing the fabrication process of the solid-state FSSC. *Source:* Reproduced with permission [76]. © 2015, Wiley-VCH. (b) Top-view SEM image of D-TiO_2/Ti; (c) Cross-sectional views of D-TiO_2/Ti; (d) A closer view showing the individual D-TiO_2 of D-TiO_2/Ti; (e) D-TiO_2 after the deposition of PEDOT for 15 s and inset shows the enlarged view of the D-TiO_2 with PEDOT. *Source:* Reproduced with permission [77]. © 2017, Wiley-VCH. (f) The FSSC consisting of two wire electrodes and an organic electrolyte of which components are presented. The inset shows an optical image of the fabricated FSSC; (g) Cross-sectional view of the layer-by-layer assembly process of MWCNTs on the Au wire; (h) Flowchart of the electrode fabrication process; (i) Capacitance retention (orange circles) and Coulombic efficiency (blue circles) of the fabricated FSSC; (j) Optical image of two FSSCs connected in series after integration into a knitted sweater to light green (3.0 V), red (2.1 V), and blue (3.2 V) LEDs. *Source:* Reproduced with permission [78]. © 2018, American Chemical Society.

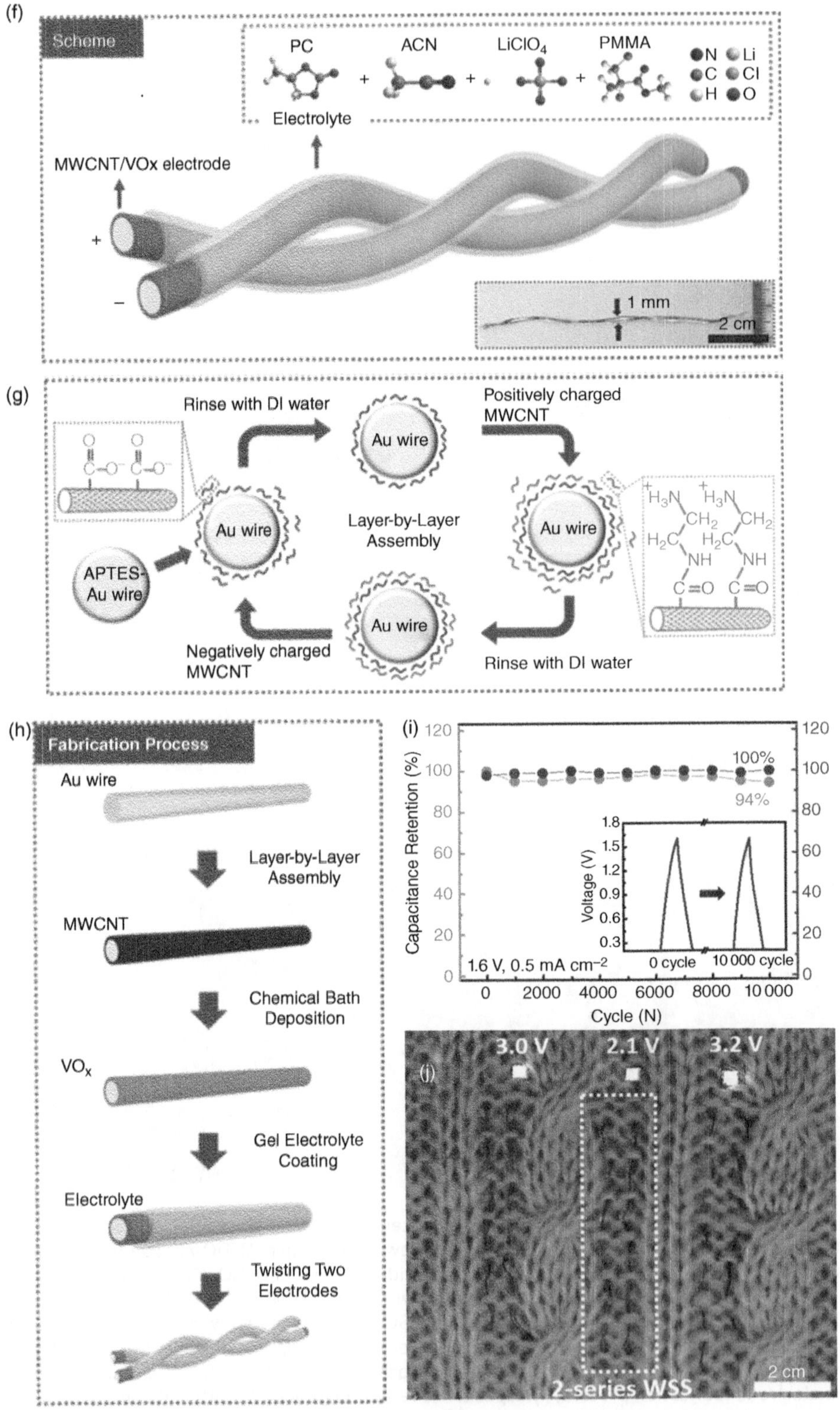

Figure 3.4 (Continued)

diffusion path) and high electrical conductivity [79, 80]. Figure 3.4e displays the SEM image of D-TiO_2 after the deposition of PEDOT, which clearly reveals the electrodeposition of PEDOT onto the inner surface of D-TiO_2/Ti. D-TiO_2/Ti could act as both current collector and support for PEDOT while PEDOT served as active material. Notably, the device showed high capacitance, ultrahigh rate capability, and fairly large potential window. Furthermore, the device also showed mechanical flexibility along with stable electrochemical performances when it was subjected to mechanical bending and straightening process, indicating its great potential in flexible electronic devices.

Weavability, knittability, and wearability are big challenges for applying conventional metallic wires in wearable electrics. However, it has been found when the size of their diameter scales down to micrometers, the situation is different for some kinds of metals. Keum et al. built a FSSC by a 100 μm thick Au wire coated with MWCNTs (multi-walled carbon nanotube) and vanadium oxide (VO_x), combining an organic electrolyte that extended the voltage window to 1.6 V (Figure 3.4f) [78]. The MWCNTs layer was deposited on the Au fiber utilizing layer by layer technique, which enabled sequential deposition of single layers of both positively charged and negatively charged MWCNTs via electrostatic interaction (Figure 3.4g). Figure 3.4h displays the whole fabrication process of FSSC which exhibited an areal capacitance of 5.23 $mF\,cm^{-2}$ with an energy density of 1.86 $\mu Wh\,cm^{-2}$ at a power density of 8.5 $mW\,cm^{-2}$. It exhibited 94% of capacitance retention and 100% Coulombic efficiency after 10 000 GCD cycles (Figure 3.4i), indicating the stable operation ability and excellent capacitance retention ability. Furthermore, two series-connected FSSC could be easily integrated into a knitted sweater owing to their overall thin thickness (~1 mm), which successfully lighted up various colored LEDs (light-emitting diodes), demonstrating its great potential in e-textile industry (Figure 3.4j).

When the diameter of stainless-steel scales down to micrometers, the stainless-steel yarn is mechanically robust and flexible enough to be weaved and knitted. 316 L stainless-steel yarns with a diameter of 180–250 μm are produced by twist-bundle-drawing multiple high-temperature-spun stainless steel 316 L thin filaments together (Figure 3.5a) [81]. The fabricated yarns are as soft and flexible as cotton yarns with low electric resistivity (40 $\Omega\,m^{-1}$), high tensile strength (over 700 Mpa) and resistance to high temperature and high vapor pressure. Therefore RGO, MnO_2,and PPy are sequentially deposited on the conductive yarn to form a thin electrocapacitive layer (Figure 3.5a). Figure 3.5b displays the SEM image of the cross-sectional PPy@MnO_2@RGO-deposited conductive yarn. The thin hierarchical structure is beneficial for ion transport during charge/discharge process. Consequently, the assembled supercapacitor demonstrated a superior capacitance of 411 $mF\,cm^{-2}$ and energy density of 1.1 $mWh\,cm^{-3}$. A 15 cm × 10 cm clothes weaved by seven yarn supercapacitors in series could light 30 LEDs (Figure 3.5c). Moreover, the yarn supercapacitor could be knitted (Figure 3.5d).

3.5.2 Carbon Materials Based Fiber Supported Electrode

FSSCs based on metal fibers display attractive performance for flexible and wearable electronics. However, the relatively heavy density increases the weight of the devices which limits their wider application. Carbonaceous materials such as carbon fiber, CNT fiber, graphene fiber with merits of low densities and good flexibility, have been used for flexible and wearable electronics. Carbon-based fibers can not only serve as physical support for active materials and excellent current collectors, but also capacitive materials.

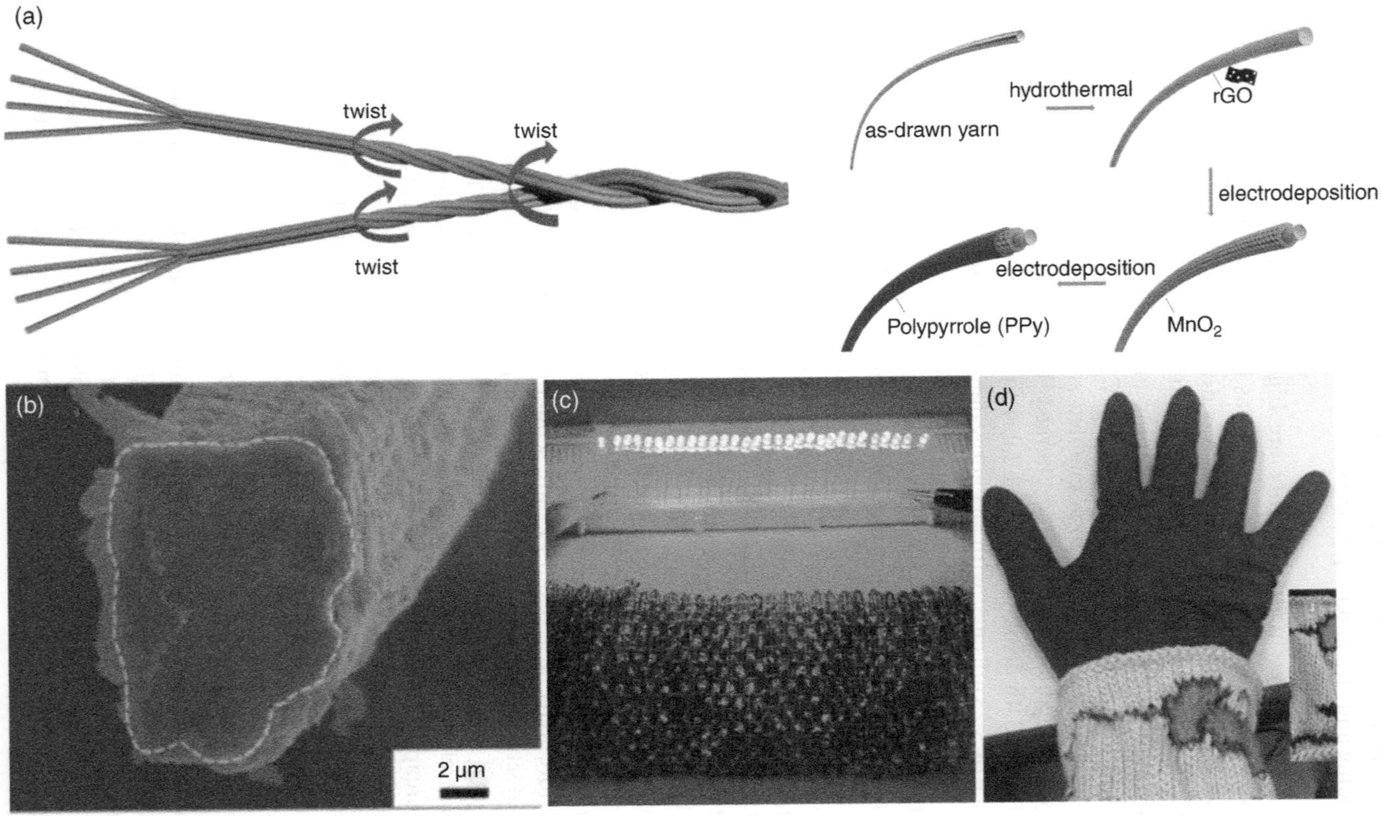

Figure 3.5 (a) Schematics of yarn fabrication and yarn modified by deposition of rGO, MnO_2, and PPy; (b) Cross section of the PPy@MnO_2@rGO-deposited conductive yarn; (c) A 15 cm × 10 cm woven clothes can light 30 LEDs; (d) A wrist band knitted with a pattern (inset shows the pattern powering a LED). *Source:* Reproduced with permission [81]. © 2015, American Chemical Society.

3.5.2.1 Carbon Fiber

Among various carbonaceous materials, carbon fiber is one of the most widely used current collectors for energy storage devices due to its one-dimensional structure, high electric conductivity and low cost [82]. An all-solid-state FSSC using $CoNiO_2$ nanowires@carbon fiber composite electrode and active carbon (AC) coated carbon fiber electrode with the length longer than 1 m was fabricated [83]. Figure 3.6a shows the schematic of whole fabrication process of the supercapacitor. It exhibited high specific capacitance of $1.68\,mF\,cm^{-1}$ and energy density of $0.95\,mWh\,cm^{-3}$ with a large voltage window of 1.8 V. The device showed excellent mechanical and electrochemical stability during bending, pulling, tying, and weaving. Furthermore, it could be easily weaved into wearable and patterned textile as a power source to power MP3 and LED flashlight as shown in Figure 3.6b–d.

Jang et al. reported a new high voltage all-solid-state fiber-shaped asymmetric supercapacitor based on carbon fiber (CF) coated by MnO_2 (CF/MnO_2) and MoO_3 (CF/MoO_3) as the positive and the negative electrode respectively [84]. The supercapacitor had a remarkable performance because of the large work function difference between the metal oxides and the high conductivity of the CF substrate. In particular, the device had a high capacitance of $4.86\,mF\,cm^{-2}$ with a wide operating voltage window of 2 V. Moreover, the electrochemical and mechanical stabilities under long-term cycling and bending tests implied its good endurance under harsh operating conditions.

3.5.2.2 CNT Fiber

CNT fibers with many merits of excellent textile properties (strong and flexible), high electric conductivity, high surface area and one-dimensional structure have been widely investigated for wearable devices [85–87]. Wei et al. have reported a high-performance CNT based FSSC by using two redox additives simultaneously: PPy for the electrodes and hydroquinone (HQ) for the electrolyte, which were used to improve the performance of the supercapacitor [9]. As shown in Figure 3.7a, PPy is first electrodeposited on the surface of the CNT strip; then they are spun into a fiber by using a motor; finally, two fibers are placed in parallel and coated with the gel electrolyte to form an all-solid-state FSSC.

In practical application of the wearable electronics, its energy density is of great importance. The energy density (E) formula of supercapacitor is $E = CV^2/2$, where C and V represent the specific capacitance and the operating voltage of the total supercapacitor. To improve the energy density of a FSSC, there are two approaches: enhancing the capacitance (C) and enlarging the voltage window (V). Generally, organic electrolytes or ionic liquids can provide higher operating voltages in comparison to aqueous electrolytes, but their applications are always constrained by safety problems or price. By integrating the respective advantages of the two different electrodes, asymmetric supercapacitors can extend the voltage window to achieve high energy density. Peng's lab created a novel type of fiber-shaped asymmetric supercapacitor with high volumetric energy densities by designing the asymmetric configuration and integrating multiple functional components as electrodes [10]. Figure 3.7b shows the schematic illustration of the fabrication of high volumetric energy density asymmetric supercapacitor. In this supercapacitor, a ternary hybrid positive electrode by growing MnO_2 nanosheets onto a PEDOT: PSS-coated CNT fiber was prepared and then it was assembled with an ordered microporous carbon (OMC)/CNT hybrid negative electrode. Figure 3.7c and d show SEM images of the bare CNT fibers at low and high magnifications. The CNT fiber has a uniform diameter of $\sim 43\,\mu m$ with a highly aligned structure, which enables high mechanical strength as well as a large surface area. The fiber supercapacitor can operate in a wide potential window from 0 to 1.8 V, achieving an energy density as high as $11.3\,mWh\,cm^{-3}$. It also exhibited high cycling stability, high flexibility, and could be woven or knitted into flexible power

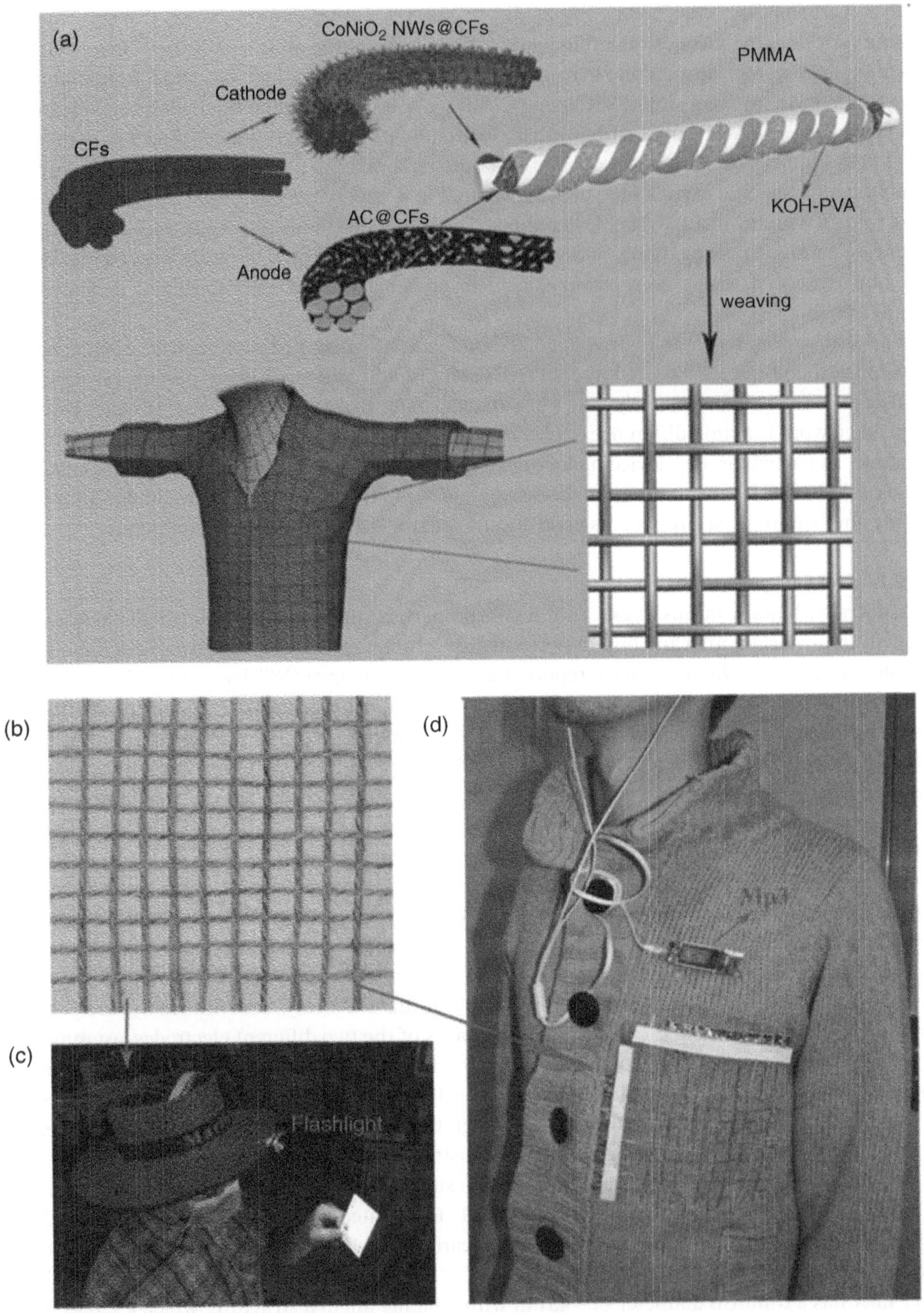

Figure 3.6 (a) Schematic illustration of the fabrication process of the asymmetric FSSC; (b–d) The FSSC can be weaved as a textile to be integrated with our daily dresses to power LED flashlight and a MP3, respectively. *Source:* Reproduced with permission [83]. © 2016, Wiley-VCH.

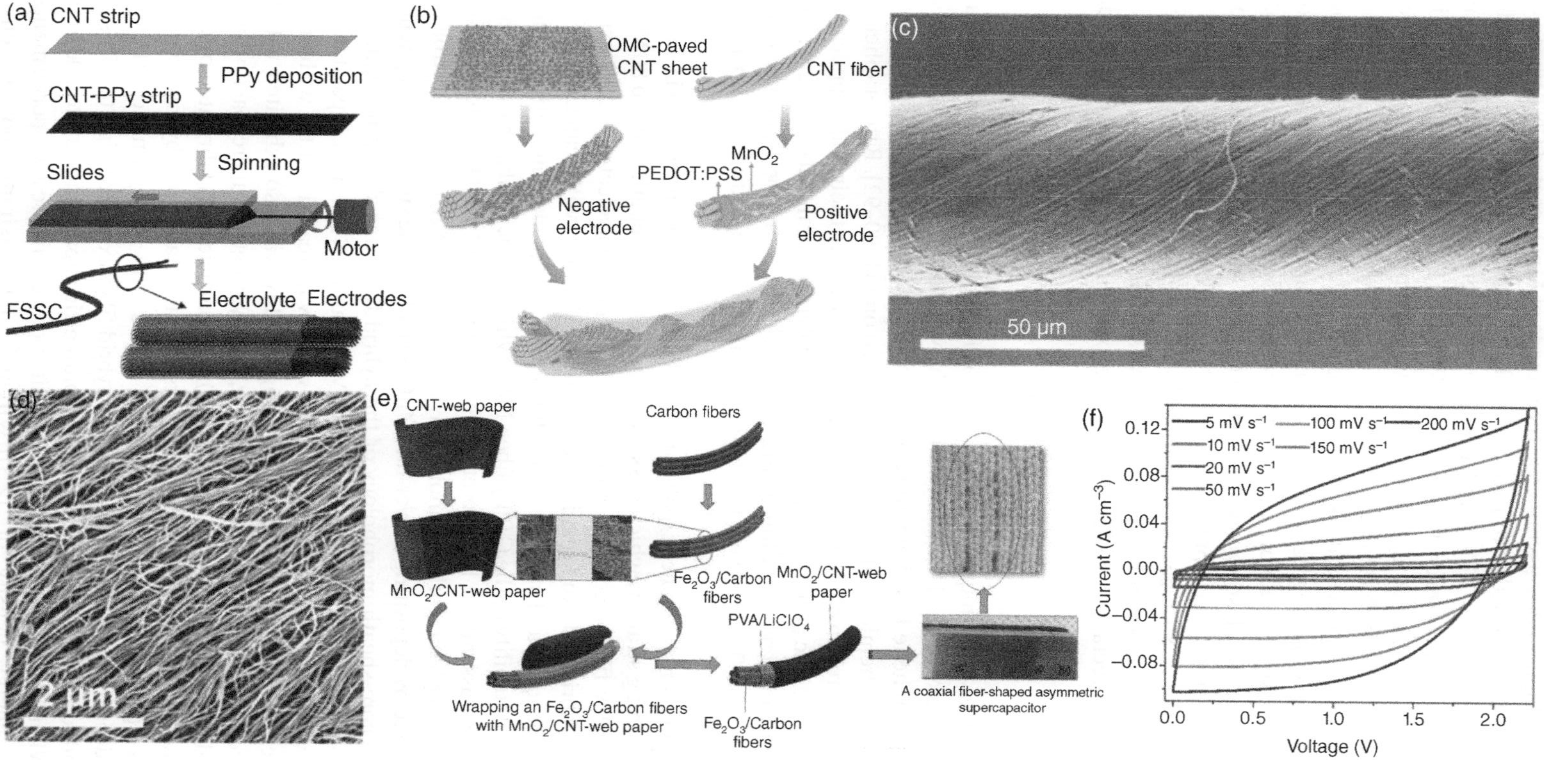

Figure 3.7 (a) Schematic illustration of the fabrication process of a symmetric parallel FSSC. *Source:* Reproduced with permission [9]. © 2015, Royal Society of Chemistry. (b) Schematic illustration of the fabrication of an asymmetric twisted FSSC; (c–d) Bare CNT yarn at low and high magnifications. *Source:* Reproduced with permission [10]. © 2016, American Chemical Society. (e) Schematic illustration of the fabrication of a coaxial asymmetric FSSC; (f) Cyclic voltammograms at different scan rates. *Source:* Reproduced with permission [88]. © 2018, Elsevier.

textiles. Similarly, another high voltage asymmetric FSSC made of MnO_2 coated CNT fiber as positive electrode and CNT fiber deposited with polyimide (PI) nanosheets as negative electrode displayed a large voltage window up to 2.1 V [89]. The supercapacitor resulted in both high energy density (36.4 μWh cm^{-2}) and high-power density (15.6 mW cm^{-2}). Furthermore, it also exhibited long cycling life, good rate performance and high flexibility, which provides a general and effective paradigm for flexible and wearable electronic devices.

For aforementioned parallel and twisted type FSSCs, the effective surface area between the two fiber electrodes is limited to some degree, which results in low capacitance. By contrast, coaxial FSSCs enable utilization of larger and more effective surface area between the electrodes and provide a short pathway for the ultrafast transport of axial electrons and ions. Ahn and co-workers developed coaxial FSSC by simply wrapping a PVA-$LiClO_4$ gel electrolyte-coated Fe_2O_3/CFs core negative electrode with MnO_2/CNT-web paper positive electrode (Figure 3.7e) [88]. The CV curves of the supercapacitor at scan rates ranging from 5 to 200 mV s^{-1} with a potential window between 0 and 2.2 V showed a quasi-rectangular shape, indicating excellent rate capability of the device (Figure 3.7f).

3.5.2.3 Graphene Fiber

Graphene fiber integrates several unique properties such as high electric and thermal conductivities, high strength, good mechanical flexibility, and light weight [90]. Various strategies were reported to fabricate 1D graphene-based macro-scaled architectures including wet-spinning methods, hydrothermal strategy, chemical vapor deposition and so on [41, 90–92]. Bare graphene-fiber based supercapacitors have relatively low capacitance which restricts their wide application. In order to improve the capacitance performance, conducting polymers, or metal oxides etc., are chosen to combine with them in consideration of the synergistic effect and performance enhancement, paving the way for the flexible, wearable and high-efficiency miniaturized supercapacitor devices.

Qu and co-workers prepared graphene/polypyrrole (G/PPy) composite fibers via facile and straightforward wet-spinning strategy, which possessed high tensile strength of up to 80 MPa [93]. By intertwining two G/PPy fiber electrodes with PVA/ H_2SO_4 gel polyelectrolyte, a flexible all-solid-state FSSC was fabricated and exhibited excellent stability with capacitance of ca. 95–105 mF cm^{-2} in the bending and straight status during 1000 cycles, demonstrating high structural and electrochemical stability of the G/PPy fiber. Meanwhile, a type of fiber-based asymmetric micro-supercapacitor was assembled by Gao et al. that composed of MnO_2-coated core-sheath graphene fiber (GF) and graphene-carbon nanotube hybrid fiber in a PVA/LiCl gel electrolyte [94]. The GMF was synthesized by depositing flower-like MnO_2 sheath on the surface of graphene fiber, and the corresponding core-sheath morphology of the hybrid fiber was shown in Figure 3.8a–c. Benefited from the synergistic effect of graphene with MnO_2 and CNT, a high capacitance up to 23.6 mF cm^{-2} was delivered when the operating voltage window increased to 1.6 V with an energy density of 11.9 mWh cm^{-2}. Meanwhile, excellent cycling stability was also exhibited with 92.7% initial capacitance retention after 8000 cycles. Similarly, an asymmetric supercapacitor based on graphene fibers coated by nickel cobalt sulfide ($NiCo_2S_4$) nanoparticles was fabricated by Cai et al. as shown in Figure 3.8d [95]. Herein, the grapheme fibers were first fabricated by low-temperature (80 °C) reduction of graphene oxide, which were then coated with $NiCo_2S_4$ nanoparticles via a simple solvothermal deposition method. The obtained hybrid fiber exhibited high tensile strength of 226 MPa, high conductivity of 39 S cm^{-1} and high volumetric capacitance up to 300 F cm^{-3} in two-electrode cell. When fabricated into a supercapacitor device, impressive performance was exhibited with high energy density of 12.3 mWh cm^{-3} and maximum power density of

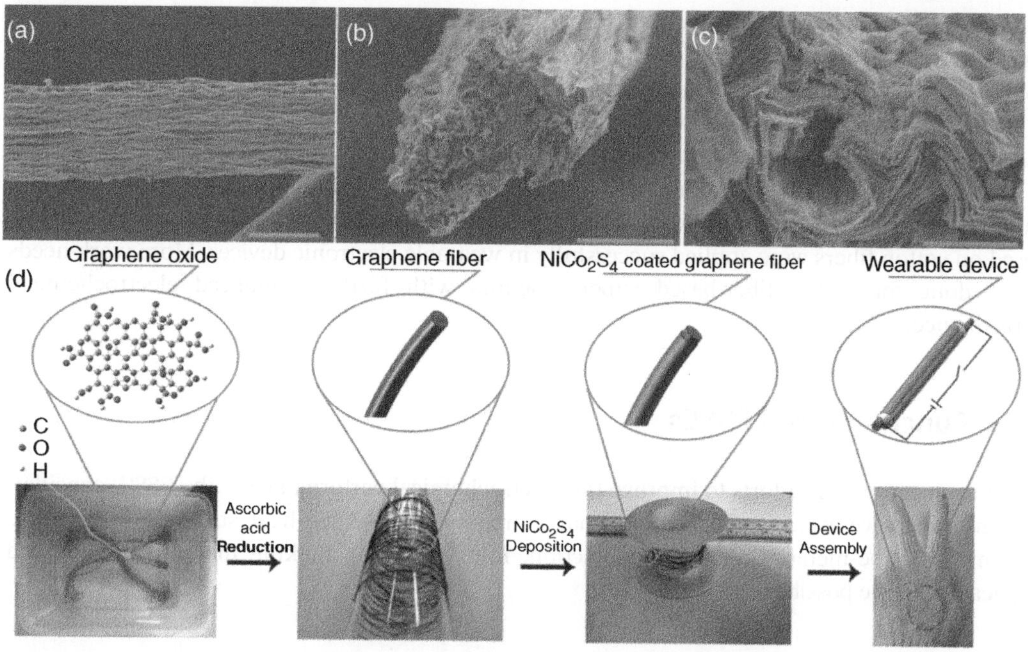

Figure 3.8 (a) Overview of GMF observed by SEM. Scale bar: 10 mm; (b) SEM cross sectional image of GMF. Scale bar: 10 mm; (c) Magnified picture of b and the boundary between MnO_2 and graphene is marked as a blue dotted line. Scale bar: 1 mm. *Source:* Reproduced with permission [94]. © 2014, Royal Society of Chemistry. (d) Schematic illustration of the fabrication process of a wearable asymmetric FSSC. *Source:* Reproduced with permission [95]. © 2016, Springer Nature.

1600 mW cm^{-3}, as well as good cyclic stability with 92% capacitance retention over 2000 cycles. In addition, the supercapacitor ring could be woven into textile and three supercapacitors connected in series could power a LED, demonstrating the potential application of FSSCs in next generation flexible and wearable electronic devices.

3.5.3 Cotton Fiber Supported Electrode

When the wearable electronics are knitted into daily clothing, the intelligent versatility, air permeability and comfortability are of great importance. In this regard, it is reasonable to select the threads derived from existing cotton-based textiles as wearable electronics substrate. Cotton thread is a macroscopic flexible and microscopic porous material that composed of countless cotton fibers [96]. The merits of high-porosity, well-hydroscopicity, and knittability as well as large quantity make it an ideal substrate to fabricate wearable electronic devices [97, 98].

It is effective to improve the conductivity of cotton fibers by introducing conductive materials (e.g. metal?carbon, CNTs) on the surface of cotton fibers via dipping or coating process [99]. For example, Liu et al. developed Ni-coated cotton yarns electrode by polymer-assisted metal deposition method, which was further coated by reduced graphene oxide (rGO) nanosheets via electrochemical deposition (Figure 3.9a and b) [100]. The unique hierarchical "graphene/metallic textile" configuration endows the assembled supercapacitor high energy and power density up to 6.1 mWh cm^{-3} and 1400 mW cm^{-3}. Furthermore, the supercapacitor yarn demonstrated good cycle stability with 82% capacitance retained after 10 000 cycles (Figure 3.9c). Additionally, the composite electrodes can be directly used as threads in commercial embroidery machines to embroider a logo on a piece of cotton fabric as shown in Figure 3.9d. The supercapacitor yarns could also be woven into fabric with other pristine cotton yarns in conventional weaving process (Figure 3.9e). Jost and coworkers fabricated a high-performance supercapacitor by using activated, carbonloaded, natural cotton yarns mixed with stainless steel fiber (Figure 3.9f) [101]. The surface of carbon-coated two-ply yarn embedded activated carbon with a mass loading of $0.6\,mg\,cm^{-1}$. And the final supercapacitor yielded a liner capacitance up to $37\,mF\,cm^{-1}$. These supercapacitors based on cotton fibers were applied successfully in wearable electronic devices. More work needs to be done for cotton fiber-based supercapacitors with further enhanced electrochemical performance.

3.6 Functionalized FSSCs

Apart from continuing efforts to improve the electrochemical performance of the FSSC, considerable interests have been attracted to add smart functions, such as self-heal, stretçhability, electrochromism, shape-memory, and photodetection, into the FSSC, making many unprecedented applications to be possible.

3.6.1 Self-Healable FSSCs

In recent decades, fascinated by nature such as cnidarian hydra and human skin which can repair damage, the impartation of the self-healing property to supercapacitors is attractive for multifunctional smart energy storage devices. At the same time, various self-healing polymers have been realized with self-healing mechanisms revealed. The most studied self-healing concept is the microencapsulation approach pioneered by White et al. which embedded a microencapsulated healing monomer into polymer matrix that contained dispersed catalyst [102]. When a crack appears and spreads to the microcapsules, the healing monomer is released into the crack through capillary action and polymerized by the catalyst distributed in the matrix, which could prevent the further growth of the cracks. This kind of self-healing requires no external stimulus, and it is considered to be “autonomous” [103].

Another self-healing mechanism is molecular interdiffusion [104]. The evolution process of self-healing involves surface rearrangement, surface approach, wetting, diffusion, and randomization. When crakes occur, the created crack surfaces have to rearrange resulting from the topographical discontinuity. Then the two crack interfaces are brought together and into contact. When the damaged surfaces approach, the cracked surfaces wet each other and form an interface before the healing process start. Then molecules diffuse and bring about the entanglement of the polymer chains and interpenetration into the unruptured matrix material. Eventually, the initial crack interfaces gradually disappear with chain randomization and the mechanical strength increased. This self-healing phenomenon is often observed in thermoplastic materials and sometimes in thermoset materials [105, 106].

Reversible covalent networks allow multiple cycles of crack healing bridged by the newly formed covalent bonds upon the application of an external stimulus such as heat [107, 108], or light [109, 110]. Apart from the aforementioned mechanisms, reversible noncovalent crosslinking is another main mechanism of supramolecular polymers based on hydrogen bonding [111, 112], ionomers [113, 114], and metal bonding [115–117], which rarely requires external stimulus. These abovementioned four self-healing mechanisms are accountable for most self-healing polymeric materials.

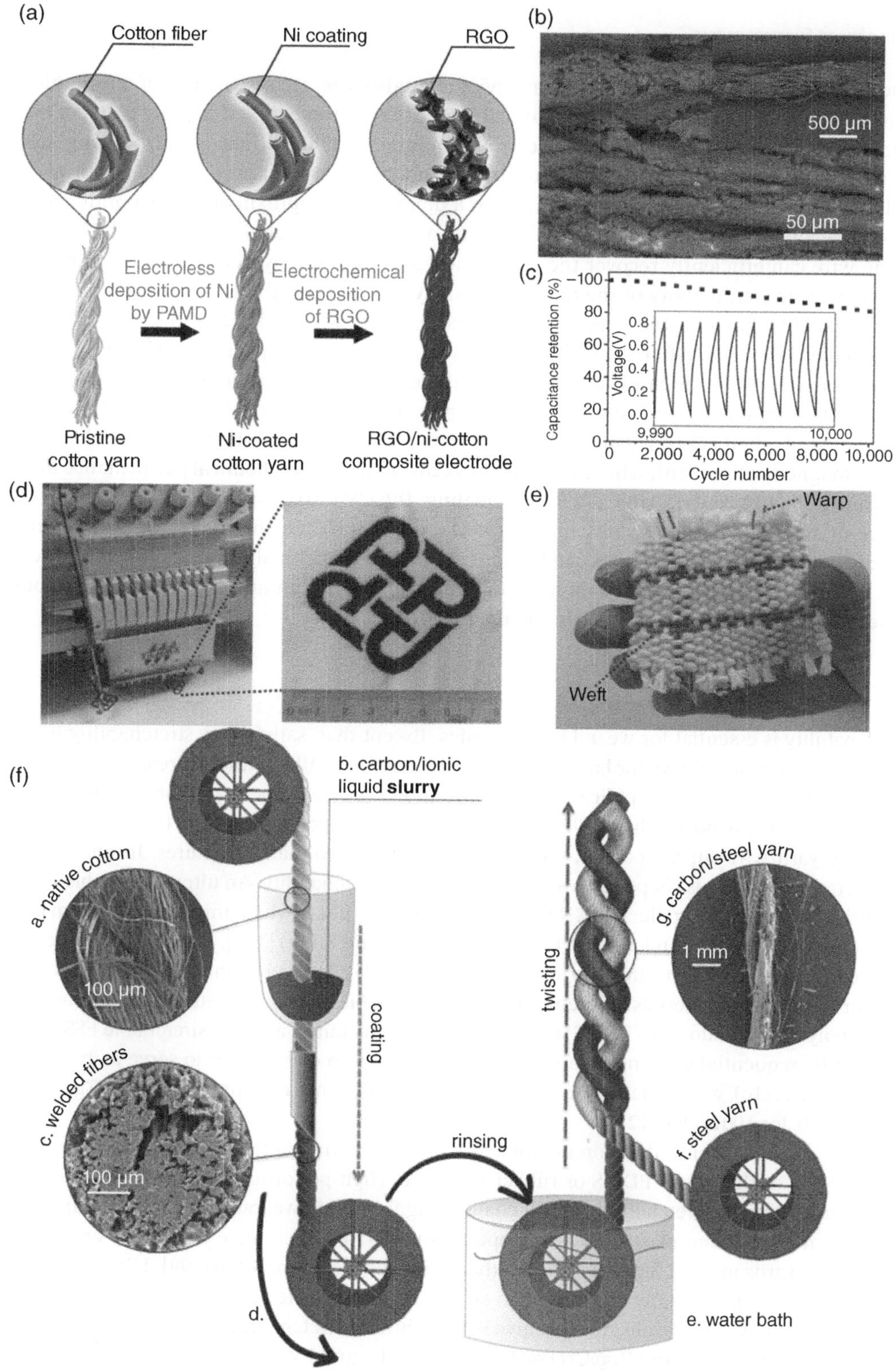

Figure 3.9 (a) Schematic illustration of the fabrication of RGO/Ni cotton yarn composite electrode; (b) SEM image of the typical RGO/Ni cotton composite electrode; (c) Cycle life of the device. The inset is GCD curves from the 9990th to 10000th cycle; (d) Digital images of embroidery logo of The Hong Kong Polytechnic University using the composite electrode yarns; (e) A woven fabric made with FSSC yarns. *Source:* Reproduced with permission [100]. © 2015, Springer Nature. (f) Yarns coated via natural fiber welding. *Source:* Reproduced with permission [101]. © 2014, Wiley-VCH.

The self-healing in the supercapacitor device contains the reconstruction of mechanical integrity and electric performance. Mechanical integrity is restored by self-healing polymer and electric performance is renovated by electrode alignment. One format of introducing self-healing polymer into supercapacitor is to use the polymer as a support of FSSC. Sun et al. developed a self-healable conducting wire with a diameter of approximately 700 μm by wrapping electrically conducting nanomaterials around a self-healing polymer fiber [118]. The conducting wire can efficiently recover the mechanical strength of polymer fiber in the core and the high electric conductivity of the nanomaterial layer in the sheath. So, a self-healable FSSC was fabricated by twisting these two wires as electrodes with PVA–H_2SO_4 gel electrolyte that also served as separator. After the fifth healing cycle, the specific capacitance was maintained at approximately 82.6%. Meanwhile, Huang et al. designed an electrically and mechanically self-healable yarn-based supercapacitor by wrapping magnetic electrodes with a self-healing carboxylated PU shell [119]. The magnetic force promotes the reconnection of broken electrode through magnetic alignment, which benefits the recovery of electrochemical performance after self-healing (Figure 3.10a). After the fourth healing, the CV curves still remained rectangular in spite of the slight decrease in capacitance (Figure 3.10b). Furthermore, the self-healing supercapacitors connected in series can light up the LED again with almost the same brightness as the normal (Figure 3.10c–e). This work enlightened the design and fabrication of various self-healable and flexible electronics in the near future.

3.6.2 Stretchable FSSCs

Good flexibility is essential for wearable electronics. Except that, satisfactory stretchability is the next necessary step to achieve for body movement. Electronic textiles without stretchability could be broken easily in practical application. In conventional planar supercapacitors, stretchability over 100% has already been achieved by "wrinkled" shapes or open mesh geometries. For a single stretchable yarn/wire supercapacitor, it can be divided into two main structures. In one case, a fiber electrode or device is shaped into a spring to achieve stretchability. An ultra-stretchable FSSC was fabricated based on two flexible CNT@graphene@MnO_2 fibers which mimicked the stem tendrils of climbing stems around a superelastic core fiber (Figure 3.11a) [120]. The obtained supercapacitor can sustain tensile strain of up to 850% with LED still effectively lighting (Figure 3.11b). In the other case, active materials are deposited on a stretchable elastic fiber substrate such as pre-strained polydimethylsiloxane (PDMS) and rubber etc. For instance, a highly stretchable FSSC was fabricated by sequentially wrapping aligned CNT sheets on an elastic fiber to serve as two electrodes (Figure 3.11c), which maintained a high specific capacitance of approximately 18 $F\,g^{-1}$ after stretch by 75% for 100 cycles [121].

For wearable electronics, the aforementioned FSSCs encountered the incompatibility to wear due to the use of the elastic PDMS or rubber substrate. High-performance yarn supercapacitors with intrinsically high stretchability is still challenging for highly wearable electronics. Sun et al. have fabricated urethane plastic fiber core spun yarns (UY), which were formed by entwining native cotton yarns around highly elastic urethane filaments (Figure 3.11d) [99]. UY were introduced for the first time to offer intrinsic high stretchability and acted as a wearable substrate for hosting conductive CNTs and electrocapacitive PPy to fabricate stretchable electrodes (Figure 3.11e). The yarn supercapacitor with PPy@CNTs@UY as electrodes and PVA/H_3PO_4 gel as electrolyte and separator, remained a high areal capacitance even at high strain of 80%. This work provides a promising strategy for large-scale, low-cost fabrication of comfortable, stretchable and wearable electronics.

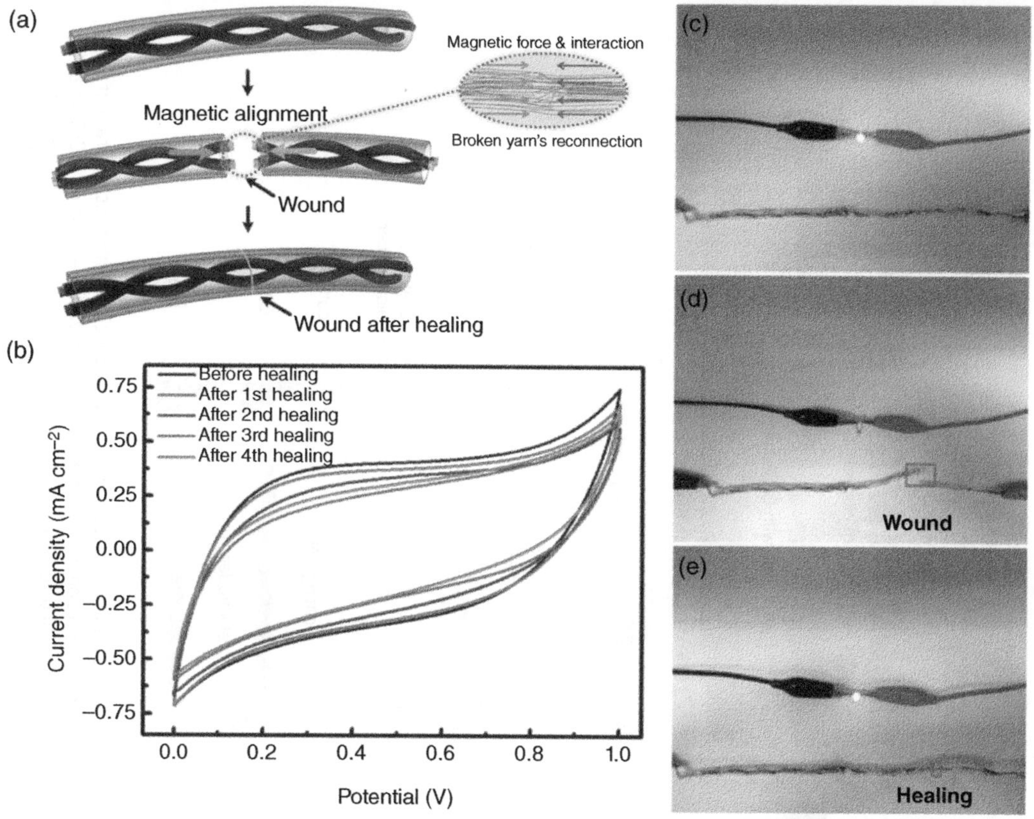

Figure 3.10 (a) Schematic illustration of the self-healable supercapacitor; (b) CVs after healing for different cycles; (c–e) Photographs of two supercapacitors connected in series to power an LED bulb. Each one is coated with a 1 mm thick self-healable PU shell: (c) before breaking, (d) after breaking, and (e) after healing. *Source:* Reproduced with permission [119]. © 2015, American Chemical Society.

3.6.3 Electrochromic FSSCs

Electrochromism is the phenomenon that active materials change their colors reversibly during ion insertion/extraction or chemical redox reaction [122–124]. The typical electrochromic device consists of supporting substrates, two transparent conducting layers, a counter electrode layer, an electrochromic electrode layer and an electrode layer (Figure 3.12a) [125].

Electrochromic materials are approximately divided into three major categories: inorganic oxides, conducting polymers, and metal coordination complexes. For electrochromic inorganic oxides, the most studied is tungsten oxides (e.g. WO_3), in which the proton insertion/extraction induces a color change [127, 128]. Similar to WO_3, Nb_2O_5, and TiO_2 are promising electrochromic materials because their crystalline vacant sites are big enough to accommodate protons or Li^+ ions insertion/extraction [129, 130]. Conducting polymers have been developed as electrochromic materials due to their low-cost, facile fabrication and remarkable electrochemical properties [131, 132]. For instance, PANI have been widely utilized in electrochromism. It has been demonstrated to be blue with the pernigraniline form, green in a slightly oxidized state, and yellow with leouco-emeraldine which is determined by the voltage applied [133]. Similarly, the dopant-modified PPy changes from pale yellow to dark blue with the increase of ionic doping [134]. The mostly used

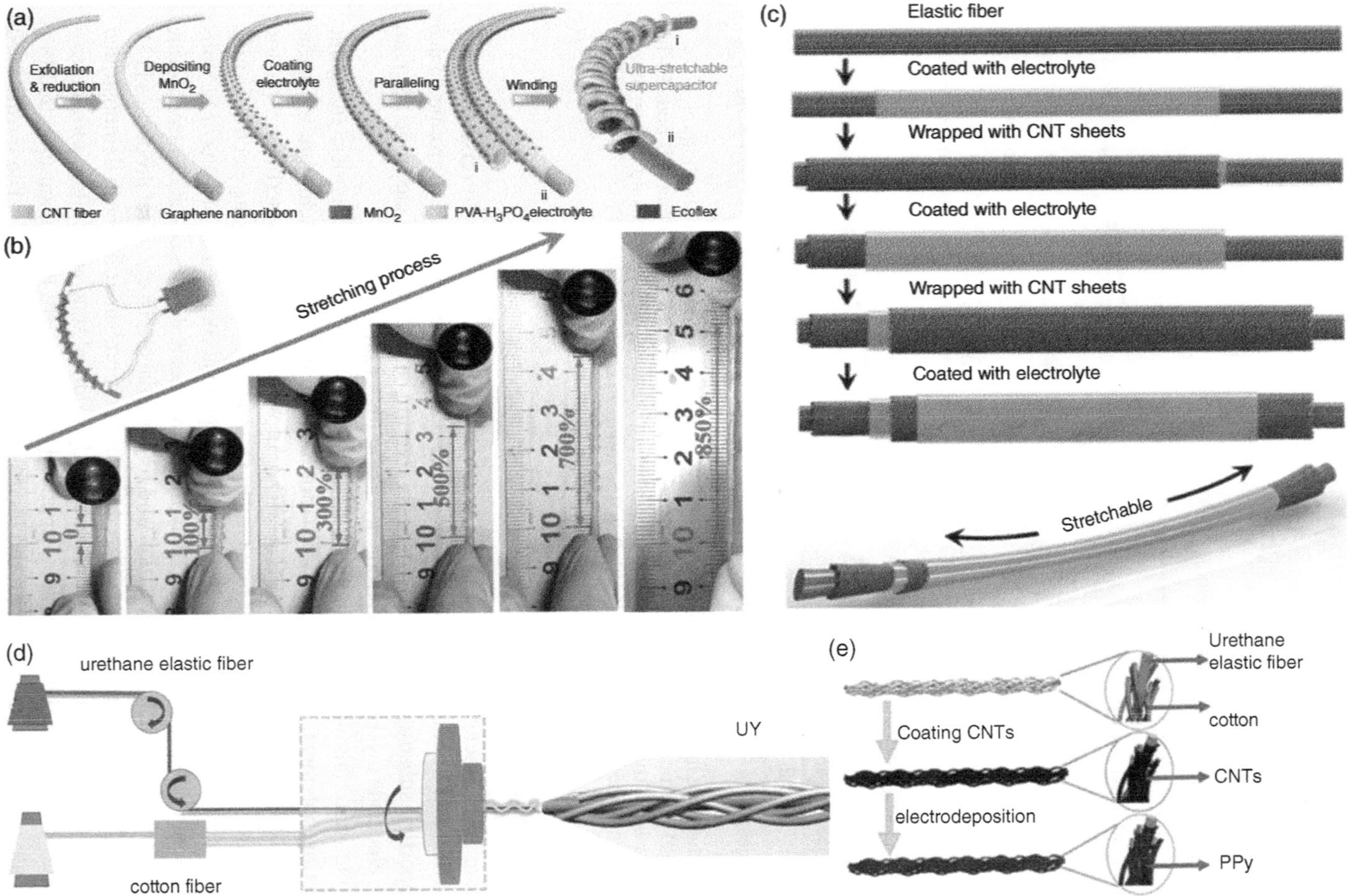

Figure 3.11 (a) Schematic illustration showing the fabrication of an ultra-stretchable FSSC; (b) Photographs of the supercapacitor with tensile strain of 0, 100%, 300%, 500%, 700%, and 850%, respectively. The insets show the LED powered by the supercapacitor. *Source:* Reproduced with permission [120]. © 2017, Springer Nature. (c) Illustration of the fabrication of a highly stretchable FSSC with a coaxial structure. *Source:* Reproduced with permission [121]. © 2013, Wiley-VCH. (d) Schematic of the spinning process of the UY; (e) Schematic illustration of the yarn modified by deposition of CNTs and PPy. *Source:* Reproduced with permission [99]. © 2016, Elsevier.

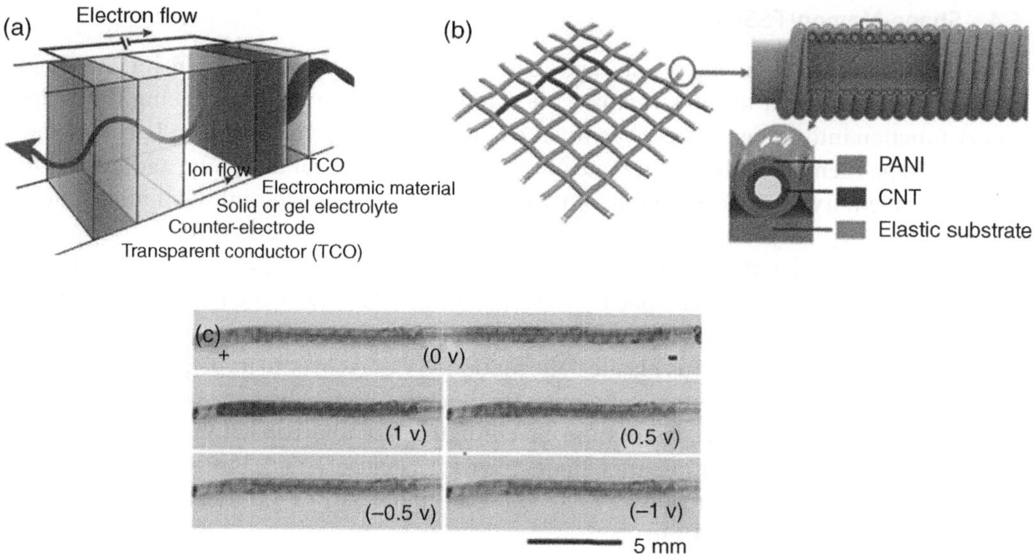

Figure 3.12 (a) Schematic illustration of the typical structure of electrochromic devices. *Source:* Reproduced with permission [125]. © 2014, Royal Society of Chemistry. (b) Schematic illustration of the structure of the electrochromic, wearable FSSC; (c) Chromatic transitions during the charge–discharge process. *Source:* Reproduced with permission [126]. © 2014, Wiley-VCH.

electrochromic material based on metal coordination complexes is Prussian blue [135, 136]. For example, the electrochromism of Prussian blue has been used for self-powered biosensors to test the concentration of ascorbic acid in orange juice depending on the rate of color change from blue to colorless, which is induced by the reduction of the ascorbic acid [137]. Apart from three major electrochromic materials, graphene-based materials are also widely applied in electrochromic devices.

Electrochromic devices have many aspects in common with a large proportion of energy storage devices with Faradic reactions in terms of similar active material, related device structure and operating mechanisms, which demonstrate their promising integration. To date, electrochromic energy storage devices have attracted significant attention. Meanwhile, FSSCs have been widely used as a power system in flexible and wearable electronic devices. The integration of electrochromism into FSSCs can remarkably improve the user experience. The users can observe the real-time display of charging/discharging process which can effectively protect the device from over-charging or unexpected powering off. For example, a flexible electrochromic FSSC has been fabricated by winding aligned CNT/PANI composite film electrodes on the elastic fiber (Figure 3.12b) [126]. Different colors of fiber electrodes, which resulted from different oxidation states of PANI, are shown under different voltages (Figure 3.12c), therefore providing dynamic and efficient communicated information during their working procedure. In this system, there are three typical colors, which are light yellow, blue, and green. Other colors can be realized by choosing other polymers or inorganic materials as electrochromic materials. These electrochromic FSSCs also show high electrochemical performance and high flexibility.

3.6.4 Shape-Memory FSSCs

During practical application, energy-storage devices unavoidably experience irreversible deformations, which may lead to the structural fracture and malfunction. Introducing the shape-memory function into energy devices can nip the potential damage in the bub and prolong their lifetimes. Shape-memory materials are one of the major elements for designing and fabricating smart shape-memory devices. Shape-memory materials can be divided into two main categories: shape-memory alloys and shape-memory polymers. They are used as the substrate or core of an electrode to endow the energy devices with shape-memory function. Huang et al. have fabricated a shape memory supercapacitor (SMSC) using NiTi wire as the main skeleton to act as current collector and substrate for active materials of MnO_2 and PPy (Figure 3.13a) [138]. Although the SMSC occurs serious plastic deformation, it can recover to its original undistorted state within a few seconds automatically once the temperature is above heat activation point (Figure 3.13b). Besides perfect restoration of configuration, the capacitive performance of SMMC has no obvious degradation after shape recovery (Figure 3.13c). Furthermore, the flexible, wire-shaped SMSC could readily weave with traditional yarns to fabricate shape memory energy storage textiles (Figure 3.13d).

Although NiTi alloy can provide much higher driving force to actuate more complex structures, its heavy weight results in low energy density of devices, while shape-memory polymers are much lighter with better flexibility. As shown in Figure 3.13e, a coaxial shape-memory FSSC was fabricated by sequentially coating thin layers of CNTs, PVA gel electrolyte, CNTs, and PVA gel electrolyte on a shape-memory polyurethane (SMPU) fiber [139]. The fiber supercapacitor could be deformed into various shapes and sizes and then recovered to the original state automatically once the temperature exceeded the thermal transition temperature. Furthermore, electronic textiles woven by such FSSCs also had shape-memory ability and could be "frozen" into different user-required shapes and sizes (Figure 3.13f). This work demonstrates a great potential for smart electronic textiles.

3.6.5 Photodetectable FSSCs

Energy storage/conversion devices as power sources are widely used to power photodetectors by connecting these two devices in series as an integrated system [140–142]. The integration of photodetection into power sources is no doubt an advance in self-powered sensors. This requires multifunctional materials that are both electrochemically active and photoresponsive. An integrated FSSC that simultaneously realizes energy storage and optoelectronic detection was reported [143]. In this integrated system, a flexible asymmetric FSSC composed of Co_3O_4 nanowires and graphene electrodes, was used as energy-storage and photon-detection device (Figure 3.14a). The integrated device improved energy storage and power delivery (at least by 1860% enhanced) by enlarging the potential window from 0–0.6 V to 0–1.5 V. Upon the absorption of light, the electron–hole pairs generated in graphene would normally recombine in tens of picoseconds. However, the electron–hole pairs became separated when an external field resulting from the fully charged supercapacitor was applied. The electrons moved to the positive Co_3O_4 electrode, and the holes to the negative graphene electrode, resulting in the improved leakage current of the supercapacitor (Figure 3.14b). The increase of the leakage current under white light irradiation demonstrated that photodetection is achieved in the integrated device (Figure 3.14c). This work indicates that flexible FSSCs can be effectively used to detect photons.

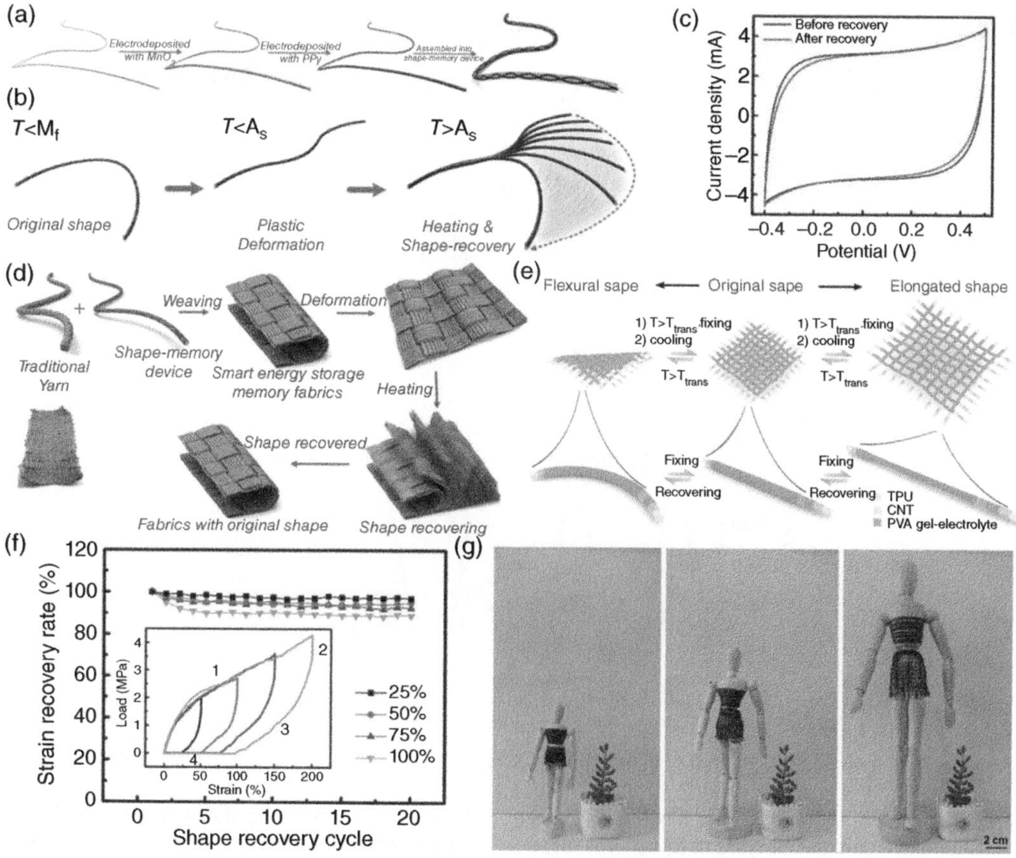

Figure 3.13 (a) Shape recovering process of a wire-shaped SMSC; (b) The reversible phase transformation of martensite and austenite; (c) CV curves of the SMSC before and after shape-recovering process; (d) Schematic demonstration of the SMSC woven with traditional fabric to fabricate a shape-recoverable smart textile. Inset is the as-prepared textile (17 × 12 cm). *Source:* Reproduced with permission [138]. © 2013, Royal Society of Chemistry. (e) Schematic design of shape-memory FSSC and the resulted textile that is reversibly transformed into flexural or elongated states and recovered to the original shape; (f) Dependence of shape recovery ratio on shape recovery cycle number. The inserted stress-strain curves compared the shape-memory effect (part 1, stretching to the maximum strain at T_{trans}; part 2, cooling down to room temperature when the maximum strain was maintained; part 3, releasing the stress; part 4, heating up to T_{trans}). (g) Photographs of the same smart clothes woven from SMSCs that were "frozen" into different shapes and sizes. *Source:* Reproduced with permission [139]. © 2015, Wiley-VCH.

3.7 Conclusion

In this chapter, we focus on the recent progresses of FSSCs with respect to their electrolyte materials, electrode materials, fiber substrate, configuration, and smart function. FSSCs are a promising energy storage candidate for flexible and wearable electronics with the merits of tiny volume, high flexibility and good deformability. Though many fundamental progresses have been achieved, some problems and challenges still remain for their practical application.

First, the electrochemical performance needs to be further improved. As energy storage devices, the core function of FSSCs is to power various flexible and wearable devices. While improving the energy density, their power density and lifetime should not be sacrificed too much. The emerging materials with high surface area, high conductivity and good mechanical properties, such as

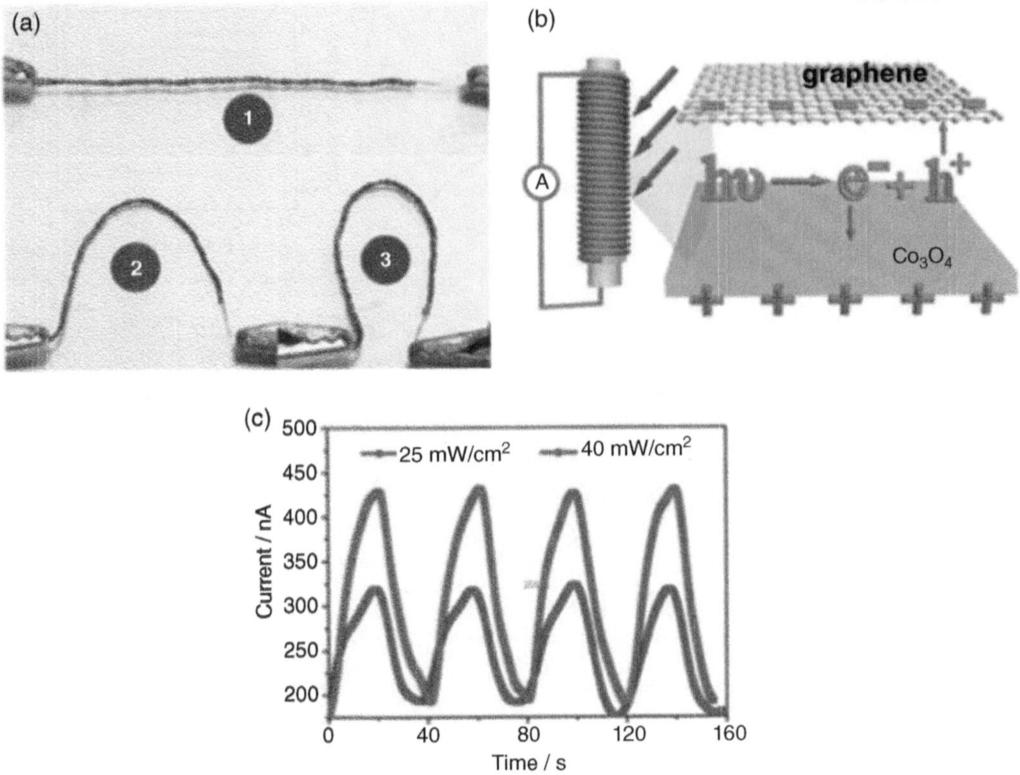

Figure 3.14 (a) Photographs of the flexible asymmetric FSSC at different bending states; (b) Schematic illustration of the integrated system; (c) Current response of the device illuminated under different incident light intensities. *Source:* Reproduced with permission [143]. © 2014, Wiley-VCH.

MXenes, MOF, provide a potential solution as electrode material candidates. In addition, suitable electrode material with high capacitance combined with suitable substrate, electrolyte with high ion conductivity and novel structure is a promising solution.

Second, light weight, high flexibility, and air permeability are essential for wearable electronics. Metal fibers are always heavy and carbon-based fibers normally have low conductivity. Meanwhile, FSSCs based on cotton yarns have low electrochemical performances. A novel fiber with light weight and high conductivity are urgently needed and offers a potential avenue to fabricate light weight, flexible and air-permeable wearable electronics.Third, aside from the basic function of storing electronic energy, multifunctional FSSCs to respond the changes of outer environment are in their infancy but will boost in the future. In this chapter, the self-healability, stretchability, electrochromic ability, shape-memory, and photodetection have been discussed. Other functions such as thermal response, force response, etc., still need to be researched.

Fourth, the aforementioned FSSCs are still limited in laboratory research, and cannot realize scale-up production. For large scale practical application, the cost-effective materials and high-throughput fabrication manners should be exploited. Additionally, although the processes of fabricating and integrating the FSSCs seem to be feasible, further development of machines for fabricating the fiber substrate, energy storage fibers/yarns and wearable electronics is extremely necessary. Keeping the balance of the energy storage capability and the processing scalability of the FSSCs is of great importance for large-scale production.

Fifth, the safety issue is always top concern when the wearable electronics are used in our daily life. Most electrode materials are highly risky when they are used in practical application, but they are indispensable for high-performance FSSCs. In this case, a proper encapsulation which can prevent the toxic electrode materials and electrolytes from leakage is a promising solution.

In summary, the research on FSSCs is an interesting and meaningful area that connects energy storage and the traditional textile industry. Although many remarkable progresses have been done in this field, they are still at their early ages and many challenges need to be overcome. In the near future, it is foreseen that significant progress will be made on high-performance electrode materials, light and flexible substrate and the integration of multifunction toward high-performance and multifunctional FSSCs.

References

1 Hu, L.B., Pasta, M., La Mantia, F. et al. (2010). Stretchable, porous, and conductive energy textiles. *Nano Lett.* 10 (2): 708–714.

2 Jost, K., Perez, C.R., McDonough, J.K. et al. (2011). Carbon coated textiles for flexible energy storage. *Energy Environ. Sci.* 4 (12): 5060–5067.

3 Jost, K., Dion, G., and Gogotsi, Y. (2014). Textile energy storage in perspective. *J. Mater. Chem. A* 2 (28): 10776–10787.

4 Huang, Q.Y., Wang, D.R., and Zheng, Z.J. (2016). Textile-based electrochemical energy storage devices. *Adv. Energy Mater.* 6 (22): 1600783.

5 Choi, C., Sim, H.J., Spinks, G.M. et al. (2016). Elastomeric and dynamic mno2/cnt core-shell structure coiled yarn supercapacitor. *Adv. Energy Mater.* 6 (5): 1502119.

6 Yang, P.H. and Mai, W.J. (2014). Flexible solid-state electrochemical supercapacitors. *Nano Energy* 8: 274–290.

7 Zeng, W., Shu, L., Li, Q. et al. (2014). Fiber-based wearable electronics: a review of materials, fabrication, devices, and applications. *Adv. Mater.* 26 (31): 5310–5336.

8 Yu, D.S., Qian, Q.H., Wei, L. et al. (2015). Emergence of fiber supercapacitors. *Chem. Soc. Rev.* 44 (3): 647–662.

9 Xu, R., Guo, F., Cui, X. et al. (2015). High performance carbon nanotube based fiber-shaped supercapacitors using redox additives of polypyrrole and hydroquinone. *J. Mater. Chem. A* 3 (44): 22353–22360.

10 Cheng, X., Zhang, J., Ren, J. et al. (2016). Design of a hierarchical ternary hybrid for a fiber-shaped asymmetric supercapacitor with high volumetric energy density. *J. Phys. Chem. C* 120 (18): 9685–9691.

11 Zhang, Z., Li, X., Guan, G. et al. (2014). A lightweight polymer solar cell textile that functions when illuminated from either side. *Angew. Chem. Int.* 53 (43): 11571–11574.

12 Ma, Y.W., Li, P., Sedloff, J.W. et al. (2015). Conductive graphene fibers for wire-shaped supercapacitors strengthened by unfunctionalized few-walled carbon nanotubes. *ACS Nano* 9 (2): 1352–1359.

13 Yu, N., Yin, H., Zhang, W. et al. (2016). High-performance fiber-shaped all-solid-state asymmetric supercapacitors based on ultrathin mno2 nanosheet/carbon fiber cathodes for wearable electronics. *Adv. Energy Mater.* 6 (2): 1501458.

14 Dubal, D.P., Chodankar, N.R., Kim, D.H. et al. (2018). Towards flexible solid-state supercapacitors for smart and wearable electronics. *Chem. Soc. Rev.* 47 (6): 2065–2129.

15 Sekhon, S.S. (2003). Conductivity behaviour of polymer gel electrolytes: role of polymer. *Bull. Mater. Sci.* 26 (3): 321–328.

16 Fei, H.J., Yang, C.Y., Bao, H. et al. (2014). Flexible all-solid-state supercapacitors based on graphene/carbon black nanoparticle film electrodes and cross-linked poly(vinyl alcohol)-h2so4 porous gel electrolytes. *J. Power Sources* 266: 488–495.

17 Huang, C.W., Wu, C.A., Hou, S.S. et al. (2012). Gel electrolyte derived from poly(ethylene glycol) blending poly(acrylonitrile) applicable to roll-to-roll assembly of electric double layer capacitors. *Adv. Funct. Mater.* 22 (22): 4677–4685.

18 Zhong, X.W., Tang, J., Cao, L.J. et al. (2017). Cross-linking of polymer and ionic liquid as high-performance gel electrolyte for flexible solid-state supercapacitors. *Electrochim. Acta* 244: 112–118.

19 Feng, E.K., Ma, G.F., Sun, K.J. et al. (2017). Superior performance of an active electrolyte enhanced supercapacitor based on a toughened porous network gel polymer. *New J. Chem.* 41 (5): 1986–1992.

20 Lewandowski, A., Zajder, M., Frackowiak, E. et al. (2001). Supercapacitor based on activated carbon and polyethylene oxide-Koh-h2o polymer electrolyte. *Electrochim. Acta* 46 (18): 2777–2780.

21 Yu, H.Y., Wu, J.H., Fan, L.Q. et al. (2011). Improvement of the performance for quasi-solid-state supercapacitor by using pva-Koh-ki polymer gel electrolyte. *Electrochim. Acta* 56 (20): 6881–6886.

22 Virya, A. and Lian, K. (2017). Li2so4-polyacrylamide polymer electrolytes for 2.0 v solid symmetric supercapacitors. *Electrochem. Commun.* 81: 52–55.

23 Virya, A. and Lian, K. (2017). Polyacrylamide-lithiumchloride polymer electrolyte and its applications in electrochemical capacitors. *Electrochem. Commun.* 74: 33–37.

24 Zhong, C., Deng, Y.D., Hu, W.B. et al. (2015). A review of electrolyte materials and compositions for electrochemical supercapacitors. *Chem. Soc. Rev.* 44 (21): 7484–7539.

25 Vaquero, S., Palma, J., Anderson, M. et al. (2013). Mass-balancing of electrodes as a strategy to widen the operating voltage window of carbon/carbon supercapacitors in neutral aqueous electrolytes. *Int. J. Electrochem. Sci.* 8 (8): 10293–10307.

26 Zhang, Y.H., Xu, S., Fu, H.R. et al. (2013). Buckling in serpentine microstructures and applications in elastomer-supported ultra-stretchable electronics with high areal coverage. *Soft Matter* 9 (33): 8062–8070.

27 Cao, L.J., Yang, M.Y., Wu, D. et al. (2017). Biopolymer-chitosan based supramolecular hydrogels as solid state electrolytes for electrochemical energy storage. *Chem. Commun.* 53 (10): 1615–1618.

28 Earle, M.J. and Seddon, K.R. (2000). Ionic liquids. Green solvents for the future. *Pure Appl. Chem.* 72 (7): 1391–1398.

29 Fan, L.Q., Zhong, J., Wu, J.H. et al. (2014). Improving the energy density of quasi-solid-state electric double-layer capacitors by introducing redox additives into gel polymer electrolytes. *J. Mater. Chem. A* 2 (24): 9011–9014.

30 Kopf, M., Eckstein, N., Pfister, D. et al. (2014). Access and in situ growth of phosphorene-precursor black phosphorus. *J. Cryst. Growth* 405: 6–10.

31 Senthilkumar, S.T., Selvan, R.K., Ponpandian, N. et al. (2012). Redox additive aqueous polymer gel electrolyte for an electric double layer capacitor. *RSC Adv.* 2 (24): 8937–8940.

32 Ma, G.F., Feng, E.K., Sun, K.J. et al. (2014). A novel and high-effective redox-mediated gel polymer electrolyte for supercapacitor. *Electrochim. Acta* 135: 461–466.

33 Yu, F.D., Huang, M.L., Wu, J.H. et al. (2014). A redox-mediator-doped gel polymer electrolyte applied in quasi-solid-state supercapacitors. *J. Appl. Polym. Sci.* 131 (2): 39784.

34 Zhang, L.L. and Zhao, X.S. (2009). Carbon-based materials as supercapacitor electrodes. *Chem. Soc. Rev.* 38 (9): 2520–2531.

35 Zhang, H., Cao, G.P., Wang, Z.Y. et al. (2008). Growth of manganese oxide nanoflowers on vertically-aligned carbon nanotube arrays for high-rate electrochemical capacitive energy storage. *Nano Lett.* 8 (9): 2664–2668.

36 Frackowiak, E., Jurewicz, K., Delpeux, S. et al. (2001). Nanotubular materials for supercapacitors. *J. Power Sources* 97-8: 822–825.

37 Chen, B.H., Bai, Y.Y., Xiang, F. et al. (2014). Stretchable and transparent hydrogels as soft conductors for dielectric elastomer actuators. *J. Polym. Sci. B Polym. Phys.* 52 (16): 1055–1060.

38 Meng, C.Z., Liu, C.H., Chen, L.Z. et al. (2010). Highly flexible and all-solid-state paper like polymer supercapacitors. *Nano Lett.* 10 (10): 4025–4031.

39 Stoller, M.D., Park, S.J., Zhu, Y.W. et al. (2008). Graphene-based ultracapacitors. *Nano Lett.* 8 (10): 3498–3502.

40 El-Kady, M.F., Strong, V., Dubin, S. et al. (2012). Laser scribing of high-performance and flexible graphene-based electrochemical capacitors. *Science* 335 (6074): 1326–1330.

41 Xu, Z. and Gao, C. (2011). Graphene chiral liquid crystals and macroscopic assembled fibres. *Nat. Commun.* 2 (1): 1–9.

42 Huang, J.Y., Wang, K., and Wei, Z.X. (2010). Conducting polymer nanowire arrays with enhanced electrochemical performance. *J. Mater. Chem.* 20 (6): 1117–1121.

43 Lipatov, A., Alhabeb, M., Lukatskaya, M.R. et al. (2016). Effect of synthesis on quality, electronic properties and environmental stability of individual monolayer ti3c2 mxene flakes. *Adv. Electron. Mater.* 2 (12): 1600255.

44 Wang, L., Feng, X., Ren, L.T. et al. (2015). Flexible solid-state supercapacitor based on a metal-organic framework interwoven by electrochemically-deposited pani. *J. Am. Chem. Soc.* 137 (15): 4920–4923.

45 Li, W.H., Ding, K., Tian, H.R. et al. (2017). Conductive metal-organic framework nanowire array electrodes for high-performance solid-state supercapacitors. *Adv. Funct. Mater.* 27 (27): 1702067.

46 Yuan, L.Y., Xiao, X., Ding, T.P. et al. (2012). Paper-based supercapacitors for self-powered nanosystems. *Angew. Chem. Int.* 51 (20): 4934–4938.

47 Zhang, C.F., Higgins, T.M., Park, S.H. et al. (2016). Highly flexible and transparent solid-state supercapacitors based on ruo2/pedot:Pss conductive ultrathin films. *Nano Energy* 28: 495–505.

48 Ferris, A., Garbarino, S., Guay, D. et al. (2015). 3d ruo2 microsupercapacitors with remarkable areal energy. *Adv. Mater.* 27 (42): 6625–6629.

49 Chodankar, N.R., Dubal, D.P., Gund, G.S. et al. (2016). A symmetric mno2/mno2 flexible solid state supercapacitor operating at 1.6 v with aqueous gel electrolyte. *J. Energy Chem.* 25 (3): 463–471.

50 Choi, D., Blomgren, G.E., and Kumta, P.N. (2006). Fast and reversible surface redox reaction in nanocrystalline vanadium nitride supercapacitors. *Adv. Mater.* 18 (9): 1178–1182.

51 Pande, P., Rasmussen, P.G., and Thompson, L.T. (2012). Charge storage on nanostructured early transition metal nitrides and carbides. *J. Power Sources* 207: 212–215.

52 Xiao, X., Peng, X., Jin, H.Y. et al. (2013). Freestanding mesoporous vn/cnt hybrid electrodes for flexible all-solid-state supercapacitors. *Adv. Mater.* 25 (36): 5091–5097.

53 Lu, X.H., Wang, G.M., Zhai, T. et al. (2012). Stabilized tin nanowire arrays for high-performance and flexible supercapacitors. *Nano Lett.* 12 (10): 5376–5381.

54 Javed, M.S., Dai, S.G., Wang, M.J. et al. (2015). Faradic redox active material of cu7s4 nanowires with a high conductance for flexible solid state supercapacitors. *Nanoscale* 7 (32): 13610–13618.

55 Shahzad, F., Alhabeb, M., Hatter, C.B. et al. (2016). Electromagnetic interference shielding with 2d transition metal carbides (mxenes). *Science* 353 (6304): 1137–1140.

56 Zhu, M.S., Huang, Y., Deng, Q.H. et al. (2016). Highly flexible, freestanding supercapacitor electrode with enhanced performance obtained by hybridizing polypyrrole chains with mxene. *Adv. Energy Mater.* 6 (21): 1600969.

57 Naguib, M., Mochalin, V.N., Barsoum, M.W. et al. (2014). 25th anniversary article: Mxenes: a new family of two-dimensional materials. *Adv. Mater.* 26 (7): 992–1005.

58 Naguib, M., Kurtoglu, M., Presser, V. et al. (2011). Two-dimensional nanocrystals produced by exfoliation of ti3alc2. *Adv. Mater.* 23 (37): 4248–4253.

59 Zhang, C.F., Anasori, B., Seral-Ascaso, A. et al. (2017). Transparent, flexible, and conductive 2d titanium carbide (mxene) films with high volumetric capacitance. *Adv. Mater.* 29 (36): 1702678.

60 Li, H.Y., Hou, Y., Wang, F.X. et al. (2017). Flexible all-solid-state supercapacitors with high volumetric capacitances boosted by solution processable mxene and electrochemically exfoliated graphene. *Adv. Energy Mater.* 7 (4).

61 Ghidiu, M., Lukatskaya, M.R., Zhao, M.Q. et al. (2014). Conductive two-dimensional titanium carbide 'clay' with high volumetric capacitance. *Nature* 516 (7529): 78–U171.

62 Wang, H.W., Naguib, M., Page, K. et al. (2016). Resolving the structure of ti3c2tx mxenes through multilevel structural modeling of the atomic pair distribution function. *Chem. Mater.* 28 (1): 349–359.

63 Hope, M.A., Forse, A.C., Griffith, K.J. et al. (2016). Nmr reveals the surface functionalisation of ti3c2 mxene. *Phys. Chem. Chem. Phys.* 18 (7): 5099–5102.

64 Zhao, M.Q., Ren, C.E., Ling, Z. et al. (2015). Flexible mxene/carbon nanotube composite paper with high volumetric capacitance. *Adv. Mater.* 27 (2): 339–345.

65 Cheng, C., Jiang, G.P., Garvey, C.J. et al. (2016). Ion transport in complex layered graphene-based membranes with tuneable interlayer spacing. *Sci. Adv.* 2 (2): e1501272.

66 Mi, B.X. (2014). Graphene oxide membranes for ionic and molecular sieving. *Science* 343 (6172): 740–742.

67 Salunkhe, R.R., Kaneti, Y.V., and Yamauchi, Y. (2017). Metal-organic framework-derived nanoporous metal oxides toward supercapacitor applications: Progress and prospects. *ACS Nano* 11 (6): 5293–5308.

68 Xuan, W.M., Zhu, C.F., Liu, Y. et al. (2012). Mesoporous metal-organic framework materials. *Chem. Soc. Rev.* 41 (5): 1677–1695.

69 Choi, K.M., Jeong, H.M., Park, J.H. et al. (2014). Supercapacitors of nanocrystalline metal-organic frameworks. *ACS Nano* 8 (7): 7451–7457.

70 Dubal, D.P., Suarez-Guevara, J., Tonti, D. et al. (2015). A high voltage solid state symmetric supercapacitor based on graphene-polyoxometalate hybrid electrodes with a hydroquinone doped hybrid gelelectrolyte. *J. Mater. Chem. A* 3 (46): 23483–23492.

71 Genovese, M. and Lian, K. (2015). Polyoxometalate modified inorganic-organic nanocomposite materials for energy storage applications: a review. *Curr. Opin. Solid State Mater. Sci.* 19 (2): 126–137.

72 Ammam, M. (2013). Polyoxometalates: formation, structures, principal properties, main deposition methods and application in sensing. *J. Mater. Chem. A* 1 (21): 6291–6312.

73 Yang, M., Hong, S.B., Yoon, J.H. et al. (2016). Fabrication of flexible, redoxable, and conductive nanopillar arrays with enhanced electrochemical performance. *ACS Appl. Mater. Interfaces* 8 (34): 22220–22226.

74 Li, L.K., Yu, Y.J., Ye, G.J. et al. (2014). Black phosphorus field-effect transistors. *Nat. Nanotechnol.* 9 (5): 372–377.

75 Hao, C.X., Yang, B.C., Wen, F.S. et al. (2016). Flexible all-solid-state supercapacitors based on liquid-exfoliated black-phosphorus nanoflakes. *Adv. Mater.* 28 (16): 3194–3201.

76 Gu, S.S., Lou, Z., Ma, X.D. et al. (2015). Cuco2o4 nanowires grown on a ni wire for high-performance, flexible fiber supercapacitors. *ChemElectroChem* 2 (7): 1042–1047.

77 Vellacheri, R., Zhao, H.P., Muhlstadt, M. et al. (2017). Rationally engineered electrodes for a high-performance solid-state cable-type supercapacitor. *Adv. Funct. Mater.* 27 (18): 1606696.

78 Keum, K., Lee, G., Lee, H. et al. (2018). Wire-shaped supercapacitors with organic electrolytes fabricated via layer-by-layer assembly. *ACS Appl. Mater. Interfaces* 10 (31): 26248–26257.

79 Vellacheri, R., Zhao, H.P., Muhlstadt, M. et al. (2016). All-solid-state cable-type supercapacitors with ultrahigh rate capability. *Adv. Mater. Technol.* 1 (1): 1600012.

80 Zhou, H. and Zhang, Y.R. (2014). Electrochemically self-doped tio2 nanotube arrays for supercapacitors. *J. Phys. Chem. C* 118 (11): 5626–5636.

81 Huang, Y., Hu, H., Huang, Y. et al. (2015). From industrially weavable and knittable highly conductive yarns to large wearable energy storage textiles. *ACS Nano* 9 (5): 4766–4775.

82 Cakici, M., Reddy, K.R., and Alonso-Marroquin, F. (2017). Advanced electrochemical energy storage supercapacitors based on the flexible carbon fiber fabric-coated with uniform coral-like mno2 structured electrodes. *Chem. Eng. J.* 309: 151–158.

83 Ai, Y.F., Lou, Z., Li, L. et al. (2016). Meters-long flexible conio2-nanowires@carbon-fibers based wire-supercapacitors for wearable electronics. *Adv. Mater. Technol.* 1 (8): 1600142.

84 Noh, J., Yoon, C.M., Kim, Y.K. et al. (2017). High performance asymmetric supercapacitor twisted from carbon fiber/mno2 and carbon fiber/moo3. *Carbon* 116: 470–478.

85 Di, J.T., Zhang, X.H., Yong, Z.Z. et al. (2016). Carbon-nanotube fibers for wearable devices and smart textiles. *Adv. Mater.* 28 (47): 10529–10538.

86 Koziol, K., Vilatela, J., Moisala, A. et al. (2007). High-performance carbon nanotube fiber. *Science* 318 (5858): 1892–1895.

87 Lu, W.B., Zu, M., Byun, J.H. et al. (2012). State of the art of carbon nanotube fibers: opportunities and challenges. *Adv. Mater.* 24 (14): 1805–1833.

88 Patil, B., Ahn, S., Yu, S. et al. (2018). Electrochemical performance of a coaxial fiber-shaped asymmetric supercapacitor based on nanostructured mno2/cnt-web paper and fe2o3/carbon fiber electrodes. *Carbon* 134: 366–375.

89 Huang, G.X., Zhang, Y., Wang, L. et al. (2017). Fiber-based mno2/carbon nanotube/polyimide asymmetric supercapacitor. *Carbon* 125: 595–604.

90 Chen, L.L., Liu, Y., Zhao, Y. et al. (2016). Graphene-based fibers for supercapacitor applications. *Nanotechnology* 27 (3): 032001.

91 Dong, Z.L., Jiang, C.C., Cheng, H.H. et al. (2012). Facile fabrication of light, flexible and multifunctional graphene fibers. *Adv. Mater.* 24 (14): 1856–1861.

92 Li, X.M., Zhao, T.S., Wang, K.L. et al. (2011). Directly drawing self-assembled, porous, and monolithic graphene fiber from chemical vapor deposition grown graphene film and its electrochemical properties. *Langmuir* 27 (19): 12164–12171.

93 Ding, X.T., Zhao, Y., Hu, C.G. et al. (2014). Spinning fabrication of graphene/polypyrrole composite fibers for all-solid-state, flexible fibriform supercapacitors. *J. Mater. Chem. A* 2 (31): 12355–12360.

94 Zheng, B.N., Huang, T.Q., Kou, L. et al. (2014). Graphene fiber-based asymmetric micro-supercapacitors. *J. Mater. Chem. A* 2 (25): 9736–9743.

95 Cai, W.H., Lai, T., Lai, J.W. et al. (2016). Transition metal sulfides grown on graphene fibers for wearable asymmetric supercapacitors with high volumetric capacitance and high energy density. *Sci. Rep.* 6 (1): 1–9.

96 Liu, N.S., Ma, W.Z., Tao, J.Y. et al. (2013). Cable-type supercapacitors of three-dimensional cotton thread based multi-grade nanostructures for wearable energy storage. *Adv. Mater.* 25 (35): 4925–4931.

97 Shim, B.S., Chen, W., Doty, C. et al. (2008). Smart electronic yarns and wearable fabrics for human biomonitoring made by carbon nanotube coating with polyelectrolytes. *Nano Lett.* 8 (12): 4151–4157.

98 Yun, Y.J., Hong, W.G., Kim, W.J. et al. (2013). A novel method for applying reduced graphene oxide directly to electronic textiles from yarns to fabrics. *Adv. Mater.* 25 (40): 5701–5705.

99 Sun, J.F., Huang, Y., Fu, C.X. et al. (2016). High-performance stretchable yarn supercapacitor based on ppy@cnts@urethane elastic fiber core spun yarn. *Nano Energy* 27: 230–237.

100 Liu, L.B., Yu, Y., Yan, C. et al. (2015). Wearable energy-dense and power-dense supercapacitor yarns enabled by scalable graphene-metallic textile composite electrodes. *Nat. Commun.* 6 (1): 1–9.

101 Jost, K., Durkin, D.P., Haverhals, L.M. et al. (2015). Natural fiber welded electrode yarns for knittable textile supercapacitors. *Adv. Energy Mater.* 5 (4): 1401286.

102 White, S.R., Sottos, N.R., Geubelle, P.H. et al. (2001). Autonomic healing of polymer composites. *Nature* 409 (6822): 794–797.

103 Wietor, J.L. and Sijbesma, R.P. (2008). A self-healing elastomer. *Angew. Chem. Int.* 47 (43): 8161–8163.

104 Zhang, M.Q. and Rong, M.Z. (2012). Theoretical consideration and modeling of self-healing polymers. *J. Polym. Sci. B Polym. Phys.* 50 (4): 229–241.

105 Yang, F. and Pitchumani, R. (2002). Healing of thermoplastic polymers at an interface under nonisothermal conditions. *Macromolecules* 35 (8): 3213–3224.

106 Wool, R.P. (2008). Self-healing materials: a review. *Soft Matter* 4 (3): 400–418.

107 Raghavan, J. and Wool, R.P. (1999). Interfaces in repair, recycling joining and manufacturing of polymers and polymer composites. *J. Appl. Polym. Sci.* 71 (5): 775–785.

108 Lee, J.Y., Buxton, G.A., and Balazs, A.C. (2004). Using nanoparticles to create self-healing composites. *J. Chem. Phys.* 121 (11): 5531–5540.

109 Yin, T., Rong, M.Z., Zhang, M.Q. et al. (2007). Self-healing epoxy composites - preparation and effect of the healant consisting of microencapsulated epoxy and latent curing agent. *Compos. Sci. Technol.* 67 (2): 201–212.

110 Chen, X.X., Dam, M.A., Ono, K. et al. (2002). A thermally re-mendable cross-linked polymeric material. *Science* 295 (5560): 1698–1702.

111 Ahn, B.K., Lee, D.W., Israelachvili, J.N. et al. (2014). Surface-initiated self-healing of polymers in aqueous media. *Nat. Mater.* 13 (9): 867–872.

112 Neal, J.A., Mozhdehi, D., and Guan, Z.B. (2015). Enhancing mechanical performance of a covalent self-healing material by sacrificial noncovalent bonds. *J. Am. Chem. Soc.* 137 (14): 4846–4850.

113 Varley, R.J. and van der Zwaag, S. (2008). Towards an understanding of thermally activated self-healing of an ionomer system during ballistic penetration. *Acta Mater.* 56 (19): 5737–5750.

114 Varley, R.J. and van der Zwaag, S. (2008). Development of a quasi-static test method to investigate the origin of self-healing in ionomers under ballistic conditions. *Polym. Test.* 27 (1): 11–19.

115 Mozhdehi, D., Ayala, S., Cromwell, O.R. et al. (2014). Self-healing multiphase polymers via dynamic metal-ligand interactions. *J. Am. Chem. Soc.* 136 (46): 16128–16131.

116 Shi, Y., Wang, M., Ma, C.B. et al. (2015). A conductive self-healing hybrid gel enabled by metal-ligand supramolecule and nanostructured conductive polymer. *Nano Lett.* 15 (9): 6276–6281.

117 Huebsch, N., Kearney, C.J., Zhao, X.H. et al. (2014). Ultrasound-triggered disruption and self-healing of reversibly cross-linked hydrogels for drug delivery and enhanced chemotherapy. *Proc. Natl. Acad. Sci. U. S. A.* 111 (27): 9762–9767.

118 Sun, H., You, X., Jiang, Y.S. et al. (2014). Self-healable electrically conducting wires for wearable microelectronics. *Angew. Chem. Int.* 53 (36): 9526–9531.

119 Huang, Y., Huang, Y., Zhu, M.S. et al. (2015). Magnetic-assisted, self-healable, yarn-based supercapacitor. *ACS Nano* 9 (6): 6242–6251.

120 Wang, H.M., Wang, C.Y., Jian, M.Q. et al. (2018). Superelastic wire-shaped supercapacitor sustaining 850% tensile strain based on carbon nanotube@graphene fiber. *Nano Res.* 11 (5): 2347–2356.

121 Yang, Z.B., Deng, J., Chen, X.L. et al. (2013). A highly stretchable, fiber-shaped supercapacitor. *Angew. Chem. Int.* 52 (50): 13453–13457.

122 Huang, Y., Zhu, M.S., Huang, Y. et al. (2016). Multifunctional energy storage and conversion devices. *Adv. Mater.* 28 (38): 8344–8364.

123 Osterholm, A.M., Shen, D.E., Kerszulis, J.A. et al. (2015). Four shades of brown: tuning of electrochromic polymer blends toward high-contrast eyewear. *ACS Appl. Mater. Interfaces* 7 (3): 1413–1421.

124 Korgel, B.A. (2013). Materials science composite for smarter windows. *Nature* 500 (7462): 278–279.

125 Runnerstrom, E.L., Llordes, A., Lounis, S.D. et al. (2014). Nanostructured electrochromic smart windows: traditional materials and nir-selective plasmonic nanocrystals. *Chem. Commun.* 50 (73): 10555–10572.

126 Chen, X.L., Lin, H.J., Deng, J. et al. (2014). Electrochromic fiber-shaped supercapacitors. *Adv. Mater.* 26 (48): 8126–8132.

127 Lee, S.H., Deshpande, R., Parilla, P.A. et al. (2006). Crystalline wo3 nanoparticles for highly improved electrochromic applications. *Adv. Mater.* 18 (6): 763–766.

128 Yan, C.Y., Kang, W.B., Wang, J.X. et al. (2014). Stretchable and wearable electrochromic devices. *ACS Nano* 8 (1): 316–322.

129 Chen, J.Z., Ko, W.Y., Yen, Y.C. et al. (2012). Hydrothermally processed tio2 nanowire electrodes with antireflective and electrochromic properties. *ACS Nano* 6 (8): 6633–6639.

130 Qiang, P.F., Chen, Z.W., Yang, P.H. et al. (2013). Tio2 nanowires for potential facile integration of solar cells and electrochromic devices. *Nanotechnology* 24 (43): 435403.

131 Sonmez, G., Sonmez, H.B., Shen, C.K.E. et al. (2004). Red, green, and blue colors in polymeric electrochromics. *Adv. Mater.* 16 (21): 1905–1908.

132 Wang, K., Wu, H.P., Meng, Y.N. et al. (2014). Conducting polymer nanowire arrays for high performance supercapacitors. *Small* 10 (1): 14–31.

133 Girotto, E.M. and De Paoli, M.A. (1998). Polypyrrole color modulation and electrochromic contrast enhancement by doping with a dye. *Adv. Mater.* 10 (10): 790–793.

134 Wang, J., Too, C.O., Zhou, D. et al. (2005). Novel electrode substrates for rechargeable lithium/polypyrrole batteries. *J. Power Sources* 140 (1): 162–167.

135 Cui, B.B., Zhong, Y.W., and Yao, J.N. (2015). Three-state near-infrared electrochromism at the molecular scale. *J. Am. Chem. Soc.* 137 (12): 4058–4061.

136 Takada, K., Sakamoto, R., Yi, S.T. et al. (2015). Electrochromic bis(terpyridine)metal complex nanosheets. *J. Am. Chem. Soc.* 137 (14): 4681–4689.

137 Zloczewska, A., Celebanska, A., Szot, K. et al. (2014). Self-powered biosensor for ascorbic acid with a prussian blue electrochromic display. *Biosens. Bioelectron.* 54: 455–461.

138 Huang, Y., Zhu, M.S., Pei, Z.X. et al. (2016). A shape memory supercapacitor and its application in smart energy storage textiles. *J. Mater. Chem. A* 4 (4): 1290–1297.

139 Deng, J.E., Zhang, Y., Zhao, Y. et al. (2015). A shape-memory supercapacitor fiber. *Angew. Chem. Int.* 54 (51): 15419–15423.

140 Yu, D.S., Goh, K., Wang, H. et al. (2014). Scalable synthesis of hierarchically structured carbon nanotube-graphene fibres for capacitive energy storage. *Nat. Nanotechnol.* 9 (7): 555–562.

141 Hou, X.J., Liu, B., Wang, X.F. et al. (2013). Sno2-microtube-assembled cloth for fully flexible self-powered photodetector nanosystems. *Nanoscale* 5 (17): 7831–7837.

142 Yang, Q., Liu, Y., Li, Z.T. et al. (2012). Self-powered ultrasensitive nanowire photodetector driven by a hybridized microbial fuel cell. *Angew. Chem. Int.* 51 (26): 6443–6446.

143 Wang, X.F., Liu, B., Liu, R. et al. (2014). Fiber-based flexible all-solid-state asymmetric supercapacitors for integrated photodetecting system. *Angew. Chem. Int.* 53 (7): 1849–1853.

4

Flexible Fiber-shaped Supercapacitors

Fabrication, Design and Applications

Muhammad S. Javed[1,2], Peng Sun[1], Muhammad Imran[3] and Wenjie Mai[1]

[1] *Siyuan Laboratory, Guangdong Provincial Engineering Technology Research Center of Vacuum Coating Technologies and New Energy Materials, Department of Physics, Jinan University, Guangzhou, 510632, PR China*
[2] *School of Physical Science and Technology, Lanzhou University, Lanzhou, 730000, PR China*
[3] *Department of Chemistry, Faculty of Science, King Khalid University, Abha, 61413, Saudi Arabia*

4.1 Introduction to Fiber-Shaped Supercapacitors

Energy storage devices support extensive applications, including electronic products, sustainable electricity supply and transport vehicles [1, 2]. Among other kinds of energy storage devices, supercapacitors (SCs) are relatively new energy storage systems with high power density and long service life [3–5]. SC has two working mechanisms: electrical double layer capacitive (EDLC) and pseudocapacitive (PC) charge storage behavior [6, 7]. In EDLC, the charge is generated electrostatically by separating at the interface between the conductive electrode surface and the electrolyte without Faradaic reactions [8]. PC is a kind of faradaic electrochemical charge storage governed by fast reversible surface redox reactions, intercalation, or electro-sorption at/on near the surface of electrode materials [9]. The energy storage in EDLC greatly depends upon the availability of electrode surface area for ion adsorption, electrode conductivity and pore-size distribution that affects electron and ion transportation [10]. Therefore, high surface porous carbon materials, such as activated carbon (AC), mesoporous carbon, carbon nanotubes (CNTs), and graphene are employed as electrodes to accomplish high performance SCs [11]. However, carbon-based electrode materials exhibiting high power density and excellent stability undergo poor energy density because of the low specific capacitance [12]. In comparison, transition metal oxides/hydroxides/sulfides exhibit pseudocapacitive type charge storage behavior and possess much higher capacitance and energy density as compared to EDLC's but they suffer from poor electronic conductivity and cycling stability [13–18]. The use of composites of pseudocapacitive and carbon materials as an electrode is a popular approach to improving the SC's performance [19–21].

With an ever-growing demand for portable electronics, lightweight and flexible energy storage devices are gaining more attention in recent years. Fiber-shaped energy storage devices present a special 1D architecture with the advantages of superior flexibility, high miniaturization efficiency, resilience to bending, and good functionality with the conventional manufacturing sector, which are especially beneficial for wearable electronics. In addition to higher efficiency, developments in multifunctional, modular, and integrable systems are also the key themes in the recent research frontier in the field of fiber-shaped energy storage devices. Traditional SCs have

Flexible Supercapacitors: Materials and Applications, First Edition. Edited by Guozhen Shen, Zheng Lou and Di Chen.

two-dimensional planar configurations [22, 23], while flexible fiber-shaped SCs (FFSCs) employ single-dimensional, cylindrical-shaped fibers as electrodes and have emerged as a special family of flexible SCs [24–26]. The schematic illustration of traditional SCs in which two metal sheets coated by active materials are served as electrodes, one separator squashed in electrolyte together (Figure 4.1a) [27]. When a potential difference across two electrodes is applied, the electrolytic cations and anions travel toward the surface of the negatively polarized electrode and the positively polarized electrode to balance the electrons and holes, respectively (Figure 4.1b) [28]. In wide stationary applications (industrial grid storage), hybrid electric cars (electric buses and trains), and mobile devices, they have uncovered disproportionate capacity (power tools) [30]. The ongoing fast growth of highly advanced consumer electronics, such as smart eyes, embedded sensors, and extendable systems, involves the production of highly flexible and mini-sized energy storage units [31]. The schematic illustration of the FFSC is demonstrated in Figure 4.1c,

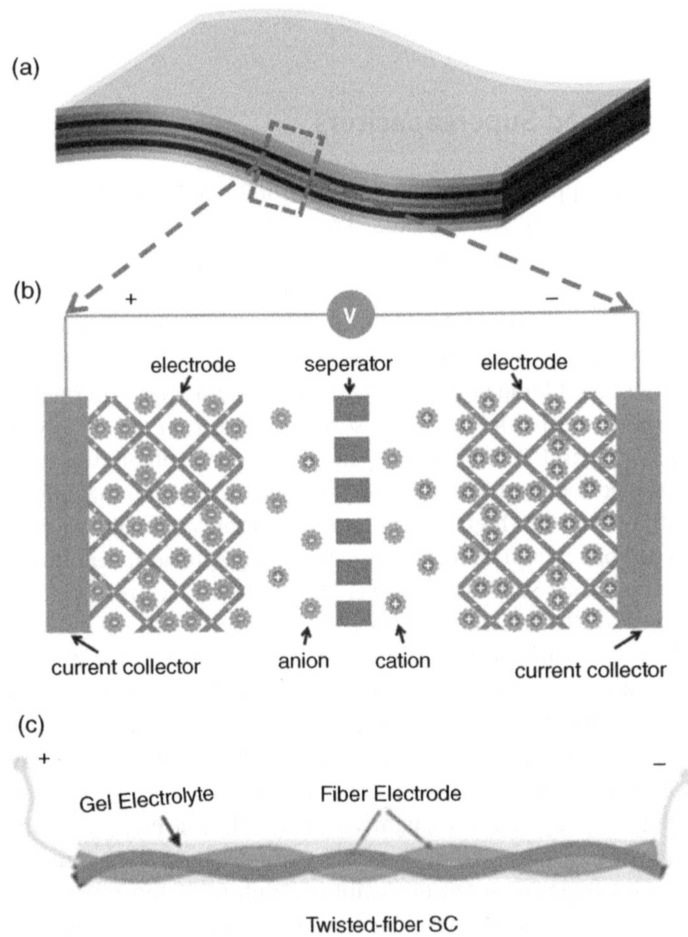

Figure 4.1 (a) Schematic description for the structure of planar supercapacitor. *Source:* Reproduced with permission [27] © 2014, American Chemical Society. (b) Discerption of charge storage mechanism of a supercapacitor [28]. (c) Schematic description for the structure of twisted-FFSC. *Source:* Reproduced with permission [29] © 2014, Royal Society of Chemistry.

in which one-dimensional (1D) fibers are used as both cathode and anode to develop the device [32]. FFSC that can be constructed from 1D wires with average diameters ranging from micrometers to millimeters and millimeters to meters in length are generally smaller in size and light in weight as compared with conventional SCs (Figure 4.1c) [29]. They are highly versatile due to their exceptional flexible fiber-shaped structure and can be woven into fabrics/textiles with superior suitability. These specific designs of FFSC also bring great flexibility in nature compared to conventional SCs, meanwhile they can be produced in numerous required shapes and placed in various locations. Therefore, FFSCs have immense promise to be conveniently combined with other fiber-shaped energy harvesting devices such as nanogenerators, solar cells, sensors into multipurpose systems [33–36].

4.2 Emerging Techniques for the Fabrication of Fiber-Shaped Electrodes

The basic parts of an FFSC are similar to conventional SC, which comprises a cathode, anode, and quasi-solid-state electrolyte [37]. The fabrication of fiber-shaped electrodes is the key issue to realize the high-performance FFSC, which generally contains two parts: obtaining the current collector for electronic/charge transfer, and decorating active materials for ion storage. At the same time, it is necessary to sufficiently facilitate multiple distortions with significantly higher conductivity and mechanical strength [38]. High electrical conductivity and mechanical flexibility are two main factors to be taken into account when choosing fiber electrodes. Additionally, consideration should also be provided to interfacial adhesion, mass density, surface area, and production costs. FFSC can be realized by growing the materials on conductive fibers through binder-free methods [39]. Recent strategies for the production of binder-free 1D fiber electrodes based on wet-spinning, electro-spinning, spray/cast-coating methods, electrodeposition method, and hydrothermal methods will be introduced in this section.

4.2.1 Wet Spinning Method

Wet spinning is the most important and popular technique for the preparation of multifunctional fibers for energy applications [40, 41]. Generally, the vicious spinning material, developed by liquifying polymer precursors (e.g. polyacrylonitrile, PAN) in an appropriate liquid and compacted through the spinnerets and forms fine streams into the clotting fibers [42]. The diffusion of solvents contributes to the condensation of continuous fibers, which are then deposited on spindles. In view of the ultimate applications, including electrodes in SCs, it is important to update nonconductive polymer precursors with conductive materials such as carbon and graphene/CNTs [43]. CNTs were ultrasonically spread into the water with the aid of a surfactant (sodium dodecyl sulfate, SDS), producing dark and diverse colloidal suspension. This CNT dispersion was slowly inserted into the polyvinyl alcohol solution (PVA, 5 wt%) via a small nozzle and entrapped with a good alignment along the flow path. The method of designing the flexible fiber-shaped SCs is illustrated schematically in Figure 4.2a [43]. Shortly, CNT yarn was formulated through the wet-spinning procedure from a commercial sizing material after which, the CNT yarn was used as a conductive substrate for the adsorption of pseudocapacitive MnO_2 and polypyrrole (PPy) through an electrosynthesis strategy. A similar method was also developed for wet spinning by Yuan et al. [44] using $CaCl_2$ and polymer composite (Figure 4.2b). The 1 m-long $CaCl_2$-polymer fibers

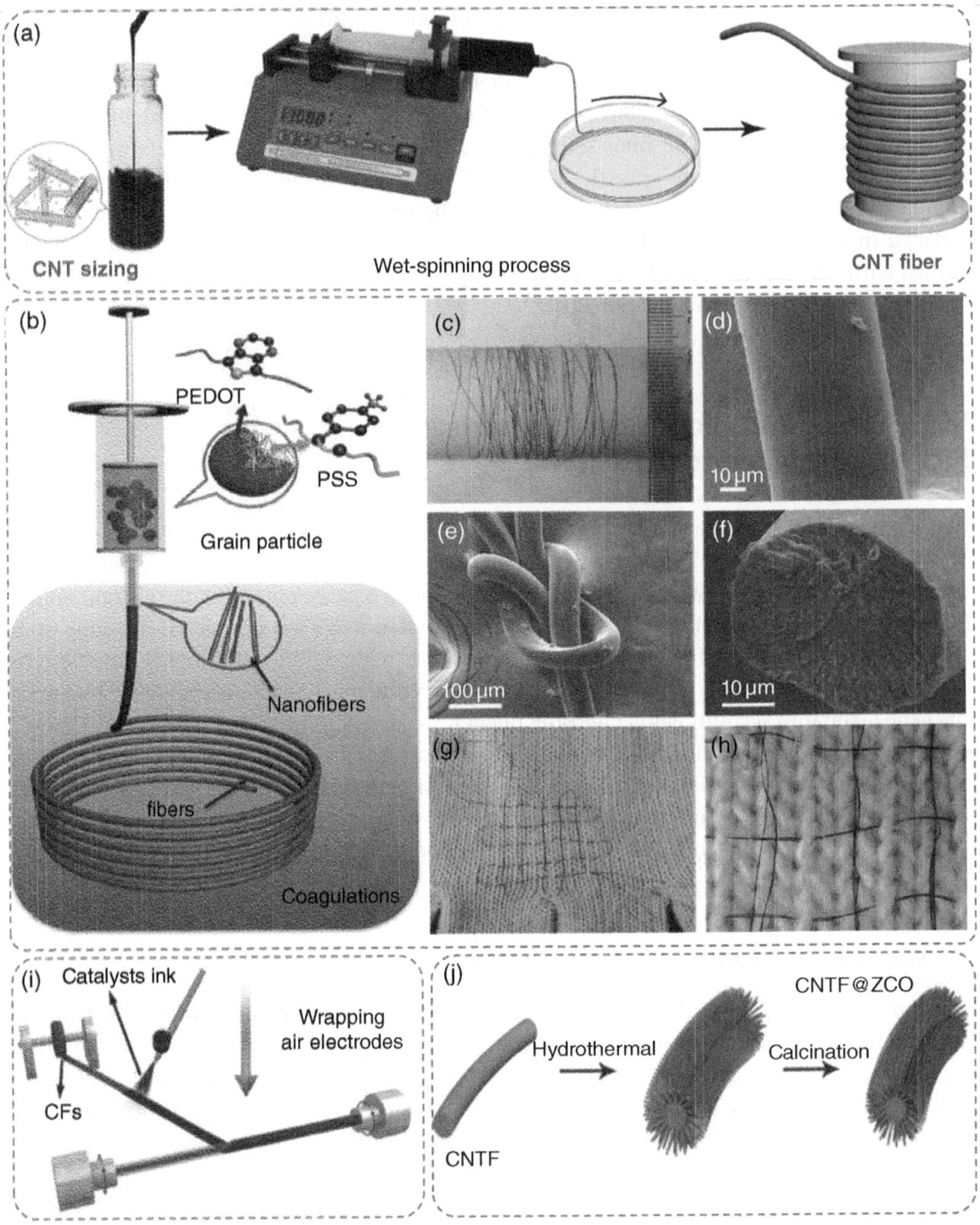

Figure 4.2 (a) Schematic description for the experimental procedure of wet-spinning process using CNTs ink. *Source:* Reproduced with permission [43] © 2020, Elsevier. (b) Schematic illustration of another kind of wet spinning process. (c) Digital photograph of few meters long PEDOT: PSS fiber around a PTFE rod. (d) SEM image of PEDOT: PSS fiber. (e) SEM image of PEDOT: PSS wet-spun fibers tied in a knot and (f) corresponding cross-sectional view. (g) Digital images of a Chinese knot tied to the cloth mask textiles. (h) Magnified image from part of Fig (g). (b–h) *Source:* Reproduced with permission [44] © 2016, Royal Society of Chemistry. (i) Schematic illustration for spray coating process. *Source:* Reproduced with permission [45] © 2018, Wiley-VCH. (j) Schematic illustration for the direct growth of nanomaterials on conductive substrate by hydrothermal synthesis process. *Source:* Reproduced with permission [46] © 2019, Elsevier.

could be bound to a plastic-rod (Figure 4.2c) and its low- and high-resolution scanning electron microscopy (SEM) images taken from different sections are shown in Figure 4.2d–f, and it also exhibits a high electrical conductivity of $38\,S\,cm^{-1}$.Furthermore, the as fabricated FFSC can be embedded and tied into the cloth mask textiles, as shown in Figure 4.2g–h.

4.2.2 Spray/Cast-Coating Method

The spray/cast-coating method is the simplest technique for the development of a flexible fiber electrode, which has also been proved to be successful for an extensive choice of 1D electrode materials for SCs. In this method, the 90 wt% active material and 10 wt% PVDF (polyvinylidene fluoride) mixture were brushed onto carbon fibers to create a flexible fiber electrode, detached by a hydro-gel electrolyte to prevent short-circuit, as shown in Figure 4.2i [45].

4.2.3 Hydrothermal Method

For the fabrication of traditional FFSCs, active material was in powder form which needs to be grounded to make a slurry by mixing a proportional amount of polymer binders and a conductive agent (carbon or acetylene black) and the slurry was further blade cast on the fibers to produce flexible electrodes. The major drawback of this method is the active material could be easily peeled off from the fiber during the electrochemical testing in the aqueous electrolyte and further in a flexible device configuration similar problem was faced during the testing of the flexible performance of the device. As a consequence, a new architecture of a binder-free and additive-free electrodes with high mechanical performance is required [47]. Therefore, the simple, cost-effective, and robust hydrothermal method was recently introduced for the fabrication of binder-free and highly flexible fiber-shaped electrodes, as demonstrated in Figure 4.2j [46].

4.3 Structures and Design/Configuration of Fiber-Shaped Electrodes

Compared to planar devices SCs, the fabrication of FFSCs devices is a more challenging task. The key parts of the FFSCs device such as electrodes, electrolytes, and separators should be consistent with the flexible fiber-shaped structure, and the integrated devices must withstand several repetitive deformations (e.g. folding, stretching, and twisting) while maintaining reasonably electrochemical performance and good mechanical strength. The accomplishment of high energy/power density and long-term stability is strongly dependent on the performance of the used active materials and their designs of manufacturing [48]. The investigation of each element and device design has encountered a development process with a progressive acquisition of knowledge. Common electrode and electrolyte designs, device specifications, and manufacturing methods are described and explained in this part to highlight the production of fiber-shaped SCs.

4.3.1 Parallel-Fiber Electrodes

Flexible fiber-shaped SCs with parallel configuration is formed by placing two electrodes (anode and cathode) sealed in a parallel configuration with the help of separator and polymer gel-based electrolytes, that is somewhat similar to the traditional flat SCs [49]. As a key electrode, transition metal oxide-based materials have been commonly employed to produce parallel-structured fiber-shaped SCs [47]. For example, Kang et al. [50] fabricated a hybrid flexible fiber-shaped SCs with a parallel structure based on nickel-cobalt layered double hydroxide (NiCo LDH)/3D Ni and 3D-manganese oxide (Mn_3O_4/3D-Ni). The fabrication process and structure configuration of (NiCo LDH)/3D Ni is schematically illustrated in Figure 4.3a [50]. Initially, 3D-Ni porous structures were constructed on Ni wire surfaces by electrodeposition using the hydrogen bubble template process. Intensely porous, flexible 3D-Ni metal with large surface area is important for the development of different nanostructures and thus considered as a cost-effective and lightweight electrode for flexible fiber-shaped SCs. The 3D-NiCo-LDH nanomaterials were then developed on the as-prepared 3D-Ni/Ni wire using the hydrothermal process and the corresponding SEM image is shown in Figure 4.3b, which shows the porous and open structure for easy ionic diffusion. To assess the potential applications of the as-synthesized NiCo-LDH/3D-Ni fiber electrode, a FFSC was manufactured using the NiCo-LDH/3D-Ni wire as an anode and the Mn_3O_4/3D-Ni as a cathode (denoted as NiCo-LDH/3D-Ni//Mn_3O_4/3D-Ni), and a pictorial diagram of the hybrid FFSC is shown in Figure 4.3c. The as-fabricated NiCo-LDH/3D-Ni//Mn_3O_4/3D-Ni FFSC exhibits excellent electrochemical performance and typical charge/discharge curves at various current densities in the potential window of 0.0–1.8 V (displayed in the inset of Figure 4.3d). The shape of charge/discharge curves demonstrated the hybrid battery/capacitor type behavior of the NiCo-LDH/3D-Ni//Mn_3O_4/3D-Ni FFSCs. The gravimetric and volumetric capacitances were calculated based on the charge/discharge curves and results are shown in Figure 4.3d. Accordingly, the as-fabricated flexible fiber-shaped NiCo-LDH/3D-Ni//Mn_3O_4/3D-Ni SC exhibited high capacitance of 331, 344.25, and 215 $F g^{-1}$ at 0.5, 2.5, and 10 $A g^{-1}$, respectively. The NiCo-LDH/3D-Ni//Mn_3O_4/3D-Ni FFSC also showed good volumetric capacitance of 9.86 $F cm^{-3}$ at 0.015 $A cm^{-3}$ with admirable rate-capability of 92.8% capacitance when the current density increased to 0.075 $A cm^{-3}$. To demonstrate the device's versatility in intelligent wearable and integrated technologies, charge/discharge curves of the NiCo-LDH/3D-Ni//Mn_3O_4/3D-Ni FFSC were recorded at various bending angles (0–150°) at 5 $A g^{-1}$, as shown in the inset of Figure 4.3e. When the NiCo-LDH/3D-Ni//Mn_3O_4/3D-Ni fiber SC unit was heavily bent at over 150° and sequentially move back to 0°, maintaining over 88% capacitance, as depicted in Figure 4.3e. As a demo of the practical potential application, two FFSCs in series could light the green light-emitting-diodes (LEDs) brightly [51–54].

4.3.2 Twisted-Fiber Electrodes

In recent years, substantial attempts have also been made to build high-performance FFSSs with twisted architectures [55, 56]. Twisted structure FFSC has recently been successfully developed and offers high-performance, ultrahigh flexibility for portable energy storage systems [57]. Nevertheless, the low energy density of the FSSC restricts the functional applications. Few recent studies concentrated on improving their basic capacitance to increase their energy efficiency, by asymmetrical architecture with twisted structures to widen the voltage windows. Fiber-shaped SCs with twisted configurations are typically built by twisting or binding together two or more fiber-based electrodes and packed them with quasi-solid polymer electrolyte or, if necessary, by inserting separators [58]. Note that a layer of quasi-solid-state gel electrolyte must be pre-coated on the exterior surface of every fiber anode/cathode to prevent the short circuits. Compared to parallel-structured flexible fiber SCs, twisted-fiber SCs do not require a flat substrate and have an architecture close to that of the cloth sheets, thus tend to be highly suitable for foldable/smart wearable applications and can be conveniently woven/sewn into scale-up energy storage textiles [59, 60]. Metal-organic frameworks (MOFs) have been established as important precursors and models for the creation of porous materials for well-tuned size and features of large surface area [59].

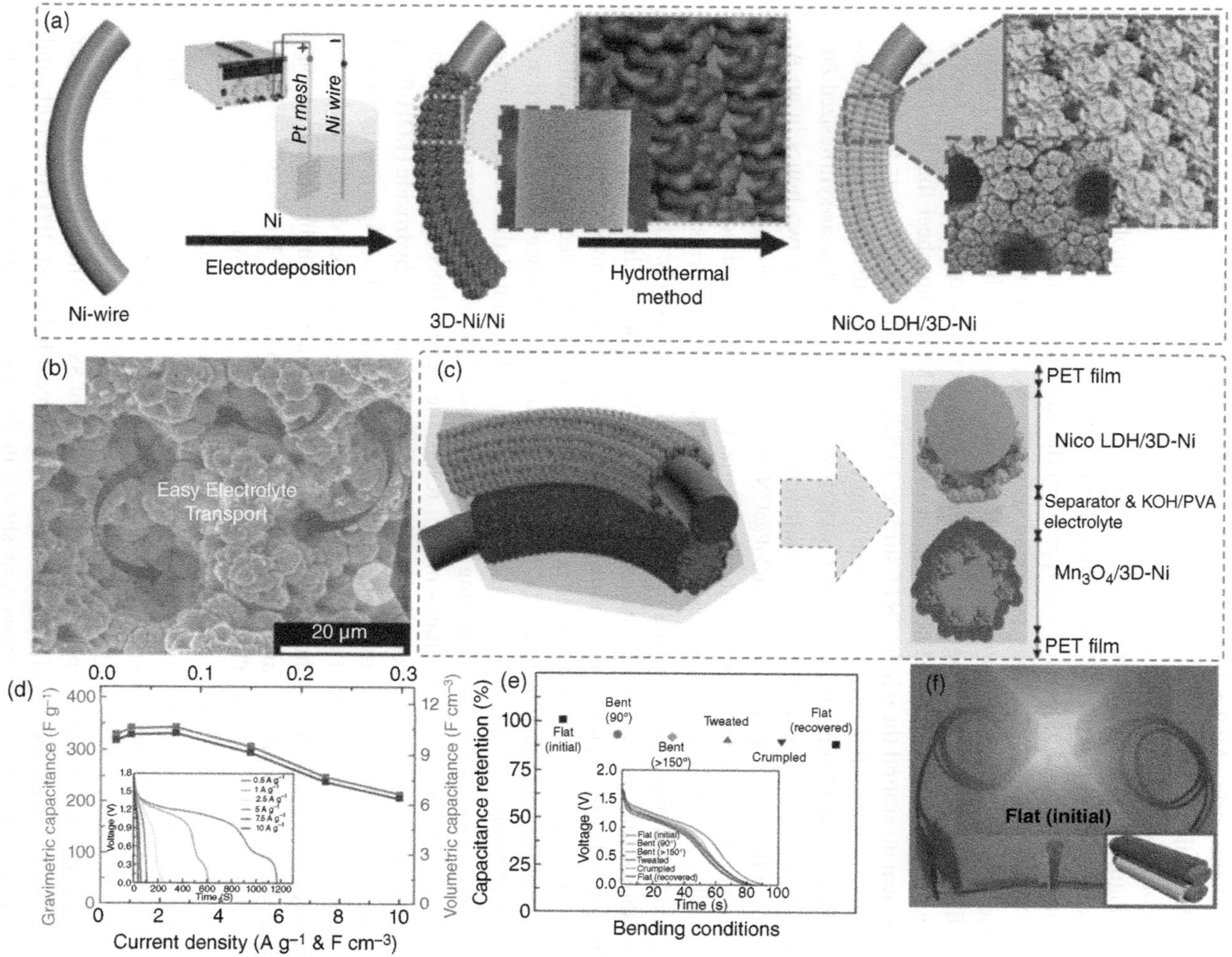

Figure 4.3 (a) Schematic depiction for the construction process of the NiCo LDH/3D-Ni electrode. (b) SEM image (c) Schematic depiction of the construction of the FFSC device with cross sectional view. (d) Gravimetric capacitance as a function of current density and inset shows the GCD curves of the device. (e) Flexible performance of the FFSC device at various bending angles and inset shows the GCD curves at various bending angles. (f) Digital photograph of lightening the green color LEDs by connecting two FFSC devices in series. *Source:* Reproduced with permission [50] ©, 2020, Springer Nature.

Noticeably, such innovative types of materials with a huge surface area and plentiful reaction spots are commonly regarded as outstanding electrochemically active materials. Nevertheless, the MOF-derivative spongy materials are typically synthesized in powder form, which includes the use of a binder and a conductor to test their electrochemical performance, resulting in reduced active sites, decline of effective materials, increased electrode/electrolyte interface resistance and inadequate flexibility [60]. The specifications to tackle these problems are the straightforward growth of MOF-derived electrode material on flexible substrates with enhanced mechanical and electrical interaction as binder-free electrodes. Recently, Zhang et al. [61] successfully fabricated the asymmetric FFSC with a twisted-fiber structure which exhibits an exceptionally high energy density and good mechanical resilience using MOFs-derived materials, which could be additionally incorporated into the flexible textile-based device to illustrate its outstanding weave-ability [62]. The manufacturing process of a twisted structured energy storage device is schematically summarized in the upper part of Figure 4.4a. First, the MOF-derived NiZnCoP nanosheet assemblies were explicitly grown on the carbon nanotube fiber (CNTF) (NiZnCoP@CNTF) as a cathode via a three-step process. At the same time, the MOF-derived Fe_2O_3 grow on the surface of oxidized carbon nanotube fiber (Fe_2O_3/OCNTF) as an anode was processed using the hydrothermal and post-annealing methods in the air atmosphere, as shown in the lower part of Figure 4.4a. They constructed a wearable flexible fiber-shaped energy storage device with a reliable potential of 1.05 V. Figure 4.4b–d shows the low- and high-resolution SEM images of the Zn-Co MOF nanosheet that have been consistently grown up around the whole CNTF surface. In response to the $Ni(NO_3)_2$ in an absolute ethanol solution, tiny nanoflakes have been uniformly established on the exteriors of the as-manufactured Zn-Co-MOF and it becomes Zn-Co-MOF@Ni-Zn-Co LDH. The as-synthesized Zn-Co-MOF@Ni-Zn-Co LDH was then annealed with NaH_2PO_2 at 350 °C for two hours in an Ar atmosphere to get the desired product of NiZnCoP@CNTF. To accomplish the good electrochemical performance of the fiber-shaped SC, an anode material was also synthesized using MOF templates (MIL-88-Fe) and directly grown at the surface of OCNTF, whose corresponding low- and high-resolution SEM images are displayed in Figure 4.4e and f, respectively. The high-resolution SEM image shows that the MIL-88-Fe has formed in regular dodecahedron-shaped material (Figure 4.4f). After the consequential annealing procedure, it was entirely converted into a spindle-like S-α-Fe_2O_3 (Figure 4.4g). The CV curves of the as-fabricated positive electrode (NiZnCoP@CNTF) and a negative electrode (S-α-Fe_2O_3/OCNTF) in the aqueous electrolyte were recorded at a fixed scan rate of $10\,mV\,s^{-1}$, as revealed in Figure 4.4h. It should be noted that both electrodes composed of prominent redox peaks indicating that full cell should possess high operating voltage with a hybrid type charge storage mechanism. As a result, the flexible fiber-shaped SC with the twisted structured prototype was successfully constructed using Potassium hydroxide (KOH-PVA) as a quasi-solid-state electrolyte and two visible redox peaks, as presented in Figure 4.4i. The CV (cyclic voltammetry) curves at several scan rates show prominent redox peaks, demonstrating the hybrid type charge storage features. The GCD curves also displayed a typical hybrid charge storage characteristic. Most significantly, a maximum capacity of $0.092\,mAh\,cm^{-2}$ has been reached at $1\,mA\,cm^{-2}$, as shown in Figure 4.4j. Thus, a total energy density of $30.61\,mWh\,cm^{-3}$ was reached along with an average power density of $3339.7\,mW\,cm^{-3}$, signifying the excellent charge storage behavior of the flexible fiber shaped energy storage devices [59, 63], as shown in Figure 4.4k. Such energy and power density values are significantly higher than those of fiber-shaped energy-storage devices mentioned earlier. The stability of fiber shaped devices was further tested at a high current density of $2\,mA\,cm^{-2}$ and it shows superior capacity retention of 88.6% of the initial capacity after 3000 cycles, as shown in Figure 4.4l. Typically, the CV curves recorded at 1st, 1000th, 2000th, and 3000th, represented the stable performance, as shown in the inset of Figure 4.4l. To show the

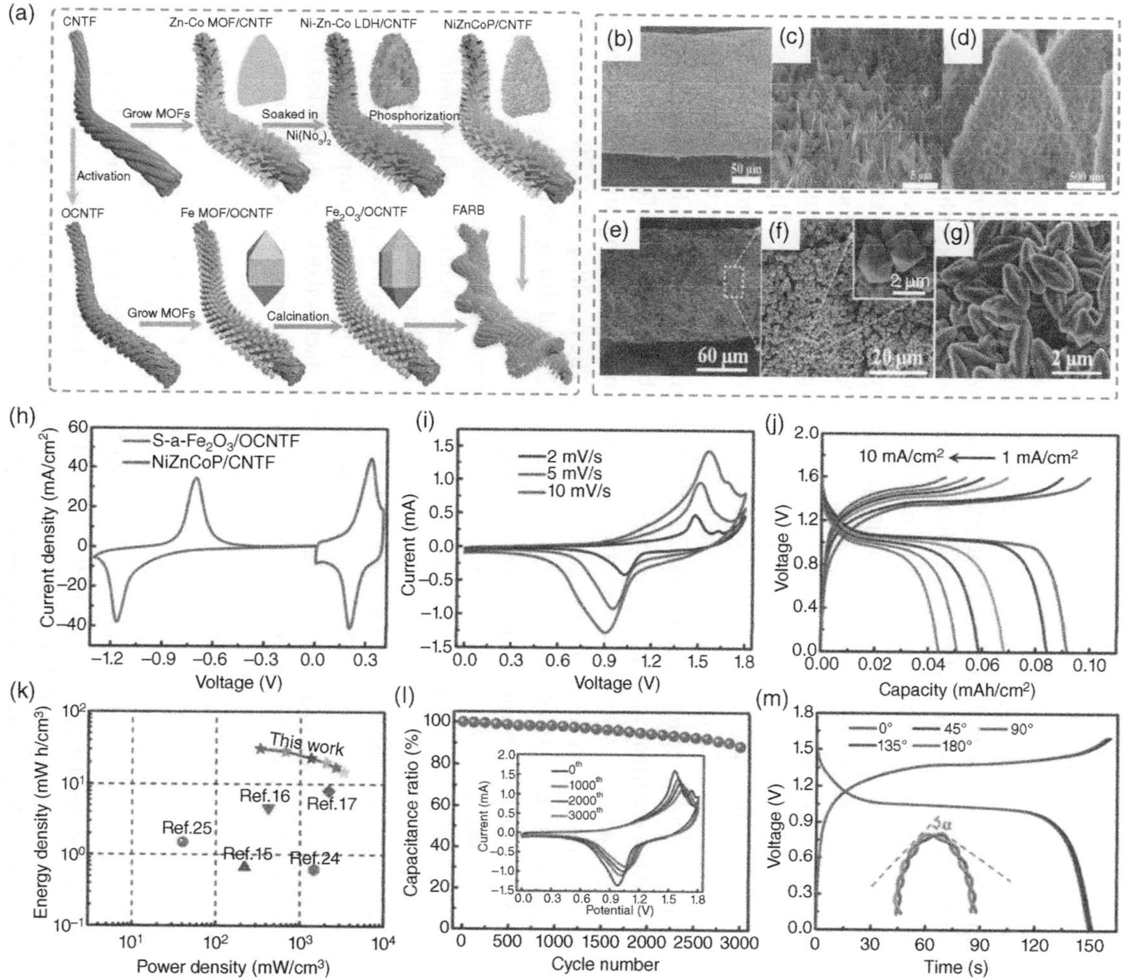

Figure 4.4 (a) Schematic illustration for the production method of the FFSC device. (b–d) Low- and high-resolution SEM images of positive and negative electrodes. (e–g), respectively. (h) CV curves of positive and negative electrodes. (i) CV curves and (j) GCD curves of as-fabricated device. (k) Ragone plot of the device along with comparison. (l) Capacitance retention and inset shows the CV curves after various number of cycles. (m) GCD curves recorded at various bending angles. *Source:* Reproduced with permission [61] © 2018, Wiley-VCH.

possible use of the fabricated fiber-shaped energy storage device in a compact and wearable system, a range of mechanical stability experiments have also carried out. The GCD graphs of the constructed device were shown in Figure 4.4m, almost entirely coincided at various bending angles extending from 0° to 180°, implying the FFSC has superior flexibility. The inset of Figure 4.4m shows the bending angle measurement for testing the flexibility of the fiber-shaped energy storage device. Throughout this end, the twist-shaped fiber devices give the relatively short path between the electrodes and the effective ion path from the central to the external electrodes, and vice versa [24, 64–66].

4.3.3 Coaxial-Fiber Electrodes

High-performance FFSCs devices cross the difference between the idea of textile energy storage and realistic use. Therefore, extensive efforts on FFSCs have been reported to date [67–69]. Up to this end, however, mechanical durability remains a challenging task. This is primarily attributed to an insufficient disruption in situations where FFSCs are exposed to different exterior constrains. Configurations such as tubular or coaxial systems have been undertaken to create wearable FFSCs with improved performance during complex mechanical pressure [70]. The coaxial design can help to sustain contact due to its dimensional integrity and provide much more reliable performance for flexible wearable electronics [71]. Nevertheless, there is a need to boost the basic capacitance of FFSCs in a coaxial configuration [72]. The constraint on such an FFSC is that it's a very small potential window even less than 1.0V, that is specified by the liquid electrolytes. As the energy density (E) of the SC followed the equation, $E = 1/2\ CV^2$, where C is the specific capacitance and V is the potential range, the upsurge of V would have high the energy density [73]. Thus, the asymmetric design is implemented to outspread the operating voltage range [74]. One conceivable way to circumvent the voltage limit in liquid electrolytes is to implement asymmetric SCs by using specific pseudocapacitive materials (metal nitrides/oxides, polymer conductors, etc.), that may improve the capacitance and potentially expand the potential window [75]. Recently, Yuan et al. [67], reported the asymmetric design for coaxial type FFSC based on the CNT@NiO@MnO_x as a positive electrode and CNT@Fe_2O_3 as a negative electrode. The coaxial configuration ensured physical supremacy throughout continuous bending conditions. The resulting FFSC offered substantial high output with a stretchability of 75% and can bear high load during the charging/discharging process and was able to operate 1.8V high voltage range.

FFSC was manufactured by utilizing CNT@NiO@MnO_x composite as the cathode (inner electrode), CNT@Fe_2O_3 composite as an anode (outer electrode), and nearly neutral PVA/LiCl hydrogel as an electrolyte. The flexible fiber was extended to a pre-defined strain of 50%, then coated with a thin layer of PVA/LiCl hydrogel electrolyte. First, the CNT@NiO@MnO_x was placed as an inner electrode on the stretched elastic fiber. Upon wrapping with CNT@Fe_2O_3 composite material, the hydrogel electrolyte was coated again. Finally, the as-manufactured device was guarded by a hydrogel electrolyte layer and the tension was released. Figure 4.5a demonstrates the alignment process of the FFSC. The thicknesses of anode and cathode are estimated to be 1.7 and 2.7 μm, depending on the number of layers of active material, respectively. The cross-sectional view of the coaxial structure of a FFSC is illustrated in Figure 4.5b. The coaxial structure could be visibly identified by the SEM images take from the cross-sectional position (Figure 4.5c and d) in which each electrode is enclosed by a hydrogel electrolyte and the width of the inner core was about 470 μm. Figure 4.5e displays the standard SEM image of the coaxial FFSC with a diameter of approximately 530 μm. Extremely overlapping CNT sheets result in a crumpled surface that offers not only excellent electronic conductivity but also flexural strength for the coaxial specification of the FFSC. The

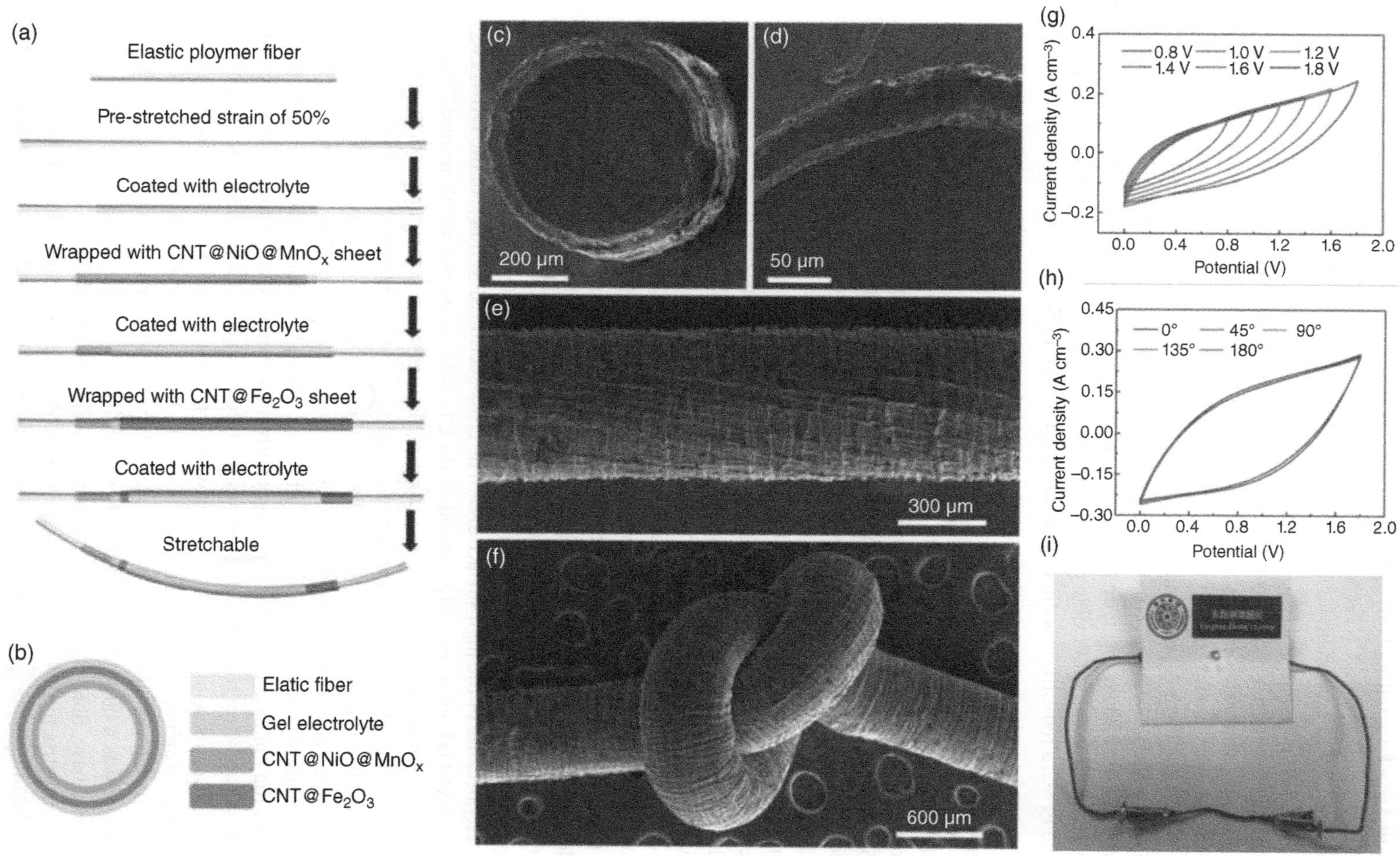

Figure 4.5 (a) Design for the production procedure of a coaxial type FFSC. (b) Illustration for the cross-sectional structure of a FFSC. (c and d) Cross-sectional view of SEM images at various resolutions. (e and f) Image of a FFSC with a crumpled structure and being tied into a knot, respectively. (g) CV curves of as-fabricated FFSC at various potential windows. (h) CV curves of FFSC at various bending angles. (i) Digital photograph of lighting the LEDs by FFSCs. *Source:* Reproduced with permission [67] ©, 2020, Springer Nature.

FFSC possibly tied in a knot without any apparent injury, as shown in Figure 4.5f. Using the robust operating potential windows of CNT@NiO@MnO_x (0–0.8 V) and CNT@Fe_2O_3 (−1.0–0 V), the operating potential 1.8 V in asymmetric configuration could be achieved [22]. The electrochemical output of the FFSCs was investigated in a two-electrode device. In order to evaluate the electrolyte stability region, a series of CV curves were recorded by varying the potential windows from 0.8–1.8 V at a constant scan rate of 50 mV s^{-1}, as shown in Figure 4.5g. There is no apparent irreversible peak current, which suggests that the asymmetric FFSC may expand the potential window up to 1.8 V. To access the flexible performance of the as-fabricated FFSC, harsh bending is important for practical applications. The electrochemical performance at several bending angles (0°, 45°, 90°, 135°, and 180°) are shown in Figure 4.5h, the similar traces of CV curves at different bent states demonstrated the excellent flexible performance of the FFSC. To illustrate the practical performance of the FFSC, a red-LED was directly illuminated by two FFSC, as depicted in Figure 4.5i.

4.3.4 Rolled-Fiber Electrodes

The most desirable architecture for FFSCs is to create a standardized and minimum gap between the two electrodes, providing a more effective route for ion/charge transfer with reduced internal resistance [76]. Nonetheless, the manufacturing method for the several coaxial sheets on the core fiber substrate may be difficult, particularly when the bonding across each coating layer and the inner material needs to be taken into account. In comparison, bulk electrodes, such as CNT nanosheets on core fiber substrate are technologically challenging and need deep proficiency. On the other side, the rolling layer-by-layer plating of flat sheets onto fiber-shaped electrodes is seen as a quick and fast approach to the manufacture of roll-type-fiber shaped SCs. As seen in Figure 4.6a, Yu and fellow workers fabricated a fiber-shaped SC based on the rolled fibers as an electrode by rolling up three alternating layers of polyaniline (PANI)/ Carbon fiber (CFs), PVA/H_2SO_4 gel electrodes, and PANI/CFs [77]. Technically, the optimization technique for the fabrication of electrode is one of the two electrodes equally spaced at a constant interval of spacing and without a short circuit. The design of the roll-type electrode can, therefore, be assumed to be the optimal strategy for FFSCs [76]. The low- and high-resolution SEM images of PANI/CFs electrode were demonstrated in Figure 4.6b and c, respectively, which displayed a porous structure that improves the electrode capacity because of the high surface area. The cross-section SEM image of the PANI/CFs revealed that the PANI is homogeneously covered on the CFs with an average width around 200 nm and the maximum CF diameter was approximately 8 μm (Figure 4.6c). A comparison between the performances of twisting-fiber and roll-fiber type electrode-based flexible fiber shaped SCs has also been made. It was observed that role-structured based fiber shaped SC had better performance over twisted-structured based SC and a comprehensive comparison of the electrochemical performances of other kinds of SCs was demonstrated in Figure 4.6d–j. The role-structured based fiber shaped SC shows the highly rectangular CV curves with the symmetric configuration at different scan rates while twisted-structured based SC shows the tilted shaped CV curves and become elliptical even at a high scan rate of 100 mV s^{-1}, as shown in Figure 4.6d and e. Further charging and discharging profiles also indicate the presence of large IR-drop for twisted-structured based SC as compared to role-structured based fiber shaped SC, as shown in Figure 4.6f and g, respectively. In consequence, there was a large difference in rate capability and energy/power densities of both kinds of structured base, fiber-shaped SCs, as shown in Ragone plots (Figure 4.6i). To prove the flexibility of the roll-structured-based fiber-shaped SC, CV curves were recorded while the FFSC bent at various bending angles (20°, 60°, 90°, 120°, and 180°) and results were summarized in Figure 4.6j. The CV curves exhibited minor changes in shape while FFSC was bent at different

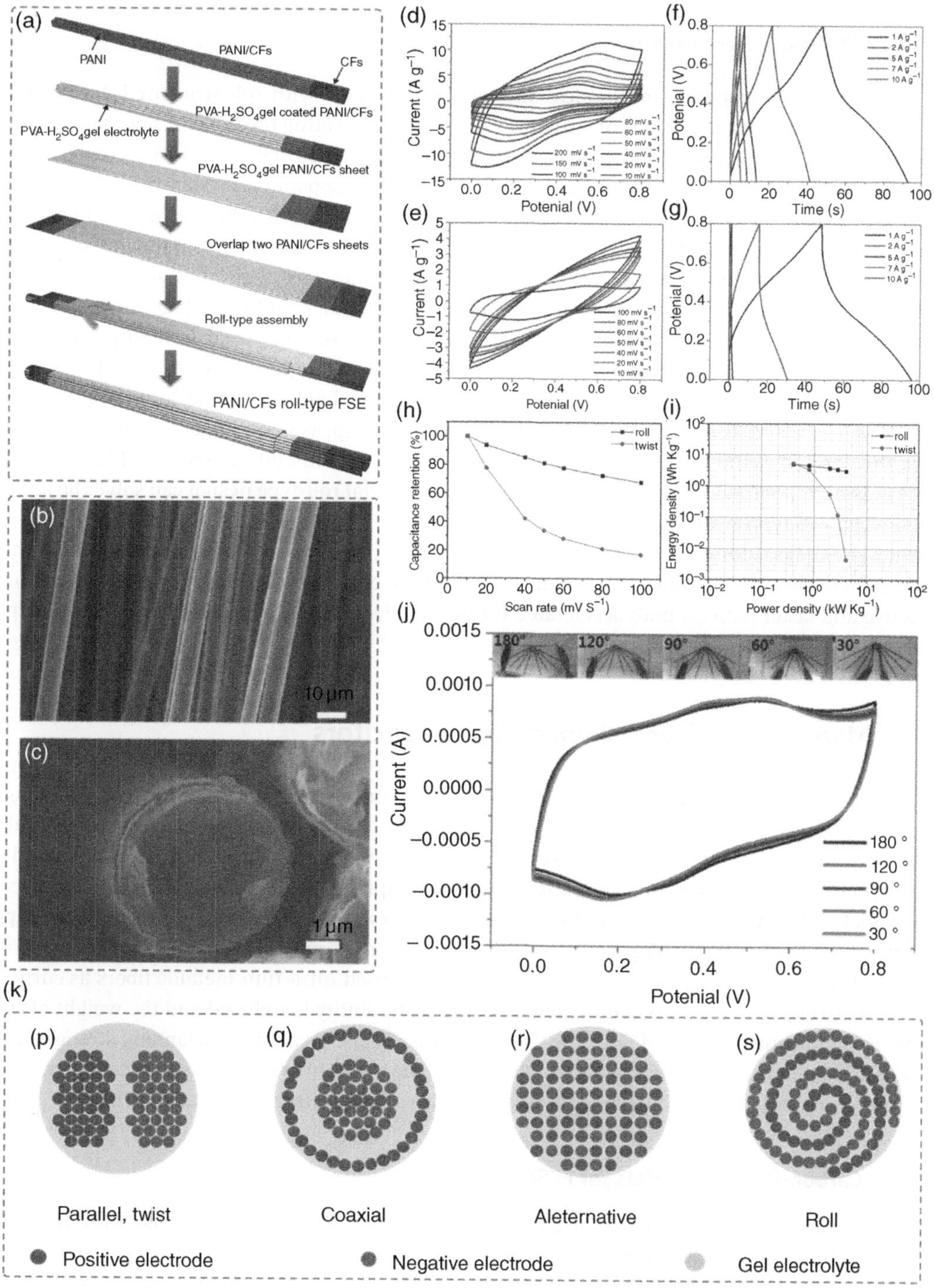

Figure 4.6 (a) Schematic diagram representing the construction process of PANI/CFs//PVA–H_2SO_4//PANI/CFs roll-structured FFSC. (b) SEM image of pristine CFs. (c) cross-sectional SEM image of PANI/CFs. (d and e) CV curves of the PANI/CFs roll- and twist-structured FSSCs. (f and g) GCD curves of the PANI/CFs roll- and twist-structured FSSCs. (h) Specific capacitance of roll- and twist-structured FSSCs as a function of scan rates. (i) Ragone plots of roll- and twist-structured FSSCs. (j) CV curves of roll- structured FSSCs at several bending angles. (k) Schematic illustration of cross-sections of FFSCs with different configurations: (p) parallel- and twist-type, (q) coaxial-type, (r) alternative-type, and (s) roll-type. *Source:* Reproduced with permission [77] © 2019, Elsevier.

angle, showing the superior mechanical strength of the FFSC with roll-structured design. The general overview of various kinds of structure designs for fabricating the flexible, fiber-shaped SCs is illustrated in Figure 4.6k. For parallel-type or twisted-type structured based FFSCs, two bundles of fiber faced one another by separating with gel electrolyte/separator, as shown in the inset of Figure 4.6k and p. The fibers are located near the middle of the FFSC, the gap between both the fibers in each electrode is the lowest. Alternatively, for fibers located next from the outside of the FFSC, the gap between the fibers in each electrode is quite far. As the distance between the electrodes increases, the mean ionic transmission diffusion reduces, resulting in high intrinsic resistance and low power density. FFSCs with a parallel- and twist-structured have large distances between the electrodes as seen in Figure 4.6k and p. However, the coaxial-type FFSC has a more balanced gap between both the electrodes, as shown in Figure 4.6k and q). The optimal setup, with a uniform space between the electrodes, is the alternative setup, as seen in Figure 4.6k and r; however, this kind of optimal, alternative-type setup is practically difficult to fabricate. Thus, the fast and convenient method for constructing FFSCs with a special roll-type design can solve the above issues. The roll-type design can be accomplished through getting rolled up two thin film-shaped fibers bundle electrodes with a solid polymer electrolyte, as shown in Figure 4.6k and s). This arrangement provides a consistent, small distance between both electrodes, providing a more effective and faster path for both accelerated transfer of electrons and ions relative to parallel-type and twist-type structures [76–78].

4.4 Materials for Fiber-shaped Supercapacitors

In order to turn SCs into fiber forms, the fiber-shaped current collectors seem to be the most important components, serving for active materials to transport electrons and fiber-shaped support substrates. Many other types of fiber current collectors have been recently developed, such as primarily plastic/polymer fibers, ultra-fine metal fibers, CNT or graphene fibers, carbon fibers, and composite fibers [79, 80]. Plastic fibers have been used for the first time in the production of fiber-shape SCs due to their excellent strength. Regrettably, the low ionic transportation restricted the performance of the manufactured FFSC. Afterward, scientists used ultra-thin metallic fibers as current collectors to enhance electron conductivity. However, durability is reduced and the weight of the FFSC is significantly increased with the use of metallic fibers. To attain outstanding strength and good electrochemical efficiency with small size, carbon-based fibers (CNT, graphene fibers, etc.) are then commonly used as current collectors and also active materials to produce FFSCs.

4.4.1 Carbon-Based Materials for FFSC

Carbon-based materials are widely used electrodes for fiber-shaped SCs, because they typically have outstanding electrical conductivity, a large specific surface area, are compact, easy to manage, and are low-cost, chemically, and thermally stable in various electrolytes [81, 82]. Fiber-shaped SCs based on the carbon materials were typically controlled by the electrical dual-layer effect, so they were referred to as electrical double-layer capacitors (EDLCs). Until now, many types of carbon materials are being used in FFSCs, such as activated carbon, single/multi-walled CNTs, PCs, and graphene, etc. [83–85].

The early research work on the flexible fiber-shaped SCs using carbon materials was carried out by Prof. Peng from Fudan University, China [81]. They introduced a simple strategy to synthesize the composite of graphene and CNTs for the fabrication of FFSCs. The schematic representation of the structural system of the composite graphene/CNTs fiber was shown in Figure 4.7a. In this case, the transport networks mentioned by the arrows were being used to highlight the cross-linked graphene which promotes the transfer of charges, even though charges may be transferred more quickly along the lengths of the CNTs, that are not shown here for the sake of clarity. The low- and high-resolution SEM images of graphene oxide/CNTs composite fibers with an average diameter of around 17 μm were shown in Figure 4.7b and c, separately. It is observed that the graphene oxide sheets have been successfully inserted between the CNTs. The SEM image of annealed composite showed the slight shrinkage in size and changed the color dark to light which signified that the π-π interactions are become stronger without hydroxyl groups, as shown in Figure 4.7d. Those same conductive fibers were further rotated to make a flexible fiber-shaped SC. The comparative CV curves of pristine CNTs and composite graphene oxide/CNT based FFSCs are depicted in Figure 4.7e. The composite graphene oxide/CNT based FFSCs shows significantly enhanced capacitive performance as compared to CNTs due to the synergistic effects of two kinds of materials. A standard rectangular shaped CV curves referring to EDLC for bare CNT fiber, whereas a redox peak in the range of at 0.3–0.6 V vs. Ag/AgCl demonstrated pseudocapacitive type storage obtained from oxygen-containing functional groups for graphene oxide/CNT composite, as shown in Figure 4.7e. The Galvanostatic charge-discharge curves for both kinds of FFSCs were also recorded and compared. The pristine CNT fiber-based FFSC displayed a symmetrical triangle for EDLC, while the graphene oxide/CNT composite fiber demonstrated a divergence from the triangle-shaped GCD curves arising from fast redox reaction in the existence of oxygen-containing groups Figure 4.7f. The graphene oxide/CNT composite-based fiber fiber-shaped SC showed high cyclic stability. No apparent decline inability was found in 5000 cycles, as shown in Figure 4.7g.

Another example of FFSCs based on the carbon materials is demonstrated by Prof. Miao from CSIRO, Australia [86]. The schematic diagram for the fabrication process of FFSC was shown in Figure 4.7h. The single CNT fiber was served as both the current collector as well as an active material. The conductive PANI polymer was being used as an electrode material due to its high energy storage capacity and flexibility. The simple CNT fiber was first rolled by spinning to produce a prolonged CNT web taken from a solid-state multi-layered CNT forest that was synthesized by the chemical vapor deposition (CVD) process. In order to optimize the electrochemical performance FFSCs, PANI nanowire assemblies were distributed onto the CNT fiber through an in-situ polymerization procedure, resulting in a hybrid fiber embodied CNT@PANI fiber, as demonstrated in step-1. Next, a slim film of polyvinyl alcohol PVA-H_2SO_4 hydrogel electrolyte being adsorbed on the surface of the CNT@PANI fiber to form a CNT@PANI@PVA composite (Step-2). At last, two CNT@PANI@PVA fibers were entangled together to form FFSCs (Step-3). Figure 4.7i show the low- and high-resolution SEM images of CNT@PANI composite fibers. The PANI nanowires are smooth and oriented perpendicular to the surface of the CNT fiber and have 50 nm in diameter and around 400 nm in length, as shown in the inset of Figure 4.7i. To access the electrochemical performance, a symmetrical configuration based on the two-electrode system with PVA-H_2SO_4 hydrogel electrolyte, and a separator was used to fabricate the FFSC. The CV curves of pristine CNT fiber and composite CNT@PANI based FFSC were recorded at a fixed scan rate of $10\,mV\,s^{-1}$ in the potential window of 0.0–0.8 V and results are presents in Figure 4.7j. All of these FFSCs exhibit the normal ECDL type charge storage properties; however, the CNT@PANI based FFSC exhibits the significantly enhanced performance as compared with pristine CNT based FFSC. The areal capacitance as a function of current density was computed based on GCD curves and the capacitance of CNT@PANI based FFSC exhibits much superior to that of pristine CNT based FFSC. Enhancement in capacitance could be directly linked to the highly arranged PANI nanowire arrays on the CNT fiber that are identified as a high-performance pseudocapacitive material. As a textile fiber, FFSC

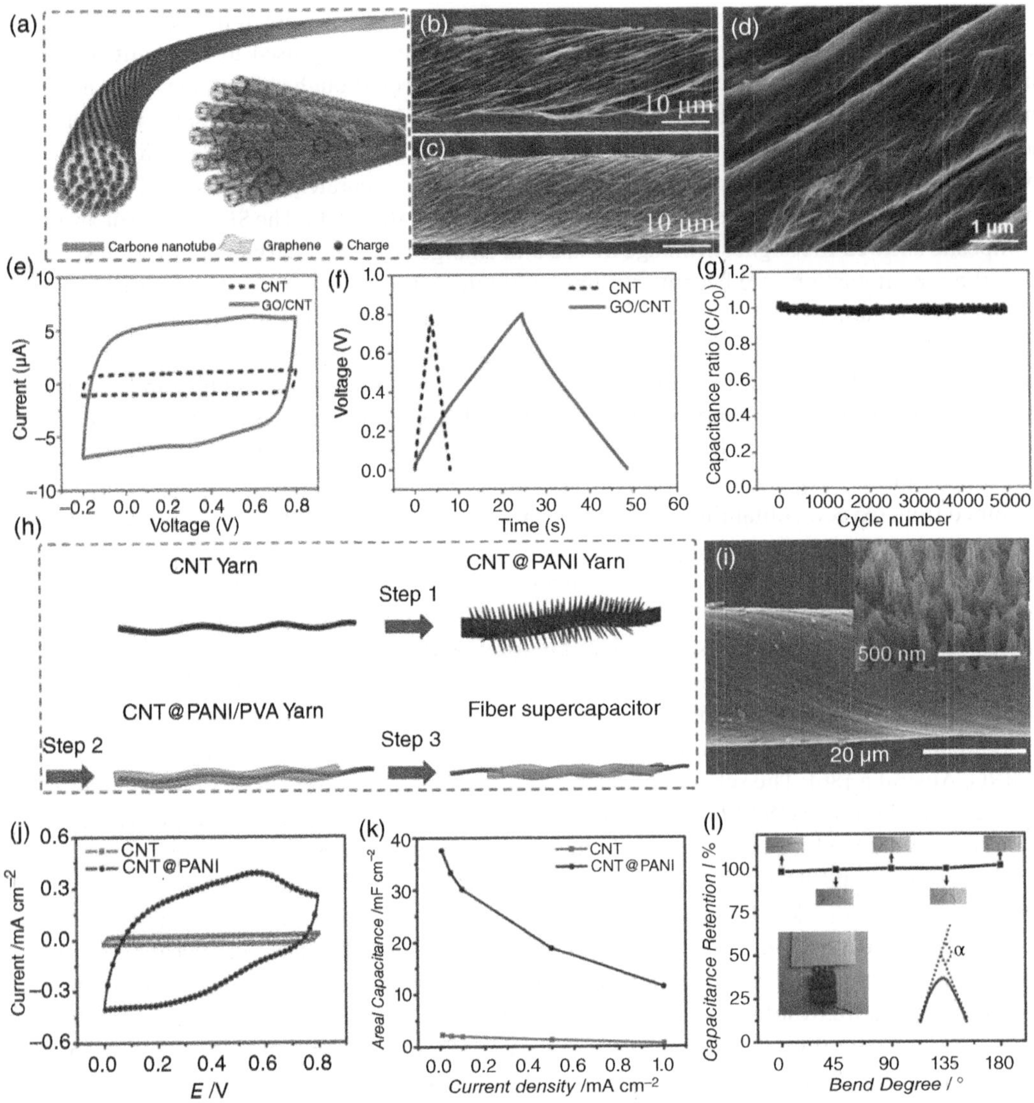

Figure 4.7 (a) Schematic diagram representing the structure of composite fiber based on the graphene/CNT. (b and c) Low- and high magnifications SEM images of graphene oxide/CNT composite fiber, respectively. (d) Low-magnifications SEM image of graphene/CNT composite fiber. (e) CV curves of FFSC with bare CNT and graphene oxide/CNT composite. (f) GCD curves of FFSC with bare CNT and graphene oxide/CNT composite. (g) Specific capacitance as a function of cycle number for graphene oxide/CNT composite. (a–g) *Source:* Reproduced with permission [81] © 2014, Wiley-VCH. (h) Diagrams for the fabrication process of two-ply yarn SC (i) Low-magnifications SEM images of CNT@PANI composite yarn and inset shows the ordered PANI nanowire arrays on the surface of CNT yarn. (j and k) Comparative CV and GCD curves of pristine CNT and CNT@PANI composite based FFSC, respectively. (l) Capacitance retention at various bending positions (Inset at bottom-left photograph shows a connector used during electrochemical testing and bottom right shows the angle measurements). *Source:* Reproduced with permission [86] © 2013, Wiley-VCH.

must have admirable flexibility without losing its electrochemical efficiency under various bending states. The electrochemical performance of the CNT@PANI fiber SC at various bending positions was assessed by an in-situ test and results are displayed in Figure 4.7l. The CNT@PANI fiber SC possesses excellent flexibility and retained high capacitance even bent at 180° (different banding states for measurements of the flexible performance of FFSC are shown in the insets of Figure 4.7l.

4.4.2 Metal Oxides and Their Composites-Based Materials for FFSC

Metal oxides are another interesting group of materials for the fabrication of high-performance FFSCs electrode [87, 88]. FFSCs based on the transition metal oxide materials as electrodes are typically categorized as pseudocapacitors [89]. The energy storage mechanism in pseudocapacitor is based on the transfer of electron during quickly reversible electrochemical reactions of the electroactive sites on the surface of the electroactive materials, which resulted in a much greater capacitance and energy density than the EDLCs based on carbon electrodes [7]. Recently, ternary transition metallic oxides have drawn intensive interest as a potential low-cost pseudocapacitive fabric-based electrode for FFSC [90, 91]. The ternary metal oxides showed high supercapacitive efficiency than single component, allowing for numerous electrochemical reactions [92]. For instance, ternary $NiCo_2O_4$ has a much higher electrical conductivity than single nickel oxide or cobalt oxide, as well as a higher electrochemical activity because of its good ability to produce rich electrochemical reactions, such as the participation of both cobalt and nickel compounds [93]. Such as Cao and coworkers [37] reported a flexible fiber-shaped asymmetric SC using the excellently aligned Mn, Ni co-substituted hydroxides nanoneedles array on carbon fibers (MnNiCo-CH/CF) as a top anode and activated carbon (AC) as a cathode. The core/shell heterostructure MnNiCo-CHs nanoneedles array on the of the CFs is manufactured using a low-cost hydrothermal synthesis method under 88 °C, as schematically demonstrated in Figure 4.8a. The fabrication procedures for FFSC and their digital photographs are representing in a schematic diagram in Figure 4.8a, respectively. The MnNiCo-CH/CF as-prepared has been treated with a PVA/KOH tunable gel polymer solution film. The activated carbon acted as a negative electrode material is then coated to the top of the above CF. Eventually, the collected CF is coated again through PVA/KOH hydrogel polymer electrolytic solution for the final coaxial FFSC unit, as shown in Figure 4.8b. Throughout this process, the diameters or volumes of FFSCs can be conveniently managed by changing the amounts of carbon fibers, which allow the effective use of fiber products. The microstructures of the pristine CF as well as the processed products of MnNiCo-CH/CF were first studied by SEM. Figure 4.8c shows the excellent flexible properties of CF, whereby two fibers could be knotted and bent spontaneously avoiding breaks. The diameter of the pristine CF is about 7 μm with a smooth surface, as shown in the inset of Figure 4.8c. Figure 4.8d. They showed that the high aspect ratio of MnNiCo-CH nanoneedles with uniformly covered the cylindrical surface of CF and tightly grown on the whole surface of CF. The as-fabricated FFSC was utilized for real-world practical application and the digital photographs of lightening the LEDs by connecting two FFSCs in series are shown in Figure 4.8e. Furthermore, the feasibility of the FFSCs was investigated as a wearable charge storage device by implementing them in a cotton fabric with combinations of parallel/series which can successfully lighten the LEDs, and wrapping at a human finger as displayed in Figure 4.8f. Seven FFSCs in embedded in a wool fabric attached to the arm of a human were used to power the blue "Q," white "D" and green "U" electronic LED screens, as shown in Figure 4.8g. Other ternary metal oxides/sulfides/selenides have also been used in the manufacture of SC fibers using similar processes [46, 56, 90, 94, 95]. Similarly, Liu and co-workers [49] recently fabricated the FFSC using asymmetrical

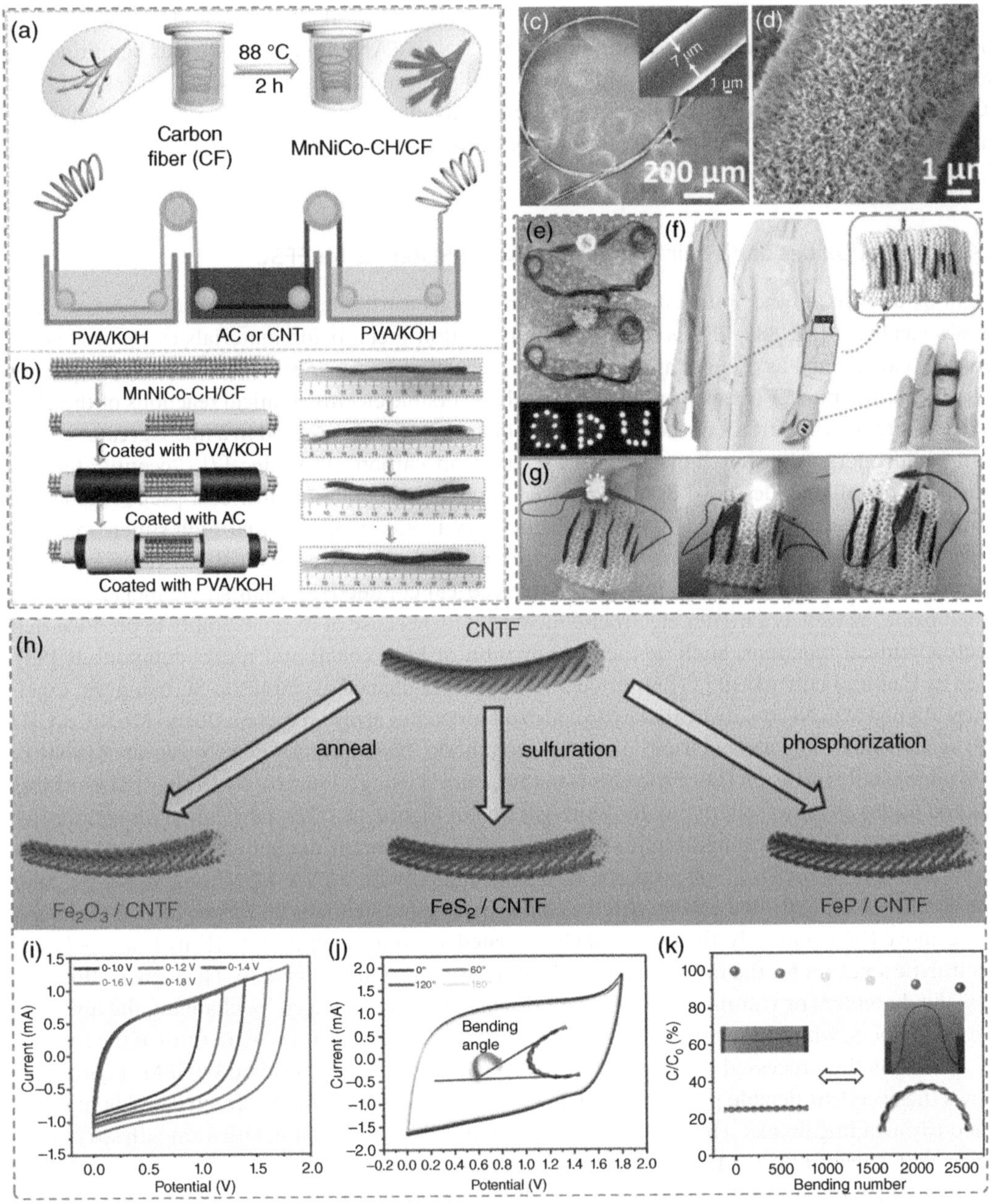

Figure 4.8 (a) Schematic diagrams for the fabrication of MnNiCo-CH nanoneedles array on carbon fiber by a hydrothermal process. (b) the construction of coaxial FFSCs device along with stepwise schemes (left) and photographs (right) of fabrication procedure for coaxial FFSCs. (c and d) Low- and high magnifications SEM images of MnNiCo-CH nanoneedles (inset shows the SEM image of pristine carbon fiber with diameter of 7 μm). (e) Real life applications of FFSCs as a flexible device by connecting two devices in series can power LEDs. (f) Seven FFSCs connected in a combination of series and parallel stitched into the cotton fabrics can power LED electronic display. (The bottom right photograph shows the flexibility of FFSC can rolled around a human finger). (g) Seven series and parallel combinations of FFSCs can lighten the "Q, D, U" LEDs cards, respectively. (a–g) *Source:* Reproduced with permission [37] © 2020, Royal Society of Chemistry. (h) Schematic description for the construction process of Fe_2O_3, FeS_2, and FeP NWAs/CNTFs, respectively. (i) CV curves of FeP/CNTF//MnO/CNTF FFSC at various potential windows and (j) bending states. (k) Capacitance retention of the FFSC while bended at angle of 180° for 2500 cycles. (The inset shows the schematic diagram and pictures of the FFSC at a bending angle of 90°). *Source:* Reproduced with permission [49] © 2020, Elsevier.

configuration based on the binder-free assembly of iron oxide/disulfide/phosphide nanoarrays on carbon fibers using hydrothermal method. The schematic illustration for the hydrothermal synthesis process was highlighted in a diagram, as shown in Figure 4.8h. Among these three samples, iron phosphide nanoarrays showed the highest performance and was further used to fabricate the FFSCs. The CV curves of as-fabricated FFSC at a fixed scan rate while varying the potential windows from 1.0 to 1.8 V are depicted in Figure 4.8i. Interestingly, the FFSC can operate at a high potential window of 0.0–1.8 V without any deformation in CV curves. Furthermore, the flexible performance of the FFSC was investigated by bending it at various bending states and recorded the CV curves, as shown in Figure 4.8j. The overlapped CV curves recorded at various bending states show the excellent performance of the FFSC device and the digital photograph for the measurement of a bending angle is displayed in the inset of Figure 4.8j. The stability in the flexibility characteristics was checked by running the FFSC over 2500 bending cycles and results are demonstrated in Figure 4.8k. The digital photographs at normal and bent state while recording the GCD curves were shown in the inset of Figure 4.8k. In addition to metal oxides, conductive polymers were also used as effective charge storage compounds including PANI [76], Ppy, and poly (3,4-ethylene dioxythiophene)/poly(styrene sulfonate) (PEDOT PSS) [96, 97]. These conductive polymers have the advantages of high capacitance, strong electrical conductivity, and easy fabrication. In several situations, such polymers are not being used individually. Instead, they can be designed into a composite material with other compounds such as CNTs/PANI, PEDOT: PSS/CNTs/graphene [96–99].

4.5 Electrolytes for Fiber-Shaped Supercapacitors

Electrolytes are another essential constituent of fiber-shaped SCs and when electrochemical reactions occur, they are correlated with the electrochemically stable potential window and the ionic transport [100–102]. Presently, electrolytes being used for energy storage devices are commonly categorized into two major classes: aqueous and non-aqueous electrolytes, the latter one also includes the ionic liquid electrolytes and organic liquid electrolytes [102]. These electrolytes have special advantages, disadvantages, and applicability. For instance, a redox reaction in aqueous conditions certainly causes a water-splitting response, such as oxygen-evolution reaction (OER) and hydrogen-evolution reaction (HER); therefore, the voltage output of aqueous SCs can hardly exceed that of organic or ionic liquid SCs [103, 104]. Nevertheless, aqueous SCs own their benefits and distinctiveness, including ease of fabrication, secure operation and low cost, even when non-aqueous SCs usually endure from safety problems exposed to harmful components and the risk of flammability that require stricter and more complex production methods. Taking the shape and state into account, electrolytes can also be grouped into three types: liquid, gel, and solid. Directly packing of liquid electrolytes along a fiber electrode is difficult in the manufacture of flexible, fiber-shaped SCs, leaking and short-circuiting problems are becoming more prevalent as these fibers SCs experience various malformations. In this situation, structurally powerful and effective separators managed to infiltrate with the liquid electrolyte with efficient formulation are extremely necessary to achieve good results of FFSCs. Besides, solid-state electrolytes, including some glasses, crystals, and ceramics, can effectively prevent unwanted short circuits between the metal conductors and minimize leakage possibility derived by liquid electrolytes [105]. However, it remains challenging to overcome their poor mechanical properties and the large interfacial resistance that stems from the limited contact among solid electrolytes and electrodes. Using gel polymer configurations as a structurally stable mechanism combined with ionic conductive additives, including all kinds of salt, has traditionally been the standard technique to meet gel electrolyte specifications for fiber-shaped SCs. Gel electrolytes are fundamentally ion-conductive compounds (e.g. acids or alkalis, ceramic salts, and ionic liquids) liquified into semi-solid polymer structures (e.g. poly(ethylene oxide), PEO and PVA) wherein ions could even drift easily [102, 106]. These electrolytes contribute to good mechanical strength and reliability; hence, they will improve the cyclic stability of the FFSCs, assembly coating, and hermetic sticking and helpful to reduce the volume of the FFSC. In addition, after coating the gel electrolytes on the surface of the fiber electrodes, it was not only act as a separator to avoid short circuits but also serve as a shielding layer to avoid the structure collapse of electrode material from the fiber surfaces while twisting, bending, and stretching during the testing of flexible performance [107].

4.6 Performance Evaluation Metrics for Fiber-Shaped Supercapacitors

Improving electrochemical performance is indeed an important step for the revelation of feasible FFSCs implementations. Specific capacity, energy density, power density, mechanical durability, and cycling reliability are some widely used output parameters.

FFSCs' energy density (E) depends on their capacity (C) and working potential window (V), which can be represented as follows: $E = 1/2CV^2$ [108]. As observed, improving efficiency or increasing the working potential window increases the energy density of FFSCs [95]. Figure 4.9 shows numerous traditional and the latest initiatives to improve the capacitance and extending the operating potential windows of FFSCs. Thus, these three factors play a critical role in a high-performance device structure [3]. A reliable FFSC needs appreciable cyclic stability over thousands of charge/discharge cycles with higher rate capability together with the optimized energy and power densities. FFSCs' power density (P) relies on their operating potential window and equivalent series resistance (R_S), as defined in the given formula: $P = V^2/4R_S$ [29]. Decreasing the total electrical impedance (i.e. the entirety of resistances of electrodes, electrolytes, and contact resistance between electrode and electrolyte) is a valuable technique for increasing the power densities. Additionally, kinetic polarization at specific voltages can also affect the power density, resulting in voltage losses at high current densities. Preceding literature reviews described the recommendations for enhancing SC's energy and power densities, as illustrated in Figure 4.9.

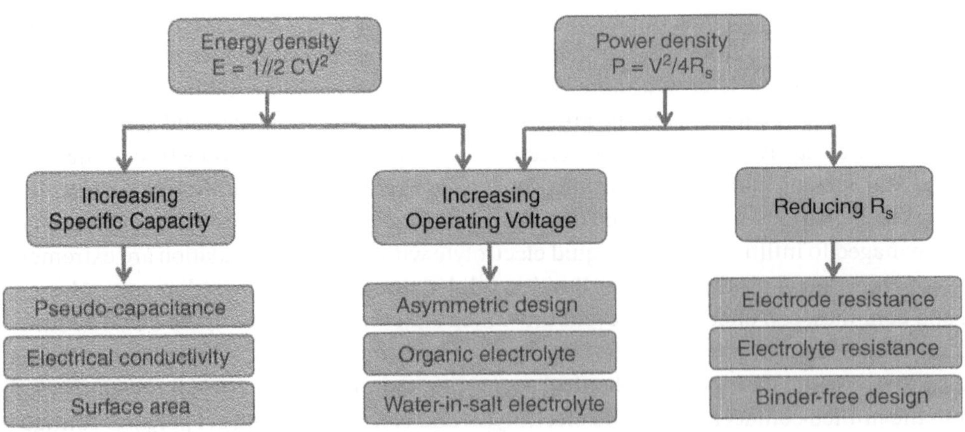

Figure 4.9 Summary of recent methodologies implemented to improve the energy and power densities of FFSCs.

Another important parameter to get high-performance is the balanced mass loading density of positive and negative electrodes for FFSC. To facilitate the robust electrochemical reaction at both ends of the electrodes, the charge balancing across the positive and negative electrodes is required. Following Eq. (4.1) is used to represent the stored charge for FFSCs [109, 110]

$$Q = C \times V \times m \tag{4.1}$$

where, Q, C, V, and m represents the charge, specific capacitance, working potential window, and mass of the active material [111]. According to the law of conservation of charge, i.e. $Q^+ = Q^-$, hence the mass density between positive and negative electrode can be calculated as follow [112]

$$C^+ \times V^+ \times m^+ = C^- \times V^- \times m^- \tag{4.2}$$

$$\frac{m^+}{m^-} = \frac{C^- \times V^-}{C^+ \times V^+} \tag{4.3}$$

where, Q^+, Q^-, C^+, C^-, V^+, V^-, $m^{+,}$ and m^- represents the stored charge, capacitance, working potential window, mass of positive and negative electrodes, respectively.

Besides, unique performance of longitudinal capacity (total capacity divides the length of devices) in FFSCs seems much important than their areal capacitance (typically, used mostly in sandwiched type 2D SCs) since the main restriction of FFSCs in the field of portable and wearable devices should be their length [113]. Thus, it should be more reasonable to provide this performance when discussing about FFSCs and even other fiber shape energy storage devices.

4.7 Applications

The wonderful advancement of FFSCs enables fully versatile embedded fiber electronic structures by bringing practical electronic devices and power supplies into one fiber [114, 115]. To date, following major techniques have been established to achieve the single multifunctional fiber network through integrating FFSC with solar or fuel cells and nanogenerators in a single fiber [33, 34, 116–118]. Integrating charge transmission and storage features into a single system is of tremendous significance to provide optimized energy wires that can eliminate redundant charging loops and external electrical connections [47, 119]. By producing dye-sensitized solar cells (DSSCs) and FFSCs on a single fiber, Peng and colleagues proposed an effective energy cable for both photoelectric conversion and electrochemical energy storage [117]. The digital photographs of the as-fabricated dual-functional device were shown in Figure 4.10a, while the low- and high-resolution SEM images of tritium nanotubes directly grown on the Ti wire and CNTs are shown in Figure 4.10b and c, respectively. Furthermore, the SEM images of both sides (FFSC unit and DSSCs unit) of the dual-functional cable (Figure 4.10e) were also depicted in Figure 4.10d and f. Figure 4.10g showed the circuit connection during the charge and discharge process. The integrated power wire would be charged upon exposure to light when the DSSC and FFSC were connected. The positively and negatively charges produced from the portion of DSSC were transferred toward the Ti wire and stored in the section of FFSC. Figure 4.10h displayed the charging and discharge curves of the built-in energy wire while subjected to AM 1.5 light radiation exposure. The voltage increased rapidly when exposed to light up to 0.6 V, close to the DSSC's open current voltage. It was determined that the overall photoelectric transmission and storage capacity efficiency was around 1.5%. The CV and GCD curves of alone FFSC are shown in Figure 4.10i and j, respectively both

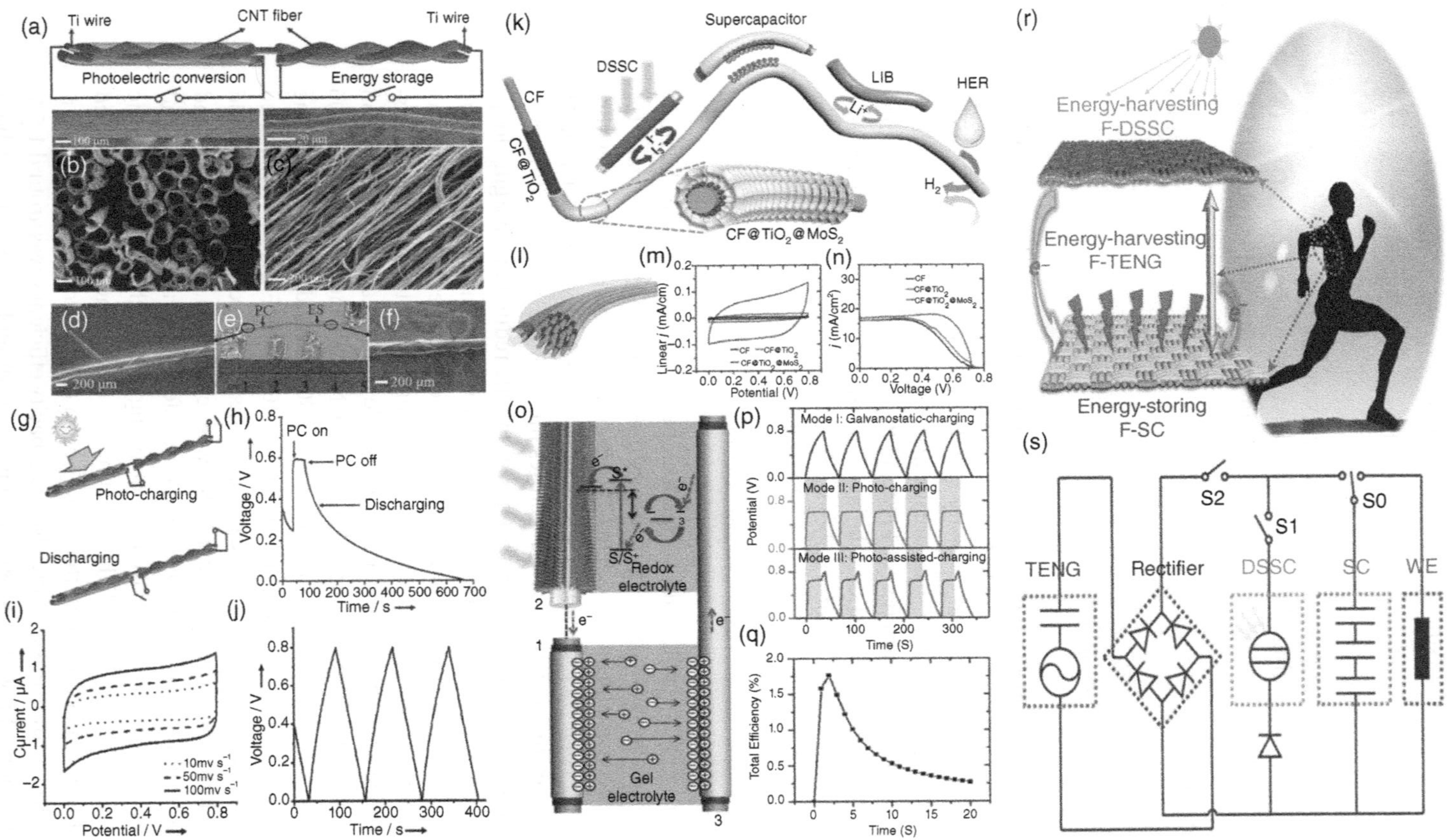

Figure 4.10 (a) Schematic design of the combined wire-shaped device for photoelectric energy conversion and storage (b) Low- and high magnifications SEM images of aligned titania nanotubes grown on a Ti wire, respectively. (c) Low- and high magnifications SEM images of a CNT fiber, respectively. (d) SEM image taken from PV part. (e) Structural photograph of a combined wire. (f) SEM image from SC part. (g) Schematic description of the circuit connection during charging and discharging processes. (h) Photo charging–discharging curve of energy wire. (i) CV and GCD curves of SC part wire, respectively. (a–j) *Source:* Reproduced with permission [117] © 2012, Wiley-VCH. (k) Schematic design of the structure of coaxial fiber CF@TiO_2@MoS_2 electrode and its multipurpose applications for energy harvesting and storage, together in one fiber-shaped DSSCs, FFSC, LIBs, and flexible HER electrode. (l) Schematic diagram of a solid-state symmetric FFSC with PVA/H_3PO_4 gel electrolyte. (m) CV curves of FFSC at a scan rate of 100 mVs^{-1} based on pristine CF, CF@TiO_2, and CF@TiO_2@MoS_2 electrodes. (n) J–V plots of fiber shaped DSSCs based on pristine CF, CF@TiO_2, and CF@TiO_2@MoS_2 counter electrodes. (o) Schematic description and working mechanism of a self-powering energy fiber created by combining a fiber-shaped DSSC and a FFSC using CF@TiO_2@MoS_2. (p) GC (Mode-I), photo-charging/GD (Mode-II) and photo-assisted-charging/GD (Mode-III) curves of the integrated energy fiber. The time periods of energy fiber under simulated AM1.5G (1 sun) light illumination are marked by light red areas. (q) Overall photochemical-electricity energy conversion efficiency of the self-powering energy fiber as a function of the photo-charging time. (k–q) *Source:* Reproduced with permission [120] © 2016. (r) Integrating solar cells and FFSCs into one fiber can collect solar energy (s) Electric model circuit diagram. *Source:* Reproduced with permission [121] © 2016 Science.

demonstrating excellent charge storage performance. Similarly, Liu and his colleagues recently proposed a MoS_2 based fiber for harvesting the energy along with a storage unit combined in one fiber [120]. The schematic illustration of the concept is presented in Figure 4.10k. Carbon fibers were first wrapped with TiO_2 nanoparticles and then processed by the development of MoS_2 nanosheets on their surface to get the active integrated energy fiber. It is verified from the experimental data that both the energy conversion portion and energy storage portion showed excellent results (Figure 4.10l–p). The DSSC section showed a 9.5% conversion efficiency and the FFSC section showed a longitudinal capacitance of 1740 $\mu F\,cm^{-1}$ and an aerial capacitance of 18.51 $mF\,cm^{-2}$. As a result, the energy conversion rate from photochemical to electrical of the integrated device was as high as 1.8% (Figure 4.10q). Furthermore, MoS_2-based fibers could also be used to fabricated fiber-based lithium-ion batteries and advanced hydrogen electrocatalytic reactions in a single fiber. This makes it easier to mount a more complex interconnected network on a single fiber.

Integrating solar cells and FFSCs into one fiber can collect solar energy efficiently. However, solar power is not available for 24 hours and also strongly depends on the environment, workplace conditions, etc. [122]. If solar energy is not available, then the system is unable to work. As a consequence, Wang and colleagues have successfully created an energy wire by combining nanogenerators, solar cells, and SCs into one fiber that will not only collect solar power but also extract mechanical energy from human motion, trying to reach the demand of flexible and wearable devices for next-generation electronics (Figure 4.10r) [121]. The proposed circuit map for the combination of three different kind devices is illustrated in Figure 4.10s, in which a bridge rectifier was employed to convert the current produced by nanogenerators from alternating current to direct current for charging the FFSC. Diodes and switches were used to block the current produced from nanogenerators which directly pass by DSSC and to control the circuits, respectively.

4.8 Conclusion and Future Prospectus

In summary, this chapter summarizes the recent developments in FFSCs. First, we discussed the emerging techniques for the fabrication of fiber-shaped electrodes and after that highlighted the three kinds of active materials for the fabrication of fiber electrodes, including carbon-based materials, single/bi-metal oxide-based materials, and composite/hybrid materials. The detailed discussion on structure/configuration, electrolytes, and performance evaluation metrics for the FFSCs were also summarized. Finally, the potential applications of the FFSCs were elaborated. Despite the recent advancement and excellent research accomplishments on the manufacture of FFSCs, many unresolved issues are remained to be addressed.

1) The electrochemical performance of FFSCs is still rather low as compared with the flexible planar SCs and also far from practical applications. Improving the energy densities while maintaining their inherently high-power densities are also critical for the production of FFSCs. Usually, enhancing the charge storage capacity and reducing the volume of the FFSCs are two major techniques for increasing energy densities.
2) Although many effective techniques have been explored for producing large-scale fiber substrates/electrodes. However, they are only suitable to quite limited materials, such as CNTs, graphene, and composite materials. Discovering and designing other flexible large-scale solutions would certainly involve a further study of the fiber substrates/electrodes.
3) Apart from CNTs or graphene fabrics, metal fabrics are another significant category of active electrode material need to expand or investigate. Nevertheless, the dimensions of most of the

metal fabrics are still too large, which significantly increases the strength but reduces the flexibility of the FFSCs. Exploring thin metal fabrics with optimal flexibility, conductivity, and mechanical strength will solve these problems and require more research attention.

4) Elastic fiber SCs are more desirable for real-world applications of FFSCs in flexible wearable electronics, as elastic fiber SCs do not only carry deformations in both directions but can also be bent and stretched without loss of output. Research on elastic fiber SCs is at an early stage and further in-depth investigations are needed. Just a few stretchable fibers substrates have been produced to date and elasticity is very restricted. It is also necessary to create new elastic fiber substrates.
5) For integrated systems with FFSCs only a few articles have been successfully proved so far. The interconnected systems should need even more attention. For the energy fibers, an alternative process should be proposed to improve the overall photoelectric conversion and storage efficiency, as the highest efficiency reported to date is still below 2%. In the near future, coupling and incorporation of FFSCs with fiber-based sensors to achieve extremely compressed optimized sensing systems for wearable electronics will be admired. This would develop a more effective production methodology to manufacture integrated systems with smaller size, weight, and increased performance.
6) One of the most appealing features of FFSCs is woven fiber SCs into cloth or other flexible substrates for large-scale applications. Presently, FFSCs will only be stitched with a very small radius by hand or by small machines but for the production of large-scale remains a challenging task. The main challenge is that the strength and mechanical durability at deformed state, as FFSCs are very far from traditional fabrics being used in industries. The production process and selection of materials (fiber-like current collector, electrode materials, electrolyte, and packing materials) should be systematically investigated to realize woven FFSCs based on the well-developed weaving technologies.

In addition to the aforementioned several critical problems that should be addressed, several other concerns should also need to be considered. Related safety problems of FFSCs that will be used to wear on human bodies should be handled seriously. Proper packing materials should also be carefully selected. Further issues, such as expenses and continued outcomes of FFSCs are also crucial for realizing the practical applications. Standard characterization procedure for assessing the performance among various types of FFSCs needs to be considered and is key for the development of next-generation FFSCs.

Acknowledgments

This work was financially supported by the Natural Science Foundation of Guangdong Province (2020A1515110537), National Natural Science Foundation of China (51772135), and the National Postdoctoral Program for Innovative Talents (BX20200150). Dr. M Imran thanks to the Deanship of Scientific Research at King Khalid University for funding through RGP. 2/3/42.

References

1 Mesbahi, T., Bartholomeus, P., Rizoug, N. et al. (2020). Advanced model of hybrid energy storage system integrating lithium-ion battery and supercapacitor for electric vehicle applications. *IEEE Trans. Ind. Electron.* 68: 3962.

2 Mastragostino, M., Arbizzani, C., Paraventi, R., and Zanelli, A. (2000). Polymer selection and cell Design for Electric-Vehicle Supercapacitors. *J. Electrochem. Soc.* 147: 407.

3 Gao, R., Tang, J., Yu, X. et al. (2020). Layered silicon-based Nanosheets as electrode for 4 V high-performance supercapacitor. *Adv. Funct. Mater.* 30 2002200.

4 García-Mateos, F.J., Ruiz-Rosas, R., Rosas, J.M. et al. (2020). Activation of electrospun lignin-based carbon fibers and their performance as self-standing supercapacitor electrodes. *Sep. Purif. Technol.* 241: 116724.

5 Bhattacharya, S., Roy, I., Tice, A. et al. (2020). High-conductivity and high-capacitance electrospun fibers for supercapacitor applications. *ACS Appl. Mater. Interfaces* 12: 19369–19376.

6 Javed, M.S., Lei, H., Shah, H.U. et al. (2019). Achieving high rate and high energy density in an all-solid-state flexible asymmetric pseudocapacitor through the synergistic design of binder-free 3D $ZnCo_2O_4$ nano polyhedra and 2D layered $Ti_3C_2T_x$-MXenes. *J. Mater. Chem. A* 7: 24543–24556.

7 Javed, M.S., Khan, A.J., Asim, S. et al. (2020). Insights to pseudocapacitive charge storage of binary metal-oxide nanobelts decorated activated carbon cloth for highly-flexible hybrid-supercapacitors. *J. Energy Storage* 31: 101602.

8 Javed, M.S., Shah, S.S.A., Najam, T. et al. (2019). Synthesis of mesoporous defective graphene-nanosheets in a space-confined self-assembled nanoreactor: highly efficient capacitive energy storage. *Electrochim. Acta* 305: 517–527.

9 Javed, M.S., Han, X., Hu, C. et al. (2016). Tracking pseudocapacitive contribution to superior energy storage of MnS nanoparticles grown on carbon textile. *ACS Appl. Mater. Interfaces* 8: 24621–24628.

10 Karaman, C., Bayram, E., Karaman, O., and Aktaş, Z. (2020). Preparation of high surface area nitrogen doped graphene for the assessment of morphologic properties and nitrogen content impacts on supercapacitors. *J. Electroanal. Chem.*: 114197.

11 Shrestha, L.K., Shrestha, R.G., Maji, S. et al. (2020). High surface area nanoporous graphitic carbon materials derived from Lapsi seed with enhanced supercapacitance. *Nanomaterials* 10: 728.

12 Zhao, J., Zhu, J., Li, Y. et al. (2020). Interfaces, graphene quantum dot reinforced electrospun carbon nanofiber fabrics with high surface area for ultrahigh rate supercapacitors. *ACS Appl. Mater. Interfaces* 12: 11669–11678.

13 Abbas, Y., Yun, S., Javed, M.S. et al. (2020). Anchoring 2D NiMoO4 nano-plates on flexible carbon cloth as a binder-free electrode for efficient energy storage devices. *Ceram. Int.* 46: 4470–4476.

14 Javed, M.S., Khan, A.J., Hanif, M. et al. (2020). Engineering the performance of negative electrode for supercapacitor by polyaniline coated Fe3O4 nanoparticles enables high stability up to 25,000 cycles. *Int. J. Hydrogen Energy*.

15 Javed, M.S., Jiang, Z., Zhang, C. et al. (2016). A high-performance flexible solid-state supercapacitor based on Li-ion intercalation into tunnel-structure iron sulfide. *Electrochim. Acta* 219: 742–750.

16 Javed, M.S., Dai, S., Wang, M. et al. (2015). Faradic redox active material of Cu_7S_4 nanowires with a high conductance for flexible solid state supercapacitors. *Nanoscale* 7: 13610–13618.

17 Shang, Y., Ma, S., Wei, Y. et al. (2020). Flower-like ternary metal of Ni-co-Mn hydroxide combined with carbon nanotube for supercapacitor. *Ionics*: 1–11.

18 Qiu, M.J., Sun, P., Shen, L.X. et al. (2016). WO_3 nanoflowers with excellent pseudo-capacitive performance and the capacitance contribution analysis. *J. Mater. Chem. A* 4: 7266–7273.

19 Liu, Y., Xu, X., and Shao, Z. (2020). Metal-organic frameworks derived porous carbon, metal oxides and metal sulfides-based compounds for supercapacitors application. *Energy Storage Mater.* 26: 1–22.

20 Chen, Y.L., Du, L.H., Yang, P.H. et al. (2015). Significantly enhanced robustness and electrochemical performance of flexible carbon nanotube-based supercapacitors by electrodepositing polypyrrole. *J. Power Sources* 287: 68–74.

21 Sun, P., Deng, Z.W., Yang, P.H. et al. (2015). Freestanding CNT-WO_3 hybrid electrodes for flexible asymmetric supercapacitors. *J. Mater. Chem. A* 3: 12076–12080.

22 Javed, M.S., Shah, S.S.A., Hussain, S. et al. (2020). Mesoporous manganese-selenide microflowers with enhanced electrochemical performance as a flexible symmetric 1.8 V supercapacitor. *Chem. Eng. J.* 382: 122814.

23 Javed, M.S., Aslam, M.K., Asim, S. et al. (2020). High-performance flexible hybrid-supercapacitor enabled by pairing binder-free ultrathin Ni–co–O nanosheets and metal-organic framework derived N-doped carbon nanosheets. *Electrochim. Acta*: 136384.

24 Yang, Z., Jia, Y., Niu, Y. et al. (2020). Wet-spun PVDF nanofiber separator for direct fabrication of coaxial fiber-shaped supercapacitors. *Chem. Eng. J.*: 125835.

25 Yang, J., Li, X.-L., Zhou, J.-W. et al. (2020). Fiber-shaped supercapacitors: advanced strategies toward high-performances and multi-functions. *Chin. J. Polym. Sci.*: 1–20.

26 Li, Y., Kang, Z., Yan, X. et al. (2017). A facile method for the preparation of three-dimensional CNT sponge and a nanoscale engineering design for high performance fiber-shaped asymmetric supercapacitors. *J. Mater. Chem. A* 5: 22559–22567.

27 Yu, C., Ma, P., Zhou, X. et al. (2014). All-solid-state flexible supercapacitors based on highly dispersed polypyrrole nanowire and reduced graphene oxide composites. *ACS Appl. Mater.* 6: 17937–17943.

28 Wang, R., Yao, M., and Niu, Z. (2020). Smart supercapacitors from materials to devices. *InfoMat* 2: 113–125.

29 Yu, D., Qian, Q., Wei, L. et al. (2015). Emergence of fiber supercapacitors. *Chem. Soc. Rev.* 44: 647–662.

30 Banerjee, S., De, B., Sinha, P. et al. (2020). Applications of supercapacitors. In: *Handbook of Nanocomposite Supercapacitor Materials II*, 341–350. Springer.

31 Chen, X., Villa, N.S., Zhuang, Y. et al. (2020). Stretchable supercapacitors as emergent energy storage units for health monitoring bioelectronics. *Adv. Energy Mater.* 10: 1902769.

32 Teng, W., Zhou, Q., Wang, X. et al. (2020). Hierarchically interconnected conducting polymer hybrid fiber with high specific capacitance for flexible fiber-shaped supercapacitor. *Chem. Eng. J.* 390: 124569.

33 Varma, S.J., Sambath Kumar, K., Seal, S. et al. (2018). Fiber-type solar cells, nanogenerators, batteries, and supercapacitors for wearable applications. *Adv. Sci.* 5: 1800340.

34 Fu, Y., Wu, H., Ye, S. et al. (2013). Integrated power fiber for energy conversion and storage. *Energy Environ. Sci.* 6: 805–812.

35 Zhang, Z., Chen, X., Chen, P. et al. (2014). Integrated polymer solar cell and electrochemical supercapacitor in a flexible and stable fiber format. *Adv. Mater.* 26: 466–470.

36 Yuan, K., Hu, T., and Chen, Y.J.F. (2020). Flexible and wearable solar cells and supercapacitors. In: *Flexible and Wearable Electronics for Smart Clothing*, 87–129.

37 Cao, X., Liu, Y., Zhong, Y. et al. (2020). Flexible coaxial fiber-shaped asymmetric supercapacitors based on manganese, nickel co-substituted cobalt carbonate hydroxides. *J. Mater. Chem. A* 8: 1837–1848.

38 Liu, S., Xu, C., Yang, H. et al. (2020). Atomic modulation triggering improved performance of MoO3 Nanobelts for fiber-shaped supercapacitors. *Small* 16: 1905778.

39 He, N., Patil, S., Qu, J. et al. (2020). Effects of electrolyte mediation and MXene size in fiber-shaped supercapacitors. *ACS Appl. Energy Mater.* 3: 2949–2958.

40 Yang, Z., Jia, Y., Niu, Y. et al. (2020). One-step wet-spinning assembly of twisting-structured graphene/carbon nanotube fiber supercapacitor. *J. Energy Chem.* 55: 434.

41 Cai, S., Huang, T., Chen, H. et al. (2017). Wet-spinning of ternary synergistic coaxial fibers for high performance yarn supercapacitors. *J. Mater. Chem. A* 5: 22489–22494.

42 Vigolo, B., Penicaud, A., Coulon, C. et al. (2000). Macroscopic fibers and ribbons of oriented carbon nanotubes. *Science* 290: 1331–1334.

43 Ren, C., Yan, Y., Sun, B. et al. (2020). Wet-spinning assembly and in situ electrodeposition of carbon nanotube-based composite fibers for high energy density wire-shaped asymmetric supercapacitor. *J. Colloid Interface Sci.* 569: 298–306.

44 Yuan, D., Li, B., Cheng, J. et al. (2016). Twisted yarns for fiber-shaped supercapacitors based on wetspun PEDOT: PSS fibers from aqueous coagulation. *J. Mater. Chem. A* 4: 11616–11624.

45 Li, Y., Zhong, C., Liu, J. et al. (2018). Atomically thin mesoporous Co3O4 layers strongly coupled with N-rGO Nanosheets as high-performance bifunctional catalysts for 1D knittable zinc–air batteries. *Adv. Mater.* 30: 1703657.

46 Yang, Z., Yang, Y., Lu, C.-x. et al. (2019). A high energy density fiber-shaped supercapacitor based on zinc-cobalt bimetallic oxide nanowire forests on carbon nanotube fibers. *New Carbon Mater.* 4: 559–568.

47 Sun, P., Lin, R., Wang, Z. et al. (2017). Rational design of carbon shell endows TiN@C nanotube based fiber supercapacitors with significantly enhanced mechanical stability and electrochemical performance. *Nano Energy* 31: 432–440.

48 Zhang, J., Seyedin, S., Qin, S. et al. (2019). Highly conductive Ti3C2Tx MXene hybrid fibers for flexible and elastic fiber-shaped supercapacitors. *Small* 15: 1804732.

49 Liu, C., Li, Q., Zhang, Q. et al. (2020). Surface-functionalized Fe2O3 nanowire arrays with enhanced pseudocapacitive performance as novel anode materials for high-energy-density fiber-shaped asymmetric supercapacitors. *Electrochim. Acta* 330: 135247.

50 Kang, K.-N., Ramadoss, A., Min, J.-W. et al. (2020). Wire-shaped 3D-hybrid supercapacitors as substitutes for batteries. *Nano Lett.* 12: 28.

51 Xu, R., Guo, F., Cui, X. et al. (2015). High performance carbon nanotube based fiber-shaped supercapacitors using redox additives of polypyrrole and hydroquinone. *J. Mater. Chem. A* 3: 22353–22360.

52 Jin, H., Zhou, L., Mak, C.L. et al. (2015). High-performance fiber-shaped supercapacitors using carbon fiber thread (CFT)@ polyanilne and functionalized CFT electrodes for wearable/stretchable electronics. *Nano Energy* 11: 662–670.

53 Lamberti, A., Gigot, A., Bianco, S. et al. (2016). Self-assembly of graphene aerogel on copper wire for wearable fiber-shaped supercapacitors. *Carbon* 105: 649–654.

54 Le, T.S., Truong, T.K., Bae, J., and Suh, D. (2020). Synergetic design of enlarged surface area and pseudo-capacitance for fiber-shaped supercapacitor yarn. *Nano Energy* 67: 104198.

55 Sun, J., Zhang, Q., Wang, X. et al. (2017). Constructing hierarchical dandelion-like molybdenum–nickel–cobalt ternary oxide nanowire arrays on carbon nanotube fiber for high-performance wearable fiber-shaped asymmetric supercapacitors. *J. Mater. Chem. A* 5: 21153–21160.

56 Wang, X., Zhang, Q., Sun, J. et al. (2018). Facile synthesis of hierarchical porous manganese nickel cobalt sulfide nanotube arrays with enhanced electrochemical performance for ultrahigh energy density fiber-shaped asymmetric supercapacitors. *J. Mater. Chem. A* 6: 8030–8038.

57 Guo, J., Zhang, Q., Sun, J. et al. (2018). Direct growth of vanadium nitride nanosheets on carbon nanotube fibers as novel negative electrodes for high-energy-density wearable fiber-shaped asymmetric supercapacitors. *J. Power Sources* 382: 122–127.

58 Liu, W., Feng, K., Zhang, Y. et al. (2017). Hair-based flexible knittable supercapacitor with wide operating voltage and ultra-high rate capability. *Nano Energy* 34: 491–499.

59 Zhou, Z., Zhang, Q., Sun, J. et al. (2018). Metal–organic framework derived spindle-like carbon incorporated α-Fe2O3 grown on carbon nanotube fiber as anodes for high-performance wearable asymmetric supercapacitors. *ACS Nano* 12: 9333–9341.

60 Zhao, J., Li, H., Li, C. et al. (2018). MOF for template-directed growth of well-oriented nanowire hybrid arrays on carbon nanotube fibers for wearable electronics integrated with triboelectric nanogenerators. *Nano Energy* 45: 420–431.

61 Zhang, Q., Zhou, Z., Pan, Z. et al. (2018). All-metal-organic framework-derived battery materials on carbon nanotube fibers for wearable energy-storage device. *ACS Nano* 5: 1801462.

62 Salunkhe, R.R., Kaneti, Y.V., and Yamauchi, Y. (2017). Metal–organic framework-derived nanoporous metal oxides toward supercapacitor applications: progress and prospects. *ACS Nano* 11: 5293–5308.

63 Zeng, Y., Meng, Y., Lai, Z. et al. (2017). An ultrastable and high-performance flexible fiber-shaped Ni–Zn battery based on a Ni–NiO heterostructured nanosheet cathode. *Adv. Mater.* 29: 1702698.

64 Zhang, K., Lv, W., Chen, J. et al. (2019). Synthesis of RGO/AC/Fe3O4 composite having 3D hierarchically porous morphology for high effective electromagnetic wave absorption. *Composites Part B* 169: 1–8.

65 Pan, Z., Yang, J., Li, L. et al. (2020). All-in-one stretchable coaxial-fiber strain sensor integrated with high-performing supercapacitor. *Energy Storage Mater.* 25: 124–130.

66 Liu, Y., Narayanasamy, M., Yang, C. et al. (2019). High-performance coaxial wire-shaped supercapacitors using ionogel electrolyte toward sustainable energy system. *J. Mater. Res.* 34: 3030–3039.

67 Hua, Y., Guang, W., Yuxing, Z. et al. (2020). A stretchable, asymmetric, coaxial fiber-shaped supercapacitor for wearable electronics. *Nano Res.*

68 Patil, B., Ahn, S., Yu, S. et al. (2018). Electrochemical performance of a coaxial fiber-shaped asymmetric supercapacitor based on nanostructured MnO2/CNT-web paper and Fe2O3/carbon fiber electrodes. *Carbon* 134: 366–375.

69 Zhang, X., Chen, X., Bai, T. et al. (2020). A simple route to fiber-shaped heterojunctioned nanocomposites for knittable high-performance supercapacitors. *J. Mater. Chem. A.*

70 Yu, Z., Moore, J., Calderon, J. et al. (2015). Coil-type asymmetric supercapacitor electrical cables. *Small* 11: 5289–5295.

71 Chen, X., Qiu, L., Ren, J. et al. (2013). Novel electric double-layer capacitor with a coaxial fiber structure. *Adv. Mater.* 25: 6436–6441.

72 Zhang, Q., Xu, W., Sun, J. et al. (2017). Constructing ultrahigh-capacity zinc–nickel–cobalt oxide@ Ni (OH) 2 core–shell nanowire arrays for high-performance coaxial fiber-shaped asymmetric supercapacitors. *Nano Lett.* 17: 7552–7560.

73 Khan, A.J., Khan, A., Javed, M.S. et al. (2020). Surface assembly of Fe3O4 nanodiscs embedded in reduced graphene oxide as a high-performance negative electrode for supercapacitors. *Ceram. Int.*

74 Zhang, Q., Sun, J., Pan, Z. et al. (2017). Stretchable fiber-shaped asymmetric supercapacitors with ultrahigh energy density. *Nano Energy* 39: 219–228.

75 Bi, S., Banda, H., Chen, M. et al. (2020). Molecular understanding of charge storage and charging dynamics in supercapacitors with MOF electrodes and ionic liquid electrolytes. *Nat. Mater.* 19: 552–558.

76 Yu, S., Patil, B., and Ahn, H.J.F. (2020). Polymers, PANI//MoO 3 fiber-shaped asymmetric supercapacitors with roll-type configuration. *Fibers Polym.* 21: 465–472.

77 Yu, S., Patil, B., and Ahn, H. (2019). Flexible, fiber-shaped supercapacitors with roll-type assembly. *J. Ind. Eng. Chem.* 71: 220–227.

78 Cho, S., Patil, B., Yu, S. et al. (2018). Flexible, Swiss roll, fiber-shaped, asymmetric supercapacitor using MnO2 and Fe2O3 on carbon fibers. *Electrochim. Acta* 269: 499–508.

79 Zhai, S., Karahan, H.E., Wang, C. et al. (2020). 1D supercapacitors for emerging electronics: current status and future directions. *Adv. Mater.* 32: 1902387.

80 Chen, D., Jiang, K., Huang, T., and Shen, G. (2020). Recent advances in fiber supercapacitors: materials, device configurations, and applications. *Adv. Mater.* 32: 1901806.

81 Sun, H., You, X., Deng, J. et al. (2014). Novel graphene/carbon nanotube composite fibers for efficient wire-shaped miniature energy devices. *Adv. Mater.* 26: 2868–2873.

82 Ren, J., Bai, W., Guan, G. et al. (2013). Flexible and weaveable capacitor wire based on a carbon nanocomposite fiber. *Adv. Mater.* 25: 5965–5970.

83 Ma, Y., Li, P., Sedloff, J.W. et al. (2015). Conductive graphene fibers for wire-shaped supercapacitors strengthened by unfunctionalized few-walled carbon nanotubes. *ACS Nano* 9: 1352–1359.

84 Meng, Y., Zhao, Y., Hu, C. et al. (2013). All-graphene core-sheath microfibers for all-solid-state, stretchable fibriform supercapacitors and wearable electronic textiles. *Adv. Mater.* 25: 2326–2331.

85 Zheng, X., Zhang, K., Yao, L. et al. (2018). Hierarchically porous sheath–core graphene-based fiber-shaped supercapacitors with high energy density. *J. Mater. Chem. A* 6: 896–907.

86 Wang, K., Meng, Q., Zhang, Y. et al. (2013). High-performance two-ply yarn supercapacitors based on carbon nanotubes and polyaniline nanowire arrays. *Adv. Mater.* 25: 1494–1498.

87 Gao, L., Surjadi, J.U., Cao, K. et al. (2017). Flexible fiber-shaped supercapacitor based on nickel–cobalt double hydroxide and pen ink electrodes on metallized carbon fiber. *ACS Appl. Mater. Interfaces* 9: 5409–5418.

88 Yin, Q., Li, D., Zhang, J. et al. (2020). Compounds, an all-solid-state fiber-type supercapacitor based on hierarchical Ni/NiO@ CoNi-layered double hydroxide core-shell nanoarrays. *J. Alloys Compd.* 813: 152187.

89 Baek, S.-H., Jeong, Y.-M., Kim, D.Y., and Park, I.-K. (2020). Phase transformation of NiCo hydroxides derived from carbonate anion and its effect on electrochemical pseudocapacitor performance. *Chem. Eng. J.*: 124713.

90 Deka, B.K., Hazarika, A., Kim, J. et al. (2019). Bimetallic copper cobalt selenide nanowire-anchored woven carbon fiber-based structural supercapacitors. *Chem. Eng. J.* 355: 551–559.

91 Abdah, M.A.A.M., Azman, N.H.N., Kulandaivalu, S., and Sulaiman, Y. (2020). Review of the use of transition-metal-oxide and conducting polymer-based fibres for high-performance supercapacitors. *Mater. Des.* 186: 108199.

92 Javed, M.S., Zhang, C., Chen, L. et al. (2016). Hierarchical mesoporous NiFe 2 O 4 nanocone forest directly growing on carbon textile for high performance flexible supercapacitors. *J. Mater. Chem. A* 4: 8851–8859.

93 Dubal, D.P., Gomez-Romero, P., Sankapal, B.R., and Holze, R. (2015). Nickel cobaltite as an emerging material for supercapacitors: an overview. *Nano Energy* 11: 377–399.

94 Zhang, J., Liu, X., Yin, Q. et al. (2019). CoNi2S4 Nanoplate arrays derived from hydroxide precursors for flexible fiber-shaped supercapacitors. *ACS Omega* 4: 11863–11870.

95 Javed, M.S., Lei, H., Li, J. et al. (2019). Construction of highly dispersed mesoporous bimetallic-sulfide nanoparticles locked in N-doped graphitic carbon nanosheets for high energy density hybrid flexible pseudocapacitors. *J. Mater. Chem. A* 7: 17435–17445.

96 Wang, Z., Zhu, M., Pei, Z. et al. (2020). Polymers for supercapacitors: boosting the development of the flexible and wearable energy storage. *Mater. Sci. Eng. R Rep.* 139: 100520.

97 Zhao, C., Jia, X., Shu, K. et al. (2020). Conducting polymer composites for unconventional solid-state supercapacitors. *J. Mater. Chem. A* 8: 4677–4699.

98 Dirican, M., Yanilmaz, M., Asiri, A.M., and Zhang, X. (2020). Polyaniline/MnO2/porous carbon nanofiber electrodes for supercapacitors. *J. Electroanal. Chem.*: 113995.

99 Wang, Y., Ding, Y., Guo, X., and Yu, G. (2019). Conductive polymers for stretchable supercapacitors. *J. Electroanal. Chem.*: 1–10.

100 Guo, L., Ma, W.-B., Wang, Y. et al. (2020). A chemically crosslinked hydrogel electrolyte based all-in-one flexible supercapacitor with superior performance. *J. Alloys Compd.*: 155895.

101 Liu, J., Huang, J., Cai, Q. et al. (2020). Design of Slidable Polymer Networks: a rational strategy to stretchable, rapid self-healing hydrogel electrolytes for flexible supercapacitors. *ACS Appl. Mater. Interfaces* 12: 20479–20489.

102 Alipoori, S., Mazinani, S., Aboutalebi, S.H., and Sharif, F. (2020). Review of PVA-based gel polymer electrolytes in flexible solid-state supercapacitors: opportunities and challenges. *J. Energy Storage* 27: 101072.

103 Du, H., Wu, Z., Xu, Y. et al. (2020). Poly (3, 4-ethylenedioxythiophene) based solid-state polymer supercapacitor with ionic liquid gel polymer electrolyte. *Polymers* 12: 297.

104 Liu, Z., Huang, Y., Huang, Y. et al. (2020). Voltage issue of aqueous rechargeable metal-ion batteries. *Chem. Soc. Rev.* 49: 180–232.

105 Yadav, N., Yadav, N., and Hashmi, S. (2020). Ionic liquid incorporated, redox-active blend polymer electrolyte for high energy density quasi-solid-state carbon supercapacitor. *J. Power Sources* 451: 227771.

106 Wang, Z., Li, H., Tang, Z. et al. (2018). Hydrogel electrolytes for flexible aqueous energy storage devices. *Adv. Funct. Mater.* 28: 1804560.

107 Lewandowski, A., Olejniczak, A., Galinski, M., and Stepniak, I. (2010). Performance of carbon–carbon supercapacitors based on organic, aqueous and ionic liquid electrolytes. *J. Power Sources* 195: 5814–5819.

108 Gao, Y., Zhang, J., Luo, X. et al. (2020). Energy density-enhancement mechanism and design principles for heteroatom-doped carbon supercapacitors. *Nano Energy*: 104666.

109 Balasubramaniam, S., Mohanty, A., Balasingam, S.K. et al. (2020). Comprehensive insight into the mechanism, material selection and performance evaluation of Supercapatteries. *Nanomicro Lett.* 12: 1–46.

110 Ye, M., Guo, H., Dou, Q. et al. (2020). Optimization of electrode potential ranges for constructing 4.0 V carbon-based supercapacitors. *ChemElectroChem* 7: 624–630.

111 Abbas, Q., Fitzek, H., Pavlenko, V., and Gollas, B. (2020). Towards an optimized hybrid electrochemical capacitor in iodide based aqueous redox-electrolyte: shift of equilibrium potential by electrodes mass-balancing. *Electrochim. Acta* 337: 135785.

112 Zhang, Y., Hu, H., Wang, Z. et al. (2020). Boosting the performance of hybrid supercapacitors through redox electrolyte-mediated capacity balancing. *Nano Energy* 68: 104226.

113 Peng, H. (2015). *Fiber-Shaped Energy Harvesting and Storage Devices*. Springer.

114 Deng, J., Zhang, Y., Zhao, Y. et al. (2015). A shape-memory supercapacitor fiber. *Angew. Chem.* 54: 15419–15423.

115 Huang, Y., Zhu, M., Pei, Z. et al. (2016). A shape memory supercapacitor and its application in smart energy storage textiles. *J. Mater. Chem. A* 4: 1290–1297.

116 Lv, T., Yao, Y., Li, N., and Chen, T. (2016). Wearable fiber-shaped energy conversion and storage devices based on aligned carbon nanotubes. *Nano Today* 11: 644–660.

117 Chen, T., Qiu, L., Yang, Z. et al. (2012). An integrated "energy wire" for both photoelectric conversion and energy storage. *ChemElectroChem* 51: 11977–11980.

118 Chai, Z., Zhang, N., Sun, P. et al. (2016). Tailorable and wearable textile devices for solar energy harvesting and simultaneous storage. *ACS Nano* 10: 9201–9207.

119 Yu, Z. and Thomas, J. (2014). Energy storing electrical cables: integrating energy storage and electrical conduction. *Adv. Mater.* 26: 4279–4285.

120 Liang, J., Zhu, G., Wang, C. et al. (2017). MoS2-based all-purpose fibrous electrode and self-powering energy fiber for efficient energy harvesting and storage. *Adv. Energy Mater.* 7: 1601208.

121 Wen, Z., Yeh, M.-H., Guo, H. et al. (2016). Self-powered textile for wearable electronics by hybridizing fiber-shaped nanogenerators, solar cells, and supercapacitors. *Sci. Adv.* 2: e1600097.

122 Chen, J., Huang, Y., Zhang, N. et al. (2016). Micro-cable structured textile for simultaneously harvesting solar and mechanical energy. *Nat. Energy* 1: 1–8.

5

Flexible Supercapacitors Based on Ternary Metal Oxide (Sulfide, Selenide) Nanostructures

Qiufan Wang[1], Daohong Zhang[1] and Guozhen Shen[2,3]

[1]Key laboratory of Catalysis and Energy Materials Chemistry of Ministry of Education & Hubei Key Laboratory of Catalysis and Materials Science, Hubei R&D Center of Hyperbranched Polymers Synthesis and Applications, South-Central University for Nationalities, 182 Minzu Road, Wuhan, 430074, China
[2]State Key Laboratory for Superlattices and Microstructures, Institution of Semiconductors, Chinese Academy of Sciences, A35 East Qinghua Road, Beijing, 100083, China
[3]College of Materials Science and opto-electronics Technology, university of Chinese academy of Sciences, 19 Yuquan Road, Beijing, 100029, China

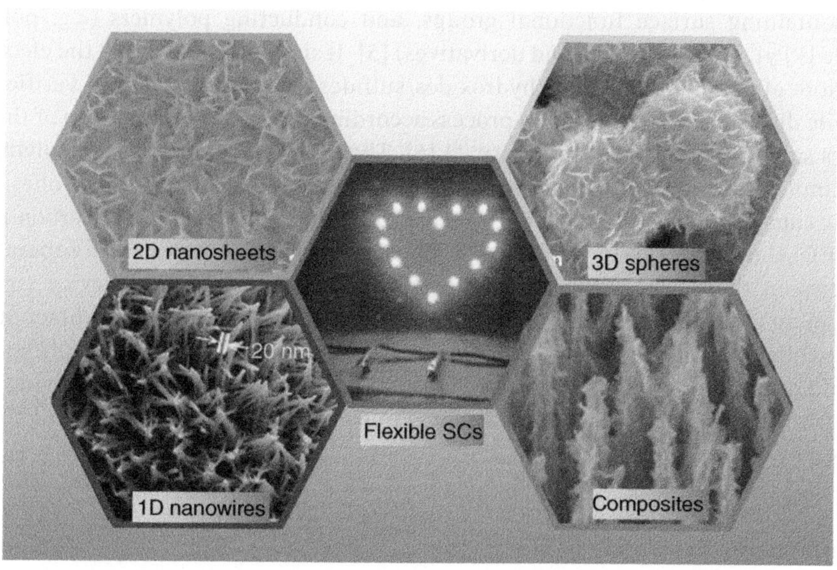

5.1 Introduction

5.1.1 Background of Electrochemical Capacitors

Owing to the limited crude oil storage and global warming situation, electrochemical devices for energy storage play a vital role in overcoming fossil fuel exhaustion in this energy-dependent world. Among various electrochemical energy storage devices, supercapacitors (SCs) have attracted significant interest as power source due to their fast charge–discharge capability, high

Flexible Supercapacitors: Materials and Applications, First Edition. Edited by Guozhen Shen, Zheng Lou and Di Chen.

power density, safe operation and long cycle life [1, 2]. Currently, SCs are used extensively in many applications, such as consumer electronics, transportation, military and aerospace, grid balancing, and power backup. Accordingly, SCs should be a promising alternative or complement to rechargeable batteries when high power delivery and/or fast energy harvesting are required.

The SC can be divided into two main types based on the difference in the storage mechanism of electric energy, e.g. electrochemical double layer capacitor (EDLC) and pseudocapacitor [3]. The former one is similar to a traditional capacitor, the mechanism of EDLCs arises from charge accumulation in the electric double layer formed at the electrode/electrolyte interface. Charge separation is a physical process without any faradaic reactions on the electrode surface. The specific capacitance of an EDLC depends on the surface properties of the electrodes, such as pore size distribution and specific surface area, which is several orders of magnitude higher than a traditional capacitor. Carbon materials including active carbon, mesoporous carbon, CNT (carbon nanotube), and graphene with a large surface area and low matrix resistivity are typical electrode materials for EDLCs [4]. The pseudocapacitors are based on fast and reversible redox reactions or Faradic charge transfer reactions at the surface of electro-active species, for example, transition metal oxides (e.g. Co_3O_4, NiO, RuO_2, and V_2O_5), transition metal sulfides (e.g. MoS_2), transition metal selenides (NiSe), carbon materials possessing oxygen- and nitrogen-containing surface functional groups, and conducting polymers (e.g. polyaniline, polypyrrole [PPy], polythiophene, and derivatives) [5]. It should be noted that the electrochemical behaviors of some metal oxides/hydroxides/sulfides/selenides have been verified to be a semi-infinite diffusion-limited reaction process according to a linear relationship of the current peak (i_p) vs square root of the scan rate ($v^{1/2}$) [6]. These electrode materials, involving typical diffusion-limited redox reactions, should be battery-type ones instead of SC-type ones. Notably, the specific capacitances of pseudocapacitors are much higher than those of carbon materials using the EDLC mechanism. In general, a SC consists of two electrodes and a separator sandwiched between them. The sandwich-like electrode/separator/electrode device structure is immersed in an aqueous or organic electrolyte. The separator prevents the contact of the two electrodes while allowing electrolyte ions freely passing through. To maintain sufficient functional liquid electrolyte in the device from leakage that is also harmful to our living environment, the whole system needs to be encapsulated in packaging such as box-like and button-like containers.

5.1.2 Performance Evaluation of SCs

The normal way to carry out material capacitance measurements is to coat the chosen material onto an inert electrode surface, then measure this electrode in the chosen electrolyte using electrochemical cyclic voltammetry (CV) to record cyclic voltammograms of the material, from which the capacitance can be calculated.

When two electrodes, positive electrode with a capacitance C_p and negative electrode with a capacitance C_n, are combined into a SC, the overall capacitance C_T of the entire cell is:

$$1/C_T = 1/C_p + 1/C_n \tag{5.1}$$

If the two electrodes are the same, $C_p = C_n$, the SC is symmetric. If $C_p \neq C_n$, the SC is asymmetric, in which case, C_T is dominated by the electrode with smaller capacitance.

The specific capacitance of a SC cell can be calculated using the voltammetric charge integrated from its CV curve:

$$C_{cell} = Q / 2m\Delta V = \int IdV / 2mv\Delta V \tag{5.2}$$

where C_{cell} is the specific capacitance of the cell, Q is the total charge obtained by integrating the positive and negative sweeps in a CV curve, m is the mass of the active materials in the two electrodes, v is the scan rate, and ΔV is the potential window between the two electrodes.

Cell capacitance can also be calculated from the galvanostatic charge–discharge curve:

$$C_{cell} = I \times \Delta t / m \times \Delta V \tag{5.3}$$

in which I is the discharge current, m is the total mass of the active materials in the two electrodes, ΔV is the applied potential, and Δt is the discharge time.

Energy density and power density are two key performance indicators for SCs. They can be calculated from the discharge curve using the following equations:

$$P = 1 / 2C_{cell} \times \Delta V \tag{5.4}$$

$$P = E / \Delta t \tag{5.5}$$

where E is the energy density, P is the power density, C_{cell} is the specific capacitance, ΔV is the applied potential, and Δt is the discharge time.

5.2 Ternary Metal Oxide

Among various functional nanomaterials, ternary transition metal oxides (designated as $A_xB_{3-x}O_4$; A, B=Zn, Ni, Co, Fe, Mn, etc.) have been widely used as high-performance electrode materials in SC and oxygen electrocatalysis due to their low-cost, environmentally friendly, rich redox chemistry and good structural stability. They can provide multiple redox reactions and they exhibit higher electrical conductivity, owing to the relatively low activation energy or electron transfer between the metal cations [7]. For example, $NiMoO_4$ [8], $NiCo_2O_4$ [9], $MnCo_2O_4$ [10], $CuCo_2O_4$ [11], and $Ni_{0.25}Mn_{0.75}O$ [12] have been previously prepared for SC electrode materials, which exhibited excellent capacitance characteristics. Moreover, ternary metal oxides can normally have various crystal structures, such as spinel and scheelite phases, which feature the electrochemical activity of synergistic effect for their SC performance. It has been previously reported that the morphology, particle size, the type of metal ions, and the crystal structure can play essential roles in determining the electrochemical performance. In this regard, selecting a suitable and effective ternary metal oxide is of enormous significance to enhance the performance of SCs.

5.2.1 1D Ternary Metal Oxide Nanostructured Electrodes

The continuous network in 1-D nanostructures including nanowires, nanorods, nanobelts, nanotubes, and nanoneedles can act as the "highway" for charge transport along their longitudinal direction and the shortened ion diffusion length in 1-D nanostructures results in the increase of rate performance when applied in electrochemical energy storage devices. Moreover, 1-D nanostructures with relatively large surface areas, can form porous network microstructure that acts as the scaffold/support of metal oxides [1], which facilitates the mass transport of the solvated ions in the electrolyte to the surface of electrodes. Li et al. reported $MnCo_2O_4$ nanowire arrays (NWAs) as the electrode materials for SC, which delivered an improved capacitance of 349.8 $F g^{-1}$ at 1 $A g^{-1}$ and high specific discharge capacity of 1288.6 $mAh g^{-1}$ at 100 $mA g^{-1}$ [13], Xu et al. synthesized porous $MnCo_2O_4$ nanowires as electrode for SC, which exhibited a remarkable specific capacitance of 1342 $F g^{-1}$ at 1 $A g^{-1}$ [14]. The Shen group has reported the growth of $ZnCo_2O_4$ nanorods on nickel foam through a polyol refluxing process followed by thermal treatment. The as-grown nanostructures have outstanding electrochemical performance in SCs with a high specific capacitance (~1400 $F g^{-1}$ at 1 $A g^{-1}$), excellent rate capability (72.5% capacity retention at 20 $A g^{-1}$) and good cycling stability (only 3% loss after 1000 cycles at 6 $A g^{-1}$) [15].

Self-branched hollow nanostructured composites is one of the most promising strategies for achieving numerous electrochemical active sites, which led to the faster diffusion of electrolyte ions, better structural and electrochemical stabilities compared to their solitary parts. Researchers have taken the effort to explore the fabrication of self-branched hollow nanostructured composites as electrode materials for high-performance SCs.

Metal-organic frameworks (MOFs)-derived nanomaterials can provide a unique set of physical and chemical properties with special morphologies/nanostructures, high specific surface areas and more abundant active sites. Yang et al. [16] reported the fabrication of ultrathin zinc-cobalt oxide nanoflakes@N-doped carbon hollow nanowall arrays from vertically aligned Co-MOF solid nanowall arrays by controllable cation ion-exchange and post annealing strategies. The synthetic strategy is schematically shown in Figure 5.1a. The ASC (asymmetric supercapacitor) device can be schematically elaborated in Figure 5.1b. Figure 5.1c shows the cross-sectional SEM (scanning electron microscope) image of the sandwich structure of $ZnCo_2O_4$@NC gel electrolyte//Fe_3O_4@r-GO ASC device with Ecoflex encapsulation. The thickness of the ASC is approximately 1 mm with 60 μm for the intermediate polymer gel electrolyte. The cycle stability was evaluated at 15 $mA cm^{-2}$ for 6000 cycles and exhibited a capacitance retention of 85.89% (see Figure 5.1d). The ASC delivers excellent flexibility at different bending angles (see Figure 5.1e), and can store a maximum volume energy density of 2.32 $mWh cm^{-3}$ at 33.3 $mW cm^{-3}$ (see Figure 5.1f).

Commercial carbon cloth woven from carbon microfibers was recently studied as a promising current collector with the advantages of high flexibility, wearability, 3D porosity, high surface area, and chemical stability. Ma et al. [17] recently reported a morphology-controlled growth of well-aligned zinc-nickel-cobalt ternary (oxy)hydroxide (Zn-Ni-Co TOH) nanostructure on carbon cloth substrate, the TOH can deliver superior capacity and rate capability to the Zn-Ni-Co oxide due to the highly open well-defined nanoarray architecture. The morphologies of the as-prepared samples depicted in Figure 5.2 reveal strong trend with the reaction temperature. At 100 °C, interconnected Zn-Ni-Co TOH nanosheets were grown on the surface of the carbon fiber (CF) substrate (see Figure 5.2b). At more elevated temperatures of 130 °C, the nanosheets disappeared and well-aligned NWA were grown (see Figure 5.2e). When the temperature increased to 150 °C, similar nanowires were also observed, with some micronsized sheets shown on top (see Figure 5.2h-i). The results indicate that the morphology evolution of Zn-Ni-Co TOH is followed by "nanosheet growth-larger nanosheet separated into nanowires-part of nanowires aggregated into microscale sheets." The ASC was assembled with Zn-Ni-Co TOH-130 cathode and the FEG (functionalized partial-exfoliated graphite) anode can deliver 5.4 $F cm^{-3}$ and a high energy density of 2.43 $mWh cm^{-3}$. Moreover, the ASCs are able to reversibly cycle with a high operating voltage of 1.8 V with no capacitive decaly in 10 000 cycle. The excellent performance can be attributed to the large loading of ordered nanoarrays allowing efficient ion diffusion with reduced diffusion length and thus resulting in a large pseudocapacitance.

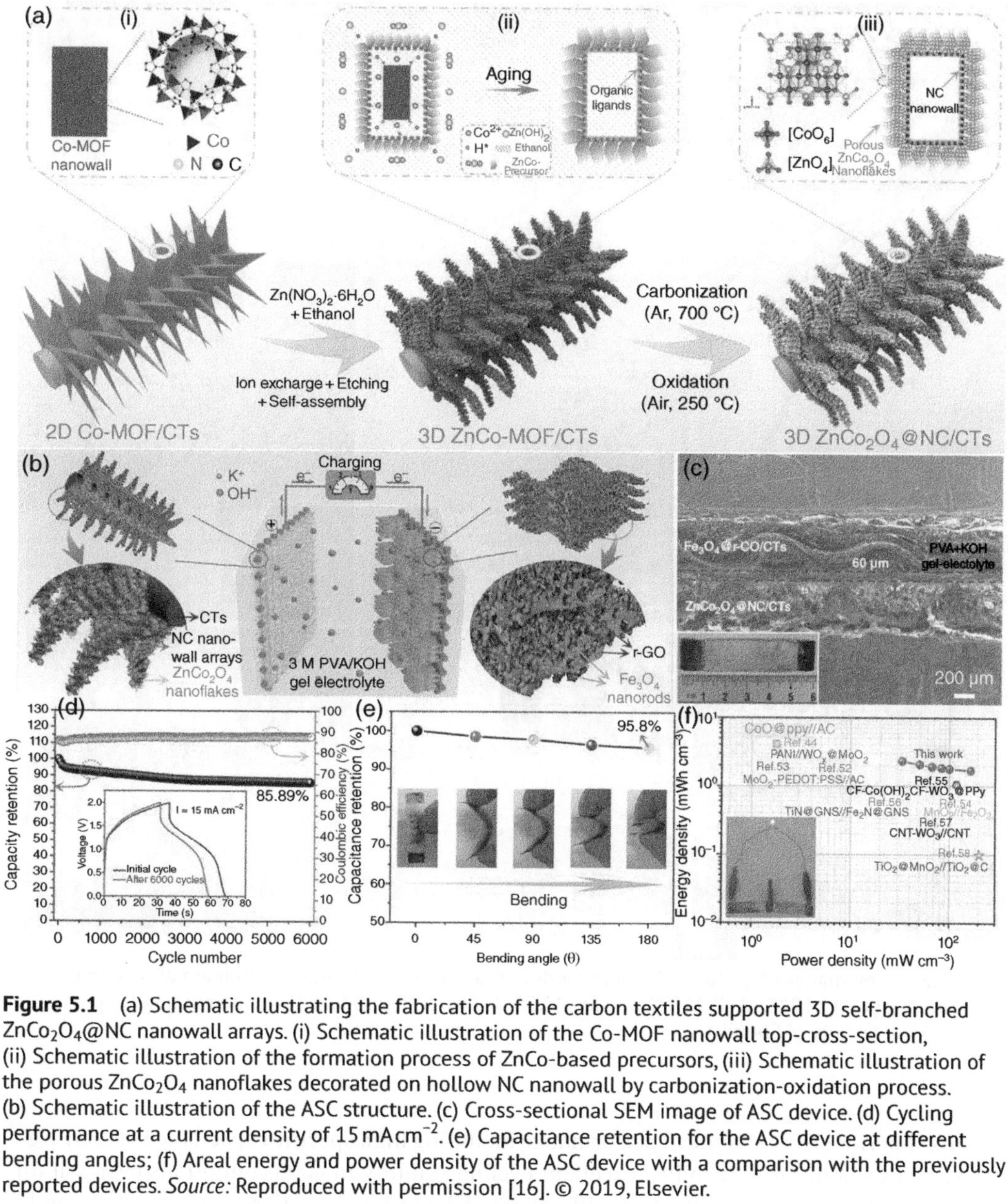

Figure 5.1 (a) Schematic illustrating the fabrication of the carbon textiles supported 3D self-branched $ZnCo_2O_4$@NC nanowall arrays. (i) Schematic illustration of the Co-MOF nanowall top-cross-section, (ii) Schematic illustration of the formation process of ZnCo-based precursors, (iii) Schematic illustration of the porous $ZnCo_2O_4$ nanoflakes decorated on hollow NC nanowall by carbonization-oxidation process. (b) Schematic illustration of the ASC structure. (c) Cross-sectional SEM image of ASC device. (d) Cycling performance at a current density of $15\,mA\,cm^{-2}$. (e) Capacitance retention for the ASC device at different bending angles; (f) Areal energy and power density of the ASC device with a comparison with the previously reported devices. *Source:* Reproduced with permission [16]. © 2019, Elsevier.

5.2.2 2D Ternary Metal Oxide Nanostructured Electrodes

Compared with 1D nanostructures, 2D nanostructures including nanosheets, nanoflakes usually exhibit better electrochemical cycling performance because the 2D feature is favorable for efficient ion and electron transport and can better accommodate the structural change during the electrochemical reaction. In particular, their ultrathin nanosheet morphology and interconnected network structure are advantageous for efficient ion/electron transport and better accommodation of volume variation [18]. Ayazpour et al. [19] has first time synthesized 2D high-ordered nanoporous $NiMoO_4$ as a high performance SC electrode material. The material

Figure 5.2 SEM images of Zn-Ni-Co TOH electrodes obtained using different hydrothermal reaction temperatures: (a–c) 110 °C, (d–f) 130 °C and (g–i) 150 °C. *Source:* Reproduced with permission [17]. © 2019, Royal Society of Chemistry.

exhibits superior electrochemical performance including high specific capacitance of 2351 F g^{-1} at 2 A g^{-1}, excellent rate capability, excellent cycling stability in 6000 continuous cycles at different current densities (only 9.3% loss after 3000 cycles at 5 A g^{-1}). Hierarchical $NiMoO_4$ architectures assembled from well-aligned uniform nanosheets are used as Faradaic capacitive electrodes [8]. An ASC is constructed using the as-prepared $NiMoO_4$ nanosheet as the positive electrode and activated carbon (AC) as the negative electrode. The optimized ASC with an extended operating voltage range of 0–1.7 V displays excellent electrochemical performance with a high energy density of 60.9 Wh kg^{-1} at power density of 850 W kg^{-1} in addition to superior rate capability. Furthermore, the $NiMoO_4$//AC ASC device exhibits remarkable cycling stability with 85.7% specific capacitance retention after 10 000 cycles. Teng and co-works [20] fabricated ASC based on 2D $CoMoO_4$ ultrathin nanosheets and, the ASC can deliver an energy density of 0.313 mWh cm^{-3} at a power density of 80 mW cm^{-3}. After 5000 cycles, 77.37% of capacitance can be remained. Yuan et al. [21] reported the fabrication of NiCo-layered double hydroxide (LDH) nanosheets on RGO (reduced graphene oxide) sheets through a facile one-pot solvothermal method, the sample exhibits ultrahigh specific capacitance up to 1911.1 $F g^{-1}$ at the current density of 2 $A g^{-1}$, and excellent rate capability with specific capacitance of 1469.8 $F g^{-1}$ at current density of 20 $A g^{-1}$. Park et al. [22] fabricated a hexagonal nanoplate-like Ni-Co-Mn hydroxide with controlled size and composition by a facile one-pot hydrothermal process. The size and composition of the hexagonal nanoplate were controlled by varying the ration of Ni to (Co + Mn) sources, the maximum capacitance of the sample is 1188 $F g^{-1}$ at 1 $A g^{-1}$ with high rate capacitance due to abundant active sites of the Ni-rich phase and rapid charge transfer kinetics of the nanoplate architecture. By integrating high-capacity Ni-rich Ni-Co-Mn hydroxide as a positive electrode with AC as a negative electrode, a full cell of a hybrid SC was configured in the enlarged voltage window of 0–1.5 V. Therefore, the cells deliver the maximum energy and power densities of 42.56 $Wh kg^{-1}$ and 17.8 $kW kg^{-1}$, respectively, showing good cycling stability with 78.8% capacity retention over 50 000 cycles and the coulombic efficiency of 100%.

5.2.3 3D Ternary Oxide Electrodes

Compared with 1D nanostructures, the complex 3D architectures constructed by 1D nanostructures show unique properties different from mono-morphological structures because they combined the features of all involved nanoscale building blocks. Chestnut-like $MnCo_2O_4$ microspheres grown on Ni foam as SC electrode was fabricated [23], the microspheres with mean diameter of 5 μm and are composed of numerous thin and long nanoneedles grown vertically outward from the center, forming an open porous network structure that can provide abundant porous surface area between the electrode and electrolyte. The nanoneedle is rounded in shape with 19 nm in tip diameter and 50 nm in bottom diameter. These unique nanoneedle structures form an open porous network for fast ion transport and ion diffusion, thus facilitating the enhancement electrochemical reaction. The $MnCo_2O_4$ nanoneedle electrodes delivered an unprecedented high capacitance of 1535 $F g^{-1}$ at 1 $A g^{-1}$ with an outstanding rate capability in 6 M KOH (potassium hydroxide) electrolyte. An ASC with 1.5 V was assembled based on $MnCo_2O_4$ as the positive electrode and graphene as the negative electrode. The ASC device showed a high energy density of 60.4 $Wh kg^{-1}$ at 375 $W kg^{-1}$. Moreover, the device demonstrate excellent cycling stability with 94.3% capacitance retention after 12 000 cycles at 30 $A g^{-1}$.

Pseudomorphic conversion of MOFs enables the fabrication of nanomaterials with well-defined porosities and morphologies for enhanced performances. However, the commonly reported calcination strategy usually requires high temperature to pyrolyze MOF particles and often results in uncontrolled growth of nanomaterials. Xiao and co-workers [24] report the controlled alkaline hydrolysis of MOFs to produce LDH while maintaining the porosity and morphology of MOF particles. Well-defined Ni/Co-LDH hierarchical microspheres with ultrathin nanosheet were obtained. These microspheres possess a 3D architecture with a large accessible surface area for high active sites and mesoporous channels for fast electrolyte and electron transport. The LDH with Ni:Co ratios of 7 : 3 show a high capacitance value of 1652 $F g^{-1}$ and outstanding long-cycling performance. The ASC based on NiCo-LDH//AC delivers a high energy density of 32.9 $Wh kg^{-1}$ at 74.3 $W kg^{-1}$. This work suggests that the alkaline hydrolysis method can be readily extended to other bimetallic LDHs from MOF templates. This facile, controllable, and scalable strategy will be widely applied in MOF-derived functional materials for electrochemical applications and beyond.

5.2.4 Core-Shell Ternary Metal Oxide Composite Electrode

5.2.4.1 Core-Shell Nanoarrays

Directly growing the core-shell nanoarrays with multi-components on the conductive substrates has been considered as a promising approach to boost the electrochemical performance of electrode materials. The core-shell nanoarrays can not only endow these hybrid electrodes with more electroactive sites, short ion and electron transportation path, but also utilize the synergistic effect of each component to remarkably boost these hybrid electrodes' electrochemical performance with respect to the single-component electrode. To date, various core-shell nanoarrays with different components and hierarchical microstructures have been reported to demonstrate notably improved electrochemical performance.

$NiCo_2O_4$ has received considerable research attention due to its high theoretical capacity and low fabrication cost. Researchers pay great efforts to boost the capacitance of $NiCo_2O_4$ by exploring various nanostructures to provide large surface area plus short ion diffusion path. A conducting polymer, like PPy, with high charge density and high electrical conductivity in doped states has also been widely explored as electrode for SCs [25, 26]. Recently, the integration of metal oxide with PPy has been demonstrated an effective method to enhance the performance by improving the electron transport and ion diffusion issues [27]. Kong and co-workers [28] fabricated $NiCo_2O_4$@PPy coaxial NWAs on carbon textiles as electrode for high-performance SCs. To optimize the thickness of PPy shell layers, the effects of growth times on the morphology and thickness of PPy are discussed. At initial deposition stage (0.5 hours), the PPy nanoparticles begin to attach to the top on the NWAs, as the deposition time increase (1.5 hours), the PPy nanospheres start to grow large. With the deposition time increased up to three hours, a tight coating of PPy nanospheres with high density was produced on the whole surface of $NiCo_2O_4$ NWAs, without an open porous structure. The $NiCo_2O_4$ nanowires act as highly capacitive "core" and uniform PPy nanospheres work as the well conductive "shell." The electrochemical properties for different hybrid $NiCo_2O_4$@PPy NWAs electrodes with different PPy polymerization times demonstrated that the current response increases with the PPy polymerization reaction times, resulting in enhancement of the capacitance. However, further increase in the polymerization reaction time up to three hours leads to the decrease of the total capacitance, which might be due to the thick and dense PPy layer could prevent the ion penetration to the inner core of $NiCo_2O_4$ NWAs. The high specific capacitance and superior cyclic stability is attributed to the remarkable synergistic contribution from the $NiCo_2O_4$ nanowires and conductive PPy, in which the PPy improve the electorn transportation and the mesoporous $NiCo_2O_4$ facilitate the ion diffusion, as a result in enhancing the rate capability. The ASC was fabricated based on $NiCo_2O_4$//AC, achieving a high energy density of 58.8 $Wh\,kg^{-1}$ at 365 $W\,kg^{-1}$. This work opens up new opportunities for design of a novel composite electrode for next generation high-performance flexible and lightweight SCs.

In order to meet the requirements of higher performance SC, many hierarchical core-shell nanostructures based on $NiCo_2O_4$ have been widely used, such as $NiCo_2O_4$@$NiMoO_4$ core-shell nanowire/nanosheet array [29, 30], $NiCo_2O_4$@polypyrrole arrays [31] and core-shell $NiCo_2O_4$@PANI nanorod arrays [32]. Que. et al. [33] reported three-dimensional $NiCo_2O_4$@$NiCo_2O_4$ core-shell nanocones arrays through a simple hydrothermal method. The unique structure is composed of $NiCo_2O_4$ nanocone array serving as the core and ultrathin $NiCo_2O_4$ nanosheets as the outer shell. Figure 5.3 illustrates the fabrication process for $NiCo_2O_4$@$NiCo_2O_4$ core-shell nanocones on Ni foam substrate. When used in SCs, the $NiCo_2O_4$@$NiCo_2O_4$ core-shell hierarchical nanostructures exhibit a remarkable capacitance of 2045.2 $F\,g^{-1}$ at 1 $A\,g^{-1}$, which are better than that of the individual components of $NiCo_2O_4$ nanosheets and $NiCo_2O_4$ nanocones. An ASC based on $NiCo_2O_4$@

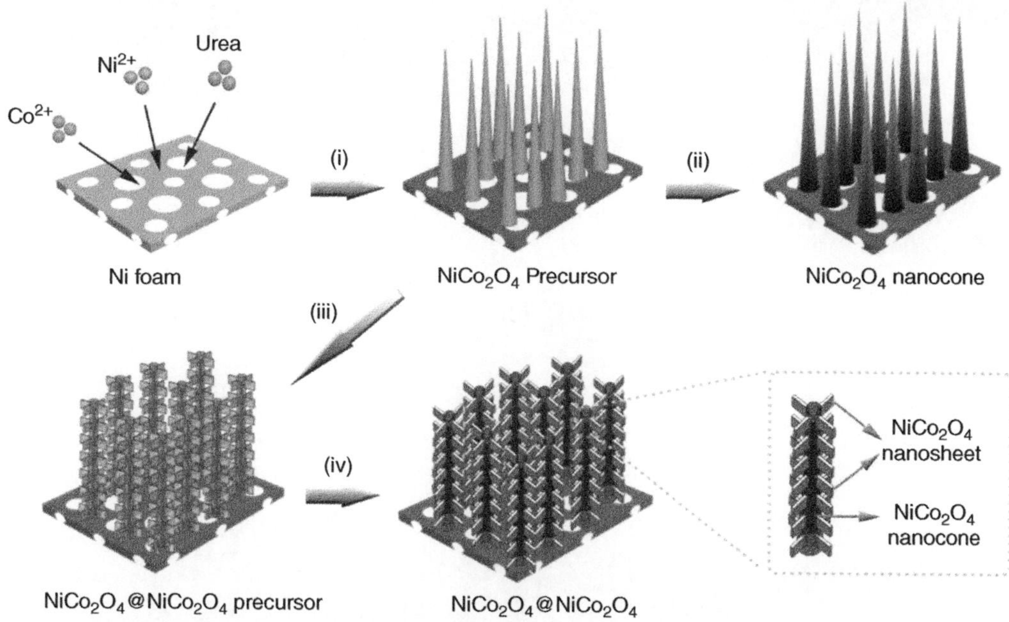

Figure 5.3 Schematic of the fabrication process for 3D $NiCo_2O_4$@$NiCo_2O_4$ hierarchical core-shell nanocones arrays on Ni foam. *Source:* Reproduced from ref. 33 with permission from the Elsevier.

$NiCo_2O_4$//AC was assembled and delivers an energy densities with a value of 82.95 Wh kg^{-1} at 0.35 kW kg^{-1}.

Pei et al. [34] adopted a two-step hydrothermal reaction to synthesize $MnCo_2O_4$@$MnMoO_4$ core-shell nanoarrays electrode for SCs, which is constructed by growing $MnMoO_4$ nanosheets as the shell onto interconnected $MnCo_2O_4$ nanoarrays as the core on Ni foam. Figure 5.4a illustrates the SEM of the $MnCo_2O_4$ nanoarrays on Ni foam. It is clear that the Ni foam is homogeneously covered by a layer of numerous $MnCo_2O_4$ nanosheets. The enlarged SEM image exhibits that these nanosheets on Ni foam are aligned perpendicularly and cross-linked with each other (see Figure 5.4b), generating an ordered and highly open network and making the architecture more stable. Afterword, the prepared $MnCo_2O_4$ nanoarrays on Ni foam were served as the backbone for growing $MnMoO_4$ nanosheets through a second hydrothermal reaction to synthesize $MnCo_2O_4$@$MnMoO_4$ core-shell nanoarrays on Ni foam. Figure 5.4c shows the low-magnification SEM image of the prepared $MnCo_2O_4$@$MnMoO_4$ core-shell nanoarrays on Ni foam after undergoing a hydrothermal reaction. The nanoarrays is still well maintained, while the outer surfaces of the prime $MnCo_2O_4$ nanosheets become very rough compared with the single $MnCo_2O_4$ nanosheets (see Figure 5.4d). To demonstrate the practical application of the $MnCo_2O_4$@$MnMoO_4$ core-shell nanoarrays electrode for SCs, an ASC was fabricated by employing the $MnCo_2O_4$@$MnMoO_4$ core-shell nanoarrays as positive electrode and the AC as a negative electrode. The specific capacity is 222 C g^{-1} at 1 A g^{-1} (see Figure 5.4e). Moreover, 91% capacity retention and 82% coulombic efficiency can be achieved for the ASC after 5000 cycles at 2 A g^{-1} (see Figure 5.4f). The ASC deliver 49.4 Wh kg^{-1} at 815 W kg^{-1} (see Figure 5.4g). Furthermore, two assembled ASC can easily power a green light-emitting diode (LED) (see Figure 5.4h). These results reveal that the prepared $MnCo_2O_4$@$MnMoO_4$ electrode has a great potential for practical application in the energy storage devices.

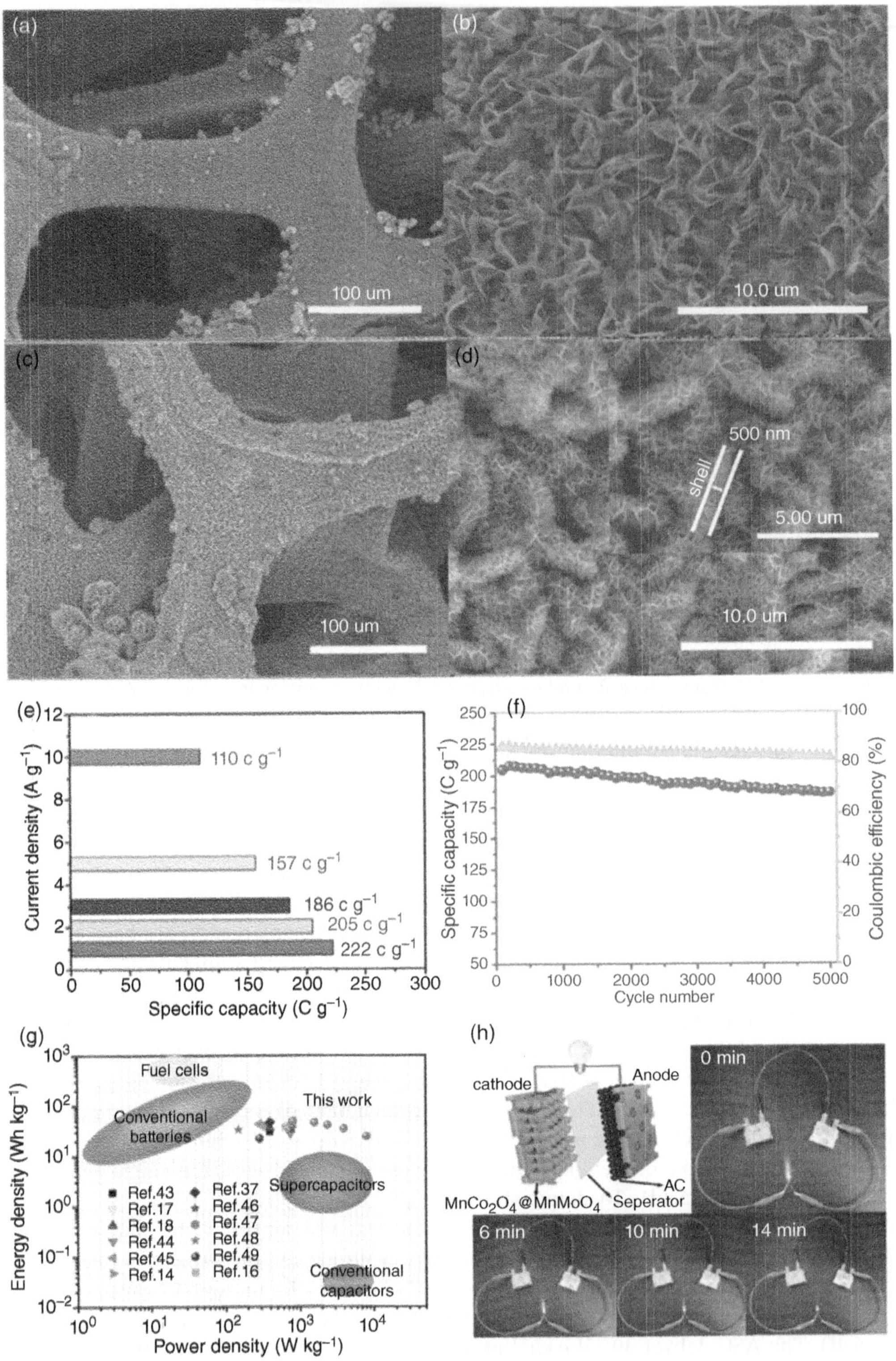

Figure 5.4 (a and b) SEM images of $MnCo_2O_4$ nanoarrays on Ni foam. (c and d) SEM images of $MnCo_2O_4$@$MnMoO_4$ core-shell nanoarrays on Ni foam. (e) The specific capacity as a function of current density and (f) cycling performance over 5000 cycles at $2\,A\,g^{-1}$ of the $MnCo_2O_4$@$MnMoO_4$//AC ASC; (g) Ragone plot of the $MnCo_2O_4$@$MnMoO_4$//AC ASC compared with other reported ASCs in the literature; (h) Schematic illustration for the assembled $MnCo_2O_4$@$MnMoO_4$//AC ASC and pictures of a green LED lighted up at different stages by two assembled ASC in series. *Source:* Reproduced with permission [34]. © 2018, Elsevier.

Researchers have proved that multi-element combination can not only overcome the shortcomings of their materials, but there may be synergies, and it can achieve unexpected results. As we know, honeycomb-like MnO_2 has large surface area, "opened" structures, and good monodispersity. When the two substances hang together, MnO_2@$NiCo_2O_4$ composite can develop their advantages. Cao et al. [35] reported a monodisperse MnO_2@$NiCo_2O_4$ core/shell nanospheres. The unique shape can increase the contact areas and facilitate the mass transport of electrolytes within the electrodes, ultimately improving the electrochemical properties. The MnO_2@$NiCo_2O_4$ composite electrode exhibits a high specific capacitance of $1127.27\,F\,g^{-1}$ at $1\,A\,g^{-1}$ and nearly 81% capacitance retention at $16\,A\,g^{-1}$. A maximum energy density as high as $26.6\,Wh\,kg^{-1}$ can be obtained at power density of $800\,Wkg^{-1}$. After 1000 cycles, capacitance retention rate is 126.8%, and only 3.7% loss after 10 000 cycles at $10\,A\,g^{-1}$, indicating a good cycling stability.

Nickel-cobalt-based LDHs attract wide attention because of their special properties of anion exchange and intercalation, high theoretical capacity and high specific surface area [36]. Chen et al. synthesized hollow nano-polyhedron LDHs by Co-based zeolitic imidazolate skeleton (ZIF-67) as sacrificial templates for SC electrode material and delivered $1203\,F\,g^{-1}$ at $1\,A\,g^{-1}$ [37]. While the MOF-derived materials have a low conductivity and low structural stability. Que et al. [38] has synthesized NiCo-LDH/CoS_2 nanocage derived from MOF on MnO_2 nanotubes and then heat-treatment sulfuration to obtained MnO_2@NiCo-LDH/CoS_2. The scan rate-dependent CV curves can be used to quantify the capacitance contribution of the capacitance control process and diffusion control process, according to the equation:

$$i(V) = k_1 v + k_2 v^{1/2}$$

$$i(V) / v^{1/2} = k_1 v^{1/2} + k_2$$

where i(V), k_1v and $k_2v^{1/2}$ represent the total current response at a fixed potential V, the current caused by the surface capacitance effect and the current caused by the diffusion effect, respectively. The dependence of $i(V)/v^{1/2}$ on $v^{1/2}$ at different potentials shows obvious linear behavior. The EIS (electrochemical impedance spectroscopy) results confirmed that the presence of CoS_2 in the nanosheets improves the conductivity of the material and promotes the transport of the electrolyte. The ASC was assembled by MnO_2@NiCo-LDH/CoS_2//AC, showing a maximum energy density of $49.5\,Wh\,kg^{-1}$ at $391.5\,W\,kg^{-1}$, excellent rate performance (retention rate 76.9% from 1 to $10\,A\,g^{-1}$). The kinetic analysis of MnO_2@NiCo-LDH/CoS_2 demonstrates that the excellent properties due to the surface capacitance control and diffusion control from MnO_2 nanotube and NiCo-LDH hollow nanocage structures.

5.3 Metal Sulfide Electrodes

Metal sulfides commonly exhibit better electrical conductivity and higher electrochemical activity. Some of them do not have a band gap in their electronic structure and present typical metallic behavior, which can offer enhanced reaction kinetics in general under high charge-discharge rates for electrodes [39, 40]. The electrochemical contributions of bimetallic sulfides can provide richer redox reactions compared with single component sulfide, resulting in higher specific capacitance [41, 42]. Such bimetallic sulfide species can also be readily reconstructed from their metal oxide/hydroxide precursors via anion exchange reactions and Kirkendall effects, and finally lead to multiform and tunable morphologies, such as nanoparticles, nanowires/nanotubes and nanosheets [43].

5.3.1 1D Metal Sulfide Electrodes

Copper antimony sulfide (Cu_3SbS_4) showed superior electrochemical performance due to their exceptional electron transport characteristics, stable in nature, large surface area and can provide better space to host smaller ions or atoms in between the layers. Kim et al. [44] reported Cu_3SbS_4 nanowires served as positive electrode for ASC. The formation of Cu_3SbS_4 nanowires can be due to the fact that Cu_3SbS_4 nanoparticles act as nucleation site for the microwave growth of Cu_3SbS_4 during the latter phase of the reaction. The Cu_3SbS_4//Cu_2MoS_4 ASC delivered a high device capacitance of 213.6 F g^{-1}. The long-term durability of Cu_3SbS_4//Cu_2MoS_4 ASC was tested at 15 mA cm^{-2} over 4000 cycles, about 90.7% of the initial capacitance can be retained.

Ternary Ni-Co sulfides, as new style of cathode materials for SC, which feature higher capacity, good electrochemical activity, and excellent conductivity, have been extensively researched. One-dimensional (1D) hollow tubular nanostructures possess robust chemical and mechanical stabilities so that active materials could void aggregation during the long charge-discharge cycle. Tube-like or hollow nanostructures can effectively eliminate the "dead" or "inactive" materials and largely increase the electrochemical active area. More importantly, nanotube structures can provide short electrolyte diffusion path through the inner cavity/channels, which will further increase the overall redox reaction efficiency. By way of the pseudocapacitive hollow nanostructures growing on conductive substrates to obtain high mass loading hierarchical structures may realize the outstanding electrochemical performance of binder-free SC electrodes. Wu and co-workers [45] developed a hierarchical $CoNi_2S_4$ nanotubes by a solvothermal route followed by sulfidation reaction for the first time, the electrode delivered a high capacitance of 2552 F g^{-1} at 1 A g^{-1} with a high rate capacity of 81% at 25 A g^{-1}. The hybrid SC device with $CoNi_2S_4$ nanotubes as the cathode and the CNTs@Fe_2O_3@C composites as the anode was assembled, the device exhibits a wide voltage range of 0 ~ 1.75 V with a maximum energy density of 90.5 Wh kg^{-1} at power density of 1.84 kW kg^{-1}. Moreover, the hybrid device has a stable cycling stability with 82.1% retention over 5000 cycles. This coordinative design strategy in this work provides a novel way to manufacture next-generation energy-storage device with high performance and safety. Liu et al. [46] designed the hollow and spinous $NiCo_2S_4$ nanotubes through a simple hydrothermal reaction using the natural silk as the template. The spinous Ni-Co precursors are grown on the natural silk through a facile hydrothermal strategy and the hollow structure is obtained by decomposing the silk via hydrothermal sulfurization. After calcination treatment, the hollow and spinous $NiCo_2S_4$ nanotubes are applied as the electrode material and exhibit better electrochemical performance than the solely vulcanized samples. The $NiCo_2S_4$ nanotube electrode shows good specific capacitance of 630 F g^{-1} at 1 A g^{-1}, low internal resistance of 0.68 Ω and high capacitance retention (91% after 3000 cycles) at 10 A g^{-1}. These might be attributed to the following superiorities: firstly, the hollow and porous structure provides an effective channel for electrolyte ion diffusion and accelerates the transfer of electron, which are beneficial to the valence transitions of Ni^{2+}/Ni^{3+}, Co^{2+}/Co^{3+} and Co^{3+}/Co^{4+} during the redox reactions; secondly, the large specific surface area can supply enormous active sites for the electrochemical reaction. To further verify the energy storage performance of the $NiCo_2S_4$ nanotube electrode for practical applications, an ASC device based on $NiCo_2S_4$ nanotube//AC was fabricated and exhibited an energy density of 52.34 Wh kg^{-1} at a power density of 2206.37 W kg^{-1} and has the ability to power a blue LED indicator. It can be envisioned that this method for fabrication of the $NiCo_2S_4$ nanotube electrode could be generalized to the preparation of other nanomaterials with similar structures for varied applications in energy storage and nanotechnology. Huang and co-workers [47] present the rational synthesis of lawn-like $FeCo_2S_4$ hollow nanoneedle array on flexible carbon nanofiber film via a two-step hydrothermal

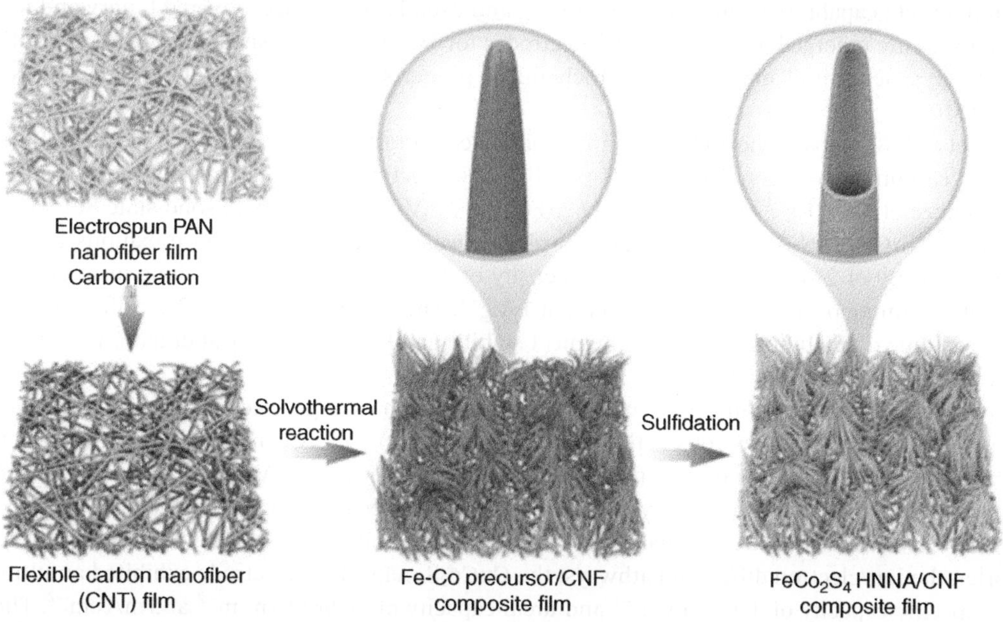

Figure 5.5 schematic for the preparation of $FeCo_2S_4$ hollow nanoneedle array/CNF composite film. *Source:* Reproduced with permission [47]. © 2019, Elsevier.

method (see Figure 5.5). Firstly, PAN (polyacrylonitrile) nanofibrous film with controlled thickness was prepared. Fresh PAN film was then stabilize in air under 250 ° C for one hour, followed by the carbonization in the N_2 atmosphere under 800 ° C for one hour to obtain carbon nanofiber film. Then, the Fe-Co precursor was prepared on CNF film with a hydrothermal method, lastly, the sulfidation treatment was performed to obtain $FeCo_2S_4$/CNF composite. The needle-like Fe-Co precursors are grown separately at first in a perpendicular orientation, while their tips start to weld together after growing to a specific length. This particular structure can prevent the nanoneedles from lodging and collapse, which is very helpful for the structure robustness. The transformation of solid Fe-Co precursor nanoneedles to $FeCo_2S_4$ hollow nanoneedles can be explained as follows. At first, Fe-Co precursor was deposited on CNT, then the Fe-Co sulfides were formed in-situ by an anion-exchange reaction between S^{2-} in Na_2S and $C_2O_4{}^{2-}$ in $FeCo_2(C_2O_4)_3$ during the sulfidation reaction. The obtained $FeCo_2S_4$ hollow nanoneedles composite electrode shows remarkable capacitive performance in terms of high specific capacitance 2476 F g^{-1} at 1 A g^{-1} and excellent cycling stability (93% after 5000 cycles). The assembled ASC based on $FeCo_2S_4$ hollow nanoneedles positive electrode and AC negative electrode delivered a high energy density of 88.5 Wh kg^{-1} at a power density of 800 W kg^{-1}, demonstrating the great potential of this approach for the exploitation on low-cost and high performance next-generation energy storage materials.

5.3.2 2D Metal Sulfide Electrodes

Recently, researcher examined various two dimensional materials (such as transition metal chalcogenide [TMC], siloxene, MXene, metalenes) beyond graphene as electrodes for SCs. Layer TMCs (such as MoS_2, $MoSe_2$, WS_2, WSe_2) have attracted much attention as high-performance electrode materials due to the presence of intercalation capacitance which can provide high capacitance with high rate capability compared to the EDLC and pseudocapacitance. Layered binary and ternary TMCs are emerged as new series of electrodes for SC due to their sheet-like structures with interlayer spacing and enhanced electrochemical properties compared to the single component TMCs.

Yao et al. [48] prepared novel Ni-Co-S interconnected nanosheet arrays on N-doped hierarchical porous carbon nanofibers (Ni-Co-S@N-pNCFs), the electrode exhibits a remarkable specific capacitance of 520.2 $F g^{-1}$ at 0.2 $A g^{-1}$ and 670 $F g^{-1}$ at 2 $mV s^{-1}$. A flexible all-solid-state SC device based on Ni-Co-S@N-pNCFs and AC@@N-pNCFs was fabricated. The device exhibits a high energy density of 21.6 $Wh kg^{-1}$ at a power density of 134.9 $W kg^{-1}$ and robust long-term cycling stability retaining over 90% after 3000 cycles at 1 $A g^{-1}$. The as-assembled ASCs have an outstanding mechanical flexibility and electrochemical stability and no significant attenuation occurred during cycling under different bending states.

MOF derived nanostructures have unique properties, such as large specific surface area, exclusive porous networks, numerous active sites, and exceptional electrochemical properties when compared to traditional nanostructures. Lee et al. [49] synthesized a MOF derived hierarchical copper cobalt sulfide nanosheet arrays. Figure 5.6 demonstrated the design and fabrication process for the electrodes. The $CuCo_2S_4$ nanosheet can provide enriched electroactive sites, as well as shorten the ion/electron diffusion pathways, the $CuCo_2S_4$ nanosheet electrode exhibited an ultrahigh specific capacity of 409.2 $mAh g^{-1}$ and areal capacity of ~0.96 $mAh cm^{-2}$ at 3 $mA cm^{-2}$. The as-fabricated ASC based on $CuCo_2S_4//Fe_2O_3$ delivered an volumetric capacity of 2.1 $mAh cm^{-3}$ at 3 $mA cm^{-2}$ and an energy density of 89.6 $W h kg^{-1}$ at 663 $W kg^{-1}$. The impressive results open a new pathway to design and fabricate sheet-like ternary metal sulfides from a MOF and construct energy storage and conversion devices for use in industrial sectors. Zhang et al. [50] designed and fabricated $NiCo_2S_4$ nanosheet supported on nitrogen-doped carbon foams for SC, which exhibited greatly improved electrochemical performance with ultrahigh capacitance (877 $F g^{-1}$ at 20 $A g^{-1}$)

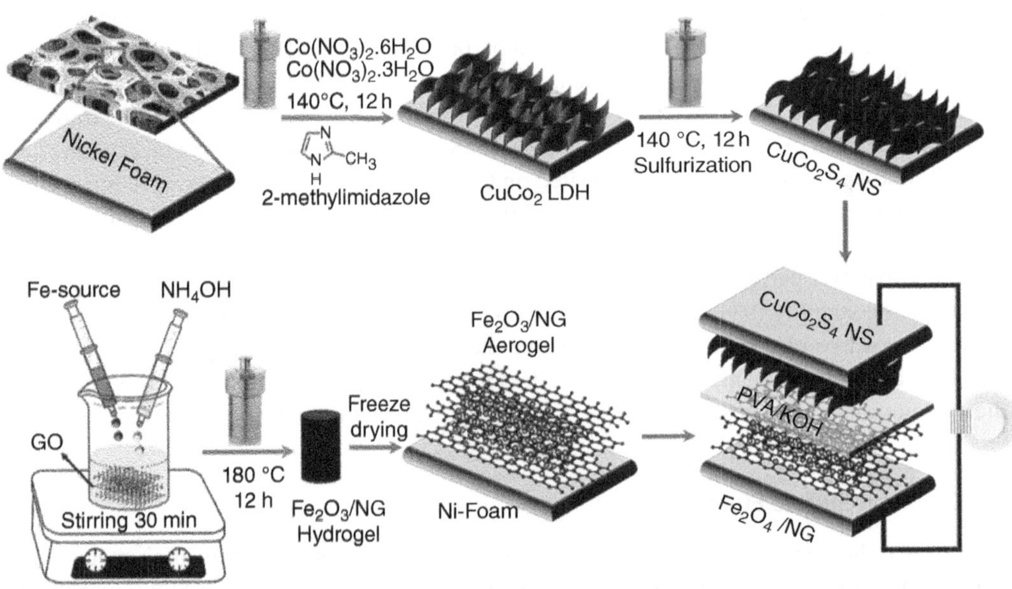

Figure 5.6 Schematic illustration of the design and fabrication of the $CuCo_2S_4$ NS and Fe_2O_3/NG aerogel and their application in all-solid-state ASCs. *Source:* Reproduced from ref. 48 with permission from the Royal Society of Chemistry.

and excellent cycling stability. The ASC was fabricated by using $NiCo_2S_4$ as the positive electrode and ordered mesoporous carbon as the negative electrode, this demonstrates high energy density (45.5 Wh kg^{-1} at 512 W kg^{-1}).

5.3.3 3D Metal Sulfide Electrodes

Highly porous and accessible nanostructures that can reduced transport lengths for ions/electrons transfer have attracted great attentions in energy storage applications. It is generally accepted that hollow cavities enhance the active sites for chemical/electrochemical reactions by mitigating the agglomeration of nanoparticles. In addition, the diffusion distance between the interior surfaces and the external electrolyte ions can be shortened by cavities which play the role of an "ion buffering reservoir." Therefore, porous hollow nanostructures can effectively enhance the electrochemical performance of electrodes by providing a competent link between the electrolyte and electrode materials. Mohammadi and co-workers [51] designed nanoporous $CuCo_2S_4$ hollow spheres by a simple self-templated method to be used in ASC. The ASC assembled by $CuCo_2S_4$ and AC electrodes delivered a power density of up to 16 kW kg^{-1} and a high energy density of 43.65 Wh kg^{-1}. Zhang et al. [52] prepared the Ni-Co-Mn sulfide yolk-shell hollow spheres via a simple solvothermal route and followed a sulfidation process. The resultant nickel cobalt manganese sulfide yolk-shell hollow spheres exhibit a large specific capacitance of 1360 F g^{-1} at 1 A g^{-1} and superior rate capability. The ASC device is assembled using the nickel cobalt manganese sulfide yolk-shell hollow spheres and AC, the device shows a high energy density of 49.8 Wh kg^{-1} at a power density of 1700 W kg^{-1} and long-term cycling life (only 1.8% loss after 6000 cycles), suggesting their potential application in high-performances electrochemical energy storage. Hierarchical hollow $NiCo_2S_4$ microspheres with a tunable interior architecture were synthesized by hydrothermal method for cathode material for ASC, the $NiCo_2S_4$//rGO/Fe_2O_3 exhibited a maximum energy density of 61.7 Wh kg^{-1} with wide operating voltage range of 0–1.75 V [53].

Zhang et al. reported a simple self-templating strategy to prepare a complex nickel cobalt manganese sulfide yolk-shell hollow spheres [52]. The sulfide yolk-shell hollow spheres exhibited a large specific capacitance of 1360 F g^{-1} at 1 A g^{-1} and superior rate capability. Furthermore, an ASC device was assembled using the nickel cobalt manganese sulfide yolk-shell hollow sphere and AC. The device showed a high energy density of 49.8 Wh kg^{-1} at a power density of 1700 W kg^{-1} and long-term cycling life (only 1.8% loss after 6000 cycles), suggesting their potential application in high-performance electrochemical energy storage. Xia and co-workers have synthesized $NiCo_2S_4$ urchin-like nanostructure through a facile precursor transformation method, these materials have shown a capacity retention of 77.3% when the current density was increased 20 times from 1 to 20 A g^{-1} [54]. Wang and co-worker have reported the synthesis of $NiCo_2S_4$ by a simple one-step hydrothermal approach. Their material exhibited a capacitance retention of 53.2% when the current density increased from 1 to 20 A g^{-1}. Moreover, a retention of 81% in specific capacitance was achieved [55].

5.3.4 Metal Sulfide Composite Electrodes

Fei et al. [56] developed a cost-effective strategy to design novel Co_9S_8 nanotubes@Ni-Co LDH nanosheets core-shell electrodes on Ni foam for high performance SCs, in which 1D Co_9S_8 nanotubes as core materials and Ni-Co LDH nanosheets as the shell can obtain the benefits from both two individual components. Numerous Ni-Co LDH nanosheets as shell tightly wrap the core Co_9S_8 nanotubes, and the Ni-Co LDH nanosheets are linked with each other in a wrinkled fashion. The hybrid Co_9S_8 nanotubes@Ni-Co LDH nanosheets electrodes achieve a high specific capacitance of 1020 C g^{-1} at 1.25 A g^{-1} with high rate capability and excellent cycling stability (90.4% capacitance after 10 000 cycles), which are due to the interlaced Ni-Co LDH nanosheets, the high conductivity of Co_9S_8 nanotubes and their synergy lead to the structural stability. Furthermore, an ASC based on Co_9S_8 nanotubes@Ni-Co LDH nanosheets as anode and AC as cathode has been fabricated. The ASC exhibits superior specific capacitance of 236 C g^{-1} at a current density of 1 A g^{-1} with a stable voltage window of 1.5 V. Furthermore, the ASC can deliver the high energy density of 50 Wh kg^{-1} at power density of 839 W kg^{-1} and an energy density of 38 Wh kg^{-1} at a high power densit of 9118 W kg^{-1}. The ASC also exhibits good cycling stability with 86.4% retention after 5000 cycles. The electrodes with remarkable electrochemical performance are promising candidates for sustainable SCs.

Despite the progress of the ternary cobalt-based system that has been achieved, the increase of specific capacity, rate performances, and cycling stability of these materials still remain as challenges. The combination with other materials can further improve electrochemical performance due to intercomponent synergy. However, little work has focus on composite electrode materials consisting of two kinds of MCo_2S_4 (M = Ni, Fe, Mn, etc.) components. Li and co-workers [57] have designed a new hierarchical structure of dendritic $NiCo_2S_4$@$NiCo_2S_4$ nanoarrays with the aim to optimize the electrochemical performances of Ni-Co-S system. In details, a facile hydrothermal route followed by a calcination process resulted in a high density of $NiCo_2O_4$ porous nanorod arrays grown vertically on a Ni foam; after that, a thin layer of $NiCo_2O_4$ nanosheets was coated on the as-grown $NiCo_2O_4$ porous nanorod arrays by a controllable hydrothermal route; finally, the S^{2-} exchange process induced the growth of dendritic $NiCo_2S_4$@ $NiCo_2S_4$ hierarchical heterostructures. Figure 5.7a shows that the $NiCo_2O_4$ nanorod coated on the Ni foam uniformly with the length of several micrometers and the width was ~100 nm. After another hydrothermal treatment to form the $NiCo_2O_4$@$NiCo_2O_4$ hierarchical nanostructures, the uniform and high density of $NiCo_2O_4$ arrays were well preserved (see Figure 5.7b). The $NiCo_2O_4$ nanorods effectively and completely covered by the $NiCo_2O_4$ nanosheets, leading to a large volume of interspace inside these structures. After S^{2-} exchange to form $NiCo_2S_4$@ $NiCo_2S_4$ hierarchical nanostructures, the dendritic structure was shown to remain intact with the $NiCo_2S_4$ nanosheets uniformly coated on the $NiCo_2S_4$ nanorods (see Figure 5.7c). To demonstrate the electrochemical superiority of the $NiCo_2S_4$@$NiCo_2S_4$ hierarchical heterostructures, a comparative study with the counterpart $NiCo_2O_4$@$NiCo_2O_4$ hierarchical heterostructures and $NiCo_2O_4$ nanorod arrays in a three-electrode configuration using 1.0 M KOH electrolyte. The CV integral area of $NiCo_2S_4$@$NiCo_2S_4$ is larger than that of $NiCo_2O_4$@$NiCo_2O_4$ and $NiCo_2O_4$, suggesting the heterostructure have the largest specific capacity (see Figure 5.7d). Such excellent properties of $NiCo_2S_4$@$NiCo_2S_4$ might be due to the unique structural and compositional features, Figure 5.7e. The $NiCo_2S_4$@$NiCo_2S_4$ electrode has a high specific capacity of 4.43 mAh cm^{-2} at 240 mA cm^{-2} with a good rate capability of 70.1% after increasing the current densities from 40 to 240 mA cm^{-2}. The void space in the interior together with the permeable thin walls provides sufficient electroactive sites and electrolyte-electrode interface for fast diffusion and reaction. The rechargeable alkaline battery based on $NiCo_2S_4$@$NiCo_2S_4$//AC on Ni foam was assembled, which has a high operating voltage of 1.6 V, and possessed the specific capacitance of 3.53, 2.74, 2.42, 2.25, 2.04 and 1.76 mAh cm^{-2} at current densities of 2, 20, 40, 80, 120, and 160 mA cm^{-2}. Figure 5.7f shows the plots of power density versus density (Ragone plot) of the device, exhibiting a high area power density of 0.26 kW cm^{-2} and high area energy density of 4.52 Wh cm^{-2}. This work delivered the design concept for hierarchical nanostructures as electrodes for rechargeable alkaline batteries.

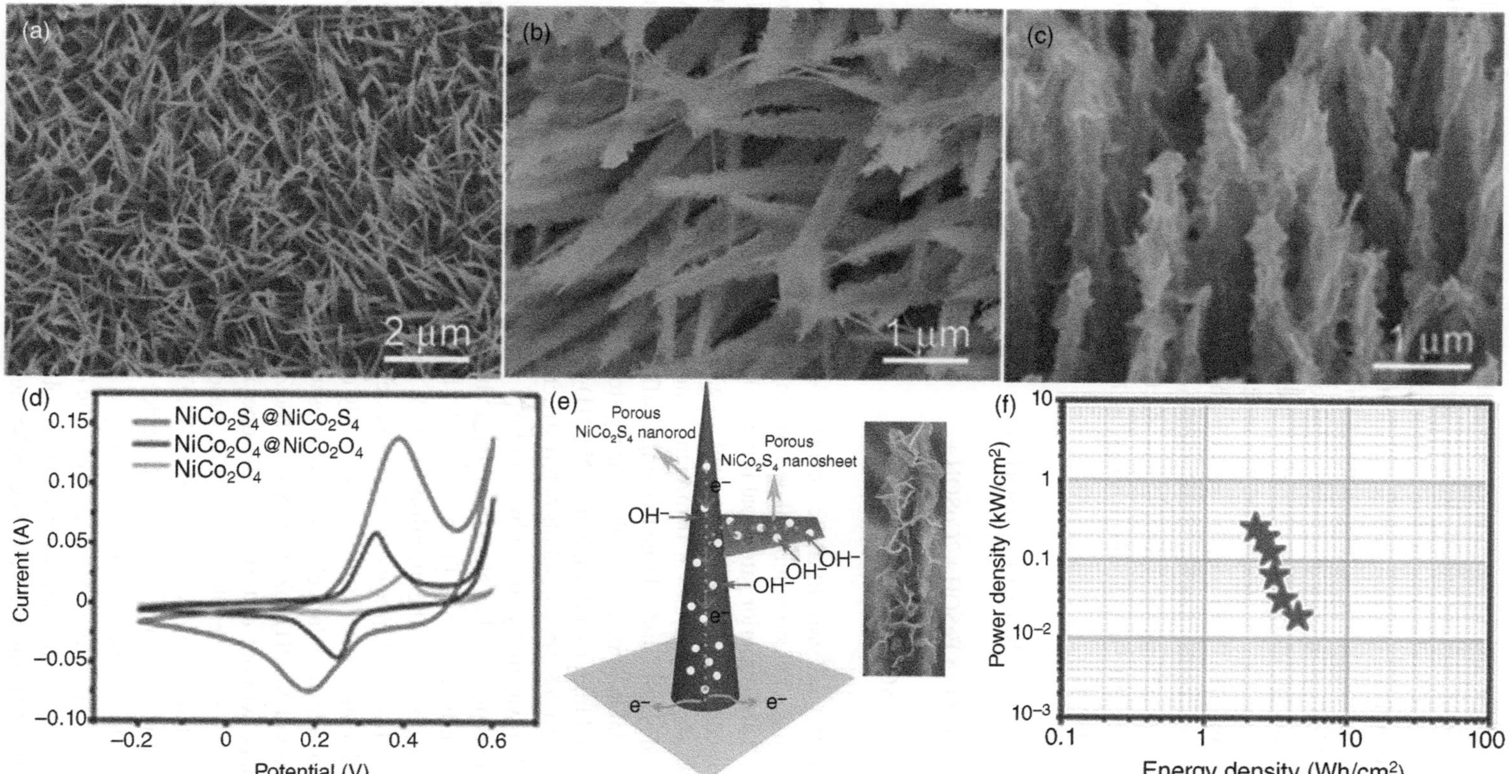

Figure 5.7 (a) SEM image of the $NiCo_2O_4$ nanorod arrays, (b) SEM image of $NiCo_2O_4$@$NiCo_2O_4$ hierarchical nanostructures, (c) SEM image of $NiCo_2S_4$@$NiCo_2S_4$ hierarchical nanostructures, (d) CV curves for as-synthesized $NiCo_2S_4$@$NiCo_2S_4$, $NiCo_2O_4$@$NiCo_2O_4$ hierarchical nanostructures and $NiCo_2O_4$ nanorod arrays, recorded at a scan rate of $5\,mV\,s^{-1}$. (e) Schematic illustration of electron paths on porous $NiCo_2S_4$@$NiCo_2S_4$ nanostructures. (f) Ragone plot of the hybrid rechargeable alkaline batteries. *Source:* Reproduced with permission [57]. © 2018, Wiley-VCH.

Hollow tubular structures have some advantages over on their inner surfaces. Meng et al. [58] reported the rational design and fabrication of a novel flexible electrode with nanotube-built multitripod architectures of ternary metal sulfides' composites ($FeCo_2S_4$-$NiCo_2S_4$) on a silver-sputtered textile cloth (SSTC). The unique multitripod architectures are highly porous but very robust, which facilitates ion and electron transfer for electrochemical reactions. In sulfuration process, sulfides are obtained while the solid precursor wires can be chemically transformed into tubes with well-defined hollow interior due to the diffusion effect. Interestingly, the resulting composites' nanostructure was found to highly depend on the S^{2-} concentration ($C_S{}^{2-}$). The saturation sulfidation occurs at around $C_S{}^{2-} = 66\times10^{-3}$ M where the precursor was completely transformed to $FeCo_2S_4$-$NiCo_2S_4$ composite. Therefore, the use of high $C_S{}^{2-}$ is unnecessary, and a suitable $C_S{}^{2-}$ not only assures a complete sulfidation but also determines the generation of desirable nanostructures. The energy storage of the $FeCo_2S_4$-$NiCo_2S_4$ electrode is ascribed to the Faradaic redox reactions of single-component $FeCo_2S_4$ and $NiCo_2S_4$ in KOH electrolyte. The reaction equations could be expressed as follows:

$$NiCo_2S_4 + OH^- + H_2O \leftrightarrow NiSOH + 2CoSOH + e^-$$

$$FeCo_2S_4 + OH^- + H_2O \leftrightarrow FeSOH + 2CoSOH + e^-$$

$$CoSOH + OH^- \leftrightarrow CoSO + H_2O + e^-$$

A comparison of GCD (galvanostatic charge-discharge) curves demonstrates the $FeCo_2S_4$-$NiCo_2S_4$ has a higher electrochemical performance, and it shows a specific capacitance of 1519 F g^{-1} at 5 mA cm^{-2} and superior rate capability (85.1% retention at 40 mA cm^{-2}). Flexible SC delivers a high specific energy of 46 Wh kg^{-1} at 1070 W kg^{-1} and exhibits high stability with capacitance retention of 92% after 3000 cycles at 10 mA cm^{-2} and outstanding stability due to no obvious degradation when the device is under large twisting configurations. This work provides a general, low-cost, green, and scalable route to wearable energy-storage devices.

Cao et al. [59] designed high pseudocapacitive $Ni(OH)_2$ and VN to form well-aligned $Ni(OH)_2$@ZnCoS-NS and VN@@ZnCoS-NS core-shell heterostructure to assemble high-performance ASC device. The carbon fiber was firstly coated by a uniform ZnCoO nanosheet array structure. After exchanging with S^{2-} ions produced by Na_2S in solution, the $ZnCo_2S_4$ nanosheets can well maintain on the carbon fiber (see Figure 5.8a). From the TEM (transmission electron microscope) image (see Figure 5.8b), the ultrathin $Ni(OH)_2$ nanoflakes are tightly and coated on the surface of ZnCoS-NSs. The TEM image of Figure 5.8c demonstrates the VN@ZnCoS-NSs possess a rough surface which consists of numerous randomly distributed tiny particles on the exposed surface, which can provide abundant electroactive sites and offer sufficient contact with the electrolyte. The GCD curves of $Ni(OH)_2$@ZnCoS-NSs//VN@ZnCoS-NSs ASC device show nearly symmetric triangular shape without distortion as the operating potential is increased to 1.6 V (see Figure 5.8d), exhibiting good reversibility. The ASC based on core-shell heterostructure delivers high energy and power densities (75 Wh kg^{-1} at 0.4 kW kg^{-1}), high specific capacitance (210 F g^{-1} at 0.5 A g^{-1}). The mechanical flexibility of the ASC device was evaluated under different bending/twisting cycles and cycle number (see Figure 5.8e and f), showing high capacitance retention. Figure 5.8g presented the operating principles, when the ASC is charged, the $Ni(OH)_2$@ZnCoS-NS positive electrode undergoes an oxidization reaction following the process of $Ni(OH)_2 + OH^- \leftrightarrow NiOOH + H_2O + e^-$ [60–62], and the VN@ZnCoS-NS negative electrode is reduced, described as $2VN_xO_y + zK^+ + ze^- + OH^-$? $VN_xO_yK_z{}^{z+} + VN_xO_y//OH^-$ [63, 64], which occurs fast at the surface

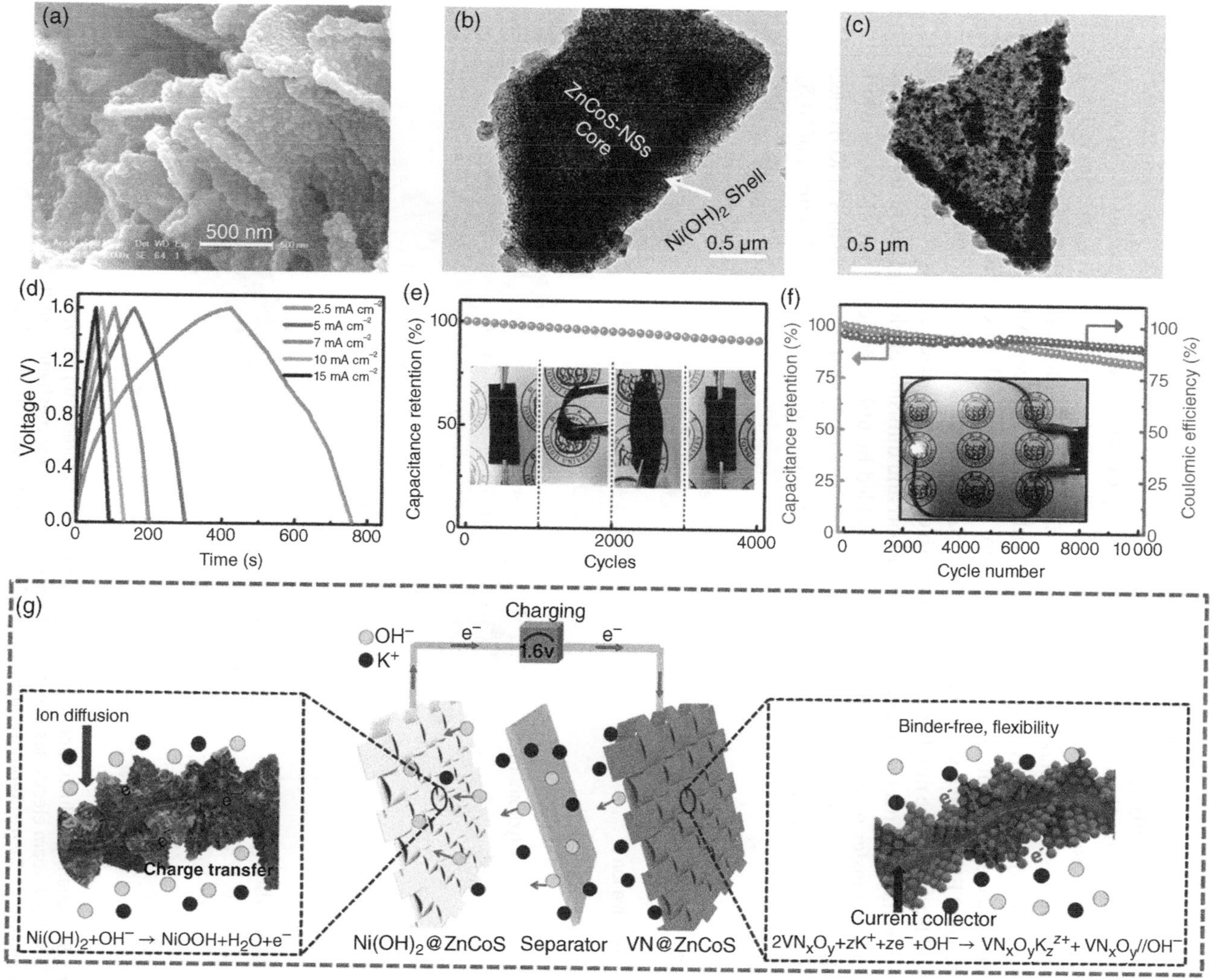

Figure 5.8 (a) SEM image of ZnCoS-NSs/CC, (b) TEM image of $Ni(OH)_2$@ZnCoS-NSs, (c) TEM image of VN@ZnCoS-NSs, (d) GCD curves of $Ni(OH)_2$@ZnCoS-NSs//VN@ZnCoS-NSs ASC device between 0 and 1.6 V at different current densities. (e) Capacitance retention of the ASC device at flat, bent and twisted states during the test over 4000 cycles, (f) Cycling stability and corresponding coulombic efficiency of the ASC device (the inset represents the digital image of the ASC lighting an LED), (g) electrode design and charge storage mechanism of the two electrodes. *Source:* Reproduced with permission [59]. © 2019, Royal Society Of Chemistry.

of the active materials leading to super high pseudocapacitance. This work offers useful and available strategy for the design of high performance heterostructure electrodes for ASC by exploiting the compositional and structural versatility of 2D MOF arrays, but also opens a new avenue for applications in high-performance electrochemical energy storage.

2D nanosheets have demonstrated their merits due to the anisotropic structure and high surface-to-bulk ratio, which grant a shorter diffusion path for electrons and ions [65, 66]. Zhai et al. [39] has successfully synthesized ultrathin and porous $Ni_3S_2/CoNi_2S_4$ network on Ni foam. Figure 5.9 illustrates the fabrication process through a two-step method. The uniform assembly of Ni-Co precursor network was directly grown on the entire surface of Ni foam with first hydrothermal synthesis step, then Ni-Co oxide and $Ni_3S_2/CoNi_2S_4$ network structure were converted from Ni-Co precursors subsequently by calcination and anion exchange reaction processes. The abundant S^{2-} ions, produced by the $Na_2S{\cdot}9H_2O$ in solution, quickly exchange with OH^- and CO_3^{2-} in Ni-Co precursor as well as accompanied by the Kirkendall effects to form $Ni_3S_2/CoNi_2S_4$ species [60, 67]. The Ni-Co precursor nanoflakes are grown on Ni foam surface successfully and interconnected with each other to form a network-like structure with plenty of spaces. After sulfurization process, the interconnected and intersected nanostructure of the original Ni-Co precursors were well retained. The TEM image reveals that $Ni_3S_2/CoNi_2S_4$ nanosheet is loaded with small nanoparticle, which will lead to a large number of edge sites derived from small-sized effects on its surfaces. The mesoporous nanoarchitecture facilitates the transport and diffusion of electrolyte, resulting in the rapid charge transfer reactions. The ASC was assembled by utilizing the $Ni_3S_2/CoNi_2S_4$ network electrode as the anode and the commercial activate carbon (AC) as the cathode in 6 M KOH electrolyte with a piece of cellulose filter paper as the separator. Figure 5.10a shows CV curves of $Ni_3S_2/$

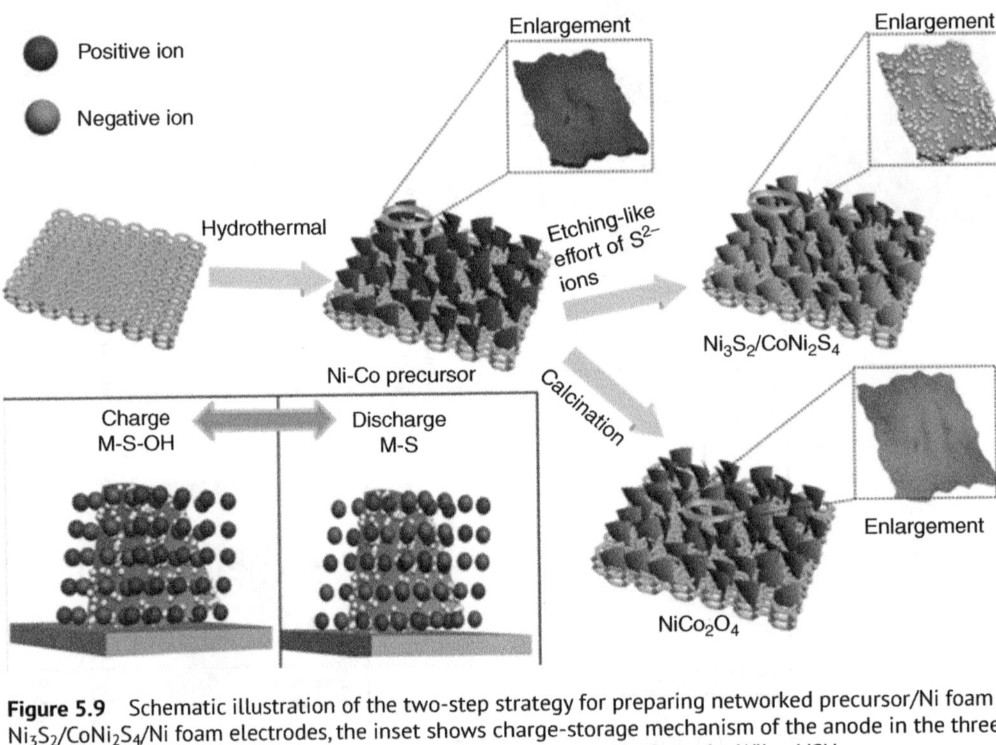

Figure 5.9 Schematic illustration of the two-step strategy for preparing networked precursor/Ni foam and $Ni_3S_2/CoNi_2S_4$/Ni foam electrodes, the inset shows charge-storage mechanism of the anode in the three-electrode system. *Source:* Reproduced from ref. 39 with permission from the Wiley-VCH.

$CoNi_2S_4$/NF (0–0.7 V) and AC/NF (0 to −1 V) electrodes with their corresponding voltage windows at 20 mV s^{-1}. According to the CV curve, the overall capacitance of the ASC device derives from the combined contribution of electrical double layer capacitance and redox pseudocapacitance. The specific capacitance is calculated to be 175 F g^{-1} at 1 A g^{-1} (see Figure 5.10b), which is still 112 F g^{-1} with the 64% of capacitance retention rate at 20 A g^{-1}, further evidencing the high capacitance and excellent rate performance of Ni_3S_2/$CoNi_2S_4$//AC/NF ASC. The ASC ahieved a high energy density of 40 Wh kg^{-1} at 17.3 kW kg^{-1} and electrochemical stabilities. Furthermore, the CV curves polarization and specific capacitance becomes more and more obvious and higher, respectively, with the increase of test temperature. These are attributed to the components' synergy assuring rich redox reactions, high conductivity as well as highly porous but robust architectures. The digital images for ASCs, the corresponding electrodes, separator, and schematic diagrams of ASCs internal structure was shown in Figure 5.10c.

It has been reported spinel-type $NiCo_2S_4$ and $CuCo_2S_4$ possess a narrower band gap and higher conductivity than the corresponding oxide/hydroxides because of the synergistic electrochemical contribution from both of the constituent metal ions [68, 69]. The electrochemical and mechanical stability of sulfides during the charge/discharge process remain to be improved. Combinating with

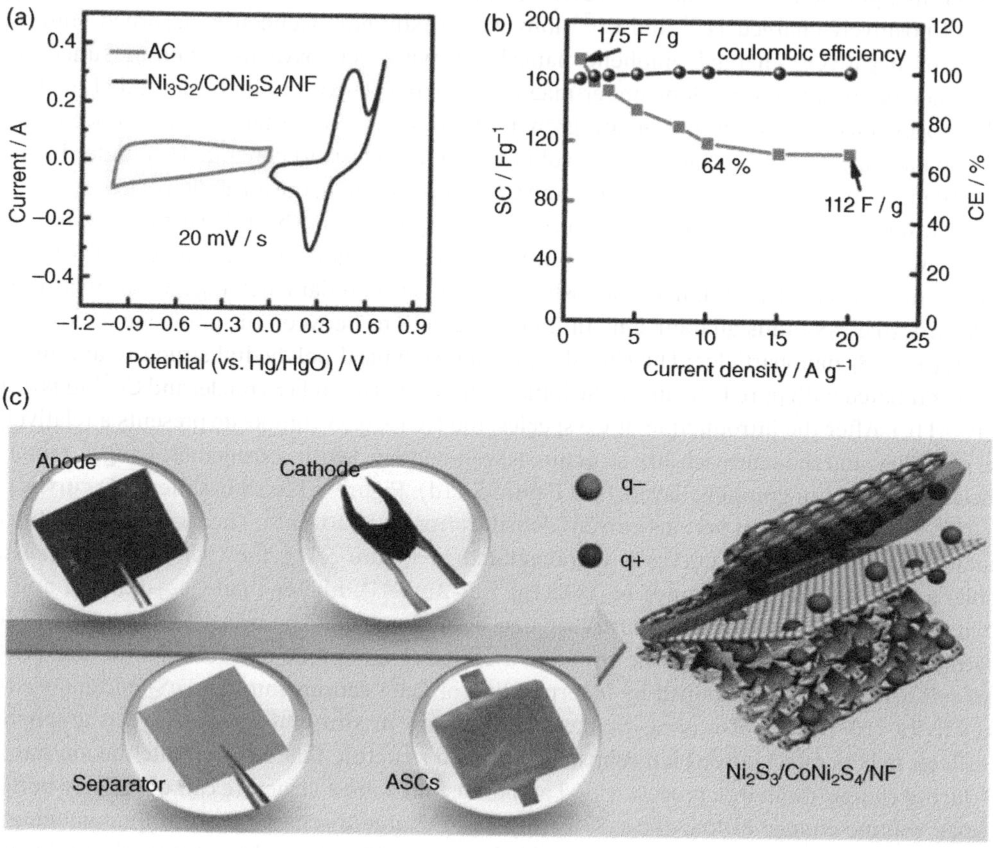

Figure 5.10 (a) Comparison of CV curves collected for Ni_3S_2/$CoNi_2S_4$/NF and AC/NF electrodes at 20 mV s^{-1}. (b) The corresponding specific capacitance and coulombic efficiency at different current densities. (c) Digital images for ASCs, the corresponding electrodes, separator, and schematic diagrams of ASCs internal structure. *Source:* Reproduced with permission [39]. © 2017, Wiley-VCH.

conductive materials, such as carbon nanotubes, graphene, conducting polymer and other hybrid inorganic compounds has been explored to enhance the structural stability of sulfides in faradaic processes. Zhou and co-workers [70] prepared $Ni(CO_3)(OH)_2$ nanosheet wrapped $CuCo_2S_4$ arrays on nickel foam via a two-step hydrothermal reaction. It is expected that by combining the highly conductive Cu/Co-based sulfides with the stable $Ni_2(CO_3)(OH)_2$, this hierarchical architecture interweaved by 1D bamboo shoot-shaped arrays and 2D nanosheets with open networks and free interspaces may efficiently create more electro-active sites and prevent the possible collapse and degradation of the electrode material during the electrochemical reactions. The obtained $CuCo_2S_4$@$Ni_2(CO_3)(OH)_2$ electrode reaches a high specific capacity of $343.9\,C\,g^{-1}$ at $0.33\,A\,g^{-1}$. The ASC was assembled using $CuCo_2S_4$@$Ni_2(CO_3)(OH)_2$ as the positive electrode and AC as the negative electrode, the energy density reaches $44\,Wh\,kg^{-1}$ at the power density of 139.3 W kg-1. Furthermore, by cladding the $CuCo_2S_4$ with $Ni_2(CO_3)(OH)_2$ nanosheets, the $CuCo_2S_4$ is effectively protected from chemical corrosion in the alkaline electrolyte during the charge/discharge process. The hierarchical structure of $CuCo_2S_4$ wrapped by $Ni_2(CO_3)(OH)_2$ nanosheets with such outstanding electrochemical performance may hold great potential in high-energy storage applications.

Yan and co-workers [71] elaborately developed a facile one-pot hydrothermal route to prepare ternary $Co_{0.33}Fe_{0.67}S_2$ nanoparticles in situ embedded between graphene nanosheets to form a sandwich-like graphene/$Co_{0.33}Fe_{0.67}S_2$ (GCFS-0.33) composite. As shown in Figure 5.11a and b, plenty of positively charged Fe^{3+} and Co^{2+} ions were firstly homogeneously adsorbed onto the surfaces of negatively charged graphene nanosheets with rich oxygen-containing functional groups via electrostatic interactions under magnetic stirring. Then, thiourea was added to allow the heterogeneous nucleation and in situ growth of $Co_{0.33}Fe_{0.67}S_2$ crystals on the surface of the graphene nanosheets. Lastly, thiourea was subjected to hydrolysis to generate H_2S under hydrothermal conditions, which reacted with Fe^{3+} and Co^{2+} ions and led to the formation of $Co_{0.33}Fe_{0.67}S_2$ nuclei. The impermeability and encapsulation effect of graphene nanosheets led to the $Co_{0.33}Fe_{0.67}S_2$ nuclei grew further and in situ crystallized into $Co_{0.33}Fe_{0.67}S_2$ nanoparticles embedded between graphene nanosheets. The graphene nanosheets provided abundant heterogeneous nucleation sites for $Co_{0.33}Fe_{0.67}S_2$, thus guaranteeing intimate interactions between the graphene nanosheets and $Co_{0.33}Fe_{0.67}S_2$ nanoparticles. The $Co_{0.33}Fe_{0.67}S_2$ shows typical polyhedral structure and much smaller compared with pure FeS_2 due to the ionic exchange between FeS_2 nuclei and Co^{2+} ions (see Figure 5.11c). After the introduction of Co species, the GCFS-0.33 composite presents a relatively loose structure, and the sandwich-like structure is well-endowed with considerable irregular nanoparticles embedded in graphene layers (see Figure 5.11d). Figure 5.11e shows the GCD curves of the GCFS-0.33 composite at various current densities from 1 to $10\,A\,g^{-1}$. The specific capacitance calculated based on the CV curves is demonstrated in Figure 5.11f, the GCFS-0.33 composite exhibits a high specific capacitance of $387.8\,F\,g^{-1}$ at $2\,mV\,s^{-1}$, higher than $Co_{0.33}Fe_{0.67}S_2$, and a capacitance retention of 61.8% at $200\,mV\,s^{-1}$ can be maintained. The excellent electrochemical performance of GCFS-0.33 composite could be ascribed to its unique and robust structure. (i) A richer redox reaction was obtained by incorporating of Co^{2+} cations into the FeS_2 and increased conductivity. (ii) Ternary $Co_{0.33}Fe_{0.67}S_2$ nanoparticles are in situ embedded between graphene nanosheets to generate a porous but robust sandwiched structure that can facilitate the ion transport during charge/discharge process. (iii) The porous but robust structure can effectively buffer the huge volume change of $Co_{0.33}Fe_{0.67}S_2$ during rapid charge/discharge process, guaranteeing a superior cycling stability. The ASC was assembled using GCFS-0.33 as the negative electrode and sulfidized graphene/CoNiAl-LDHs as the positive electrode with well-separated potential windows. The fabricated ASC delivered an excellent energy density of $66.8\,Wh\,kg^{-1}$ at $300.5\,W\,kg^{-1}$ and still retained $13.1\,Wh\,kg^{-1}$ even at a high power density of $29.4\,kW\,kg^{-1}$. The ASC displays

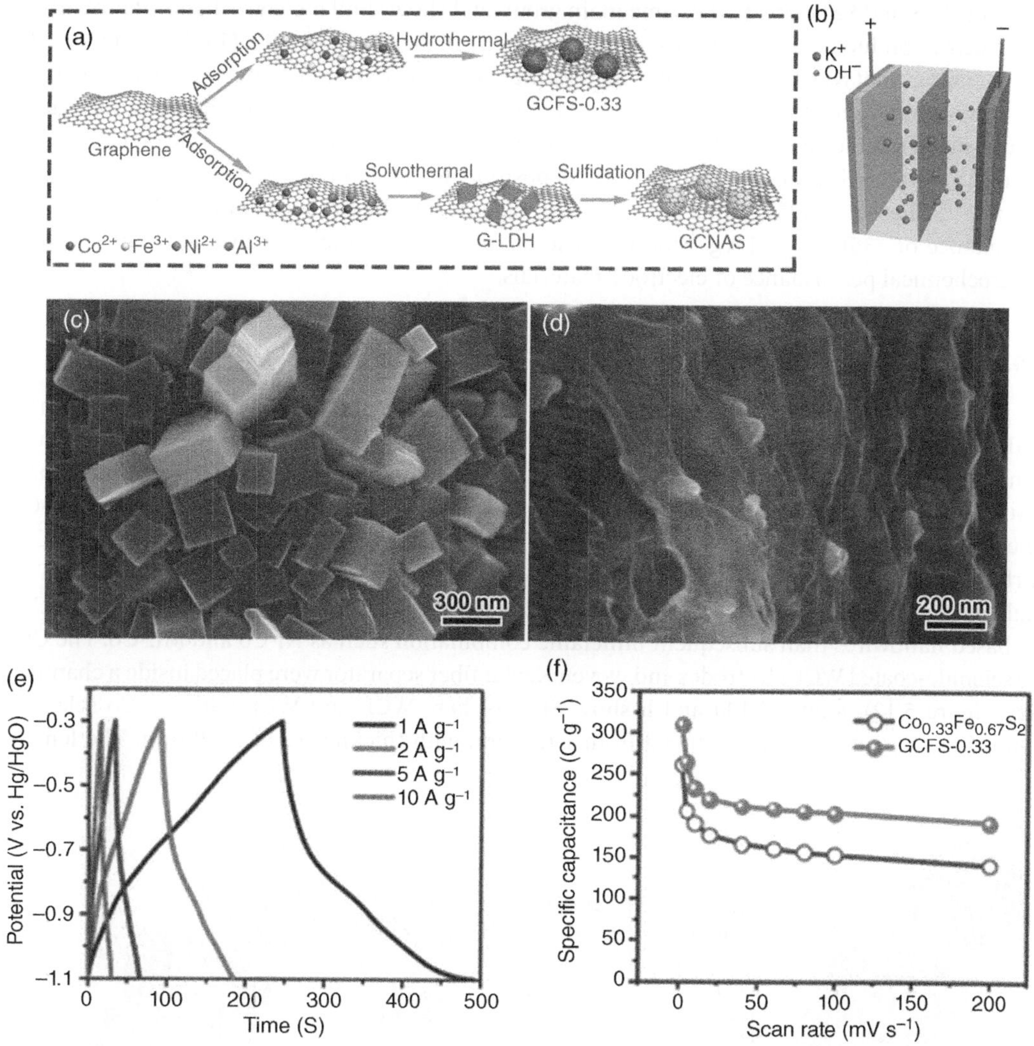

Figure 5.11 (a) Schematic illustration of the synthesis process of the composite and fabrication of ASC (b). (c) SEM image of Graphene/FeS_2, (d) SEM image of GCFS-0.33 composite. (e) GCD of the GCFS-0.33 composite electrode at different current densities. (f) Specific capacitance of $Co_{0.33}Fe_{0.67}S_2$ and the GCFS-0.33 composite at different scan rates. *Source:* Reproduced with permission [71]. © 2018, Royal Society Of Chemistry.

impressive long-term cycling stability with a capacitance retention of 102.2% relative to the initial capacitance after 1000 cycles. This versatile synthetic strategy can be readily extended to synthesize other transition-metal-sulfide-based composites with excellent electrochemical performance.

5.4 Metal Selenide Electrodes

Among many ternary Ni-Co based compounds, Ni-Co oxides/sulfides have been synthesized and used as high-performance electrode materials of SCs. In general, Ni-Co sulfides show largely improved electric conductivity compared to that of the corresponding oxides [72]. As we known, elements O, S and Se belong to the same main group and show similar electrochemical properties. The electric conductivity of Se ($1 \times 10^{-3}\,S\,m^{-1}$) is greatly higher than that of S ($1 \times 10^{-28}\,S\,m^{-1}$) [73], which implies that Ni-Co selenides may possess higher conductivity and better electroactivity than the corresponding oxides and sulfides. Recently, researchers are focused on the development of unique metal selenide electroactive materials for SCs. Wang et al. [74] prepared a $GeSe_2$ nanomaterial on a graphite substrate by chemical vapor deposition, which exhibited a capacitance of $300\,F\,g^{-1}$ at $1\,A\,g^{-1}$. Chen et al. [75] synthesized a Ni-Co selenide by a hydrothermal method, delivering a capacitance of $535\,C\,g^{-1}$ at $1\,A\,g^{-1}$ for SCs. The design of nanostructure can highly enhance the electrochemical performance of electrode materials.

5.4.1 1D Metal Selenide

1D nanostructures possess some distinct superiorities for application in the field of SCs compared with other nanostructures. For example, ternary $Ni_{0.8}Co_{0.2}Se$ NWAs//AC device exhibits an energy density of $17\,Wh\,kg^{-1}$ at $1526.8\,W\,kg^{-1}$ [76]. $(NiCo)_{0.85}Se$ nanotube arrays grown on carbon fabric result in an areal capacitance of $2.33\,F\,cm^{-2}$ at $4\,mA\,cm^{-2}$ [77]. Park et al. [78] reported Cu-Co selenide nanowires containing various amount of Cu and Co grown on woven carbon fiber (WCF) surfaces as a means of enhancing both the electrochemical and mechanical properties of a SC. The high electrical conductivity of Cu over other metals such as Ni, Mn offers better candidate of Cu-Co based nanowires than subsequent bimetallic combination such as Ni-Co and Mn-Co. The Cu-Co selenide-coated WCF electrodes and woven Kevlar fiber separator were placed inside a chamber (see Figure 5.12). Figure 5.13a and b show SEM of bare WCF and WCF with Cu-Co selenide nanowires. Nanowires' length ranged from 1 to 5 μm, with thicknesses of 50–90 nm. The length

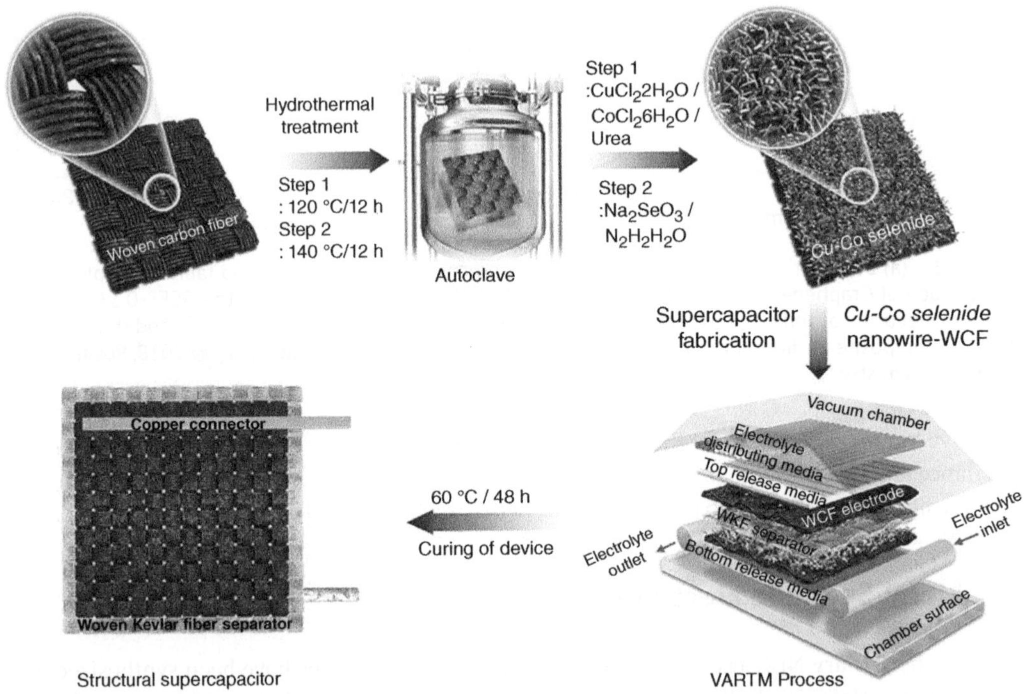

Figure 5.12 SC fabrication. *Source:* Reproduced with permission [78]. © 2019, Elsevier.

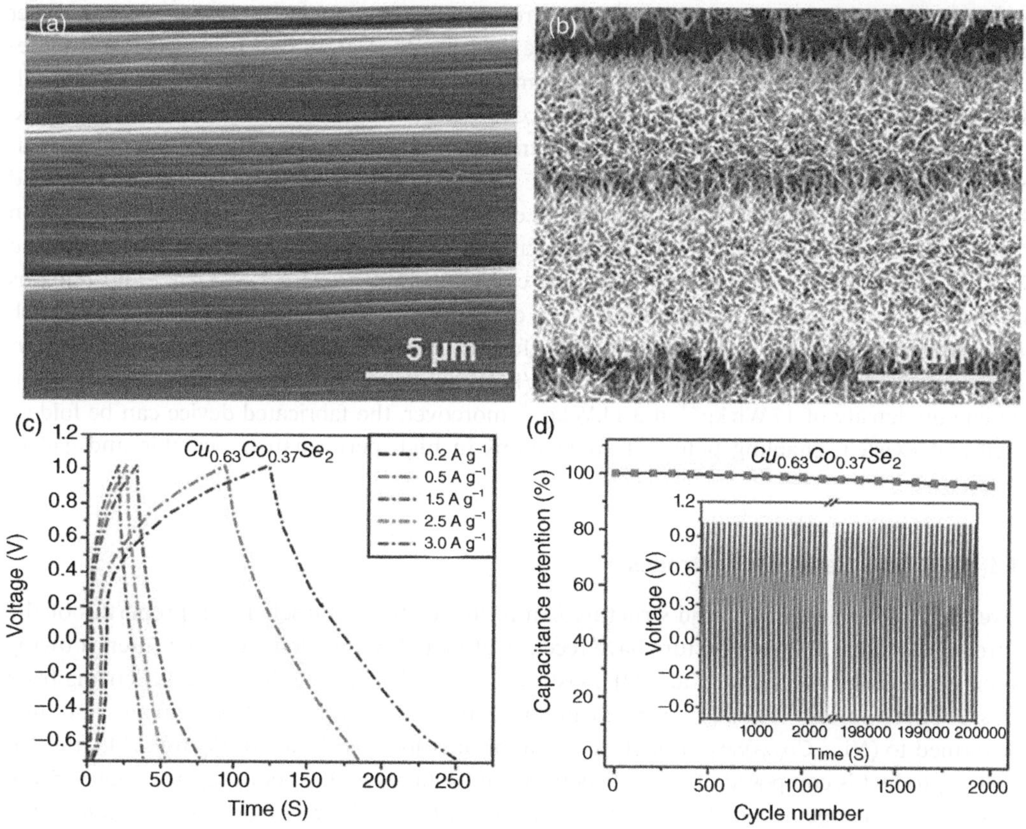

Figure 5.13 (a and b) SEM images. (c) Charge-discharge cycles as a function of discharge current. (d) Capacitance retention as a function of cycle number is given for $Cu_{0.63}Co_{0.37}Se_2$ nanowire-based SC. *Source:* Reproduced with permission [78]. © 2019, Elsevier.

and size distributions of the nanowires increase with enhancing the copper content as the copper ion predominant the nanowires growth process. The GCD curves from 0.2 to 3 A g^{-1} demonstrate stable characteristics (see Figure 5.13c). The overall cycling stability of the device was further evaluated over 2000 continuous charge-discharge cycles at 1 A g^{-1} (see Figure 5.13d). The potential was slightly improved during the charge-discharge cycles due to high energy density and good electrochemical reversibility of this device. The device achieved 100% capacity retention after up to 700 successive cycles and ended with 96.5% of its initial capacity after 2000 cycles.

5.4.2 2D Metal Selenide Electrodes

Zhang et al. [79] firstly reported $(Ni_xCo_{1-x})_{0.85}Se$ nanosheet arrays via a two-step method. ASC have been assembled $(Ni_xCo_{1-x})_{0.85}Se$ nanosheet arrays as positive electrode and nitrogen-doped porous carbon as negative electrode. The specific capacities of the ASC device are 88.22, 81.33, 78.22, 70.13, 64.02 and 58.67 mAh g^{-1} at 0.4, 0.8, 1.6, 2.4, 3.2 and 4 A g^{-1}, respectively. Such high specific capacities can be attributed to the large capacity to store tons of charges within the working area. Furthermore, the ASC can maintain 91.88% capacitance after 8000 cycles at 4 A g^{-1}, providing excellent cyclic stability. The $(Ni_xCo_{1-x})_{0.85}Se$//nitrogen-carbon ASC can achieve a high energy density of 70.58 Wh kg^{-1} at 320.02 W kg^{-1}. These results evidently prove $(Ni_xCo_{1-x})_{0.85}Se$ nanosheet arrays have excellent prospect for energy storage devices. Zhao et al. [80] have successfully prepared $NiCo_{2.1}Se_{3.3}$ nanosheets/graphene/Ni foam electrode via chemical vapor deposition, followed by a facile two-step hydrothermal process. The $NiCo_{2.1}Se_{3.3}$ nanosheets are arranged vertically on graphene/Ni foam with a uniform and stable structure without using chemical binders. The electrode exhibits an excellent specific capacitance of ~742.4 F g^{-1} at 1 mA cm^{-2}, and retains 471.78 F g^{-1} at 10 mA cm^{-2}. Hu et al. [81] developed a one-step growth of $(Ni_xCo_{1-x})_9Se_8$ in various conductive substrates as advanced electrodes for flexible, foldable SC. The optimized $(Ni_{0.1}Co_{0.9})_9Se_8$ nanodendrites deliver a specific capacitance of 3762 F g^{-1} at 5 A g^{-1} and remains 94.8% of the initial capacitance after 5000 cycles, owing to the advantage from fractal feature with numerous exposed surface as well as fast ion diffusion. The ASC based on $(Ni_{0.1}Co_{0.9})_9Se_8$ i@carbon fiber cloth//PVA (polyvinyl alcohol)/KOH//reduced graphene oxide exhibits an ultra-high energy density of 17 Wh kg^{-1} at 3.1 kW kg^{-1}. moreover, the fabricated device can be folded, pulled and twisted, enabling potential applications in high performance wearable and gadget devices.

5.4.3 3D Metal Selenide Electrodes

As we know, the morphology and structure can influence the electrochemical properties of the electrode materials, hollow structure have recently attracted tremendous research interest owing to their unique features. Zhan et al. [82] construct hollow $(Ni_{0.33}Co_{0.67})Se_2$ spheres. Starting from MOFs, NCO (nickel cobalt oxide) hollow spheres are synthesized as templates and then transformed to $(Ni_{0.33}Co_{0.67})Se_2$ through selenization method. As shown in Figure 5.14a, the Ni-Co-MOF product is composed of monodispersed solid spheres with smooth surface. Figure 5.14b shows the SEM image of the NCO derived from the Ni-Co-MOF precursor. After annealing, the spherical morphology is well maintained. The morphology of $(Ni_{0.33}Co_{0.67})Se_2$ has been drastically changed after the selenization reaction (see Figure 5.14c and d), and the unique hollow inner structure was obtained. The ASC device based on $(Ni_{0.33}Co_{0.67})Se_2$ spheres//AC display a high energy density of 29.1 Wh kg^{-1} at 800 W kg^{-1} and still retains 23 Wh kg^{-1} at high power density of 6400 W kg^{-1}. The high energy and power density confirms that the $(Ni_{0.33}Co_{0.67})Se_2$ spheres are promising for practical energy storage applications. Furthermore, Yuan and co-workers [83] fabricated monodisperse hollow $NiCoSe_2$ sub-microspheres using a solvothermal strategy. A detailed time-dependent morphology evolution investigation was carried out to gain insight into the formation process of hollow $NiCoSe_2$ sub-microspheres. When selenylation for one hour, partially hollow interior is visibly observed. With the extension of selenylation time up to three hours and even five hours, a core-shell structure can be discerned, and very large gap space can be found between the well-defined shell and nanocore with the selenylation further prolonging. Then a hollow spherical structure with completely void cavity can be obtained when the selenylation time is further increased to seven hours. The as-fabricated hollow $NiCoSe_2$ sub-microspheres possess appealing structural and compositional superiorities as potential electrode for SCs. To elaborate the potential application of the hollow $NiCoSe_2$ sub-microspheres as positive electrode for SCs, an ASC device was assembled using the commercial AC as negative electrode and hollow $NiCoSe_2$ sub-microspheres as positive electrode in 6 M KOH aqueous electrolyte. The ASC with the operating potential window extended to 1.5 V is able to provide a maximum specific energy density as high as 35 Wh kg^{-1} coupled with a specific power density of 188 W kg^{-1}. The long-term cycling performance is explored at 1 A g^{-1} over 5200 cycles, and 82.3% of SC retention forcefully supports good electrochemical stability of the ASC for practical applications. The electrode design and new

Figure 5.14 SEM images of the (a) Ni-Co-MOF solid spheres, (b) $NiCo_2O_4$, (c-d) hierarchical $(Ni_{0.33}Co_{0.67})Se_2$ hollow spheres. *Source:* Reproduced with permission [82]. © 2018, Elsevier.

insights hold huge promise in enriching inorganic synthesis methodologies, and pave an avenue to achieve in-depth understanding of the complex electrochemical charge storage process for next-generation SCs in the future.

The similar $CoFe_2Se_4$ has also been estimated owing excellent electrochemical properties. Meanwhile, combinating different materials can efficiently enhance the SC performance due to synergistic effect [84]. So far, little attention has been paid to synthesize composite electrode materials consisting of two kinds of ternary metal selenide components. Kim et al. [85] synthesized microsphere-like $CoFe_2Se_4$-$CoNiSe_2$ nanostructures by a simple hydrothermal method on copper/nickel coated polyester fabric (CNF), the unique architectures can facilitate ion and electron transfer for electrochemical reactions due to the porous and high surface area. The flexible $CoFe_2Se_4$-$CoNiSe_2$ electrode exhibited a high specific capacity of $183.4\,mAh\,g^{-1}$ at $1\,A\,g^{-1}$, excellent rate capability and long cycling lifetime (99.2% after 3000 cycles).

5.5 Fiber-Shaped SCs

Personal electronics with multifunctionality and wearability will become an advanced area and important trend. The subsequently increasing demand for lightweight and highly flexible power modules is essentially one of the critical challenges for the progress of miniaturized portable and wearable electronics. 1D fiber supercapacitor (FSC), as a promising candidate among various electrode materials, has been intensively investigated. The FSCs can be easily integrated into extremely small devices with tiny sizes and various kinds of shapes due to their small volume and high flexibility. Consequently, the construction of solid-state flexible energy storage systems is of great significance for the new generation of electronic devices.

The fundamental components of the fiber SC are the same as a traditional capacitor, which consists of two separated electrodes electrically connected via electrolyte. In contrast to conventional planar SCs, fiber SCs are linear devices, therefore usually smaller in size and lighter in weight. Fiber SCs take one of four types of configurations, parallel ply, twisted ply, coaxial and series. In general, a parallel ply fiber SC is composed of two separated linear electrodes connected via gel-electrolyte which also serves as a separator. Twisted ply fiber SCs are constructed by twisting two linear electrodes together with a separator infiltrated with an electrolyte or a gel polymer electrolyte serves as both separator and electrolyte. Coaxial fiber SCs are assembled by wrapping a gel-electrolyte coated core electrode with an outer electrode layer. In series fiber SCs, two non-contacting electrodes are placed on a common linear substrate in an end-to-end configuration and a continuous coat of gel-polymer electrolyte provides the electrical connection between the two electrodes.

Wang et al. reported the growth of $NiMoO_4$ nanowires radially on carbon cloth with good electrochemical properties [86]. Typically, $NiMoO_4$ nanowires directly grown on carbon cloth as electrodes in SCs possess obvious advantages: the open space between nanowires allowing for easy diffusion of the electrolyte, numerous fast electron-transport accesses, avoiding the use of a polymer binder, and excellent structural stability leading to the cycling stability and high rate capability evidently during long-term cycling at high current densities. Thus, the as-synthesized $NiMoO_4$ nanowires yielded high capacitance performance with a high specific capacitance of $1.27\,F\,cm^{-2}$ ($1587\,F\,g^{-1}$) at a current density of $5\,mA\,cm^{-2}$, which was higher than the reported binder enriched $NiMoO_4$ powders ($1517\,F\,g^{-1}$ at $1.2\,A\,g^{-1}$) [87]. The asymmetric SC devices with a maximum voltage of 1.7V have been fabricated, delivering both high energy density ($70.7\,Wh\,kg^{-1}$) and power density ($16\,000\,W\,kg^{-1}$ at $14.1\,Wh\,kg^{-1}$). Such high ASCs at large current densities further proved the great advantages of the binder-free $NiMoO_4$ nanowires. Shen group fabricated fiber SC by growing $CuCo_2O_4$ on Ni wires [88]. The device has a high capacitance of $11.09\,F\,g^{-1}$ at a current of 2 mA and an outstanding cycling stability with about 93.5% retention at 4 mA after 4000 cycles.

Nanotube arrays with well-defined pore structures have recently drawn great attention in the design of electrode materials due to the high surface area. Huang used low-cost and environment-friendly ZnO nanorod arrays grown on carbon fiber cloth (CFC) as a template to design and fabricate Ni-Co@Ni-Co LDH nanotube arrays with separated tubes and hierarchical top for efficient utilization of high surface area [89]. The well-aligned Ni-Co@Ni-Co LDH nanotube arrays designed in this work represent a new example of nanostructure engineering for better electrochemical performance, owing to the following features: (i) the Ni/Co metallic core provides a pathway for efficient charge transport, which effectively overcomes the weakness of Ni-Co LDH (relatively poor conductivity), (ii) the binary metal ions of Ni^{2+}/Ni^{3+} and Co^{2+}/Co^{3+} enhance capacitive properties since the mixed valence states can increase the number of active sites to trigger more redox reactions, (iii) the nanotube structure with a hierarchical top in addition to the separated tubes not only facilitates the transport of ions but also provides numerous electrochemical reaction sites, and (iv) well-aligned NTAs (nanotube arrays) growing homogenously on the CFC current collector enable almost all the nanotubes to participate in the electrochemical reactions without any dead volume. The electrochemical properties for the Ni:Co with different ratio depend on a proper amount of Ni and Co, which is also reflected in the gradual changes of crystal structure and morphology of the hybrid as a function of Ni:Co ratio. The sample with Ni:Co = 5:5 demonstrates the highest specific capacitance at all the current densities, which decreases from 2414 to 1980 F g^{-1} when the current density is increased from 1 to 20 A g^{-1}, about 82.1% retention rate of its initial value, indicating a high rate capability. The rate capability and capacitance imply the importance of doping the proper amount of cobalt into the active material. The ASC using Ni-Co@Ni-Co LDH nanotubes/CFC and CNFs/CFC as the positive and negative electrodes, respectively, was fabricated. The device shows good flexibility, high energy density (100 Wh kg^{-1} at 1500 W kg^{-1}), good cycling stability (96.9% after 3000 cycles) at a voltage window of 1.5 V. This work shows a good example of the combination of the electrode structural design and composition optimization for SC applications.

Portable electronic devices with roll-up and wearable characters attracted great attention in recent years. To fullfully flexible electronic devices, it is vital to design flexible and wearable energy storage units. Fiber SCs or planar SCs with fiber-type electrodes were thus recently developed. Figure 5.15a shows the schematic drawing of a flexible planar device with fiber-type electrodes, which consists of several $ZnCo_2O_4$ NWAs/carbon fibers, PET (Plythylene terephthalate) substrates, the electrolyte and two silver electrodes [90]. The photograph of the as-fabricated device in bending state is shown in Figure 5.15b, consisting of 14 $ZnCo_2O_4$ NWAs/carbon composite bers. Figure 5.15c shows the equivalent circuit of the planar-integrated ber SCs, where the enhanced distributed capacitance effect can be theoretically derived. It is evident that this flexible device based on the planar-integrated fiber electrodes can largely reduce the size of the device and achieve superior utilization efficiency and maximum functionality. Shen

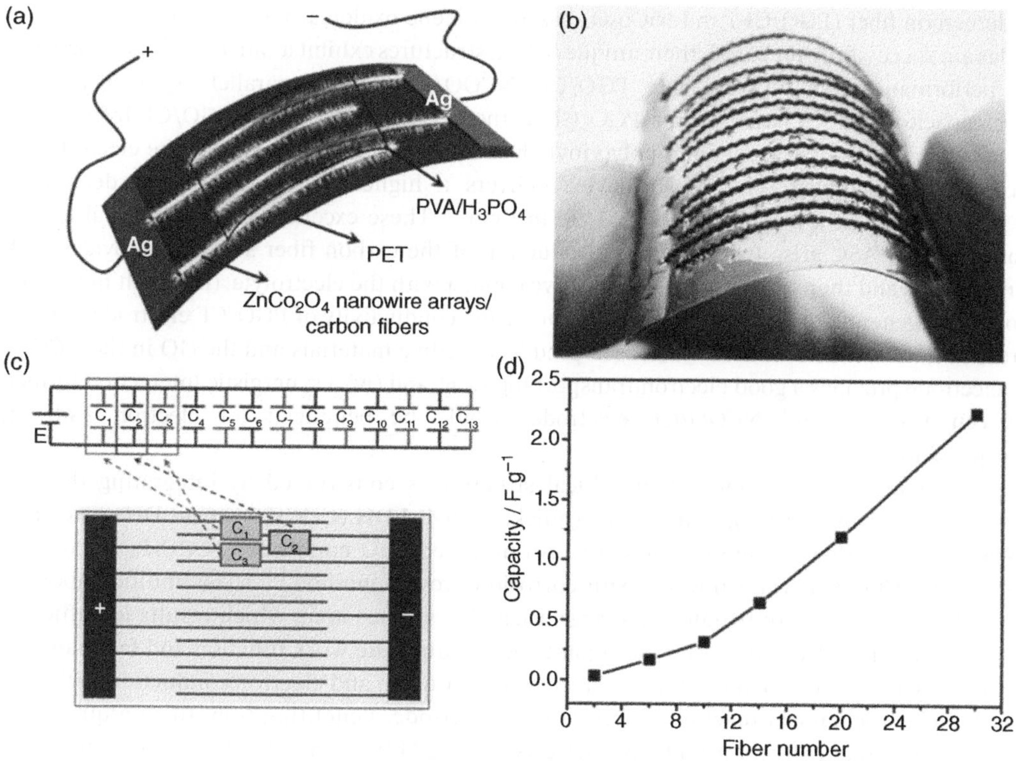

Figure 5.15 (a) Schematic illustration, (b) photograph, (c) equivalent circuit and (d) electrochemical performance of the flexible microsupercapacitors with fiber-type electrodes. *Source:* Reproduced with permission [90]. © 2013, Wiley-VCH.

has successfully developed a coaxial-type flexible FSC based on $NiCo_2O_4$ nanosheets/Ni wire as the fiber electrodes [91]. The volume capacitance of the fiber-shaped SCs reached $10.3\,F\,cm^{-3}$ at a current of 0.08 mA with outstanding cycling stability. An energy density of $1.44\,mW\,h\,cm^{-3}$ and a power density of up to $17\,W\,cm^{-3}$ were also obtained for the ber SCs, which are almost 48-fold higher than the previous values. Furthermore, the coaxial-type ber SCs do not show any apparent degradation in the bending test, illustrating the promise for use as electrodes for portable and wearable energy storage.

Wang et al. presented a novel type of thread-shaped ASC by depositing $NiCo_2Se_4$ nanosheets and $NiCo_2O_4$@PPy on Ti threadlike materials to obtain a large work function difference between positive and negative electrodes [92]. The fabrication procedure of the thread-shaped ASC is schematically illustrated in Figure 5.16a. Hydrothermal reactions were used to deposit ternary $NiCo_2Se_4$ and $NiCo_2O_4$@PPy on Ti wires to produce two types of electrode materials that were later used as positive and negative electrodes, respectively. High performance coaxial ASCs show stable output voltage of 1.7 V and provide very high volumetric capacitance of $14.2\,F\,cm^{-3}$ and excellent energy density of $5.18\,mWh\,cm^{-3}$ at a power density of $51.6\,W\,cm^{-3}$ together with superior flexibility under different bending conditions. These encouraging findings can open up the possibilities for selenide composites for applications in ASC with high voltage, high energy, and power densities to meet the diverse demands for wearable electronic devices.

Xiao et al. designed microscale fiber electrode by using carbon fibers as substrates and capillary channels as microreactors to space-confined hydrothermal assembling [93]. As P-doped graphene oxide/carbon fiber (PGO/CF) and $NiCo_2O_4$-based graphene oxide/carbon fiber (NCGO/CF) electrodes are successfully prepared, their unique hybrid structures exhibit a satisfactory electrochemical performance. An all-solid-state PGO/CF//NCGO/CF flexible parallel asymmetric fiber supercapacitor (AFSC) based on the PGO/CF as the negative electrode, NCGO/CF hybrid electrode as the positive electrode, and poly(vinyl alcohol)/potassium hydroxide as the electrolyte is successfully assembled. The AFSC device delivers a higher volumetric energy density of $36.77\,mW\,h\,cm^{-3}$ at a power density of $142.5\,mW\,cm^{-3}$. These excellent electrochemical performances of the ASC arise from: (i) the introduction of the carbon fiber substrate provides high surface areas and thereby increases the effective contact with the electrolyte, (ii) P-rich functional groups can generate high capacitance and improve the conductivity of PGO/CF electrode, (iii) the direct contact between the $NiCo_2O_4$-based pseudocapacitive materials and the GO in the NCGO/CF electrode provides a good electron transport highway, and (iv) a synergistic interaction formed between PGO/CF and NCGO/CF electrodes would be beneficial for electron and ion transportations.

An efficient multicomponent hierarchical structure is constructed by integrating the Cu-doped cobalt copper carbonate hydroxide@nickel cobalt LDH (CCCH@NiCo-LDH) core-shell NWAs on Cu fibers with highly conductive Au-modified CuO nanosheets (CCCH@NiCo-LDH NWAs@Au–CuO/Cu) via a novel in situ corrosion growth method.[101] This multicomponent hierarchical structure contributes to a large accessible surface area, which results in sufficient permeation of the electrolyte. The Cu dopant could reduce the work function and facilitate fast charge transfer kinetics. Therefore, the effective ion diffusion and electron conduction will facilitate the electrochemical reaction kinetics of the electrode. Benefiting from this unique structure, the electrode delivers a high specific capacitance ($1.97\,F\,cm^{-2}/1237\,F\,g^{-1}/193.3\,mAh\,g^{-1}$) and cycling stability (90.8% after 30 000 cycles), exhibiting superb performance compared with most oxide-based fiber electrodes. The hybrid FSC of CCCH@NiCo-LDH NWAs@Au-CuO/Cu// VN/carbon fibers is fabricated (see Figure 5.17a). As shown in Figure 5.17b, the potential

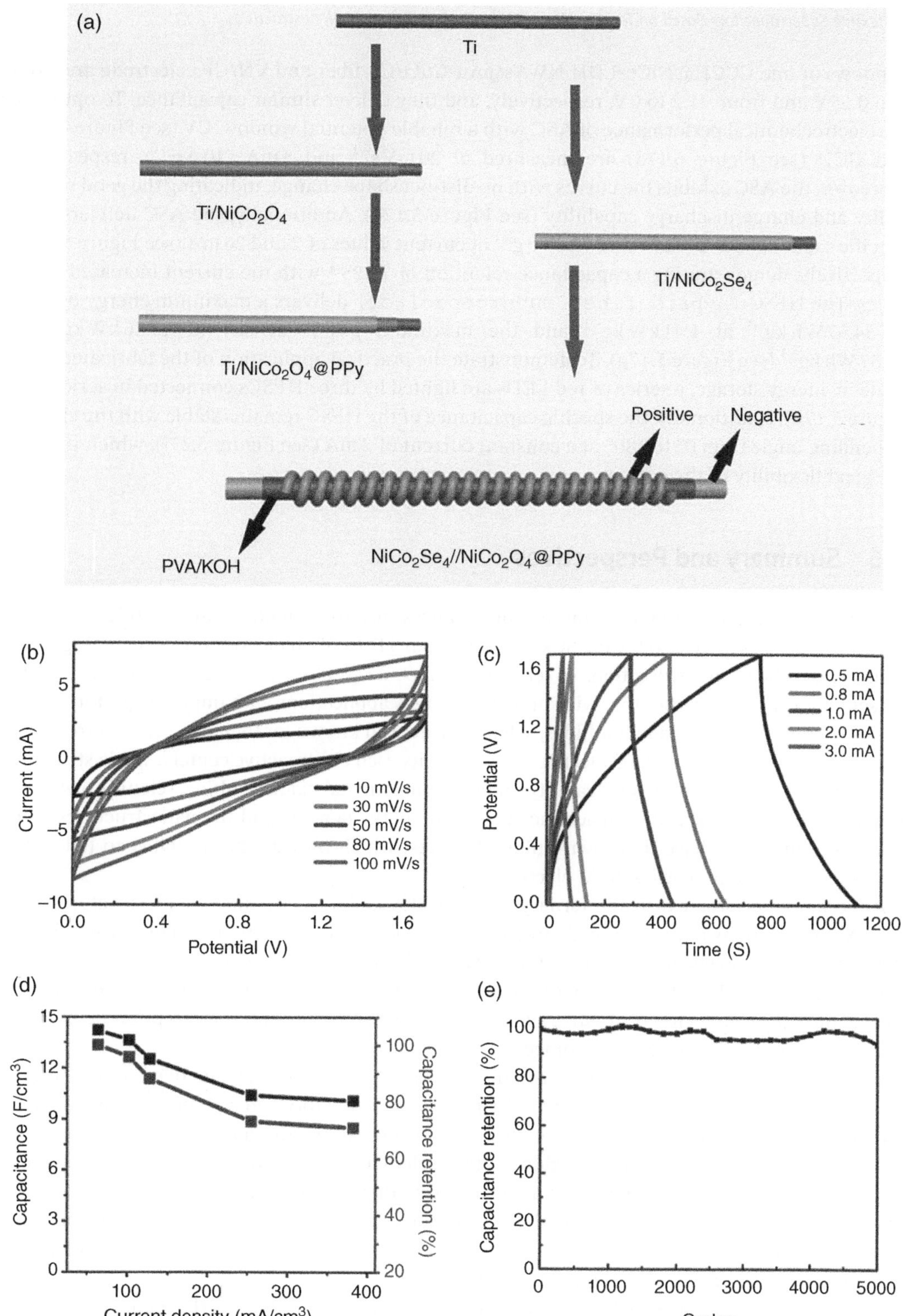

Figure 5.16 (a) Fabrication of thread-like ASC based on $NiCo_2Se_4$ nanosheets//$NiCo_2O_4$//PPy. (b) CV curves for the ASC device at different scan rates, (c) GCD curves at different currents, (d) volumetric capacitance and capacitance retention as a function of charge–discharge current, (e) capacitance retention of the flexible ASC device after up to 5000 cycles. *Source:* Reproduced from ref. 92 with permission by Wiley-VCH.

windows of one CCCH@NiCo-LDH NWAs@Au-CuO/Cu fiber and VN/CFs electrode are from 0 to 0.55 V and from −1.2 to 0 V, respectively, and they deliver similar capacitance. To optimize the electrochemical performance of ASC with a reliable potential window, CV (see Figure 5.17c) and GCD (see Figure 5.17d) are measured at $20\,mV\,s^{-1}$ and 5 mA ($10\,A\,g^{-1}$), respectively. Moreover, the ASC exhibits the curves with no distinct shape change, indicating the good reversibility and charge/discharge capability (see Figure 5.17e). Additionally, the ASC delivers high specific capacitances of 92.47 and $72.39\,F\,g^{-1}$ at current values of 2 and 20 mA (see Figure 5.17f), respectively, demonstrating a capacitance retention of 78.28% with the current increased by 10 times. The HFSC (hybrid fiber supercapacitor) delivers a maximum energy density of $34.97\,Wh\,kg^{-1}$ at $1.41\,kW\,kg^{-1}$ and the maximum power density of $13.86\,kW\,kg^{-1}$ at $27.37\,Wh\,kg^{-1}$ (see Figure 5.17g). To demonstrate the practical application of the fabricated electrode in energy storage, a series of red LEDs are lighted by three HFSCs connected in series (see Figure 5.17h). Additionally, the specific capacitance of the HFSC remains stable with the change of bending angle from 0° to 180° at a constant current of 2 mA (see Figure 5.17i), which implies the good flexibility of the device.

5.6 Summary and Perspectives

This chapter seeks to summarize the recent advances in ternary metal oxide (sulfide, selenide) nanostructure based electrodes for SCs, ranging from 1D to 3D nanostructures, including their composites. The future generations of ternary metal oxide (sulfide, selenide) nanostructures will present many new properties that will surely lead to new applications with improved performance. Generally ternary metal oxides (sulfides, selenides) are well known for their superior electrochemical characteristics, such as good electrical conductivity, rich electroactive centers, stable structure and environmentally benign applicability. In fact, ternary metal oxides (sulfide, selenide) can provide a source of high specific capacitance and high energy density, and the nanostructures with different dimensions have their own unique features, which provide the flexibility to tailor the microstructure and properties of SC electrodes.

Although great progress in the ternary metal oxides (sulfides, selenides) nanostructures and their SC applications have been made in recent years, some of the current challenges and future directions for the practical realization of ternary metal oxide (sulfide, selenide) are as follows:

Designing 1D and 3D architectures with high surface area and special structures, an effective way to maximize the electrochemical active sites for redox reactions or chemical adsorption-desorption and to further increase the energy storage density.

Ternary metal oxides (sulfides, selenides) are restricted by their poor conductivity, in the meantime, the energy density and capacitance of carbon is unsatisfactory and conductive polymers suffer from swelling and shrinking of the structure during charge/discharge or doping/dedoping processes. Combining ternary metal oxide (sulfide, selenide) with carbon or electroactive conducting polymers to form a core-shell structure by utilizing their synergistic effects.

Developing the composite ternary metal oxide (sulfide, selenide) nanostructured directly on conducting substrates as binder-free electrodes for electrochemical application can deliver fast diffusion of electrolyte and simplify the electrode processing.

Some engineering aspects should be focused on the development of high performance electrodes. It is still difficult to make electrodes or devices that can tolerate mechanical strains and to prepare ternary metal oxide (sulfide, selenide) nanostructured electrodes with tailored architectures in a cost-effective manner or on a large scale.

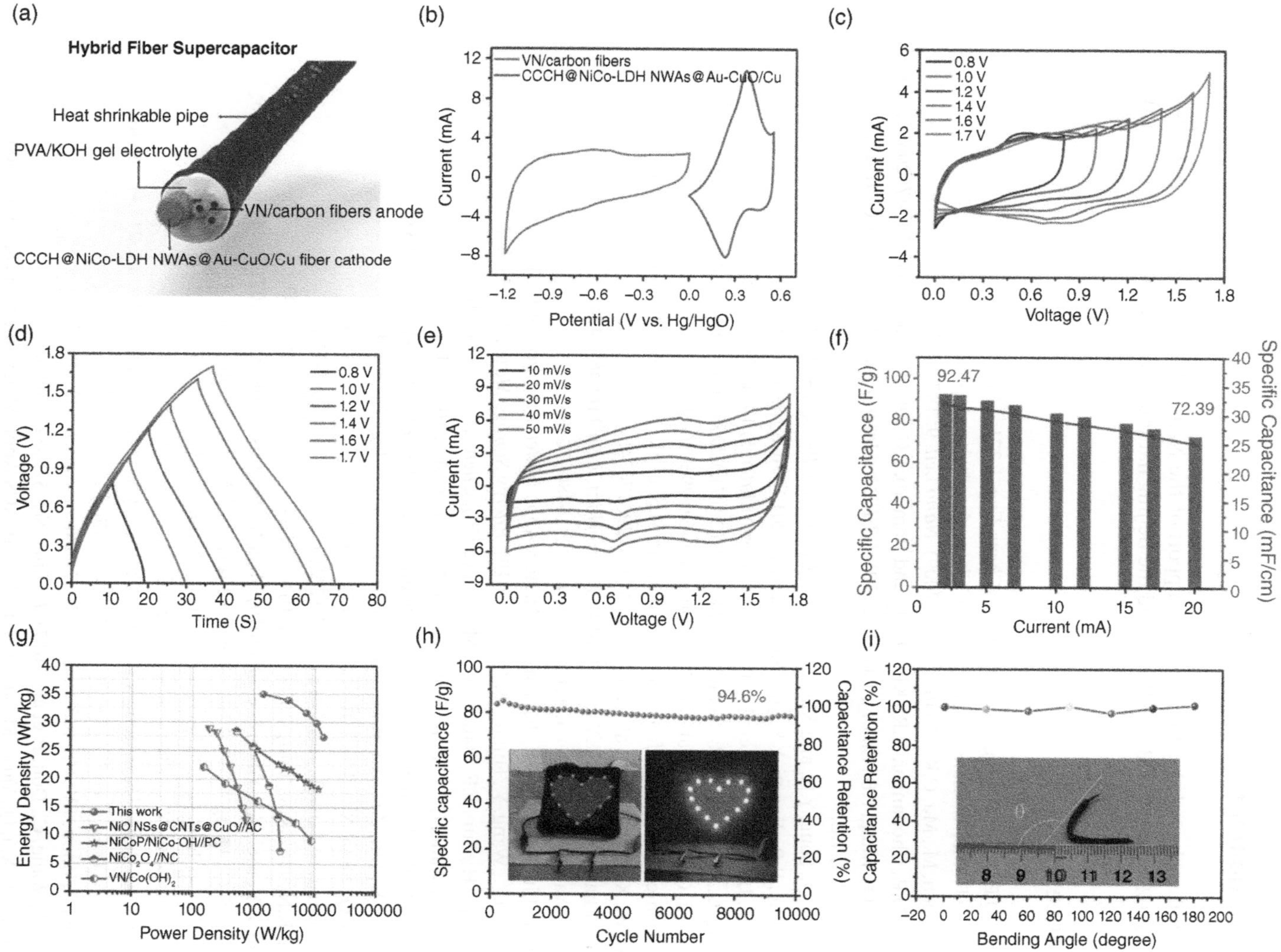

Figure 5.17 (a) Schematic diagram of the assembly of solid-state flexible device. (b) CV curves of CCCH@NiCo-LDH NWAs@Au–CuO/Cu fiber and VN/CFs. (c) CV and (d) GCD curves of the ASC operated at different voltages with a constant scan rate of $20\,mV\,s^{-1}$ and current of 5 mA, respectively. (e) CV curves, (f) specific capacitance values, (g) Ragone plots, and (h) long-term cycling performance of the ASC device and three serially connected ASCs for powering LED. (i) Capacitance retention at different bending angles. *Source:* Reproduced with permission [94]. © 2019, Wiley-VCH.

Declaration of Competing Interest

The authors declare that they have no known competing financial interests or personal relationships that could have appeared to influence the work reported in this paper.

Acknowledgments

We gratefully acknowledge the financial support of the National Natural Science Foundation of China (51702369, 51873233, 61625404) and the Beijing natural Science Foundation (4162062).

References

1 Wang, Q.F., Ma, Y., Liang, X. et al. (2019). *Chem. Eng. J.* 371: 145–153.
2 Wang, Q.F., Ma, Y., Liang, X. et al. (2018). *Chem. A* 6: 10361–10369.
3 Kaempgen, M., Ma, C.K., Chan, J. et al. (2009). *Nano Lett.* 9: 1872–1876.
4 Dubal, D.P., Chodankar, N.R., Kim, D.H., and Romero, P.G. (2018). *Chem. Soc. Rev.* 47: 2065–2129.
5 Jiang, Y.Q., Wang, P.B., Zhang, J. et al. (2010). *IEEE 23rd International Conference on IEEE*, 1171–1174. IEEE.
6 Wang, Y.G., Song, Y.F., and Xia, Y.Y. (2016). *Chem. Soc. Rev.* 45: 5925–5950.
7 Salunkhe, R.R., Kaneti, Y.V., and Yamauchi, Y. (2017). *ACS Nano* 11: 5293–5308.
8 Peng, S., Li, L., Wu, H.B. et al. (2015). *Adv. Energy Mater.* 5: 1401172.
9 Young, C., Salunkhe, R.R., Alshehri, S.M. et al. (2017). *Chem. A* 5: 11834–11839.
10 Li, F., Li, G., Chen, H. et al. (2015). *J. Power Sources* 296: 86–91.
11 Pendashteh, A., Moosavifard, S.E., Rahmanifar, M.S. et al. (2015). *Chem. Mater.* 27: 3919–3926.
12 Zuo, W., Xie, C., Xu, P. et al. (2017). *Adv. Mater.* 29: 1703463.
13 Li, L., Zhang, Y.Q., Liu, X.Y. et al. (2014). *Electrochim. Acta* 116: 467–474.
14 Xu, Y., Wang, X., An, C. et al. (2014). *Chem. A* 2: 16480–16488.
15 Liu, B., Liu, B.Y., Wang, Q.F. et al. (2013). *ACS Appl. Mater. Interfaces* 5: 10011.
16 Kong, D.Z., Wang, Y., Huang, S.Z. et al. (2019). *Energy Storage Mater.* 23: 653–663.
17 Huang, Z.H., Sun, F.F., Batmunkh, M. et al. (2019). *Chem. A* 7: 11826–11835.
18 Chen, D., Wang, Q.F., Wang, R.M. et al. (2015). *Chem. A* 3: 10158–10173.
19 Moosavifard, S.E., Shamsi, J., and Ayazpour, M. (2015). *Ceram. Int.* 41: 1831–1837.
20 Zhao, Y.X., Liu, Z.L., Gu, W.H. et al. (2016). *Nanotechnology* 27: 505401.
21 Cai, X.Q., Shen, X.P., Ma, L.B. et al. (2015). *Chem. Eng. J.* 268: 251–259.
22 Sivakumar, P., Jana, J., Jung, M.G. et al. (2019). *Chem. A* 7: 11362–11369.
23 Hui, K.N., Hui, K.S., Tang, Z.K. et al. (2016). *J. Power Sources* 330: 195–203.
24 Xiao, Z.Y., Mei, Y.J., Yuan, S. et al. (2019). *ACS Nano* https://doi.org/10.1021/acsnano.9b02106.
25 Shang, C.Q., Dong, S.M., Wang, S. et al. (2013). *ACS Nano* 7: 5430–5436.
26 Huang, L., Chen, D.C., Ding, Y. et al. (2013). *Nano Lett.*: 3135–3139.
27 Yu, M.H., Zeng, Y.X., Zhang, C. et al. (2013). *Nanoscale* 5: 10806–10810.
28 Kong, D.Z., Ren, W.N., Cheng, C.W. et al. (2015). *ACS Appl. Mater. Interfaces* 7: 21334–21346.
29 Cheng, D., Yang, Y.F., Xie, J.L. et al. (2015). *Chem. A* 3: 14348–14357.
30 Zhang, P., Zhou, J.Y., Chen, W.J. et al. (2017). *Chem. Eng. J.* 307: 687–695.
31 Wei, X., Wang, C., Dou, P. et al. (2017). *J. Mater. Sci.* 52: 3893–3905.
32 Jabeen, N., Xia, Q., Yang, M., and Xia, H. (2016). ACS Appl. *Mater. Interfaces* 8: 6093–6100.

33 Wang, X.H., Fang, Y., Shi, B. et al. (2018). *Chem. Eng. J.* 344: 311–319.
34 Lv, Y.M., Liu, A.F., Che, H.W. et al. (2018). *Chem. Eng. J.* 336: 64–73.
35 Zhou, Y., Ma, L., Gan, M.Y. et al. (2018). *Appl. Surf. Sci.* 444: 1–9.
36 Gao, Z., Wang, J., Li, Z. et al. (2011). *Chem. Mater.* 23: 3509–3516.
37 Jiang, Z., Li, Z., Qin, Z. et al. (2013). *Nanoscale* 5: 11770–11775.
38 Wang, X.H., Huang, F.F., Rong, F. et al. (2019). *Chem. A* 7: 12018–12028.
39 He, W.D., Wang, C.G., Li, H.Q. et al. (2017). *Adv. Energy Mater.* 7: 1700983.
40 Zhou, J., Wang, L., Yang, M. et al. (2017). *Adv. Mater.* 29: 1702061.
41 Yang, K., Yu, C., Fan, X. et al. (2016). *Energy Environ. Sci.* 9: 1299.
42 Yu, X., Yu, L., Shen, L. et al. (2014). *Adv. Funct. Mater.* 24: 7400.
43 Bai, D., Wang, F., Lv, J. et al. (2016). *ACS Appl. Mater. Interfaces* 8: 32853.
44 Mariappan, V.K., Krishnamoorthy, K., Pazhamalai, P. et al. (2019). *Nano Energy* 57: 307–316.
45 Li, M., Li, X.Y., Li, Z.Z., and Wu, Y.H. (2018). *ACS Appl. Mater. Interfaces* 10: 34254–34264.
46 Huang, W.G., Zhang, A.T., Liang, H. et al. (2019). *J. Colloid Interface Sci.* 549: 140–149.
47 Huang, Y.P., Zhao, Y., Bao, J. et al. (2019). *J. Alloys Compd.* 772: 337–347.
48 Liu, Y.K., Lu, Q.L., Huang, Z. et al. (2018). *J. Alloys Compd.* 762: 301–311.
49 Bahaa, A., Balamurugan, J., Kim, N.H. et al. (2019). *Chem. A* 7: 8620–8632.
50 Shen, L.F., Wang, J., Xu, G.Y. et al. (2015). *Adv. Energy Mater.* 4: 1400977.
51 Mohammadi, A., Moosavifard, S.E., Tabrizi, A.G. et al. (2019). *ACS Appl. Energy Mater.* 2: 627–635.
52 Wei, C.Z., Chen, Q.Y., Cheng, C. et al. (2019). *Inorg. Chem. Front.* https://doi.org/10.1039/C9QI00173E.
53 Wang, Y., Chen, Z., Lei, T. et al. (2018). *Adv. Energy Mater.*: 1703453.
54 Chen, H., Jiang, J., Zhang, L. et al. (2013). *Nanoscale* 5: 8879.
55 Pu, J., Cui, F., Chu, S. et al. (2014). *ACS Sustain. Chem. Eng.* 2: 809.
56 Jia, H.N., Wang, Z.Y., Zheng, X.H. et al. (2018). *Chem. Eng. J.* 351: 348–355.
57 Li, W.Y., Zhang, B.J., Lin, R.J. et al. (2018). *Adv. Funct. Mater.*: 1705937.
58 Zhu, J., Tang, S.C., Wu, J. et al. (2016). *Adv. Energy Mater.*: 1601234.
59 Sun, W., Du, Y.C., Wu, G.M. et al. (2019). *Chem. A* 7: 7138–7150.
60 Ma, M., Hua, Y., Chen, R. et al. (2016). *Nano Energy* 24: 139.
61 Xie, M.J., Xu, Z.C., Duan, S.Y. et al. (2018). *Nano Res.* 11: 216.
62 Xiong, X.H., Ding, D., Chen, D.C. et al. (2015). *Nano Energy* 11: 154–161.
63 Choi, D., Blomgren, G.E., and Kumta, P.N. (2006). *Adv. Mater.* 18: 1178.
64 Lu, X.H., Yu, M.H., Zhai, T. et al. (2013). *Nano Lett.* 13: 2628–2633.
65 Chen, J., Guan, C., Gui, Y., and Blackwood, D.J. (2017). *ACS Appl. Mater. Interfaces* 9: 496.
66 Zhao, M., Wang, Y., Ma, Q. et al. (2015). *Adv. Mater.* 27: 7372.
67 Hu, W., Chen, R., Xie, W. et al. (2014). *ACS Appl. Mater. Interfaces* 6: 19318.
68 Xu, T., Li, G., Yang, X. et al. (2019). *Chem. Eng. J.* 362: 783–793.
69 Tian, Z., Wang, X., Li, B. et al. (2019). *Electrochim. Acta* 298: 321–329.
70 Zhou, D.Y., Huang, J.P., Li, C.Y. et al. (2019). *New J. Chem.* 43: 5904–5913.
71 Liu, W., Niu, H., Yang, J. et al. (2018). *Chem. Mater.* 30: 1055–1068.
72 Xiao, J., Wan, L., Yang, S. et al. (2014). *Nano Lett.* 14: 831–838.
73 Cai, Q., Li, Y., Wang, L. et al. (2017). *Nano Energy* 32: 1–9.
74 Wang, X.F., Liu, B., Wang, Q.F. et al. (2013). *Adv. Mater.* 25: 1479.
75 Chen, H., Chen, S., Fan, M. et al. (2015). *Chem. A* 3: 23653–23659.
76 Guo, K., Cui, S., Hou, H. et al. (2016). *Dalton Trans.* 45: 19458–19465.
77 Xia, C., Jiang, Q., Zhao, C. et al. (2016). *Nano Energy* 24: 78–86.
78 Deka, B.K., Hazarika, A., Kim, J. et al. (2019). *Chem. Eng. J.* 355: 551–559.

79 Du, L.L., Du, W.M., Zhao, Y.P. et al. (2019). *J. Alloys Compd.* 778: 848–857.

80 Wang, Y.X., Zhang, W.J., Guo, X.L. et al. (2019). *ACS Appl. Mater. Interfaces* 11: 7946–7953.

81 Yang, P.Y., Wu, Z.Y., Jiang, Y.C. et al. (2018). *Adv. Energy Mater.* 8: 1801392.

82 Quan, L., Liu, T.Q., Yi, M.J. et al. (2018). *Electrochim. Acta* 281: 109–116.

83 Hou, L.R., Shi, Y.Y., Wu, C. et al. (2018). *Adv. Funct. Mater.* 13: 1705921.

84 Zhu, J., Tang, S., Wu, J. et al. (2017). *Adv. Energy Mater.* 7: 1601234.

85 Gopi, C.V.V.M., Reddy, A.E., Kim, H.J., and Mater, J. (2018). *Chem. A* 6: 7439–7448.

86 Guo, D., Luo, Y.Z., Yu, X.Z. et al. (2014). *Nano Energy* 8: 174–182.

87 Senthilkumar, B., Sankar, K.V., Selvan, R.K. et al. (2013). *RSC Adv.* 3: 352.

88 Gu, S.S., Lou, Z., Ma, X.D., and Shen, G.Z. (2015). *ChemElectroChem* 2: 1042–1047.

89 Liu, Y., Fu, N.Q., Zhang, G.G. et al. (2017). *Adv. Funct. Mater.* 27: 1605307.

90 Liu, B., Tan, D.S., Wang, X.F. et al. (2013). *Small* 9: 1998–2004.

91 Wang, Q.F., Wang, X.F., Wang, P. et al. (2014). *Nano Energy* 8: 44–51.

92 Wang, Q.F., Ma, Y., Wu, Y.L. et al. (2017). *ChemSusChem* 10: 1427–1435.

93 Zhou, C.Z., Gao, T.T., Wang, Y.J. et al. (2019). *Small* 15: 1803469.

94 Guo, Y.Q., Hong, X.F., Wang, Y. et al. (2019). *Adv. Funct. Mater.* 29: 1809004.

6

Transition Metal Oxide Based Electrode Materials for Supercapacitors

Xiang Wu

School of Materials Science and Engineering, Shenyang University of Technology, China

In recent decades, the increasing energy consumption has posed a huge threat to human society. It is very necessary to develop sustainable energy with high performance, low cost, and environment friendly energy conversion/storage characteristics. Supercapacitor devices are emerging as one of the promising energy devices for the future energy technology. In this chapter, we focus on metal oxide based nanomaterials as the electrodes for supercapacitors. Different synthesis methods and characteristics of electrode materials and energy storage mechanism of supercapacitors have been presented. The disadvantages of SCs based on metal oxide materials are also discussed.

6.1 Introduction

The supercapacitors, also named as electrochemical capacitors, are deemed as one of the most promising energy storage devices compared with the traditional batteries because they possess various intriguing properties such as high power density, fast charge-discharge speed, long cycle life and low temperature performance [1–6]. In addition, they can bridge the gap between conventional capacitors and batteries/fuel cells [7]. According to the charge storage mechanism, the supercapacitors can be divided in two kinds, i.e. electric double layer capacitors and pseudocapacitors [8–10]. The electric double layer capacitors are formed at the electrode/electrolyte interface due to the accumulation of electrostatic charge. The charge accumulation can be achieved by positive and negative charges residing on two interfaces separated by a vacuum or a molecular dielectric (the double layer) [11, 12]. While the capacitance of pseudocapacitor is stored by an under-potential deposition or redox reaction that occurs between electrode and electroactive ions. The charge storage is achieved by an electron transfer that produces chemical or oxidation state changes in the electroactive materials according to Faraday's law related to electrode potential [13–15].

In general, a pseudocapacitor possesses higher specific capacitance than an electric double layer capacitor. At the same time, the performance of supercapacitors mainly depend upon the characteristic of electrode materials [16–20]. The active materials in electroactive species contain hydroxides, transition metal oxides, and conducting polymers [21–28]. Highly efficient energy storage requires a device to store large quantities of electrical energy in very small space and can release them rapidly. Metal oxides, especially on nanoscale, have been research focuses as potential

Flexible Supercapacitors: Materials and Applications, First Edition. Edited by Guozhen Shen, Zheng Lou and Di Chen.

electrode materials for pseudocapacitor due to large scale and facile preparation and their rich redox reactions involving different ions, which contribute to high specific capacitances [29–33]. The size, morphology and orientation of metal oxide nanostructures can also be easily tuned, which makes it possible to systematically investigate the relationships between the structure and electrochemical properties. Meanwhile, most metal oxides are of great interest because of their richness in nature and low cost. Here, we have mainly discussed metal oxides including Co_3O_4, NiO, Fe_2O_3, MnO_2, V_2O_5.

6.2 Co_3O_4 Electrode Materials

Co_3O_4 nanomaterials have received considerable attention as electrode materials due to their excellent electrochemical performance, high theoretical specific capacitance, environment benign feature, and simple synthesis with versatile morphologies. Co_3O_4 electrode has been found to possess high efficiency, long term performance, and good corrosion resistance [34, 35]. The capacitive reactions of Co_3O_4 can be described as follows [36]:

$$CoO + OH^- + H_2O \rightarrow 3CoOOH + e^- \tag{6.1}$$

$$CoOOH + H^- \rightarrow CoO_2 + H_2O + e^- \tag{6.2}$$

To obtain large specific capacitance, Co_3O_4 nanostructures with various morphologies have been prepared. Co_3O_4 nanoflakes can be grown on nickel foam (NF) through a simple one-step hydrothermal route without using any surfactants at 200 °C for six hours. The as-synthesized products showed an initial discharge capacitance of 518 mF cm^{-2} at a current density of 1 mA cm^{-2} and can maintain 75% capacitance retention even after 6000 charge-discharge cycles [37]. However, Co_3O_4 material, when used as single electrode, possesses poor electrode stability, lower specific capacity (much lower than the theoretical value) and cyclic stability, which limit their practical applications [38, 39]. Its large volume expansion/contraction associated with alkali ion insertion and extraction process leads to electrode pulverization and a large irreversible capacity loss and poor cycling stability [40]. Thus, to enhance the electrode properties of Co_3O_4 nanomaterials, the combination of two metal oxide nanomaterials shows better electrochemical performances. Recently, Zheng et al. synthesized Co_3O_4 nanostructures with high yield on flexible Ni foam through a facile hydrothermal method (See Figure 6.1). Co_3O_4 products with high specific area are used as the scaffolds to support MnO_2 materials. Then 3D Co_3O_4@MnO_2 heterostructures exhibit a discharge

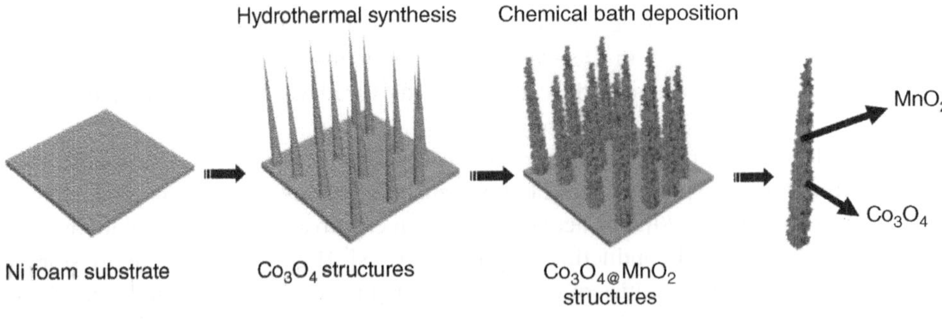

Figure 6.1 The schematic of 3D Co_3O_4@MnO_2 heterostructures grown on Ni foam.

capacitance of 1397.2 mF cm^{-2} at a current density of 1 mA cm^{-2}. Compared with a single material, it obviously improves the capacitive characteristic. After 6000 charge/discharge cycles, the hybrid structure can maintain its discharge capacitance of 86.1% [41]. Shen and coworkers fabricated nanowires assembled Co_3O_4@$NiCo_2O_4$ architectures through a two-step hydrothermal process. Owing to the synergistic effect between Co_3O_4 scaffolds and $NiCo_2O_4$ nanowires, the hierarchical electrodes exhibit an areal specific capacitance of 9.12 F cm^{-2} at the current density of 2 mA cm^{-2}. The device shows a maximum voltage of 1.0 V and delivers a specific capacitance of 1343.7 mF cm^{-2} at 2 mA cm^{-2}, and excellent cycling stability with no obvious capacitance attenuation after 5000 cycles at the current density of 10 mA cm^{-2} and high energy density of 75.6 Wh kg^{-1} with power density of 1053 W kg^{-1} [42]. Wu's group reported Co_3O_4@$NiCo_2O_4$ core-shell structures fabricated on flexible carbon cloth using a facile hydrothermal method. The as-fabricated core-shell structures present an areal capacitance of 4.35 F cm^{-2} at 1 mA cm^{-2} and specific capacitance of 1450 F g^{-1} at 1 A g^{-1} with a stable operational voltage of 0–0.4 V and outstanding cycling stability (~4.2% loss after 6000 cycles at 10 mA cm^{-2}, See Figure 6.2). They contribute the enhanced electrochemical behaviors to the rational design of $NiCo_2O_4$ nanoflakes adhering on Co_3O_4 nanowires, which promotes two electroactive materials to supply more pathways for accelerating fast electron and ion transfer [43].

The 3D Co_3O_4 nanowires@NiO nanosheet arrays were synthesized using a simple hydrothermal method. Co_3O_4 nanowires served as the support plate for NiO to increase the cycling stability of the material. NiO nanosheets connected with Co_3O_4 can shorten the ion diffusion path and enlarge the specific surface area to ensure that the active material and electrolyte possess effective contact and thus can show better cycle stability and conductivity. The electrochemical tests showed that the specific capacitances reached 2018 mF cm^{-2} and 608 F g^{-1} at a current density of 2 mA cm^{-2} and 2 A g^{-1}, and exhibited excellent cycle stability. An asymmetric supercapacitor was assembled using Co_3O_4 nanowire@NiO nanosheet as the cathode and activated carbon (AC) as the anode materials. It was observed that the fabricated supercapacitor exhibited capacity retention rate of 73.5% after 10 000 cycles at a current density of 10 mA cm^{-2} (See Figure 6.3) [44].

Cobalt oxides and hydroxides possess a very large theoretical pseudocapacitance of 3500–4600 Fg^{-1} [45]. Many cobalt oxide/hydroxide nanomaterials and their electrochemical behaviors have been investigated in the past few years [46–48]. However, highly conductive cobalt oxides usually show small surface area, while cobalt hydroxides with large surface area present low conductivity. Such disadvantages limit the practical application of cobalt oxides/hydroxides as electrode materials. Pang et al. developed a controllable one-pot hydrothermal methodology to prepare heterogeneous Co_3O_4 nanocube/$Co(OH)_2$ nanosheet hybrids. Electrochemical measurements showed that the electrode possessed a large specific capacitance of 1164 Fg^{-1}. The as-assembled flexible asymmetric all solid-state supercapacitor indicated the maximum energy density of 9.4 mWh cm^{-3} with excellent cycle stability even after 5000 cycles (See Figure 6.4) [49].

Co_3O_4@$Ni(OH)_2$ nanosheet arrays with well-defined morphologies were obtained by hydrothermal reaction without additive surfactants followed by calcination in air. The two-dimensional core-shell architecture exposes numerous active sites for fast faradic reaction. Such architecture results in an increased reaction area and a close contact of electrolyte. The electrodes showed enhanced electrochemical performance with high cycling stability of 90% after 3000 cycles. For an assembled supercapacitor, it delivers a specific capacitance of 98.4 F g^{-1} at 5 mA cm^{-2} in a potential range of 0–1.7 V and high energy density of 40.0 Wh kg^{-1} at a power density of 349.6 W g^{-1} [50].

To improve chemical reactivity for fast and efficient faradaic reactions, the surface of Co_3O_4 electrode can be functionalized with phosphate ions [51]. Phosphate ion functionalized Co_3O_4 nanosheet arrays showed enhanced specific capacitance and cycling stability compared with the

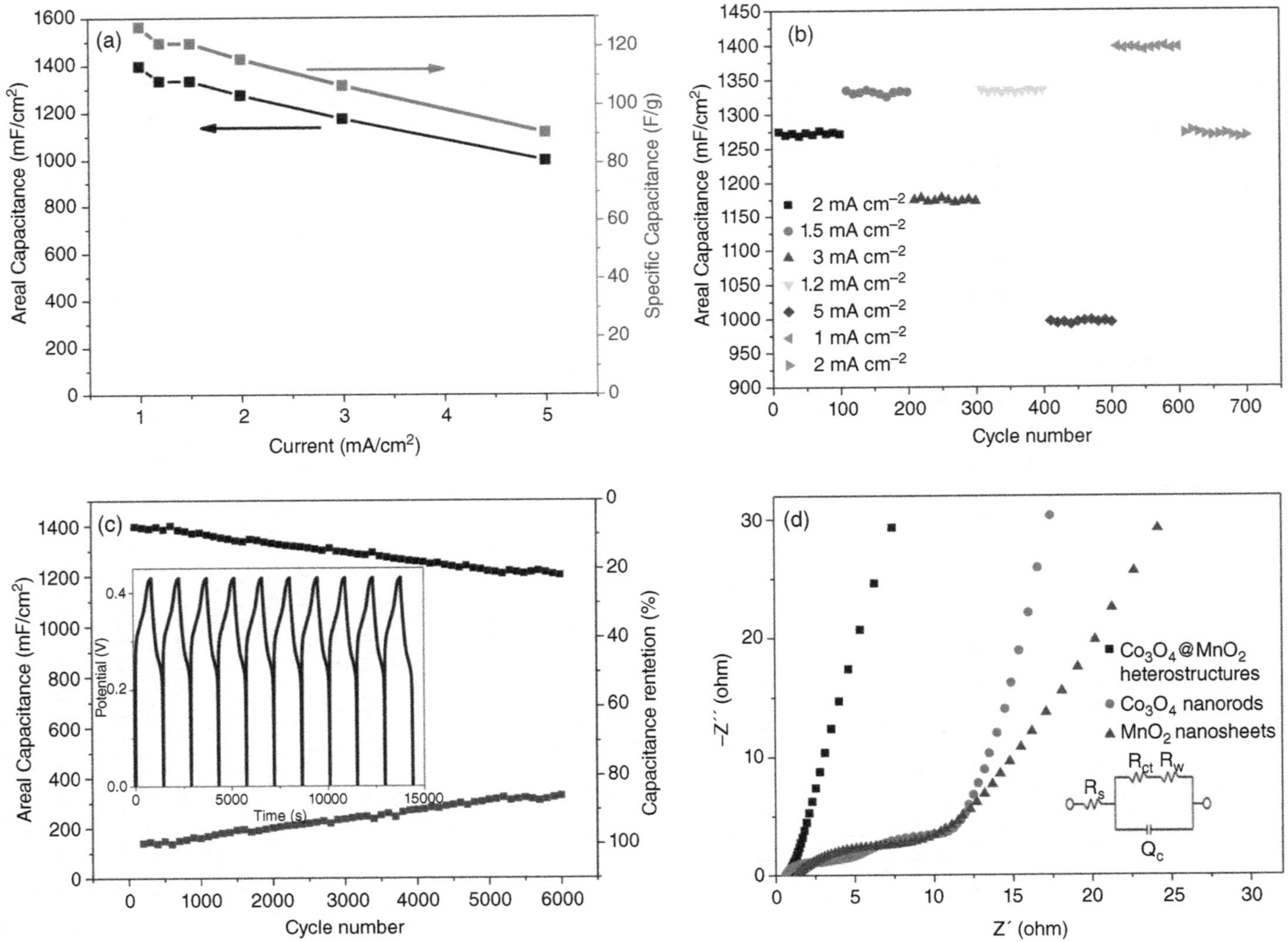

Figure 6.2 (a) Areal capacitance and specific capacitance of the hybrid electrode material as a function of current densities (b) Current density dependent cycling performance (c) Cycling performance and capacitance retention of 3D Co_3O_4@MnO_2 heterostructures at a current density of $1\,mA\,cm^{-2}$, the inset is 10 cycles of the hybrid electrode for charge-discharge curve (d) Nyquist plots of 3D Co_3O_4@MnO_2 heterostructures, individual MnO_2 and Co_3O_4 nanostructures, the inset is AC impedance equivalent matching circuit diagram.

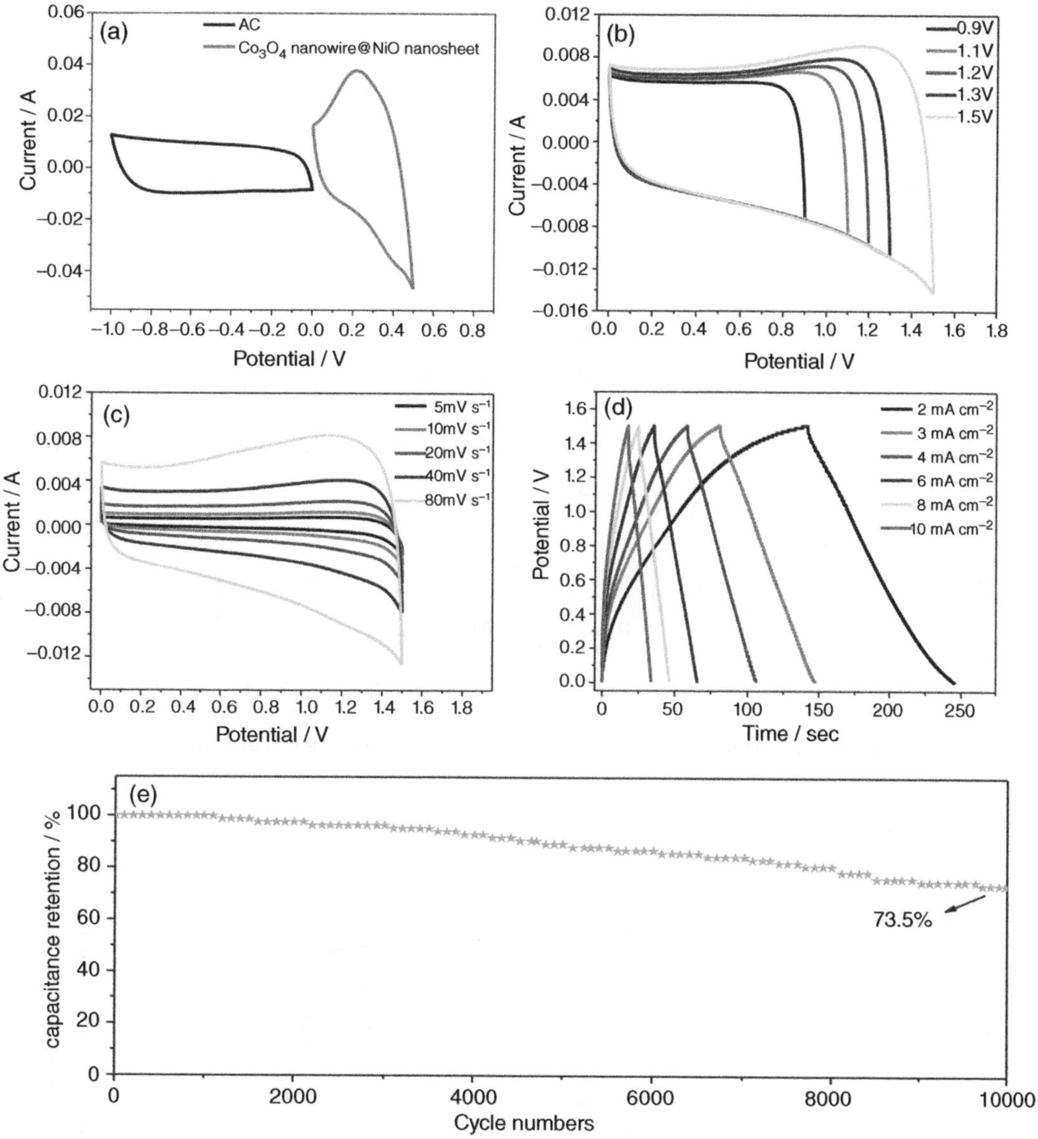

Figure 6.3 (a) CV curves of the electrodes (b) CV curves of ASC device within various operation voltage windows (c) CV curves of ASC device at various scan rates (d) charge-discharge curves of ASC device collected at various current densities (e) cycling performance of ASC at constant current density of $10\,mA\,cm^{-2}$.

as-prepared Co_3O_4 nanosheet array electrode. The functionalizing electrode can deliver a large specific capacitance of $1716\,F\,g^{-1}$ at a scan rate of $5\,mV\,s^{-1}$ in 6 M KOH electrolyte. It was found that phosphate ion functionalization on Co_3O_4 might significantly reduce the charge transfer resistance and increase the active reaction sites, resulting in improved reactivity and pseudocapacitive performance. The as-assembled asymmetric supercapacitor achieved an energy density of $71.6\,Wh\,kg^{-1}$ at an average power density of $1500\,W\,kg^{-1}$. The schematic of the synthesis process of phosphate ion functionalized Co_3O_4 nanosheet arrays (See Figure 6.5).

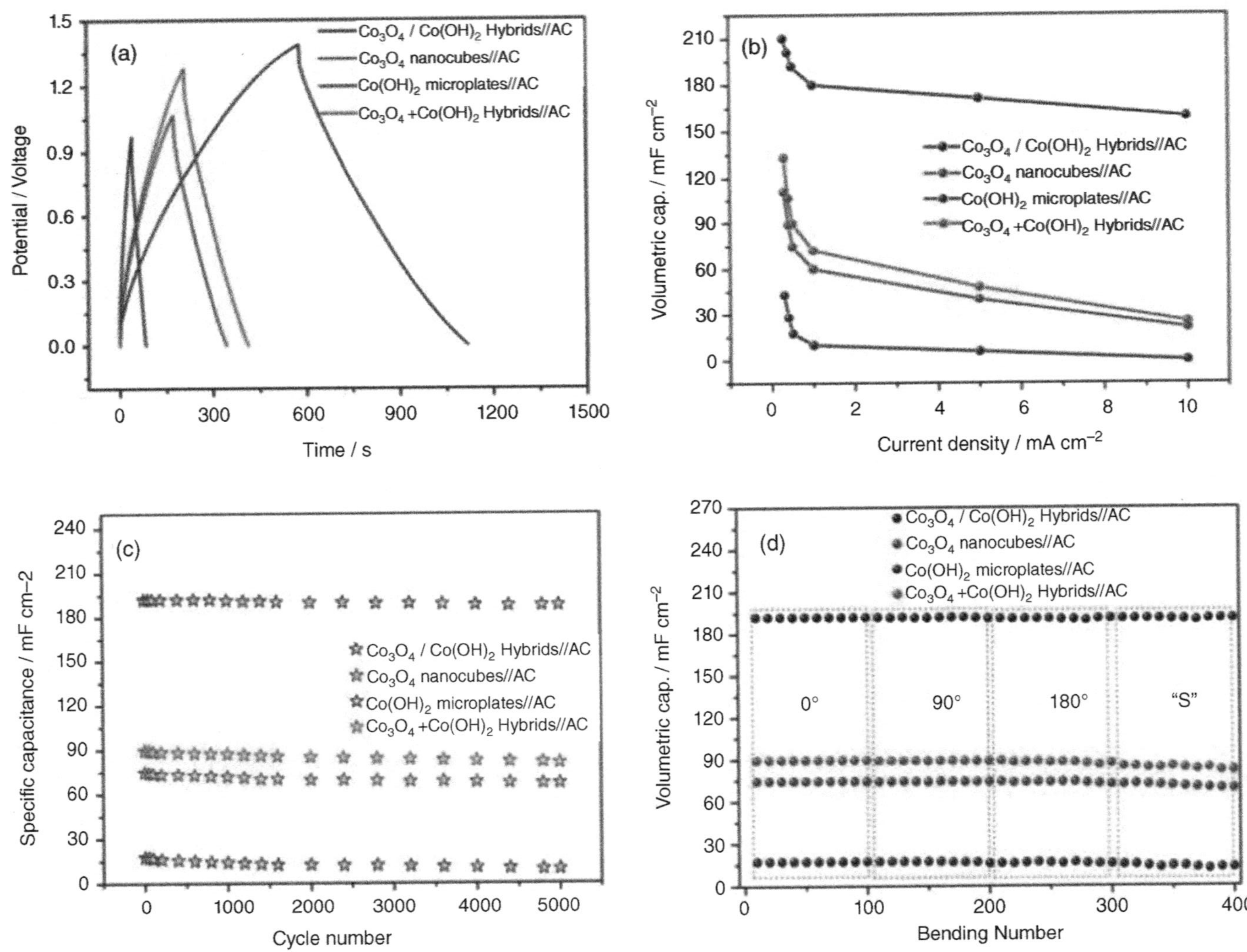

Figure 6.4 electrochemical characterization of ASC devices (a) charge-discharge curves (b) the specific capacitance (c) cycling performance at $0.5\,mA\,cm^{-2}$ (d) ASC devices under different bending degrees (0°, 90°, 180° and "S"), and each bending degree with 100 cycles.

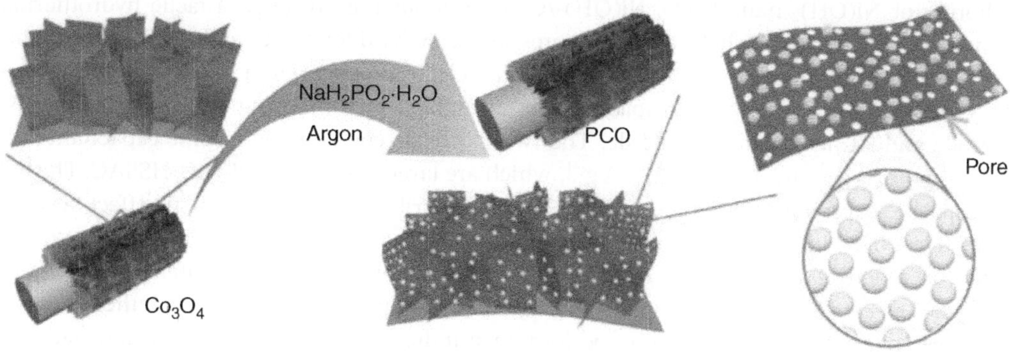

Figure 6.5 Schematic of phosphate ion functionalized Co_3O_4 nanosheet arrays.

6.3 NiO Electrode Materials

Among transition metal oxides, NiO is promising candidate for pseudocapacitor due to its easy availability, high theoretical capacitance, good chemical stability and pseudocapacitive behavior [52]. The performance of nickel oxide electrode materials for pseudocapacitor is determined by the redox reaction of NiO or $Ni(OH)_2$ in alkaline electrolytes. To study the electrochemical performance of NiO based electrode materials, the electrical conductivity and the degree of crystallinity need be taken into account [53, 54]. It has been generally believed that the electrochemical performance of NiO nanostructure largely depends on its morphology, surface area and the presence of dopants [55–57]. In particular, hollow nanosphere structures present several advantages, which can effectively reduce mass weight and enhance spatial dispersion, resulting in high surface area and better mass transportation. Cao et al. used a template-free microwave-assisted gas/liquid interfacial method to synthesize flowerlike NiO hollow nanospheres. The flowerlike structures show excellent electrochemical properties with specific capacitance of 585 F g^{-1} at a discharge current of 5 A g^{-1} and superior cycling performance [58].

Hierarchically porous NiO spheres were also prepared by a disodium citrate assisted method. The as-obtained products as the electrode for supercapacitor showed charge discharge performance of 463 F g^{-1} at 4.5 A g^{-1} and long cyclic stability with 95% capacitance retention after 1000 charge discharge cycles at 0.5 A g^{-1} [59]. Hu and coworkers reported a three-dimensionally nanoporous NiO film through an electrochemical route. The as-prepared NiO film showed a highly nanoporous structure and surface area of 264 m^2 g^{-1}. The electrochemical results showed a specific capacitance of 1776 F g^{-1}, power density of 89 Wh kg^{-1} at 16.5 kW kg^{-1} [60].

To improve the conductivity of electrode materials, Gan's group reported encapsulation of NiO nanoparticles in mesoporous carbon nanospheres through a hydrothermal route, followed by a carbonization and template removal procedure. The as-synthesized NiO/carbon sample as a supercapacitor electrode showed a specific capacitance of 406 F g^{-1} at 1.0 A g^{-1}. Furthermore, the electrode shows high rate capability (315 F g^{-1}) at 20.0 A g^{-1}. In addition, NiO(7.20%)/MCN electrode presented superior electrochemical stability after 10 000 cycles [61]. Yu et al. used semi-sacrificial template-assisted hydrothermal and calcination methods to prepare a hierarchical NiO/C hollow sphere composite (NiO/C-HS). They firstly prepared SiO_2/polydopamine (SiO_2/PDA) spheres with uniform diameters by the hydrolysis of tetraethoxysilane and the polymerization of dopamine. SiO_2/PDA spheres were then transformed into SiO_2/C solid spheres through calcination under N_2 atmosphere condition. Finally, SiO_2/C spheres were used as semi-sacrificial templates for the anchoring of $Ni(OH)_2$ nanosheets. $Ni(OH)_2$/C-HS was obtained through a facile hydrothermal method, where the carbon shells and SiO_2 cores served as hard templates and self-sacrificial templates, respectively. To evaluate the composite as an electrode material in practical application, a NiO/C-HS||AC asymmetric supercapacitor was assembled by employing NiO/C-HS as positive electrode and AC as negative electrode. The two-electrode cell showed specific capacitances of $99\,F\,g^{-1}$ at $0.25\,A\,g^{-1}$ and $47\,F\,g^{-1}$ at $2.0\,A\,g^{-1}$, which are larger than those of NiO-HS||AC. The elevated specific capacitance of NiO/C-HS||AC could be attributed to the synergistic effect between conductive C-HS and NiO nanosheets. For asymmetric supercapacitors, stability is an important factor for practical applications. NiO/C-HS||AC ASC almost didn't capacitance decay after 5000 cycles at $1\,A\,g^{-1}$. They thought the integrity of the composite provided some buffer space for volume change. The excellent cycling performance indicated that the composite can maintain good mechanical integrity and high electrochemical stability [62].

In addition, Wang et al. developed a novel ASC based on carbon nanotubes@nickel oxide nanosheets (CNT@NiO) core-shell composites as positive electrode and porous carbon polyhedrons (PCPs) as negative electrode in aqueous KOH solution. The CNT@NiO core-shell hybrids were prepared by a facile chemical bath deposition method followed by thermal annealing. PCPs were obtained by direct carbonization of Zn-based metal-organic frameworks (MOFs). The as-fabricated CNT@NiO//PCPs ASC showed maximum energy density of $25.4\,Wh\,kg^{-1}$ at a power density of $400\,W\,kg^{-1}$ and still remains $9.8\,Wh\,kg^{-1}$ at $16000\,W\,kg^{-1}$ in a wide voltage region of 0–1.6 V. Especially, the CNT@NiO//PCPs asymmetric supercapacitor showed ultralong cycling stability with 93% capacitance retention after 10 000 cycles [63].

Recently, graphene based transition metal oxides have attracted one's interest a due to their enhanced electronic conductivity and the unique graphene 2D structure. Jiang's group used a facile method to fabricate monolayer graphene/NiO nanosheets with a two-dimensional structure. The process is based on self-assembly due to mutual electrostatic interactions of two species. All Ni^{2+} ions are uniformly dispersed on the surface of graphene oxide (GO) sheets. This novel 2D structure improves the conductivity of the composite and prevents the aggregation of NiO nanoparticles. The composite obtained capacity retention of 95.4% after 1000 cycles due to the volume change of graphene during charge/discharge process [64].

6.4 Fe_2O_3 Electrode Materials

Hematite (α-Fe_2O_3) is an environmentally friendly, low cost, non-toxic and stable electrode material. It possesses a high theoretical capacitance ($3265\,F\,g^{-1}$) and a suitable negative potential working voltage (−1.0 V–0 V) [65, 66]. So far, α-Fe_2O_3 has received extensive attention as high performance anode material for ASCs. However, α-Fe_2O_3 possesses a lower conductivity, what makes the reported α-Fe_2O_3 electrode material lower capacitance. Therefore, some strategies have been taken to improve the capacitance, including designing the microstructure of materials to shorten the transmission path of ion electrons and coating the surface with conductive polymers or carbon materials to improve the conductivity of the materials. Oxygen vacancies are introduced or doped with other metal elements [67–70]. However, the obtained specific capacitance is still far below the theoretical capacitance. Therefore, improving the capacity of α-Fe_2O_3 electrode materials while still maintaining its capacity remains is a huge challenge. A carbon-coated α-Fe_2O_3 (α-Fe_2O_3@C) core/shell nanowire arrays (NWAs) grown on a flexible carbon cloth were reported by using a simple hydrothermal method. The conductivity of α-Fe_2O_3 can be significantly improved by coating carbon shell. The as-prepared α-Fe_2O_3@C nanostructures show a capacitance of

$280\,F\,g^{-1}$ at a current density of $1\,A\,g^{-1}$, and an ASC are assembled, which shows the maximum energy density of $30.625\,Wh\,kg^{-1}$ and the maximum power density of $5000\,W\,kg^{-1}$. To show the practical application of the ASC, the prototype device is connected to a 3.2V blue light emitting diode (LED) and can successfully lighten it for three minutes [71]. The electrochemical performances of two electrode materials are also shown (See Figure 6.6).

Yu et al. synthesized large-scale CC/Fe_2O_3 electrode material ($14.50\times31.00\,cm^2$) by a simple, efficient and low-energy cost hydrothermal route followed by a post annealing treatment. They constructed symmetrical supercapacitors with CC/Fe_2O_3 as electrode and 2.0 M Li_2SO_4 as electrolyte, revealing an operational voltage window of 2.0V. A energy density of $11.0\,mWh\,cm^{-3}$ was achieved, which is much higher than most of the previously reported carbon cloth-based asymmetric supercapacitors. Moreover, this kind of supercapacitor can be amplified from 1.50 to $100.00\,cm^2$ without obvious areal capacitance loss [72]. Xia et al. developed a novel strategy to engineer ultrafine Ni nanotube arrays as robust conductive backbones for depositing ultrafine Fe_2O_3 active material. They fabricated the ultrafine Ni nanotube arrays with ultrathin tube walls by a modified template-assisted method. The α-Fe_2O_3 nanoneedles are then controllably electrodeposited on ultrafine Ni nanotube arrays (NiNTAs@Fe_2O_3 nanoneedles) through a bottom-up strategy, which can effectively enhance specific capacitance of $418.7\,F\,g^{-1}$ at $10\,mV\,s^{-1}$ and rate capability of $215.3\,F\,g^{-1}$ at a current density of $64\,A\,g^{-1}$. Moreover, this effective strategy was also adopted for cathode design, and NiNTAs@MnO_2 nanosheets were successfully fabricated, revealing excellent electrochemical performance as cathode for supercapacitors. At the same time, a 1.6V ASCs integrated based on NiNTAs@Fe_2O_3 nanoneedle anode and NiNTAs@MnO_2 nanosheet cathode were demonstrated. A schematic of the synthesis procedure for NiNTAs@Fe_2O_3 nanoneedles and NiNTAs@MnO_2 nanosheets is shown (See Figure 6.7). The inset indicates different Ni films formed on ZnO nanorod with and without Au layer [73].

Lu and coworkers reported the rational design and fabrication of Ti-doped Fe_2O_3@poly (3,4-ethylenedioxythiophene) (denoted as Ti-Fe_2O_3@PEDOT) core/shell NR arrays grown on flexible carbon cloth as high-performance anode for ASCs. Ti^{4+} has been reported to be an electron donor by replacing Fe^{3+} and reducing Fe^{3+} to Fe^{2+}. It was proposed that the donor density of Fe_2O_3 could be significantly enhanced after Ti doping, and could boost its capacitive performance. In addition, PEDOT is an ultrahigh stable and conductive polymer. It can not only improve the conductivity of nanomaterials, but also act as a protective layer to prevent the architectures from destruction/degradation. Such unique core/shell architectures can offer a high electrical conductivity of the overall electrode for charge transport, a large reaction interfacial area and many channels for rapid diffusion of electrolyte ions within electrode. They enable the designed Ti-Fe_2O_3@PEDOT electrode with an excellent capacitive performance. The as-prepared electrode showed a remarkably large areal capacitance ($311.6\,F\,g^{-1}$ and $28.8\,F\,cm^{-3}$ at $1\,mA\,cm^{-2}$). They also assembled a flexible ASC device with a maximum energy density of $0.89\,mWh\,cm^{-3}$ and a maximum power density of $0.44\,W\,cm^{-3}$. The Ti-Fe_2O_3@PEDOT electrode also exhibited ultrahigh cycling durability with more than 96% capacitance retention after 30000 cycles [23]. Li et al. fabricated α-Fe_2O_3 NWAs on carbon cloth by electrodeposition followed by an annealed process in air. Subsequently, a thin PANI was electrodeposited on α-Fe_2O_3 nanowires. They obtained highly ordered 3D α-Fe_2O_3@PANI core-shell NWAs. Compared with pure α-Fe_2O_3 electrode, the composite electrode shows better electrochemical performance with a twice increase in C_{sp} from 33.93 to $103\,mF\,cm^{-2}$. In addition, a high-performance ASC was designed with α-Fe_2O_3@PANI nanowires as anode and PANI nanorods as cathode, and it showed a volumetric capacitance of $2.02\,mF\,cm^{-3}$, an energy density of $0.35\,mWh\,cm^{-3}$ at the power density of $120.51\,mW\,cm^{-3}$ based on volume of whole device, and very good stability with C_{sp} retention of 95.77% after 10, 000 cycles [74].

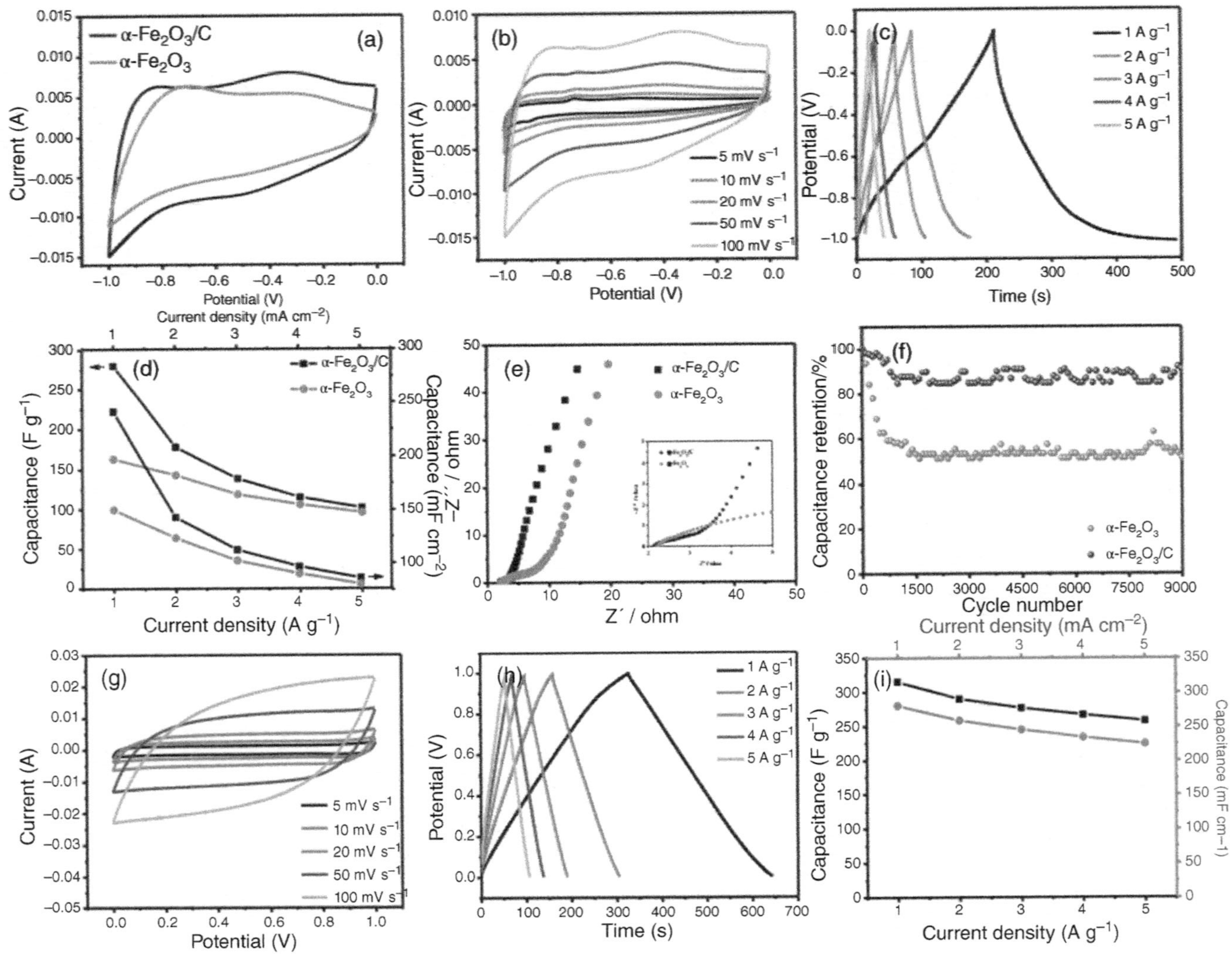

Figure 6.6 (a) CV curves at a scan rate of $100\,mV\,s^{-1}$ (b) CV curves of electrode at different scan rates (c) GCD curves of α-Fe_2O_3@C electrode at varied current densities (d) The capacitance from CGD curves (e) Nyquist plots (f) Cycle performance at $5\,A\,g^{-1}$ for 9000 cycles (g) CV curves of MnO_2 electrode (h) GCD curves of MnO_2 electrode (i) The capacitance of MnO_2.

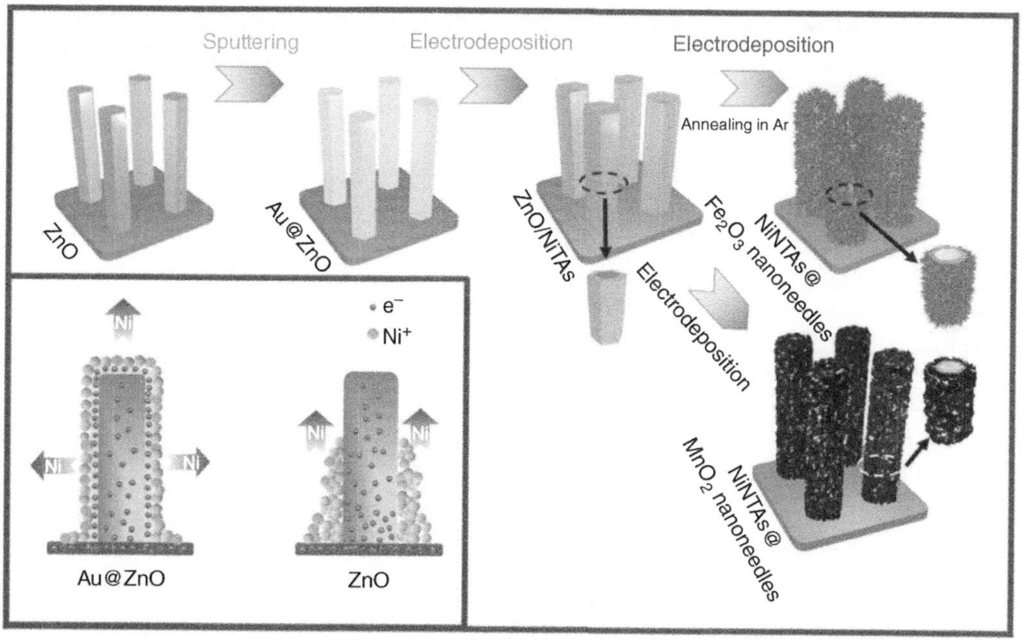

Figure 6.7 Schematic of the synthesis procedure, the inset shows different Ni films formed on ZnO nanorod with and without Au layer.

In addition, one chose various flexible substrates for the growth of electrode materials. Hu et al. reported on the fabrication of hierarchical T-Fe_2O_3/polypyrrole (PPy) nanoarrays (NAs) grown on conductive carbon cloth through a self-sacrificing template and in situ vapor-phase polymerization route under mild conditions. The electrode made of T-Fe_2O_3/PPy NAs showed an areal capacitance of 382.4 mF cm^{-2} at a current density of 0.5 mA cm^{-2} and good reversibility with capacitance retention of 97.2% after 5000 cycles. The as-assembled solid-state ASC consisting of T-Fe_2O_3/PPy NAs and MnO_2 electrodes achieved an energy density of 0.22 mWh cm^{-3} at a power density of 165.6 mW cm^{-3}. The work further demonstrated that 3D hierarchical architectures enables high charge-storage capability [75]. Arbiol et al. synthesized core-branch honeycomb-like Fe_2O_3 nanoflakes@PPy nanoleaves arrays grown on three-dimensional nickel foam (NF) scaffold. The as-prepared core-branch Fe_2O_3@PPy electrode showed a remarkably large specific capacitance (1167.8 F g^{-1} at 1 A g^{-1}) and a good rate capability [76]. Li et al. constructed aligned and evenly distributed Fe_2O_3 NNAs on SiC NWs skeletons for the first time. Meanwhile, a SiC@$NiCo_2O_4$/$Ni(OH)_2$ hybrid with intriguing electrochemical behaviors was also prepared by a simple technique. Based on the above designed two electrodes, an ASC device with superior performances including both high energy/power density and desired cycling lifespan was assembled, revealing the potential application of the advanced hybrid nanoarchitectures in exploiting highly efficient energy storage systems. The electrochemical performance between SiC NWs@Fe_2O_3 NNAs electrode and single components SiC NWs and Fe_2O_3 NNAs electrodes are presented (See Figure 6.8) [77].

Liu's group fabricated thumb-ring-like α-Fe_2O_3 structures for the first time through the introduction of phosphate ions and sulfate ions to effectively tune the size and shape of nanorings and control well the growth of hematite crystals. Then, they synthesized the composites with reduced

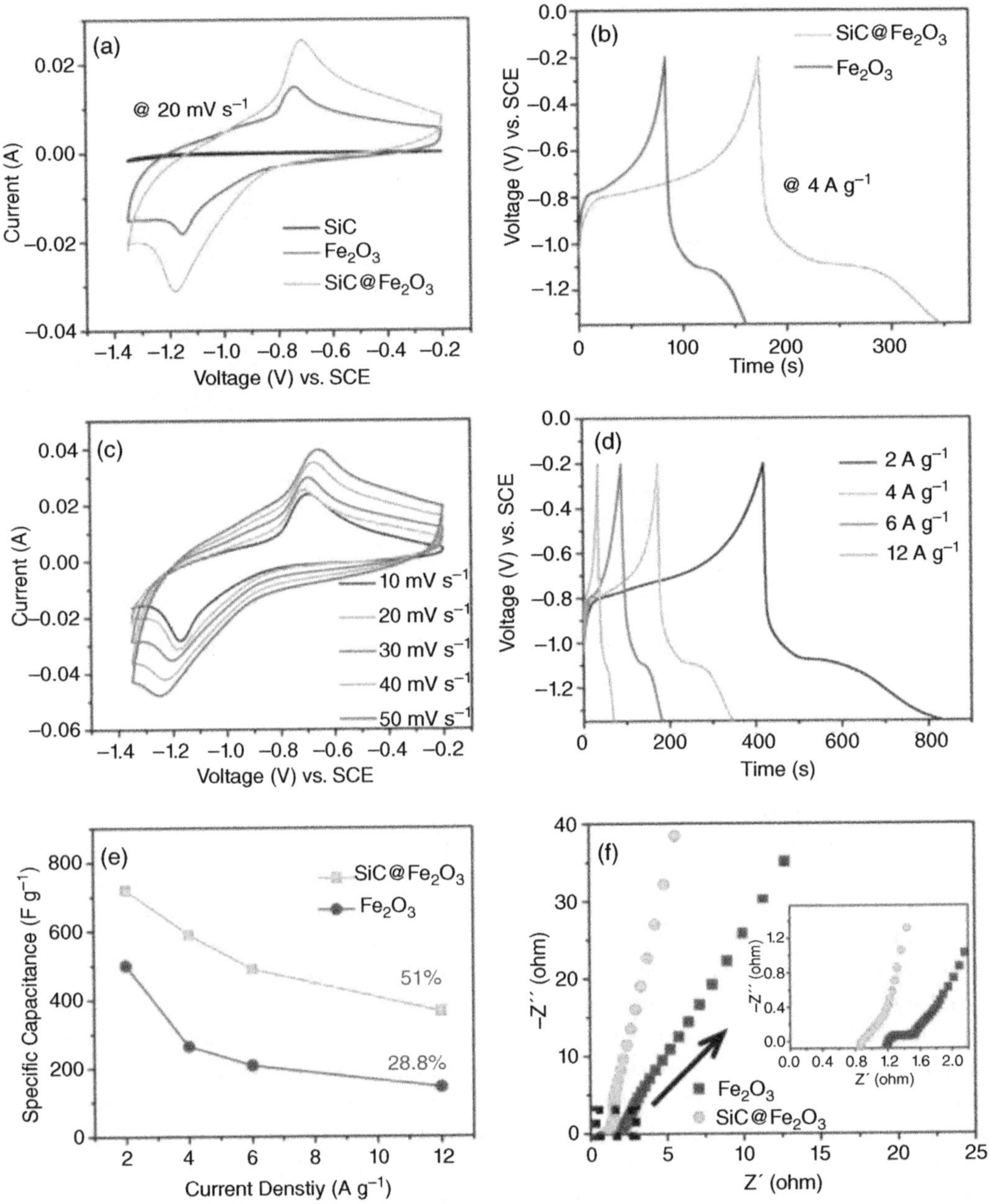

Figure 6.8 electrochemical performances of electrodes.

graphene oxide (rGO) by a two-step hydrothermal procedure and surfactant modification. The sample showed a good rate performance with 75% of capacitance retention with scan rate from 5 to 200 mV s^{-1}, and a high cycling stability with about 10% decay of the initial capacitance after 11 000 cycles at a wide potential window (−1 to 0.2 V vs Ag/AgCl). In addition, it was found that α-Fe_2O_3 began to be gradually wrapped by rGO during cycling test, while pure α-Fe_2O_3 electrode emerged with much lower specific capacitances without redox pairs under a same test condition.

Hence, it can be speculated that rGO not only served as conductive pathway but also effectively promoted the Faradaic reactions of α-Fe_2O_3 and protected α-Fe_2O_3 from possible mass loss or aggregation [78].

6.5 MnO_2 Electrode Materials

To overcome high cost drawbacks of RuO_2 for supercapacitor, people shift their focus to MnO_2 for its high theoretical specific capacitance ($1370\,F\,g^{-1}$), low cost, environmental safety, and natural abundance [79–81]. However, the capacitive performance and power density of MnO_2 electrode are restricted by its poor conductivity. Electrochemical performance of MnO_2 materials are especially influenced by various crystal structures. It is noted that the capacitance of Mn oxides comes mainly from pseudocapacitance. The energy storage mechanism in this material is associated mainly with reversible redox reactions and transition between different oxidation states [82]. There are two mechanisms to explain charge storage behavior. The first one shows the insertion of electrolyte cations ($C^+ = Li^+$, Na^+, and K^+) in the bulk of electrode:

$$MnO_2 + C^+ + e^- \leftrightarrow MnOOC \tag{6.3}$$

The second one is the surface adsorption of cations on MnO_2 electrode:

$$(MnO_2)_{surface} + C^+ + e^- \leftrightarrow (MnOOC)_{surface} \tag{6.4}$$

In order to achieve high specific capacitance and rate capability, it is vital to control the electrical conductivity of MnO_2 electrode. But it is very difficult to improve the capacitive performance of MnO_2 electrode due to low electrical conductivity (~10^{-5}–$10^{-6}\,S\,cm^{-1}$). Therefore, one of the ways to increase the electrical conductivity and charge-storage capability is to design hybrid electrode materials. Recently, MnO_2/nitrogen-doped ultra-microporous carbon nanospheres (MnO_2/N--UCNs) for supercapacitor have been synthesized by a template-free polymerization of resorcinol/formaldehyde on the surface of colloids in the presence of hexamethylenetetramine, followed by carbonization and then a redox reaction between carbons and $KMnO_4$. The as-prepared MnO_2/N--UCNs showed regular ultramicropores, high surface area and high content of MnO_2. A typical MnO_2/N-UCNs with 57 wt.% MnO_2 doping content (MnO_2(57%)/N-UCNs) produces the synergistic effect between carbons and metal oxides. MnO_2(57%)/N-UCNs as a supercapacitor electrode delivers a specific capacitance of $401\,F\,g^{-1}$ at $1.0\,A\,g^{-1}$ and charge/discharge stability with 86.3% of the initial capacitance after 10 000 cycles at $2.0\,A\,g^{-1}$ in $1.0\,mol\,l^{-1}$ Na_2SO_4 electrolyte [83].

Incorporation of pseudocapacitive materials onto MWCNT fiber is a promising strategy to improve the capacitance of fiber based supercapacitor. However, because of the hydrophobic nature of MWCNTs, the electrochemical deposition of MnO_2 nanocrystals primarily occurred at the outer surface of MWCNT fiber, and the poor electrical conductivity of MnO_2 nanocrystals was not much improved [84]. Therefore, high pseudocapacitance of MnO_2 cannot be efficiently utilized, resulting in low specific capacitance, rate capability, and poor cycling stability. Moreover, the as-prepared hybrids were sensitive to mechanical deformation such as folding and twisting, which result in poor mechanical stability.

To overcome the above mentioned disadvantages, Huang et al. reported a kind of amorphous MnO_2@ MWCNT fiber. The incorporated amorphous MnO_2 nanoparticles and the well-aligned characteristic of interconnected MWCNT scaffold enable the amorphous MnO_2@ MWCNT fiber outstanding mechanical robustness, highly electrical conductivity and fast ion-diffusion. The as-assembled solid-state supercapacitor based on the hybrid fibers delivered a specific capacitance of 8.5 $F cm^{-3}$ at a current density of 1 $A cm^{-3}$ (nearly 13 times higher than that of MWCNT fiber-based device, 0.66 $F cm^{-3}$), high energy density (1.5 $mWh cm^{-3}$), and exceptional cycling stability with capacity retention of over 90% after 15000 cycles. Furthermore, the amorphous MnO_2@MWCNT fiber-based supercapacitor presented excellent flexibility and mechanical reliability [85].

It is well known that GO possesses abundant functional groups, large surface area (400–1500 $m^2 g^{-1}$) and outstanding dispersion in water medium [86–88]. By combining the advantages of GO and MnO_2, manganese ions can transport more easily onto the surface of GO, which will effectively increase the active surface of MnO_2 and enhance the energy storage property [89]. Zhao and coworkers synthesized MnO_2/GO composites through a facile chemical precipitation method. They used the as-synthesized product as electrode materials for supercapacitor. Electrochemical results indicate that the asymmetric supercapacitor can be cycled reversibly between 0 and 2 V in 1 M Na_2SO_4 solution and shows a maximum specific capacitance of 84 $F g^{-1}$ at a current density of 0.1 $A g^{-1}$, and the energy density of 46.7 $Wh kg^{-1}$ at a power density of 100 $W kg^{-1}$. In addition, such asymmetric supercapacitor shows cycle life along with 100% specific capacitance retention after 1000 cycles and 93% after 4000 cycles [90].

As ternary transition metal oxide, spinel structured $NiCo_2O_4$ nanomaterials have been extensively studied for supercapacitors due to its high theoretical capacity and great electrochemical activity [91, 92]. Moreover, spinel $NiCo_2O_4$ possesses much better electrical conductivity and higher electrochemical activity than nickel oxides or cobalt oxides. It contributes to provide richer redox reactions, including contributions from both nickel and cobalt ions, than those of the monometallic nickel oxides and cobalt oxide [91, 93]. In addition, $NiCo_2O_4$ materials often exhibit different shapes and orientations [92]. Lou et al. prepared hierarchical $NiCo_2O_4$ tetragonal microtubes by annealing nickel cobalt layered double hydroxide microtubes. Benefiting from the unique structural features, these hierarchical $NiCo_2O_4$ microtubes manifest an excellent electrochemical performance in terms of high specific capacitance and remarkable cycle life [94]. The images and structures of $NiCo_2O_4$ microtubes are shown (See Figure 6.9). However, the crystal structure of $NiCo_2O_4$ nanomaterials often expand or shrink significantly during charge–discharge process due to high amount of charge storage. Combining two or more electrode materials is considered to be an effective method, which can further show the synergetic effect between active materials [95]. Asymmetric supercapacitors reasonably combine the merits of the battery-type capacitor and an electrochemically double-layered capacitor, which not only enhances the working potential significantly but also improves the capacitance storage ability. Wu's group prepared hybrid MnO_2@$NiCo_2O_4$ nanosheets through a simple hydrothermal route. The as-prepared products are employed directly as electrode material for the supercapacitors. It shows that MnO_2@$NiCo_2O_4$ electrode delivers a specific capacitance of 3086 $mF cm^{-2}$ at a current density of 2 $mA cm^{-2}$. The as-assembled asymmetric supercapacitor presents an energy density of 74.1 $Wh kg^{-1}$, an a power density of 12.05 $kW kg^{-1}$ at 22.89 $Wh kg^{-1}$, and an enhancement of cyclic stability with a capacitance retention of 97.3% over 6000 cycles at 10 $mA cm^{-2}$. This work provides a general strategy to the design of electrode materials in asymmetric supercapacitors with a high energy and power density [96]. The electrochemical performance of MnO_2, $NiCo_2O_4$ and MnO_2@$NiCo_2O_4$ electrode materials are shown (See Figure 6.10).

Recently, $FeCo_2O_4$ has been investigated as electrode materials for supercapacitors since it possesses high electrical conductivity and electrochemical activity [97]. Moreover, two-dimensional nanosheets can provide high efficient electrochemical active sites, reduce the ions and electron diffusion path and improve the structural stability [98]. Therefore, $FeCo_2O_4$ nanosheets can be an

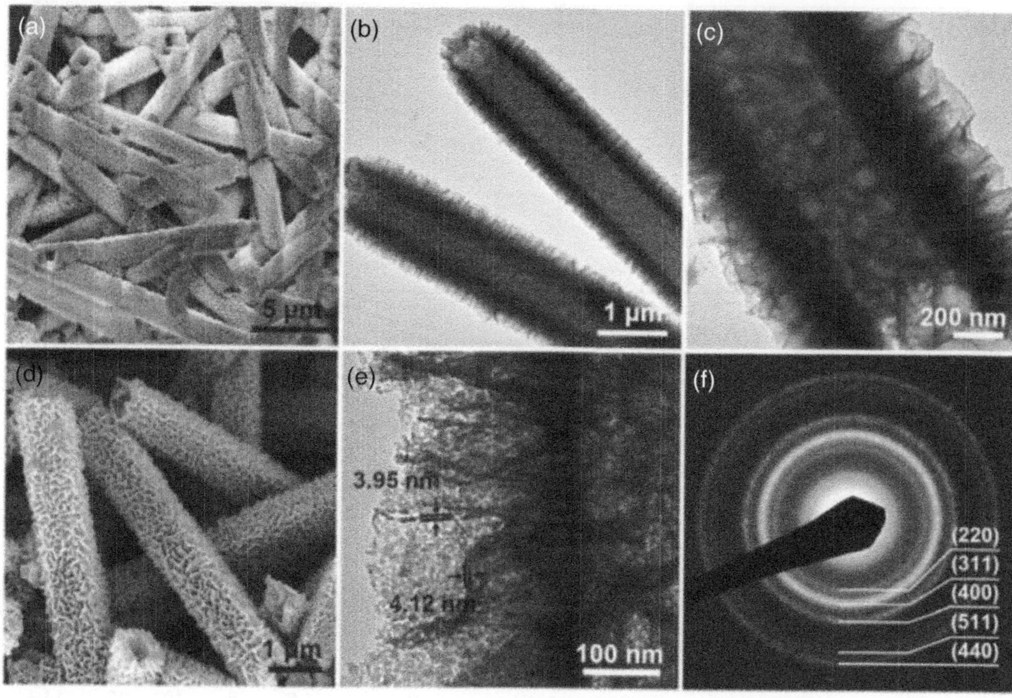

Figure 6.9 (a) FESEM and (b and c) TEM images of NiCo-LDH tetragonal microtubes (d) FESEM, (e) TEM image and (f) SAED pattern of hierarchical $NiCo_2O_4$ microtubes after thermal treatment.

excellent backbone material for supercapacitors. Sun and coworkers synthesized hierarchical $FeCo_2O_4$@MnO_2 core-shell nanosheet arrays on CFs through a facile two-step hydrothermal method for supercapacitors. The network structure composed of $FeCo_2O_4$ nanosheets arranging in the pattern offers large surface area and reliable electrical connections to the outer MnO_2 nanosheets. The hybrid electrode showed an energy density of 22.68 Wh kg^{-1} at 406.01 W kg^{-1} and an energy density 7.06 Wh kg^{-1} at a power density of 1802.5 W kg^{-1}. At the same time, it presents an excellent cycling performance with 90.1% retention of the initial specific capacitance after 5000 cycles in 3 M KOH solution [99].

$ZnCo_2O_4$ has attracted much interest due to its better electrochemical activity. Furthermore, $ZnCo_2O_4$ is well known with low cost, abundant resources and environmental friendliness [100–102]. On the other hand, $ZnCo_2O_4$ shows a unique crystal structure. Zinc ions of $ZnCo_2O_4$ occupy the tetrahedral sites in cubic spinel structure, and cobalt ions occupy the octahedral sites [103]. Simultaneously, various morphologies of $ZnCo_2O_4$ nanostructure have been prepared through simple synthesis methods. Such as nanosheets, nanowires and nanotubes etc. [102, 104, 105] Recently, Wang et al. prepared $ZnCo_2O_4$ nanowires directly grow on nickel foam by a facile hydrothermal method. Electrochemical measurements show that $ZnCo_2O_4$/nickel foam electrode exhibits a high specific capacitance (1625 F g^{-1} at 5 A g^{-1}), excellent rate capability (59% capacitance retention at 80 A g^{-1}) and good cycling stability (94% capacitance retention over 5000 cycles) [106]. Wang et al. reported hierarchical porous $ZnCo_2O_4$ microspheres synthesized on nickel foam through a solvothermal method. $ZnCo_2O_4$ electrode shows a specific capacitance of 647.1 F g^{-1} at 1 A g^{-1} and 440.6 F g^{-1} at 10 A g^{-1} in 2 M KOH. After 2000 cycles, the capacity loss is 8.5% [101]. Nanostructures grown on conducting substrates like

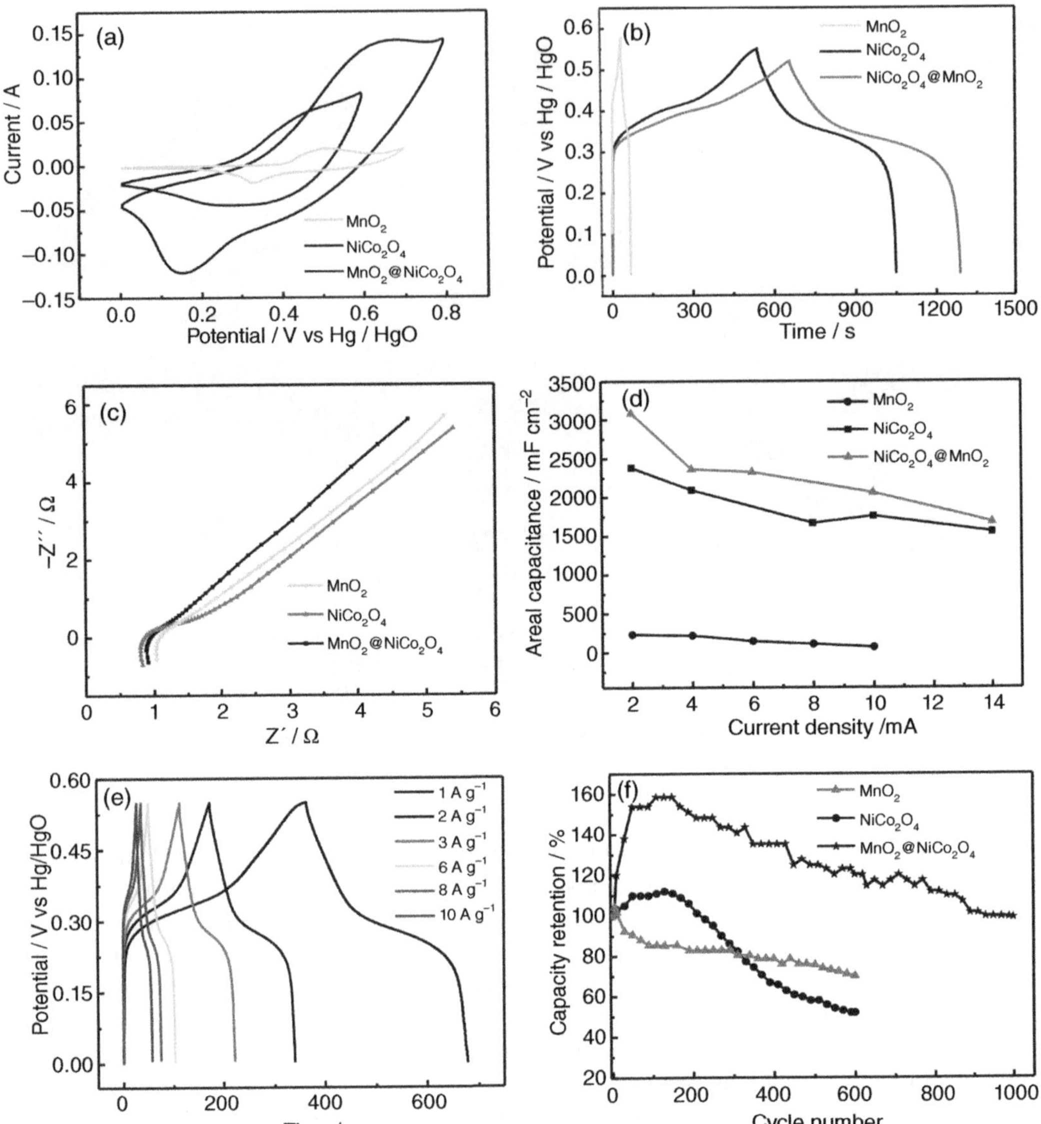

Figure 6.10 (a) CV curves (b) Galvanostatic charge-discharge curves (c) Nyquist plots (d) Areal capacitance (e) Galvanostatic charge-discharge curves of hybrid $MnO_2@NiCo_2O_4$ structure (f) Capacitance retention at a current density of $10\,mA\,cm^{-2}$.

Ni-foam and carbon cloth can be used directly as binder-free electrodes for hybrid supercapacitor. This method avoids the occurrence of an inactive surface which result from the use of conductive agents and polymer binders in the normal slurry coating process and hinder the accessibility of electrolyte ions.

Metal oxides/hydroxides and conducting polymers store charges in a Faradic or redox-type process similar to batteries. It gives rise to high energy densities. But the reactions are kinetically unfavorable. Thus, tremendous efforts have been devoted to rational synthesize superior electrode materials, especially in the form of core@shell hetero-structures. Luo's group fabricated hierarchical $ZnCo_2O_4/MnO_2$ composite nanocone forests. The supercapacitor electrodes prepared from the unique structure exhibits exceptional specific capacitances of 2339 and 1526 $F g^{-1}$ at current densities of 1 and 10 $A g^{-1}$ and a capacity retention of 95.9% after 3000 cycles at 2 $A g^{-1}$ and 94.5% after 8000 cycles at 10 $A g^{-1}$ [107] The schematic of the fabrication processes of $ZnCo_2O_4/MnO_2$ structure is illustrated (See Figure 6.11).

The binder-free film/array electrodes can provide an exciting platform for high and fast energy storage. Zhang et al. reported $ZnCo_2O_4@MnO_2$ nanotube arrays electrode by a facile two-step method [108]. The electrode exhibits a specific capacitance of 1981 $F g^{-1}$ (2.38 $F cm^{-2}$) at a current density of 5 $A g^{-1}$ and excellent cycling stability (5000 cycles). Furthermore, a low-cost, high performance asymmetric supercapacitor (ASC) with $ZnCo_2O_4@MnO_2$ core-shell nanotube arrays on Ni foam (as positive electrode) and 3D porous a-Fe_2O_3 on Fe foil (as negative electrode) was successfully designed. The device with an extended operating voltage window of 1.3 V achieved a specific capacitance of 161 $F g^{-1}$ at 2.5 $mA cm^{-2}$ with a maximum energy density of 37.8 $W h kg^{-1}$ and excellent stability with a capacitance retention of 91% after 5000 cycles. SEM images of $ZnCo_2O_4$ nanowire arrays and hierarchical $ZnCo_2O_4@MnO_2$ core-shell nanotube are shown (See Figure 6.12).

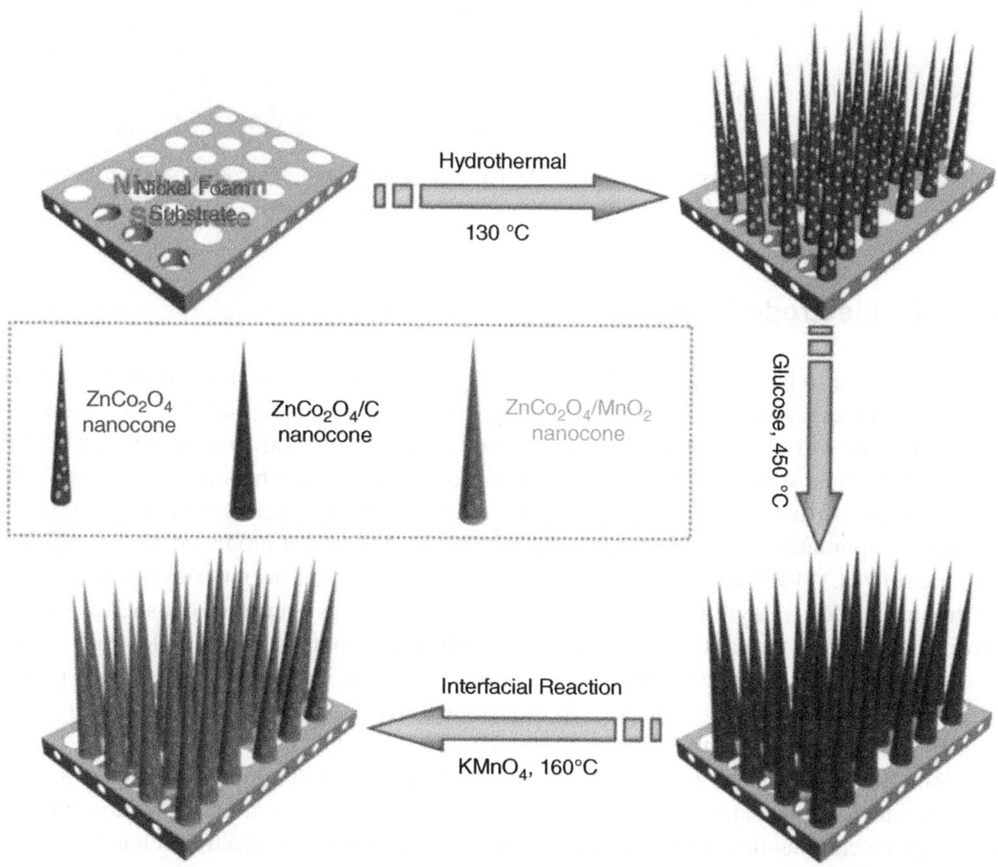

Figure 6.11 Schematic of $ZnCo_2O_4/MnO_2$ composite nanocone forests (NCFs).

Figure 6.12 Typical SEM images of (a and b) $ZnCo_2O_4$ NW arrays and (c and d) hierarchical $ZnCo_2O_4$@ MnO_2 core-shell nanotube arrays grown on Ni foam.

6.6 V_2O_5 Electrode Materials

V_2O_5 is considered to be potentially active cathode materials for electrochemical energy storage owing to the high capacitance, high working voltage, and superior cycling stability. V_2O_5 possesses a layered structure with an easy cleavage plane perpendicular to c-axis and a band gap of 2.8 eV. Thus far, various V_2O_5 nanostructures have been synthesized by some facile preparation methods. Yolk-shelled V_2O_5 nanostructures composed of the ultrathin nanosheets were synthesized through a hydrothermal approach. As the electrode material, the as-synthesized product delivers a specific capacitance of 704.17 $F\,g^{-1}$ at 1.0 $A\,g^{-1}$ and a capacity retention of 89% over 4000 cycles at 3.0 $A\,g^{-1}$. Then they assembled an asymmetric supercapacitor by using the as-prepared yolk-shelled V_2O_5 as positive electrode and Osmanthus-derived coralline-shaped carbon as negative electrode, and it presents an energy density of 29.49 $Wh\,kg^{-1}$ at a power density of 800 $W\,kg^{-1}$ with a cycling performance with 90.6% capacity retention after 2000 cycles at 3.0 $A\,g^{-1}$. Furthermore, two cells in series can easily light up the light-emitting diode (3 V) [109].

Using simple capping-agent-assisted precipitation technique, V_2O_5 nanoporous network as electrode materials for supercapacitor was synthesized. The effect of annealing temperature on the morphology, electrochemical and structural properties, and stability of electrodes has been studied. The results showed the interconnected V_2O_5 nanoporous network possesses a specific capacitance of 316 $F\,g^{-1}$. The nanoporous network presents facile nanochannels for ion diffusion and facilitates the accessibility of the ions [110].

To improve the conductivity and mechanical properties of electrode materials, V_2O_5 nanostructures were constructed on the activated carbon cloth (ACC) through an ion-exchange column, and the as-prepared V_2O_5 is in favor of the in situ oxidation polymerization process of indole monomer, resulting in the fabrication of bamboo-like V_2O_5/polyindole@activated carbon cloth (V_2O_5/PIn@ACC) materials. The introduction of PIn shell with close contact with V_2O_5 is beneficial for not only preventing V_2O_5 dissolution into electrolyte by avoiding the direct contact between them during cycling, but also affording a facile electron transport to ensure electrochemical activity. By matching V_2O_5/PIn@ACC with the negative electrode of reduced graphene oxide@activated carbon cloth (rGO@ACC), an asymmetric supercapacitors device was assembled, which obtained a maximum energy density of 38.7 W $(h\,kg)^{-1}$ at a stable potential window of 1.8 V. Besides, the device showed a superior cycling stability with capacitance retention of 91.1% after 5000 charging discharging cycles [111].

Nanocomposites have received great attention because of their structure-dependent properties and relative synergistic effects. Constructing electrically conductive backbone and directly attaching nanoscale V_2O_5 onto it can enhance the electrochemical performance of V_2O_5-based electrode materials. At the same time, the unique properties of nanostructures show advantages over conventional electrode materials by exploiting the synergistic effects of different dimensions in the structures. Wu et al. reported a facile strategy to fabricate hierarchical nanocomposites of V_2O_5 NWAs on three-dimension N-doped graphene aerogel (N-GA). They found that the introduction of N atoms into graphene lattice plays an important role in the formation of vertical VOx NWAs on graphene nanosheets. The resultant nanocomposite exhibits excellent capacitance characteristic with 710.5 F g^{-1} at a discharge current density of 0.5 A g^{-1}, outstanding rate performance and good cycling behavior [112]. The possible growth mechanism of V_2O_5 NWAs is proposed (See Figure 6.13).

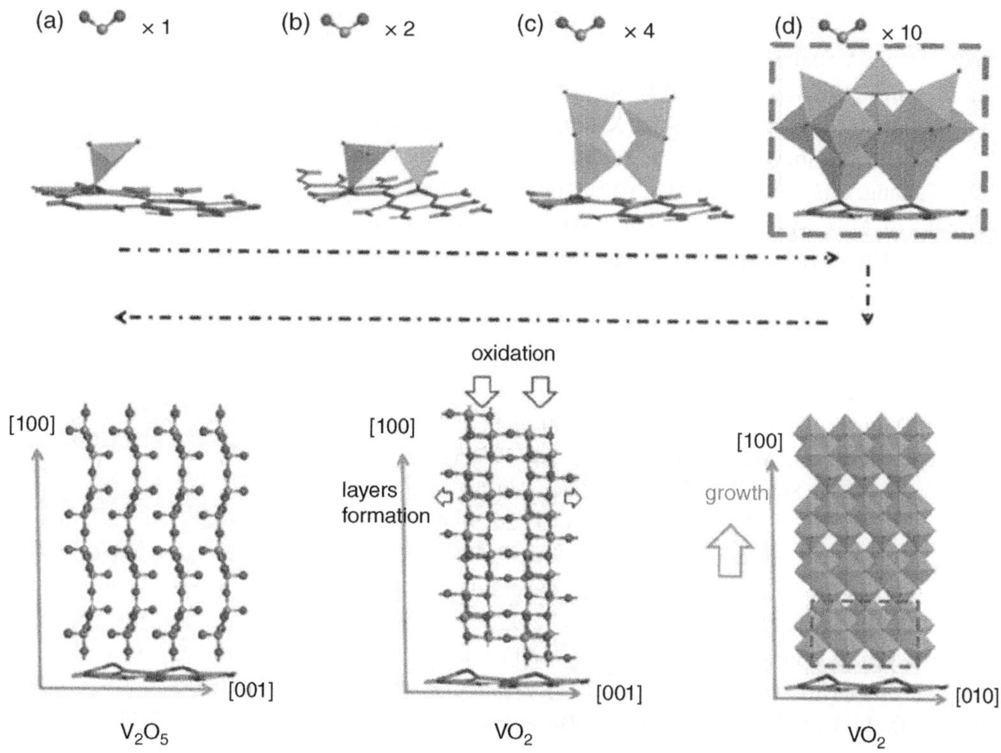

Figure 6.13 Schematic of growth mechanism of V_2O_5 NWAs.

In a summary, we have outlined the recent developments in design and fabrication of transition metal oxide nanostructures as electrodes for supercapacitors. It is very important to integrate active/synergetic nanostructures into flexible substrates for improving energy density and power density of the devices. Moreover, integration of diverse conducting agents can also improve the overall electrical conductivity of metal oxide films/arrays. Finally, it is desired to understand the electrochemical mechanisms of transition metal oxide-based hybrid structures and to investigate the physical/chemical properties of the interface within hybrid structure and their effects on the electrochemical performance.

References

1 Liu, T.Y., Zhang, F., Song, Y. et al. (2017). *Chem. A* 5: 17705–17733.
2 Zuo, W.H., Li, R.Z., Zhou, C. et al. (2017). *Adv. Sci.* 4: 1600539.
3 Yang, P.H. and Mai, W.J. (2014). *Nano Energy* 8: 274–290.
4 Hu, C.G., Song, L., Zhang, Z.P. et al. (2015). *Energy Environ. Sci.* 8: 31–54.
5 Huang, Y., Shan, W., Lau, Y.Y. et al. (2017). *ACS Nano* 11: 8953–8961.
6 Yoo, J.J., Balakrishnan, K., Huang, J. et al. (2011). *Nano Lett.* 11: 1423–1427.
7 Sk, M.M., Yue, C.Y., Ghosh, K., and Jena, R. (2016). *J. Power Sources* 308: 121–140.
8 Xiao, H.H., Qu, F.Y., and Wu, X. (2016). *Appl. Surf. Sci.* 360: 8–13.
9 Chen, Z., Qin, Y.C., Weng, D. et al. (2009). *Adv. Funct. Mater.* 19: 3420.
10 Zakaria, M.B., Hu, M., Salunkhe, R.R. et al. (2015). *Chem. Eur. J.* 21: 3605.
11 Chen, H., Hu, L.F., Yan, Y. et al. (2013). *Adv. Energy Mater.* 3: 1636–1646.
12 Xiong, S.L., Yuan, C.Z., Zhang, X.G. et al. (2009). *Chemistry* 15: 5320–5326.
13 He, X.Y., Li, R.M., Liu, J.Y. et al. (2018). *Chem. Eng. J.* 334: 1573–1583.
14 Pan, Y., Gao, H., Zhang, M.Y. et al. (2017). *J. Colloid Interface Sci.* 497: 50–56.
15 Jiao, Y., Liu, Y., Yin, B.S. et al. (2015). *Sci. Adv. Mater.* 7: 1395–1399.
16 Yao, S.Y., Qu, F.Y., Wang, G., and Wu, X. (2017). *J. Alloys Compd.* 724: 695–702.
17 Jiang, W., Hu, F., Yan, Q.Y., and Wu, X. (2017). *Inorg. Chem. Front.* 4: 1642–1648.
18 Tang, Z., Tang, C.H., and Gong, H. (2012). *Adv. Funct. Mater.* 22: 1272–1278.
19 Lu, X.H., Yu, M.H., Wang, G.M. et al. (2014). *Energy Environ. Sci.* 7: 2160–2181.
20 Jabeen, N., Hussain, A., Xia, Q.Y. et al. (2017). *Adv. Mater.* 29: 1700804.
21 Wang, H., Yi, H., Chen, X. et al. (2014). *Chem. A* 2: 1165–1173.
22 Liu, J., Zheng, M., Shi, X. et al. (2016). *Adv. Funct. Mater.* 26: 919–930.
23 Zeng, Y.X., Han, Y., Zhao, Y.T. et al. (2015). *Adv. Energy Mater.* 5: 1402176.
24 Wang, K., Meng, Q.H., Zhang, Y.J. et al. (2013). *Adv. Mater.* 25: 1494–1498.
25 Yang, P., Ding, Y., Lin, Z. et al. (2014). *Nano Lett.* 14: 731–736.
26 Su, F.H., Lv, X.M., and Miao, M.H. (2015). *Small* 11: 854–861.
27 Choi, C., Kim, S.H., Sim, H.J. et al. (2015). *Sci. Rep.* 5: 9387.
28 Ferris, A., Garbarino, S., Guay, D., and Pech, D. (2015). *Adv. Mater.* 27: 6625–6629.
29 Jiang, J., Li, Y.Y., Liu, J.P. et al. (2012). *Adv. Mater.* 24: 5166–5180.
30 Ji, L.W., Lin, Z., Alcoutlabi, M., and Zhang, X.W. (2011). *Energy Environ. Sci.* 4: 2682–2699.
31 Cheng, F.Y., Liang, J., Tao, Z.L., and Chen, J. (2011). *Adv. Mater.* 23: 1695.
32 Choi, J.Y., Lee, D.J., Lee, Y.M. et al. (2013). *Adv. Funct. Mater.* 23: 2108–2114.
33 Fan, Y.Q., Shao, H.B., Wang, J.M. et al. (2011). *Chem. Commun.* 47: 3469–3471.
34 Zhang, Y.Q., Li, L., Shi, S.J. et al. (2014). *J. Power Sources* 256: 200–205.

35 Kate, R.S., Khalate, S.A., and Deokate, R.J. (2018). *J. Alloys Compd.* 734: 89–111.
36 Xia, X., Tu, J., Zhang, Y. et al. (2012). *RSC Adv.* 2: 1835–1841.
37 Han, Z.C., Zheng, X., Yao, S.Y. et al. (2016). *Appl. Surf. Sci.* 365: 240–244.
38 Ning, F., Shao, M., Zhang, C. et al. (2014). *Nano Energy* 7: 134–142.
39 Zhou, J.J., Li, Q., Chen, C. et al. (2018). *Chem. Eng. J.* 350: 551–558.
40 Wang, G.X., Shen, X.P., Horvat, J. et al. (2009). *J. Phys. Chem. C* 113: 4357–4361.
41 Zheng, X., Han, Z.C., Yang, W.J. et al. (2016). *Dalton Trans.* 45: 16850–16858.
42 Lu, Y., Li, L., Chen, D., and Shen, G.Z. (2017). Nanowires-assembled Co_3O_4@$NiCo_2O_4$ architectures for high performance all-solid-state asymmetric supercapacitors. *J. Mater. Chem. A* 5: 24981–24988.
43 Wu, X., Han, Z.C., Zheng, X. et al. (2017). *Nano Energy* 31: 410–417.
44 Xing, L., Dong, Y.D., Hu, F. et al. (2018). *Dalton Trans.* 47: 5687–5694.
45 Gao, S., Sun, Y., Lei, F. et al. (2014). *Angew. Chem. Int. Ed.* 53: 12789.
46 Zhou, C., Zhang, Y.W., Li, Y.Y., and Liu, J.P. (2013). *Nano Lett.* 13: 2078–2085.
47 Hosono, E., Fujihara, S., Honma, I. et al. (2006). *J. Power Sources* 158: 779–783.
48 Zhang, F., Yuan, C.Z., Lu, X.J. et al. (2012). *J. Power Sources* 203: 250–256.
49 Pang, H., Li, X.R., Zhao, Q.X. et al. (2017). *Nano Energy* 35: 138–145.
50 Bai, X., Liu, Q., Liu, J.Y. et al. (2017). *Chem. Eng. J.* 315: 35–45.
51 Zhai, T., Wan, L.M., Sun, S. et al. (2017). *Adv. Mater.* 29: 1604167.
52 Sato, Y., Nomura, T., Tanaka, H., and Kobayakawa, K. (1991). *J. Electrochem. Soc.* 138: L37–L39.
53 Lee, J.W., Ahn, T., Soundararajan, D. et al. (2011). *Chem. Commun.* 47: 6305–6307.
54 Mai, L., Yang, F., Zhao, Y. et al. (2011). *Nat. Commun.* 2: 381–386.
55 Xia, Y., Wang, B., Wang, G., and Wang, H. (2015). *RSC Adv.* 5: 98740–98746.
56 Sun, W.P., Rui, X.H., Zhu, J.X. et al. (2015). *J. Power Sources* 274: 755–761.
57 Wei, T.Y., Chen, C.H., Chien, H.C. et al. (2010). *Adv. Mater.* 22: 347–351.
58 Cao, C.Y., Guo, W., Cui, Z.M. et al. (2011). *J. Mater. Chem.* 21: 3204–3209.
59 Han, D.D., Xu, P.C., Jing, X.Y. et al. (2013). *J. Power Sources* 235: 45–53.
60 Liang, K., Tang, X.Z., and Hu, W.C. (2012). *J. Mater. Chem.* 22: 11062–11067.
61 Liu, M.X., Wang, X., Zhu, D.Z. et al. (2018). *Chem. Eng. J.* 308: 240–247.
62 Liu, T., Zhang, L.Y., Cheng, B. et al. (2018). *Chem. Commun.* 54: 3731–3373.
63 Yi, H., Wang, H.W., Jing, Y.T. et al. (2015). *J. Power Sources* 285: 281–290.
64 Zhao, B., Song, J.S., Liu, P. et al. (2011). *J. Mater. Chem.* 21: 18792–18798.
65 Tian, W., Wang, X., Zhi, C. et al. (2013). *Nano Energy* 2: 754–763.
66 Zheng, X., Han, Z.C., Yao, S.Y. et al. (2016). *Dalton Trans.* 45: 7094–7103.
67 Sun, S., Zhai, T., Liang, C.L. et al. (2018). *Nano Energy* 45: 390–397.
68 Jiao, Y., Liu, Y., Yin, B.S. et al. (2014). *Nano Energy* 10: 90–98.
69 Xie, K., Li, J., Lai, Y. et al. (2011). *Electrochem. Commun.* 13: 657–660.
70 Huang, J., Yang, S., Xu, Y. et al. (2014). *J. Electroanal. Chem.* 713: 98–102.
71 Dong, Y.D., Xing, L., Chen, K.F., and Wu, X. (2018). *Nano* 8: 487.
72 Chen, L., Yu, Z., Ma, X. et al. (2014). *Nano Energy* 9: 345.
73 Li, Y., Xu, J., Feng, T. et al. (2017). *Adv. Funct. Mater.* 27: 1606728.
74 Lu, X., Chen, X., Zhou, W. et al. (2015). *ACS Appl. Mater. Interfaces* 7: 14843.
75 Wang, L., Yang, H., Liu, X. et al. (2016). *Angew. Chem. Int. Ed.* 55: 1.
76 Tang, P., Han, L., Genc, A. et al. (2016). *Nano Energy* 22: 189.
77 Zhao, J., Li, Z., Yuan, X. et al. (2018). *Adv. Energy Mater.* 8: 1702787.
78 Zhu, Y., Cheng, S., Zhou, W. et al. (2017). *ACS Sustain. Chem. Eng.* 5: 5067.
79 Chang, J.K., Lee, M.T., and Tsai, W.T. (2007). *J. Power Sources* 166: 590–594.

80 Pan, Z., Qiu, Y., Yang, J. et al. (2016). *Nano Energy* 26: 610–619.
81 Yin, B.S., Zhang, S.W., Liu, Y. et al. (2014). *CrystEngComm* 16: 9999–10005.
82 Messaoudi, B., Joiret, S., Keddam, M., and Takenouti, H. (2001). *Electrochim. Acta* 46: 2487–2498.
83 Lu, W.J., Huang, S.Z., Miao, L. et al. (2017). *Chin. Chem. Lett.* 28: 1324–1329.
84 Choi, C., Lee, J.A., Choi, A.Y. et al. (2014). *Adv. Mater.* 26: 2059–2065.
85 Shi, P.P., Li, L., Hua, L. et al. (2017). *ACS Nano* 11: 444–452.
86 Yang, X.Y., Zhang, X.Y., Ma, Y.F. et al. (2009). *J. Mater. Chem.* 19: 2710–2714.
87 He, Y.M., Chen, W.J., Li, X.D. et al. (2013). *ACS Nano* 7: 174–182.
88 Liao, Q.Y., Li, N., Cui, H. et al. (2013). *Chem. A* 1: 13715–13720.
89 Yang, H.P., Jiang, J., Zhou, W.W. et al. (2011). *Nanoscale Res. Lett.* 6: 531–538.
90 Zhao, Y.F., Ran, W., He, J. et al. (2015). *Small* 11: 1310–1319.
91 Hu, J., Li, M.C., Lv, F.C. et al. (2015). *J. Power Sources* 294: 120–127.
92 Wang, Q.F., Wang, X.F., Liu, B. et al. (2013). *Chem. A* 1: 2468–2473.
93 Zhang, J.N., Liu, P., Jin, C. et al. (2017). *Electrochim. Acta* 256: 90–99.
94 Ma, F.M., Yu, L., Xu, C.Y., and Lou, X.W. (2016). *Energy Environ. Sci.* 9: 862–868.
95 Zhao, D.P., Liu, H.Q., and Wu, X. (2019). *Nano Energy* 57: 363–370.
96 Zhao, D.P., Wu, X., and Guo, C.F. (2018). *Inorg. Chem. Front.* 5: 1378–1385.
97 Ervin, M.H., Miller, B.S., Hanrahan, B. et al. (2012). *Electrochim. Acta* 65: 37–43.
98 Pandolfo, A.G. and Hollenkamp, A.F. (2006). *J. Power Sources* 157: 11–27.
99 Sun, W., Gao, G.H., Du, Y.C. et al. (2018). *J. Colloid Interface Sci.* 512: 419–427.
100 Lin, C., Li, Y.Y., Yu, M. et al. (2007). *Adv. Funct. Mater.* 17: 1459–1465.
101 Wang, Q., Zhu, L., Sun, L. et al. (2015). *J. Mater. Chem. A* 3: 982–985.
102 Guan, B., Guo, D., Hu, L. et al. (2014). *J. Mater. Chem. A* 2: 16116–16123.
103 Sharma, Y., Sharma, N., Subba Rao, G.V., and Chowdari, B.V.R. (2007). *Adv. Funct. Mater.* 17: 2855–2861.
104 Wu, H., Lou, Z., Yang, H., and Shen, G.Z. (2015). *Nanoscale* 7: 1921–1927.
105 Huang, T., Zhao, C., Zheng, R. et al. (2015). *Ionics* 21: 3109–3115.
106 Wang, S., Pu, J., Tong, Y. et al. (2014). *J. Mater. Chem. A* 2: 5434.
107 Qiu, K.W., Lu, Y., Zhang, D.Y. et al. (2015). *Nano Energy* 11: 687–696.
108 Ma, W.Q., Nan, H.H., Gu, Z.X. et al. (2015). *Chem. A* 3: 5442–5448.
109 Xing, L.L., Zhao, G.G., Huang, K.J., and Wu, X. (2018). *Dalton Trans.* 47: 2256–2265.
110 Saravanakumar, B., Purushothaman, K.K., and Muralidharan, G. (2012). *ACS Appl. Mater. Interfaces* 4: 4484–4490.
111 Zhou, X., Chen, Q., Wang, A.Q. et al. (2016). *ACS Appl. Mater. Interfaces* 8: 3776–3783.
112 Sun, W., Gao, G.H., Du, Y.C. et al. (2018). *J. Mater. Chem. A* 6: 9938–9947.

7

Three-Dimensional Nanoarrays for Flexible Supercapacitors

Jing Xu[1,2]

[1]*School of Materials Science and Engineering, Nanjing University of Science and Technology, Xuanwu District, Nanjing, 210094, China*
[2]*Herbert Gleiter Institute of Nanoscience, Nanjing University of Science and Technology, Xuanwu District, Nanjing, 210094, China*

List of Abbreviations

Supercapacitors	(SCs)
electrical double-layer capacitors	(EDLCs)
three-dimensional	(3D)
polyethylene terephthalate	(PET)
polyethylene naphthalate	(PEN)
polyimide	(PI)
polycarbonate	(PC)
polyethersulfone	(PES)
polydimethylsiloxane	(PDMS)
glass transition temperature	(T_g)
indium tin oxide films	(ITO)
carbon nanotubes	(CNTs)
chemical vapor deposition	(CVD)
metal–organic framework	(MOF)
ion concentrations	($[M^{n+}]$)
anodized aluminum oxide	(AAO)
porous carbon nanotube arrays	(PCNTAs)
cyclic voltammetry	(CV)
potassium hydroxide	(KOH)
polyvinyl alcohol	(PVA)
saturated calomel electrode	(SCE)
bending angle	(θ)
length of the device	(L)
end-to-end distance along the bending direction	(LEE)
bending radius	(R)

Flexible Supercapacitors: Materials and Applications, First Edition. Edited by Guozhen Shen, Zheng Lou and Di Chen.

American Society for Testing and Materials	(ASTM)
International Organization for Standardization	(ISO)
twist angle	(θ)
applied torque	(T)

7.1 Introduction

The development of flexible electronics has attracted considerable attention recently, as great R&D efforts are devoted to facilitate the practical applications of various flexible electronics ultimately, such as flexible displays, electronic skin, electronic papers, wearable electronics, conformable sensors, adaptable robots, and implantable medical devices [1]. In order to functionalize these devices, flexible power sources with high performance are required to be integrated in the system. In this regard, flexible power sources must be capable of undergoing deformation and at the same time the electronic and electrochemical parameters must be unaffected by the strain process [2].

Supercapacitors (SCs) are one of the most common power sources that are widely used for commercial electronic, hybrid electric vehicles and stationary power grids, etc. with the features of high power density (>10 kW kg^{-1}), high energy density, ultralong cycle life (>10^5 cycles), and high rate capability with fast charge–discharge capability within seconds. As elucidated in Chapter 1, an SC is composed of an anode, cathode, separator, and electrolyte. The conventional method to fabricate SCs includes coating slurry containing the active electrode materials, carbon black, and polymer binder on planar metal current collectors, drying, pressing, layer-by-layer crimping, filling with liquid electrolytes, and packaging [3]. The working mechanism of SCs can be classified into electrical double-layer capacitors (EDLCs) and pseudocapacitors [4]. Since both the EDLCs and pseudocapacitors are surface electrochemical related devices, electrode materials with high surface area are necessary in SCs. At the same time, specified porosity, high electrical conductivity, abundant electro-active sites, and high thermal and chemical stabilities are also desired [5].

However, the active electrode materials coated on current collectors are easily detached from the substrates when the device is bent or twisted, making it difficult to be integrated in the flexible systems. In addition, the insulating polymer binder in the electrode will inevitably increase the "dead volume" of the electrode, due to the electrochemical inertness of the binder during the charge–discharge cycles [6]. Therefore, subtle structure engineering is necessary to improve the flexibility of the SC electrodes. Till now, many attempts have been reported to achieve high-performance flexible SCs [7]. Typically, replacing the rigid components with flexible ones and designing additive/binder-free electrode architectures are thought as the major two strategies. Particularly, recent years have witnessed significant progress in the design of additive/binder-free electrode architectures to avoid the "dead surface" in the traditional slurry-derived electrode and allow for more efficient charge and mass exchange, which includes the three-dimensional (3D) nanowire/nanosheet/nanotube arrays [8], mesoporous/macroporous foams [9], free-standing textiles [10], and so on. Among them, 3D nanowire/nanosheet/nanotube arrays aligned on flexible and conductive substrates have been extensively developed as promising flexible electrodes for SCs, due to the robust structure to alleviate the volume change of electrode materials, open structure for efficient electronic/ionic transport paths, and abundant surface active sites for charge transfer. Taking nanowire arrays as an example, direct one-dimensional (1D) pathways can be accessed because of the direct connection between the nanowire and the current collectors, which are more favorable for the enhancement of the specific capacitance and rate capability of SCs. In this chapter, we will review the design and fabrication methods of 3D nanoarray flexible electrodes for SCs. In the first part, the fabrication methods of 3D nanoarrays are introduced. Following to this part, typical 3D topologies based on nanoarrays for flexible supercapacitors are reviewed. Thirdly, the evaluation methods of flexible supercapacitors are discussed. In the last section, the remaining challenges and future opportunities for this emerging field are concluded.

7.2 Fabrication of 3D Nanoarrays

7.2.1 Selection of Substrates

Selection of substrates is critically important as their mechanical properties play a key role in the flexibility of the electrodes. The materials of substrates can be generally divided into three categories: metal foils, plastic films, and fibrous materials (including paper and textiles).

7.2.1.1 Metal Foils

Metal foils (e.g. copper foils and aluminum foils) are conventional current collectors utilized in the commercialized SCs, which are also commonly used as the substrates for the growth of 3D nanoarrays due to their high temperature resistance and excellent electrical conductivity. In high-performance supercapacitors, the conductivity of metallic films ($\sigma \sim 10^4$–$10^6\,S\,cm^{-1}$) is more advantageous than other current collectors. Among them, stainless steel, titanium, copper, nickel, and aluminum are proper options because of their modest flexibility and low cost. However, the high weight density of metal foils is a drawback and the flexibility of the electrode is limited when the loading of active materials is high. Moreover, effective barriers are necessary to restrict the corrosion of metal foils by water and oxygen when they are used as substrates in the SCs involving aqueous electrolytes.

7.2.1.2 Polymeric Films

Polymeric films are also one of the most common flexible substrates that have been utilized in the flexible electronics, due to the various material species, light weight, and low cost, making them well applicable to large-scale manufacturing [11]. Till now, some key plastic materials such as polyethylene terephthalate (PET), polyethylene naphthalate (PEN), polyimide (PI), polycarbonate (PC), polyethersulfone (PES) are widely utilized. If the stretchability is important, other polymer substrates such as polydimethylsiloxane (PDMS) and polyurethanes are better choice. Because of the different compositions, the thermal stability, mechanical stability, resistance to oxygen and water penetration, surface energy, chemical stability, optical clarity, and surface smoothness of these polymeric materials are various, giving rise to different selection criteria for the growth of 3D nanoarrays. The main challenge for synthesizing binder-free 3D nanoarray electrode on polymeric substrates is their low thermal stability. For instance, the glass transition temperature (T_g) of PET, PEN, and PC are 60–80 °C, 120–125 °C, and 140–150 °C, respectively [12]. Particularly, PI exhibits the highest T_g in the range of 310–410 °C, which has been utilized in many flexible electronics [12]. When using the polymeric films as substrates in SCs, conductive layers such as indium tin oxide films (ITO) [13], carbon nanotubes (CNTs) [14], graphene films [15], and some metal nanowires [16] are required to be deposited on the substrates.

7.2.1.3 Textile-Like Materials

In order to improve the flexibility of the substrates and the loading mass of the active materials in the electrodes, fibrous materials (e.g. paper and textiles) are employed as substrates for flexible energy storage devices due to their ultrahigh capability of withstanding different bending-mode deformations and the high surface area compared with the planar metal foils and polymeric films [2a, 10b, 17]. Paper and some traditional textiles are inexpensive and light, while the insulated property makes them necessary to composite with conductive components, including CNTs and graphene, etc [10b, 18]. In recent years, free-standing conductive textiles such as carbon cloth, CNT, and graphene papers are emerged as flexible current collectors [19]. These substrates can be chemically and physically stable in most fabrication conditions of 3D nanoarrays, which are promising as the expense of these materials is decreasing gradually and the large-scale manufacturing is ultimately expected. Moreover, the fibrous feature of these substrates makes it possible to integrate the flexible electronics with human's clothes for wearable electronics.

7.2.2 Synthesis Methods of Flexible 3D Nanoarrays

Various synthesis methods have been explored for the growth of flexible 3D nanoarrays, including the hydrothermal method, chemical vapor deposition (CVD), sputtering, anodic oxidation, electrochemical deposition and so on. According to the species of materials, the reaction conditions, and the choice of substrates, a wide range of 3D nanostructure arrays with complex 3D topologies and desired energy storage properties can be further constructed.

7.2.2.1 Flexible 3D Nanoarray Electrodes Fabricated by Hydrothermal Methods

Hydrothermal methods are one of the most commonly used methodology to fabricate 3D nanoarrays. In general, most metal oxides and some metal-organic framework (MOF) nanoarrays can be fabricated by this method [19a, b, 20]. In particular, due to the circumstance of high temperature and high pressure during the hydrothermal reaction, as-obtained metal oxides usually have good crystallinity compared with some other methods (e.g. anodic oxidation and electrochemical deposition). In addition, the production scale of hydrothermal method is easy to be expanded, leading to great production flexibility. In this decade, Fe_2O_3 nanorod arrays [21], MnO_2 nanosheet/nanorod arrays [22], $Ni(OH)_2$ nanoflake arrays [23], TiO_2 nanorod arrays[8c, 24], WO_3 nanowire arrays [25], and Co_3O_4 nanowire arrays [26] etc. fabricated by hydrothermal method have been reported as the most popular pseudocapacitive SC electrodes. To further improve the specific capacitance, some binary metal oxide-based nanostructures were also developed as pseudocapacitive electrodes, including $NiCo_2O_4$ [27], $CuCo_2O_4$ [28], and $MnCo_2O_4$ [29], etc.

In general, the precursor solution, the temperature, the reaction time, and the pressure are key factors in the hydrothermal system. Since the crystallization and phases play a key role in the growth process of 3D nanoarrays and the pseudocapacitive behavior, it is necessary to disclose the growth mechanism of these metal oxides in the hydrothermal system. For instance, as one of the most popular electrodes with high theoretical specific capacity, MnO_2 is widely studied [30]. Among all of the MnO_2 phases, α-MnO_2 and δ-MnO_2 can exhibit higher specific capacities compared with the others (e.g. β-MnO_2 and R-MnO_2) [31]. In particular, 3D α-MnO_2 nanorod and δ-MnO_2 nanosheet arrays can be directly fabricated by the hydrothermal method. However, the diverse polymorphism of MnO_2 makes hydrothermal synthesis a challenge, as reaction products are often sensitive to subtle variations in synthesis parameters. Particularly, δ-MnO_2 is a bulk-metastable phase that is difficult to be stable in the bulk size. Recently, Chen et al. investigated the crystallization pathways leading to manganese oxide polymorph formation via in situ X-ray scattering under varying solution potassium ion concentrations ([K^+]) in the hydrothermal system, providing a guiding principle to design rational synthesis pathways [20a]. Bulk and size-dependent phase diagrams at varying K^+ chemical potential were calculated to reveal the influence of K^+ ion concentration on MnO_2 crystallization. At the redox potential (E) of 1.2V, β-MnO_2 is the equilibrium phase at low [K^+], and α-K_xMnO_2 can be stabilized against β-MnO_2 under high [K^+] (Figure 7.1a). While a more complicated phase stability landscape was shown by the size-dependent phase diagram (at $E = 1.2$ V and fixed PH = 2; Figure 7.1b), indicating that the low surface energies of δ-MnO_2 and R-MnO_2 can stabilize these bulk-metastable phases at small sizes and low [K^+], and a δ-$K_{0.33}MnO_2$•$0.66H_2O$ can be stabilized at higher [K^+]. These diagrams can provide a foundation toward designing nanostructured α-MnO_2 and δ-MnO_2 electrodes for supercapacitors. Based on these principles, Zhang's group developed series of 3D MnO_2 core-shell nanoarrays, such as the β-MnO_2/δ-MnO_2 (Figure 7.1c) and β-MnO_2/α-MnO_2 (Figure 7.1d) core-shell structures by tuning the PH value and the concentration of K^+.

7.2.2.2 Flexible 3D Nanoarray Electrodes Fabricated by CVD/Sputtering Methods

3D nanostructured materials obtained via CVD method can also have the advantages of high crystallinity and great production flexibility. Apart from this, the CVD synthesis is easy because no foreign templates or other rigid requirements are needed, leading to great production flexibility. Some three-dimensional nanoarrays made from carbonaceous materials can be easily fabricated by CVD methods. There is some important progress in this field recently. For instance, Lin et al. developed a "two-step" CVD method for the synthesis of aligned graphene/CNT carpets on the substrate [34]. Firstly, few-layer graphene was grown on the Si/SiO_2 substrates with the adhesion layer and catalyst layer. After that, CNT carpets were synthesized on the graphene layer by CVD method following by the deposition of catalyst particles. Based on this method, patterned 3D microsupercapacitor can be fabricated, as shown in Figure 7.2a and b. Very recently, vertical graphene nanoflake arrays on flexible carbon cloth are also fabricated and reported by Deng et al. via the plasma-enhanced CVD method (Figure 7.2c) [35]. Similar with the traditional synthesis method for planar graphene, hydrogen and methane are utilized as the precursor gases. Since CNTs and graphene are popular carbonaceous materials for double-layer electrochemical supercapacitors, these 3D nanoarrays are promising electrodes for flexible SCs with high power density and fast response.

Magnetron Sputtering is also one of the important techniques for the fabrication of nanostructured thin films. Recently, some research demonstrated that 3D nanoarrays on the flexible substrates can also be fabricated via this method. For instance, Xia et al. found that lower substrate temperature is beneficial to the growth of 3D nanostructures due to the relatively low energy and tend to evolve into 3D islands via the Volmer-Weber growth mode (Figure 7.2d) [36]. In contrast, when the substrate temperature is high, the adatoms have enough energy to diffuse on the substrate and thin film growth favors Frank-van der Merwe model (2D layer-by-layer growth) [37]. Based on this method, 3D nanoarrays on flexible stainless steel foils such as Li_2MnO_4 nanosheet arrays and α-MoO_{3-x} nanoflake arrays were successfully fabricated [36b, c, 37].

7.2.2.3 Flexible 3D Nanoarray Electrodes Fabricated by Electrochemical Deposition Methods

Electrochemical deposition is another popular method for the synthesis of 3D nanostructure arrays. Typically, there are three strategies for the 3D structure engineering by this technique, including the template-free electrodeposition, the template-assisted electrodeposition, and the anodic oxidation method. Among all of the capacitive electrode materials, most of them can be synthesized via the electrochemical deposition method, such as MnO_2 [38], RuO_2 [39], Co_3O_4 [40], Fe_2O_3 [41], polyaniline (PANI) [42], Poly(3,4-ethylenedioxythiophene) (PEDOT) [43], Polypyrrole (PPY) [44], TiO_2 [24], and so on.

During the electrochemical deposition process, a stable precursor solution, appropriate deposition potential/current and conducting temperature are required, which will critically influence the

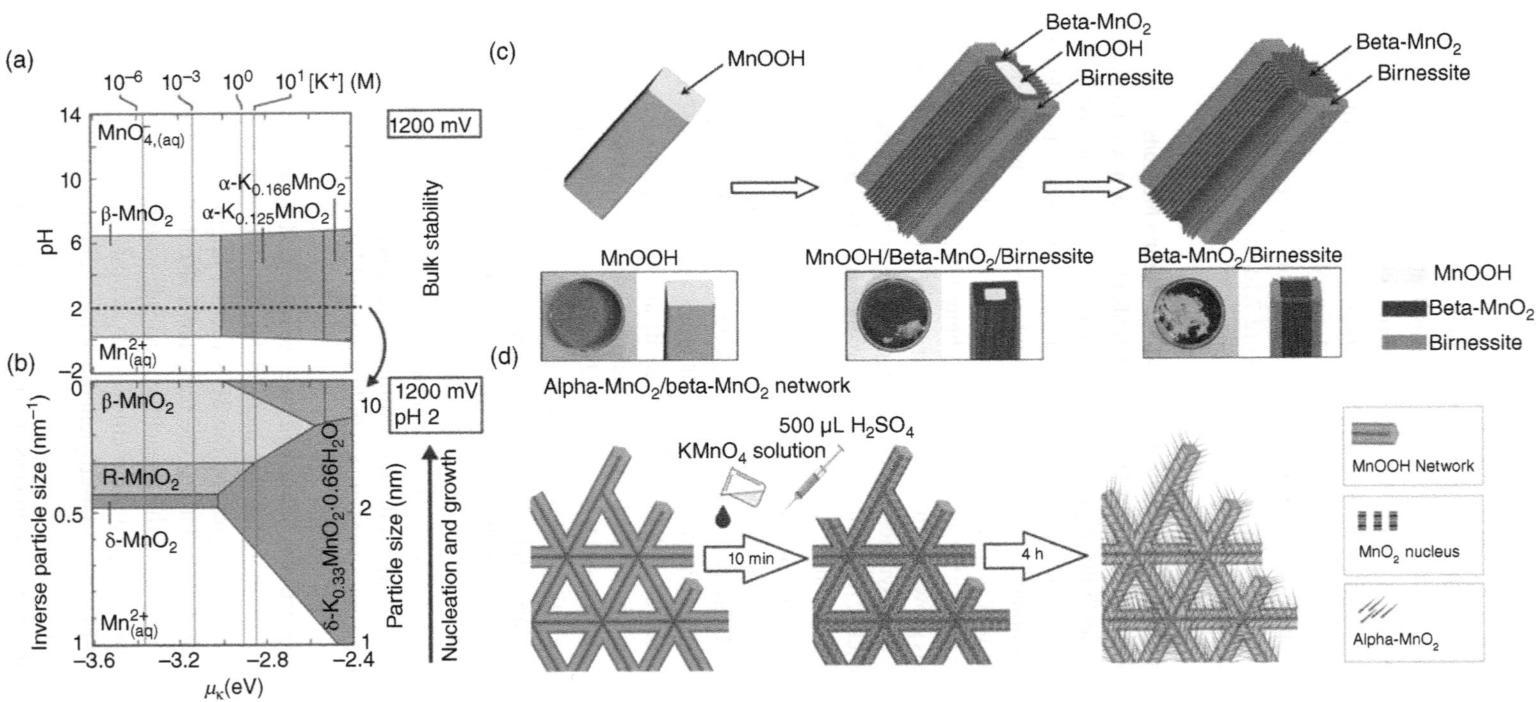

Figure 7.1 Phase diagrams of K-Mn oxide systems at E = 1.2 V. (a) PH-μ_K slice at 1/R = 0 indicating bulk stabilized phases. (b) 1/R-μ_K slice at PH = 2 illustrating the size-stabilized phases [20a]. *Source:* Reproduced with permission, © 2018, Springer Nature. (c) Schematic representation of β-MnO_2/δ-MnO_2 core/shell nanorod [32]. *Source:* Reproduced with permission, © 2018, American Chemical Society. (d) Schematic representation of β-MnO_2/α-MnO_2 core/shell network [33]. *Source:* Reproduced with permission, © 2019, American Chemical Society.

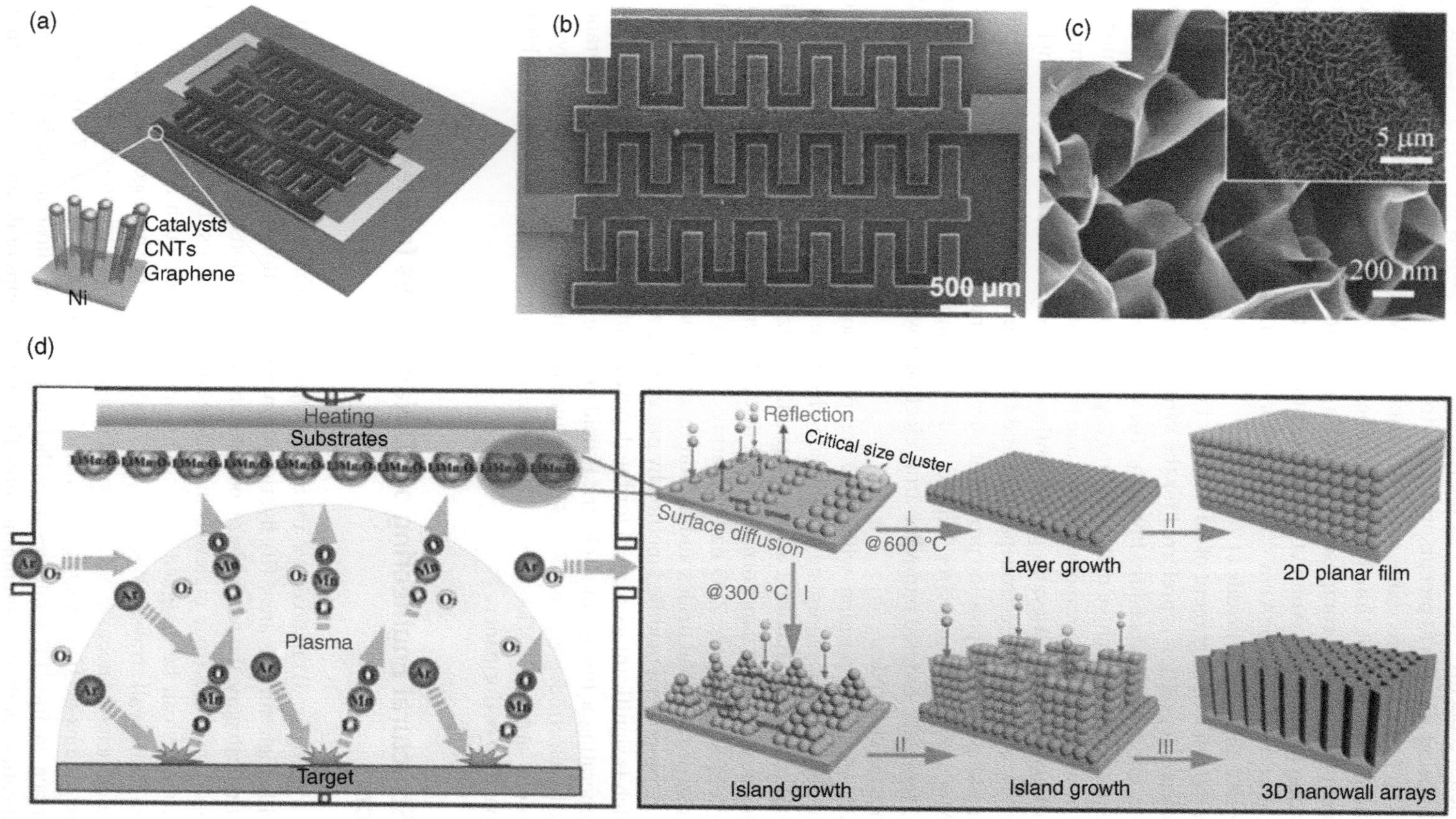

Figure 7.2 (a and b) Schematic and field-emission scanning electron microscopy (FESEM) image of the aligned graphene/carbon nanotube carpets on the substrate for micro-supercapacitors [34]. *Source:* Reproduced with permission, © 2013, American Chemical Society. (c) FESEM image of graphene nanosheet arrays on carbon cloth [35b]. (d) Schematic illustration of the DC sputtering process and the depositions of 3D $LiMn_2O_4$ nanowall arrays and 2D planar $LiMn_2O_4$ thin flm via different growth modes [36b]. *Source:* Reproduced with permission, © 2018, Wiley-VCH.

morphology and pseudocapacitive behavior of the films deposited. Taking MnO_2 as an example, the earliest report of using acetates as precursors for electrodeposition of MnO_2 is from Tench and Warren in 1980s [45]. Later, a more stable solution containing $MnSO_4$ and sodium acetate was obtained by Broughton et al. [46] The thin films of carambola-like γ-MnO_2 nanoflakes was reported by Chou et al. with another solution composed of 0.1 M Na_2SO_4 and 0.1 M $Mn(CH_3COO)_2$ at pH 6.0 by using potentiostatic method [47]. Yuan et al. reported the electrodeposition of MnO_2 nanorod arrays at a constant current of $0.5\,mA\,cm^{-2}$ in a solution of 0.01 M $Mn(CH_3COO)_2$ and 0.02 M NH_4Ac containing 10% dimethyl sulfoxide at 70 °C [48].

Comparing with the random alignment of nanoarrays fabricated by template-free methods, the template-assisted electrodeposition is preferred for the synthesis of ordered hybrid nanostructure arrays. For example, one typical procedure involving the anodized aluminum oxide (AAO) template allows the deposition of capacitive materials into aligned channels of AAO [49]. Commonly, a layer of Au/Ag/Cu was thermally evaporated onto one planar surface side of the AAO template to serve as the working electrode. Ajayan's group synthesized arrays of multi-segmented hybrid CNT/Au nanowires by combining electrodeposition and CVD method (Figure 7.3a–e) [50]. Lee's group reported a one-step electrodeposition process of pure MnO_2 nanowires, pure PEDOT nanowires, and coaxial MnO_2/PEDOT nanowires (Figure 7.3f) into aligned AAO template [38]. After removing the AAO template, ordered nanowire-array electrodes were obtained and demonstrated as high-performance SC electrodes.

Besides the inside-electrode position with AAO template, some other nanorod arrays and nanosheet arrays are also effective templates for outside-electrode position. For instance, as a low-cost transition metal oxide, ZnO nanorod arrays are usually utilized as a sacrificial template [21, 51]. Fan's group developed nickel nanocorn arrays as three-dimensional and conductive support by electrodeposition of nickel on the ZnO nanorod arrays, which efficiently boosted the supercapacitive performance of the metal oxides and carbon materials-based electrodes [51a]. Based on this strategy, Li et al. proposed a modified method by pre-coating a thin Au layer on the surface of ZnO nanorod, which is significantly essential for the deposition of homogeneous and continuous Ni thin film by generating a finely tuned electrical field (Figure 7.3g). Two hybrid core-branch nanoarchitectures by electrodeposition of Fe_2O_3 nanoneedles and MnO_2 nanosheet arrays on ultrafine Ni nanotube arrays (NiNTAs@Fe_2O_3 nanoneedles) were then developed, exhibiting excellent supercapacitive performance.

7.3 Typical Structural Engineering of 3D Nanoarrays for Flexible Supercapacitors

From the view of topologies, 3D nanoarrays are mainly classified into two categories: (i) intrinsic nanostructures with high surface area; (ii) hybridization with functional species, as shown in Figure 7.4. In the first category, the particle size, pore size and distribution, and morphology dominate the basic properties of electrode materials in flexible supercapacitors. Derived from the bulk materials, nanostructures can be designed into various morphologies, such as nanowire arrays, nanotube arrays, and nanowall arrays (NWAs) (Figure 7.4a). Based on the homogeneous nanostructures, subtly designed heterogeneous structures have been demonstrated as more effective surface engineering configurations of 3D nanoarrays for flexible supercapacitors (Figure 7.4b). Typically, three kinds of surface engineering strategies onto the nanostructured host are devised. The first strategy is a modification method by anchoring some functional groups or doping elements on the surface of active materials, aiming at modified surface chemical and physical properties. The

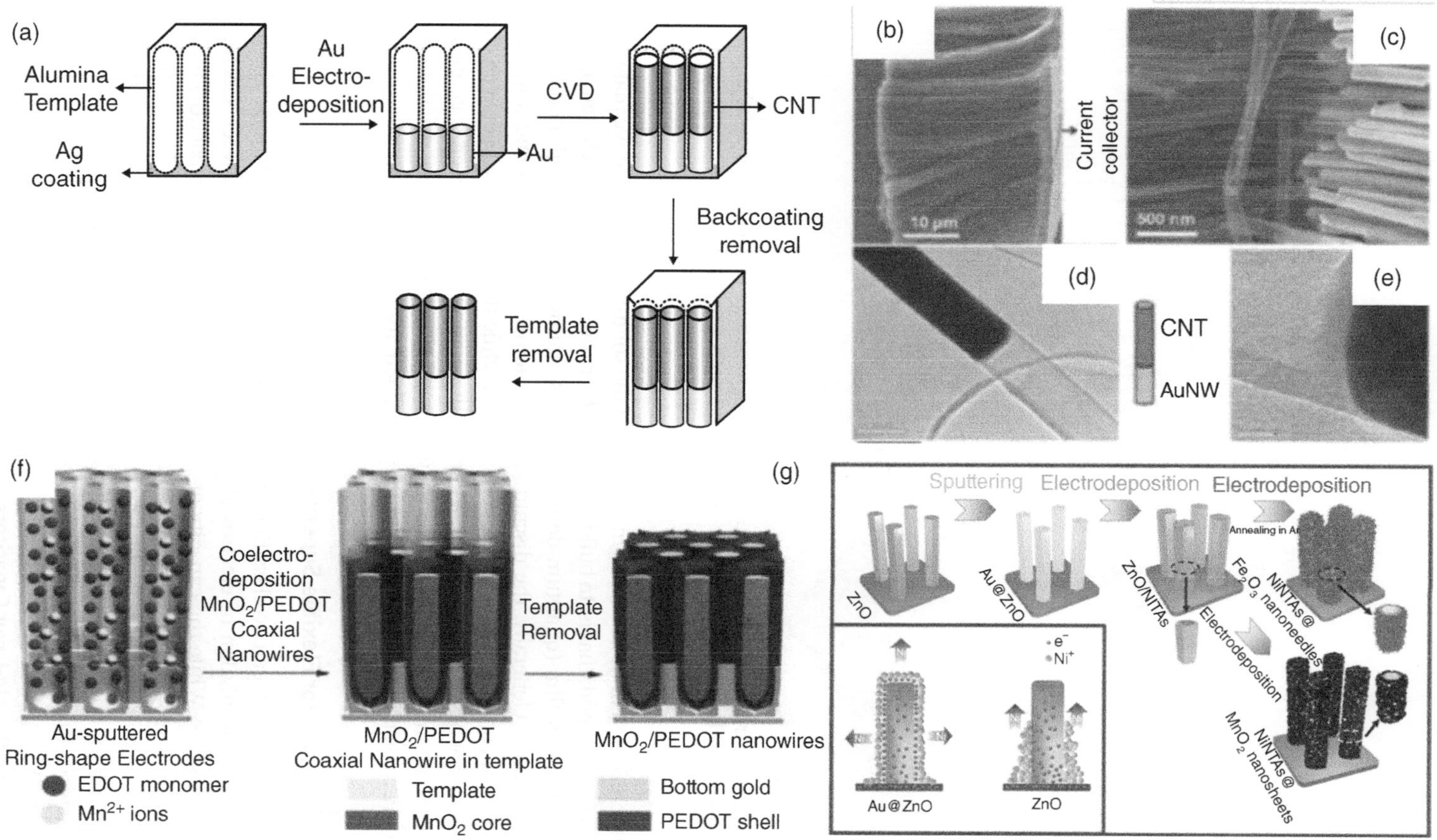

Figure 7.3 Schematic (a), FESEM (b and c), and TEM (d and e) images showing the fabrication of CNT/Au nanowires hybrid structures inside the AAO template [50]. *Source:* Reproduced with permission, © 2008, The Royal Society of Chemistry. (f) Schematic showing the growth mechanism of coaxial MnO_2-PEDOT nanowire arrays [38]. *Source:* Reproduced with permission, © 2008, American Chemical Society. (g) Schematic illustration of the synthesis procedure for the NiNTAs@Fe_2O_3 nanoneedles and NiNTAs@MnO_2 nanosheets. The inset shows the different Ni flms formed on the ZnO nanorod with and without Au layer [21]. *Source:* Reproduced with permission, © 2017, Wiley-VCH.

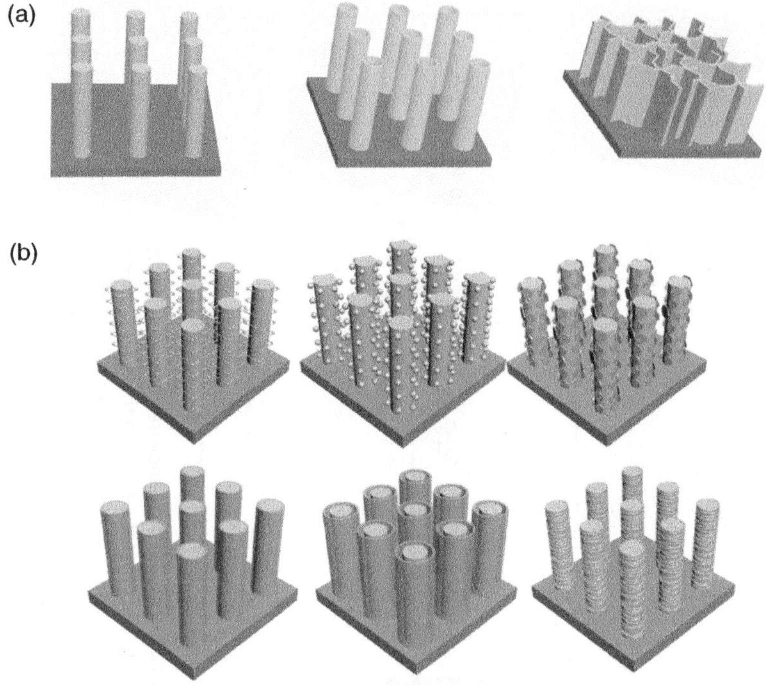

Figure 7.4 (a) Schematic illustration of typical 3D nanowire arrays, nanotube arrays, and nanowall arrays. (b) Schematic illustration of 3D hybrid nanostructures based on nanorod arrays.

second one is pre-intercalation of heteroatoms in the bulk of the solid-phase of the active materials. The third one is a coating method, that is, to build core-branch or core-shell structures based on host materials with discontinuous film (quantum dots, nanowires, nanosheets, etc.) and a conformal film (solid or gapped films). Generally, the discontinuous structure can exhibit some functional effects for better electrical conductivity and ionic conductivity, while the contact area between core materials and electrolyte won't be diminished. Whereas, the conformal film covered core-shell structure may prevent core materials from the electrolyte, suppressing the secondary reactions and structure deformation, as well as the electrical and ionic conductivity enhancement.

7.3.1 Basic 3D Nanoarrays for Flexible Supercapacitors

The working mechanism of SCs can be classified into EDLCs and the pseudocapacitors. The former one works on the accumulation of electrostatic charges at the interface between electrode materials and the electrolyte. While the capacity of pseudocapacitor (Faradaic capacitor) results from the redox reaction of electro-active species on the surface of the electrode. Since both the EDLCs and the pseudocapacitors are controlled by surface electrochemical phenomena, the SCs can reach high rate capability with fast charge-discharge within seconds, long cycle life ($>10^5$ cycles) and high power density ($>10\,\mathrm{kW\,kg^{-1}}$).

7.3.1.1 Flexible Electrical Double-Layer Capacitors

In general, electrode materials with high electrical conductivity, large ion-accessible surface area, fast ionic transport and excellent electrochemical stability can achieve high capacitances and desired lifetime in EDLCs. Carbon materials with high specific surface area, hierarchical morphology, tunable porosity, and appropriate texture are regarded as the most appropriate candidates for EDLCs. Various carbon-based materials such as activated carbon, carbide-derived carbon, CNTs, graphene, carbon nanocages have been explored extensively. Compared to conventional electrolytic capacitors (on the order of micro-or pico-farads), activated carbon-based commercial EDLCs have a much higher theoretical capacitance (100–300 $F g^{-1}$). Conventional activated carbon electrodes, however, are not appropriate for flexible energy storage devices due to the fragile nature of particulate film coated on the metal current collectors. In this context, 3D nanoarrays grown on flexible current collectors can exhibit higher electrical conductivity and better mechanical contact for flexible EDLCs compared with the conventional electrodes. Till now, researchers developed several kinds of carbon material-based flexible supercapacitors by constructing aligned CNTs or graphene nanosheet arrays on the carbon cloth. For instance, Zhang et al. reported a flexible EDLC by constructing hierarchically aligned porous carbon nanotube arrays (PCNTAs) on flexible carbon fibers [52]. The PCNTAs are obtained by catalytic conversion of ethanol on ZnO nanorod arrays and then reduction-evaporation of ZnO nanorods, resulting in uniform and controllable wall thicknesses of the final PCNTAs. The important parameters such as 3D arrangement, the diameters, and the lengths of the PCNTAs can be tuned by adjusting the synthesis protocols of the ZnO nanorod arrays. The electrode exhibits a high specific capacitance of 182 $F g^{-1}$ at 40 $A g^{-1}$ in 6 M KOH. And the symmetric supercapacitor (Figure 7.5a) shows an excellent cycling stability with only 0.0016% loss per cycle after 10 000 cycles at the current density of 12 $A g^{-1}$. In addition, as shown in Figure 7.5b, the nearly rectangular cyclic voltammetry (CV) shape at different scan rates indicates the ideal electric double-layer capacitive behavior and fast ion transport/diffusion at high charge/discharge rates. Meanwhile, under different bending geometries, the CV curves show clearly the almost same capacitive behavior (Figure 7.5c and d). The excellent rate performance and long cycle life can be ascribed to multiple channels in the unique 3D hierarchical structure of the PCNTAs, which facilitate the diffusion and migration of the electrolyte ion, as schematically illustrated in Figure 7.5e.

7.3.1.2 Flexible Pseudocapacitors

Since redox reactions usually take place in metal oxides where metal ions have multiple valence states, the pseudocapacitive materials take the advantages from both mechanisms. Progress in pseudocapacitor technology has benefited greatly by moving from conventional to 3D-nanostructured electrodes. The large surface area created by 3D nanoarrays significantly enhance the efficiency in utilizing the electrode materials and therefore improve the electrode performance.

Great progress have been achieved on the cathode materials for the high-performance supercapacitors, such as MnO_2 [53], Co_3O_4 [54], RuO_2 [55], $NiCo_2O_4$ [56]. For the 3D structure engineering of cathode materials, Shen's group reported a fiber-based flexible all-solid-state asymmetric supercapacitors by growing Co_3O_4 nanowires on nickel fibers (Figure 7.6a) as positive electrode and employing graphene as negative electrode, which can also act as a flexible photodetector (Figure 7.6b) [57]. With an operating potential of 1.5 V (Figure 7.6c), this device achieved a maximum volumetric energy density of 0.62 $mWh\,cm^{-3}$ and power density of 1.47 $W cm^{-3}$. Spinel $NiCo_2O_4$ with both nickel and cobalt metal elements shows higher electrochemical activity than single component nickel or cobalt oxides. A flexible electrode based on $NiCo_2O_4$ nanowire arrays on nickel foam was also reported by Shen's group (Figure 7.6d and e), which exhibited a high specific capacity of 2305 $F g^{-1}$ at the current density of 8 $A g^{-1}$ (Figure 7.6f) [56a].

Active efforts have been attracted to explore novel pseudocapacitive anode materials to achieve large specific capacitance and good rate performance, including VO_x [58], MoO_x [59], VN [60],

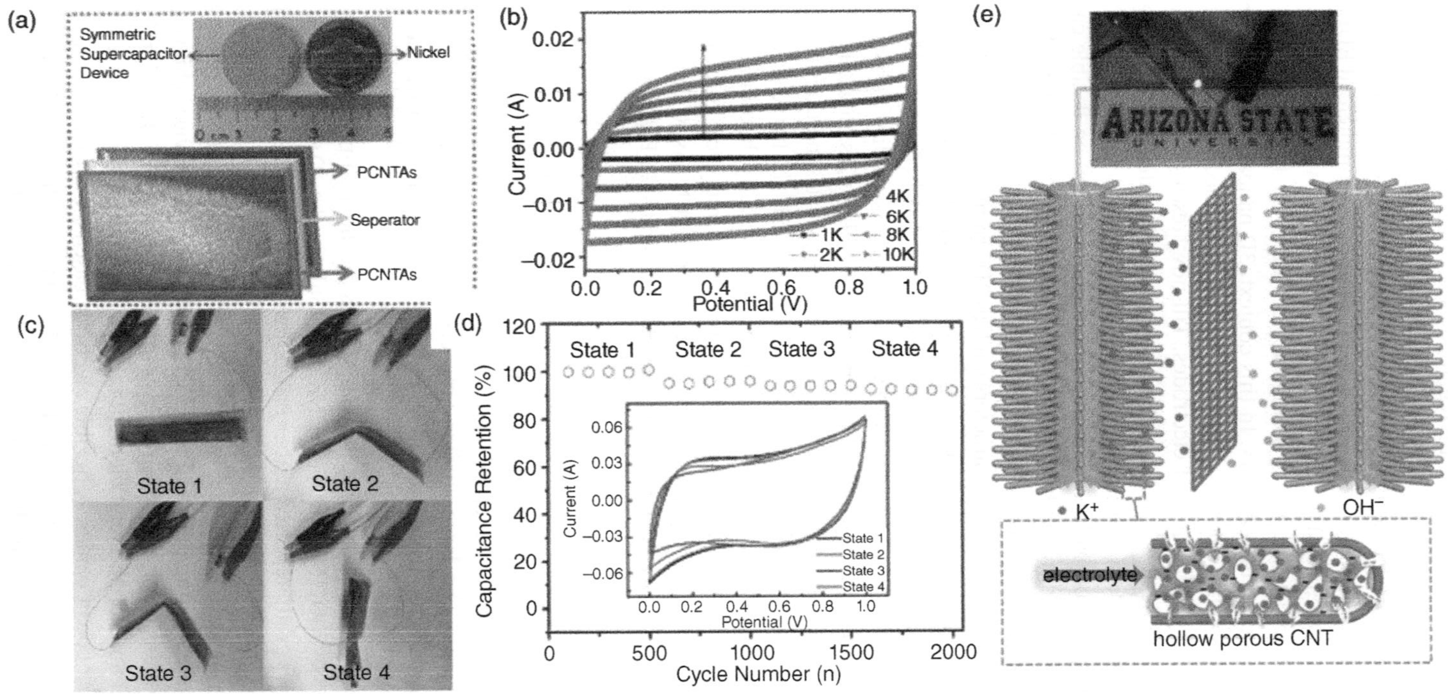

Figure 7.5 (a) Digital photograph and schematic illustration of the symmetric supercapacitor configuration; (b) CV curves of the device at different scan rates; (c) Photographs of flexible symmetric device at different bending states; (d) Cycle stability tested at different bending states with the inset showing CV curves obtained at different bending states at a scan rate of 500 $mV s^{-1}$. (e) A schematic illustration of multi-transport pathways for the electrons and ions in the porous carbon nanotubes [52]. *Source:* Reproduced with permission, © 2016, Wiley-VCH.

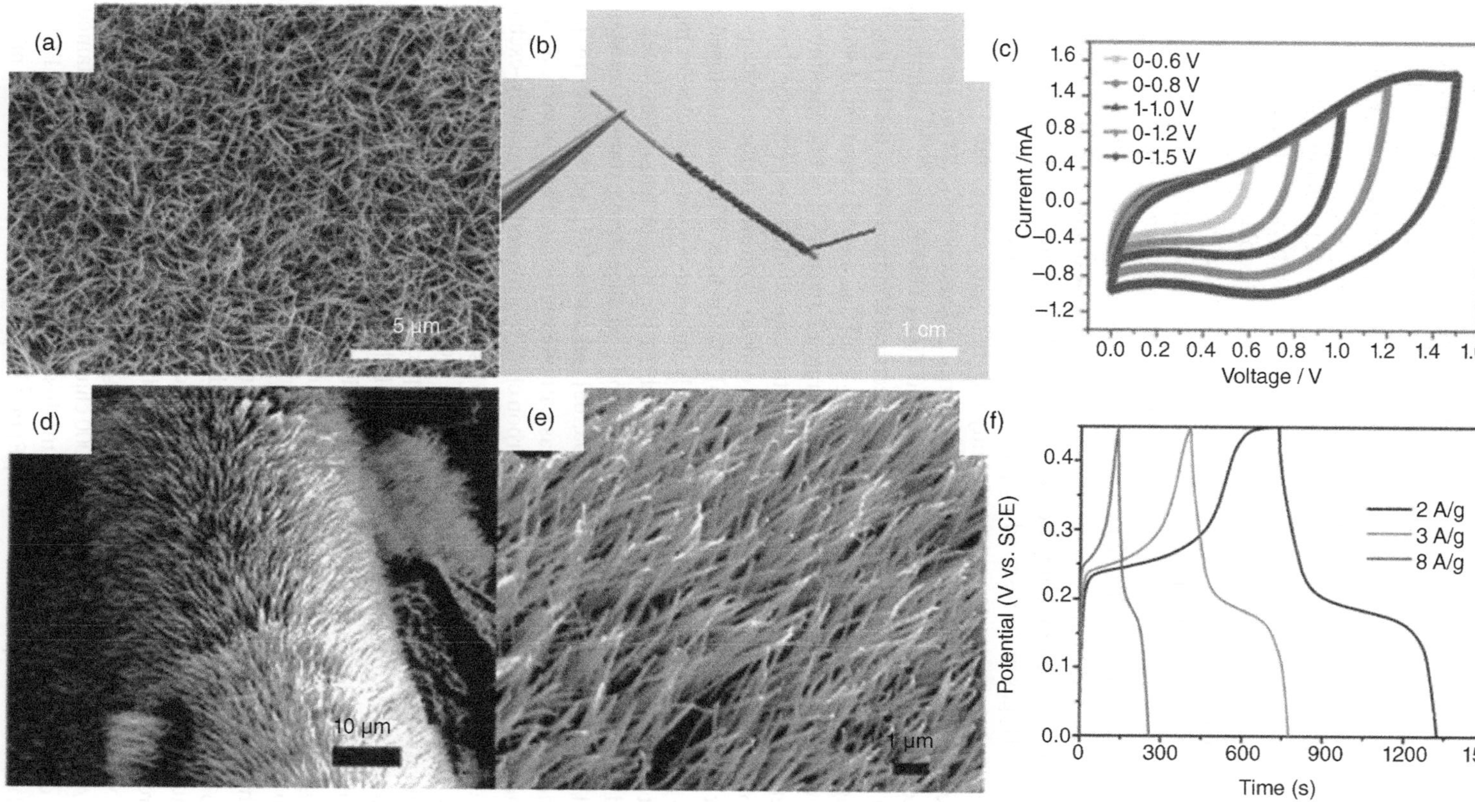

Figure 7.6 (a) SEM image of Co_3O_4 nanowire arrays on nickel fibers. (b) The fiber-based flexible all-solid-state asymmetric supercapacitors based on the cathode of Co_3O_4 nanowire arrays and anode of graphene fiber. (c) CV curves of the fiber-based flexible supercapacitor in different voltage windows [57]. *Source:* Reproduced with permission, © 2014, Wiley-VCH. (d and e) SEM image of the $NiCo_2O_4$ nanowire arrays. (f) Charge-discharge curves of the flexible $NiCo_2O_4$ electrode [56a]. *Source:* Reproduced with permission, © 2013, The Royal Society of Chemistry.

TiO_2 [61], FeO_x [21, 62], and FeOOH [63]. Among them, iron oxides/hydroxides (α-Fe_2O_3, β-Fe_2O_3, γ-Fe_2O_3, Fe_3O_4, α-FeOOH, β-FeOOH, γ-FeOOH, and δ-FeOOH) have received tremendous research interest as promising anode materials because of the multiple valences of iron element, rich redox chemistry in the negative potential window, low cost, and abundant resource. Particularly, α-Fe_2O_3 is regarded as one of the most promising pseudocapacitive materials owing to its stable crystal structure, environment compatibility, and high theoretical capacitance (3625 $F\,g^{-1}$). However, the currently reported specific capacitance of α-Fe_2O_3 (100–300 $F\,g^{-1}$) is still far below the theoretical value, which is mainly attributed to the inferior conductivity and insufficient ionic diffusion rate of electrodes. Researchers found that building 3D nanoarrays on flexible current collectors can address these problems. For instance, a high specific capacity of 418.7 $F\,g^{-1}$ was achieved by Li et al. by constructing Fe_2O_3 nanoneedle arrays on nickel nanotube arrays [21]. Lei's group reported a highly porous Fe_2O_3 nanosheet arrays with a high specific capacity of 627 $F\,g^{-1}$ [62].

However, for the transition metal oxides-based pseudocapacitive materials, the electrochemical performance is mainly impeded by the low electrical conductivity. Rational surface vacancies in nanocrystals enhance the electron mobility, by treating the pseudocapacitive metal oxides (TiO_2, VO_2, etc.) with reductive H_2 gas or corresponding plasma.[8c, 61b, 64] For example, Lu et al. reported that the hydrogen treatment improve the carrier mobility TiO_2 nanotube arrays from $3.4\times10^{20}\,cm^{-3}$ to $1.4\times10^{23}\,cm^{-3}$, delivering a 124~ and 40~ fold capacitance enhancement (3.24 $mF\,cm^{-2}$) compared to the untreated TiO_2 (0.026 $mF\,cm^{-2}$) and air treated TiO_2 (0.08 $mF\,cm^{-2}$).[8c] Additionally, the small IR drop (0.02 V) again confirmed the superior electrical conductivity of the H-TiO_2 electrode. For the cycling capability, the hydrogen treated TiO_2 exhibited a good retention of 96.9% of the initial capacitance after 10 000 cycles scanning at the rate of 100 mVs^{-1}, while the untreated TiO_2 and air-TiO_2 retained only 65.2% and 55.7% of the initial capacitances respectively. Based on this strategy, a flexible hydrogen-treated TiO_2 electrode on carbon cloth was further developed by them, as an efficient anode in flexible supercapacitors [61b]. Meanwhile, Wu et al. developed a H_2 plasma treated TiO_2 nanotube arrays for SCs, which not only induced oxygen vacancies for higher carrier mobility, but also increased the surface area by etching the wall of nanotubes [64a]. This treatment generated 6.2 times of increase in capacitance and a smaller charge transfer resistance.

Besides the metal oxides, some metal sulfides [19a, 65], selenides [66], phosphides [67], carbides [68], and nitrides [60, 69] are also developed as cathode materials. For instance, Xu et al. obtained a high capacitive Co_9S_8 nanorod arrays on carbon cloth by a hydrothermal anion-exchange reaction from Co_3O_4 nanorod arrays (Figure 7.7a) [19a]. A significantly enhanced specific capacitance (783.3 $F\,g^{-1}$ for active materials and 113.5 $F\,g^{-1}$ for the overall electrode at 5 $mV\,s^{-1}$) was obtained by the sulfuration reaction. Asymmetric supercapacitors in liquid potassium hydroxide (KOH) electrolyte and in polyvinyl alcohol (PVA)/KOH gel electrolyte (Figure 7.7b) based on the C_9S_8 cathode and $RuO_2@Co_3O_4$ nanoarray anode were fabricated respectively, which can be cycled reversibly in the range of 0–1.6 V and exhibit superior electrochemical performance with an energy density of 1.21 $mWh\,cm^{-3}$ at a power density of 13.29 $W\,cm^{-3}$ in aqueous electrolyte and an energy density of 1.44 $mWh\,cm^{-3}$ at the power density of 0.89 $W\,cm^{-3}$ in solid-state electrolyte. A simple test on the flexibility of the solid-state device was evaluated by testing the CV curves of the device under normal, bent, and twisted conditions (Figure 7.7c). Via a simple thermal evaporation method, a 3D hierarchical structure consisted of pyramid-like bases and wire-like branches grown epitaxially on them was successfully fabricated [66a]. A high specific capacitance of 300 $F\,g^{-1}$ at a current density of 1 $A\,g^{-1}$ and 99.3% retention after 2000 charge/discharge cycles were obtained. By stacking the $GeSe_2$ coated electrodes into a thin-film all-solid-state supercapacitor on PET substrates (Figure 7.7d), an areal capacitance of 240 $\mu F\,cm^{-2}$ was delivered at the 100 $mA\,m^{-2}$, and about 77.5% of the capacitance at 10 $mA\,m^{-2}$ was still retained (Figure 7.7e). Excellent cycling life

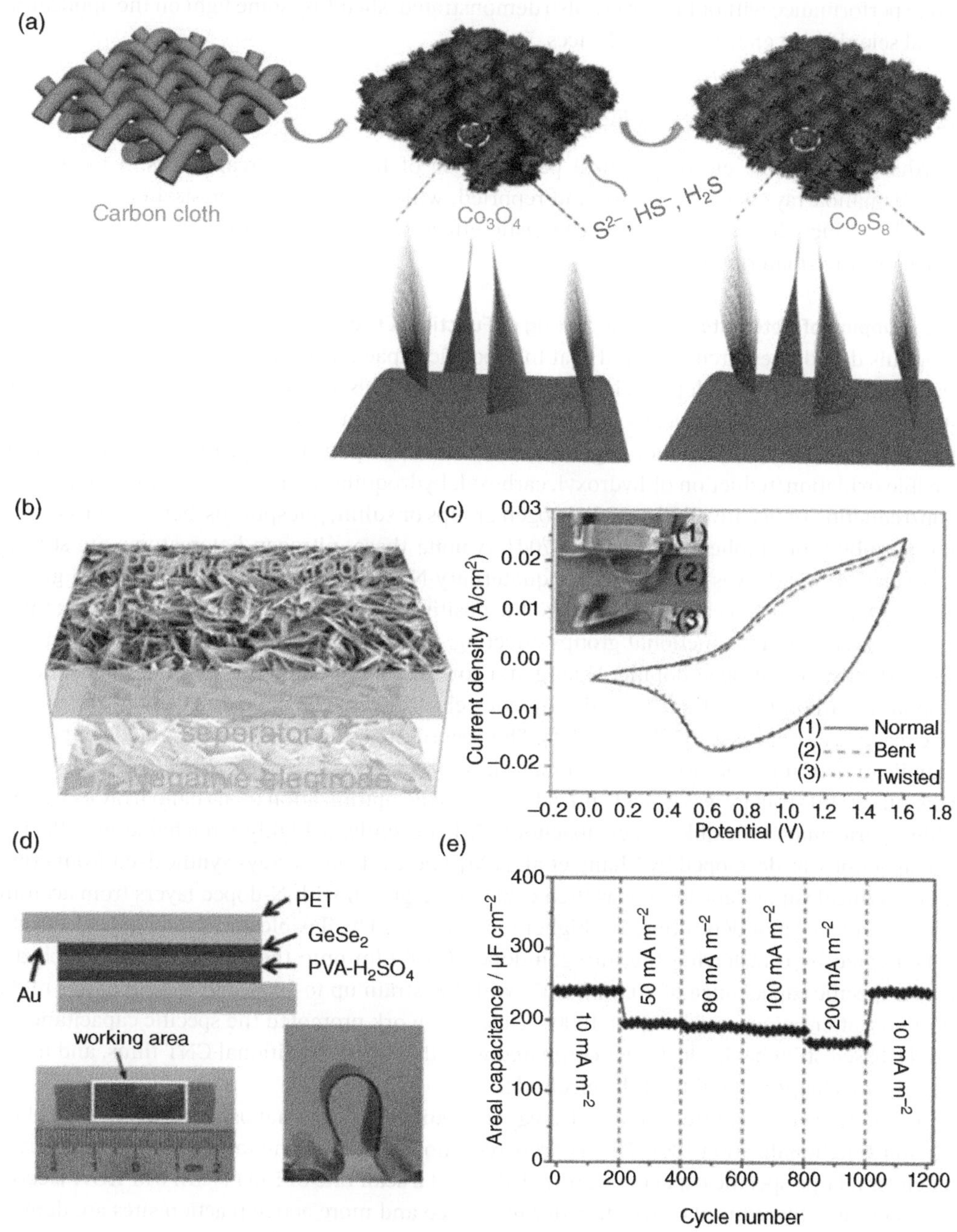

Figure 7.7 (a) Schematic illustration of the synthesis process of the Co_9S_8 nanoarrays on carbon cloth. (b) Schematic illustration of the as-assembled asymmetric supercapacitor. (c) CV curves collected at the scan rate of $100\,mV\,s^{-1}$ for the solid-state ASC device under normal, bent, and twisted conditions. Insets are the device photographs under different test conditions [19a]. *Source:* Reproduced with permission, © 2013, American Chemical Society. (d) Schematics and photographs of the stacked $GeSe_2$ supercapacitor. (e) Rate performance at progressively varying current densities [66a]. *Source:* Reproduced with permission, © 2012, Wiley-VCH.

and rate performance without loss were also demonstrated, shedding some light on the application of metal selenides in energy storage devices.

7.3.2 Hybrid 3D Nanoarrays for Flexible Supercapacitors

To further improve the electrochemical performance of flexible supercapacitors, a number of hybrid 3D nanoarrays were developed and reported, which enhance the physical properties (e.g. electrical conductivity) and take the synergistic effect of electrical double-layer capacitive and pseudocapacitive materials efficiently.

7.3.2.1 Doping of Heteroatoms and Anchoring of Functional Groups

During this decade, researchers found that the specific capacitance can be improved by introducing surface redox groups or doping element to carbon materials, including –OH, N, B, P, S, etc. [70] Post treatment process, chemical vapor deposition and biomass derivation method are popular synthetic routes for these functional groups. The dominant part of oxygen functionalities is the reversible oxidation/reduction of hydroxyl, carboxyl, hydroquinone/quinone groups [71]. The oxidation/reduction of negatively charged nitrogen groups or sulfur, phosphorus, boron atoms locates at the periphery of graphene-like layers [70d]. Among these, nitrogen heteroatoms are studied extensively for SCs. It was noted that the quaternary-N and pyridinic-N-oxides nitrogen groups have enhanced effects on capacitance due to the positive charge, and pyrrolic and pyridinic nitrogen are most important functional groups affecting energy storage performance at high current loads. Recently, via nitrogen doping, Huang and coworkers promoted the specific capacitance of mesoporous carbon to 855 $F g^{-1}$ in acidic aqueous electrolytes. The energy density of the packed device reached as high as 23.0 $Wh kg^{-1}$ [70f]. This record-breaking value demonstrated the promising application of nitrogen-doped carbon materials.

This successful strategy was quickly introduced into the optimization of 3D nanoarray electrodes for high-performance flexible supercapacitors [72]. Recently, a highly stretchable and flexible supercapacitor was developed by Zhang et al. [72a] The CNT nanoarrays synthesized from ethylene by chemical vapor deposition was then coaxially re-grown with N-doped layers from acetonitrile by chemical vapor deposition at a higher temperature. The flexible and stretchable electrodes was fabricated by transferring the nitrogen-doped CNT film onto the polyurethane, which displayed a specific capacitance of 30.8 $mF cm^{-2}$ with the strain up to 400% and a capacitance retention of 96% after being stretching for 1000 cycles. This work promoted the specific capacitance of the CNT-based stretchable electrodes compared with that of the traditional CNT films, and maintained the mechanical stability of the electrode.

Recently, Zhai et al. reported a novel strategy to modify the Co_3O_4 nanosheet arrays by the phosphate ion functionalization [26]. The as-prepared phosphate ion functionalized-Co_3O_4 electrode can deliver a large specific capacitance of 1716 $F g^{-1}$ at a scan rate of 5 $mV s^{-1}$ in 6 M KOH electrolyte. A significantly reduced charge transfer resistance and more active reaction sites are demonstrated by this strategy. Therefore, it is intriguing to develop more effective functional groups for higher pseudocapacitive performance.

7.3.2.2 Pre-Intercalation of Heteroatoms

For the typical intercalated-pseudocapacitive materials, such as MnO_2 and VO_2, the low ionic diffusion ability inside the solid phase usually limits the rate capability of the electrodes, which is a fundamentally unresolved problem. Mai et al. found that pre-intercalation of Na^+ into the layered MnO_2 by the hydrothermal reaction can improve the Na^+ diffusion behaviors [73]. The as-prepared

Na_xMnO_2 nanoflake based supercapacitors exhibit faster ionic diffusion with enhanced redox peaks, 10-fold-higher energy densities up to 110 Wh·kg^{-1} and higher capacitances over 1000 F·g^{-1} in aqueous sodium system compared with traditional MnO_2 supercapacitors. Recently, Xia's group proposed that it is also an efficient strategy to extend the voltage window of the intercalated-pseudocapacitive materials by pre-intercalating Na^+ into the layered MnO_2 [22b]. The common working voltage window of 0–0.8 V vs. saturated calomel electrode (SCE) for MnO_2 electrode limits the overall cell voltage to 1.0–2.0 V. To address this problem, high Na content Birnessite $Na_{0.5}MnO_2$ nanosheet assembled nanowall arrays were in situ formed on carbon cloth substrate via electrochemical oxidation on the Mn_3O_4 nanowall arrays (Figure 7.8a). It was found that the electrode potential window of the $Na_{0.5}MnO_2$ nanowall arrays electrode can be extended to 0–1.3 V (vs Ag/AgCl) by using the carbon cloth substrate (Figure 7.8b), providing the possibility to construct aqueous asymmetric supercapacitors beyond 2.0 V (Figure 7.8c). A $Na_{0.5}MnO_2$//Fe_3O_4@C asymmetric supercapacitor exhibited excellent cycling stability (93% capacity retention after 10 000 cycles) in such a large voltage window of 0–2.6 V (Figure 7.8d).

7.3.2.3 Coaxial Branched and Core-Shell 3D Hybrid Nanostructures

Coaxial branched and core-shell nanostructures are a hybrid 3D structure engineer mainly based on the 1D or 2D axle and the branches in diverse dimensions, which have been vigorously designed as the electrode materials for electrochemical energy storage devices. These typical 3D nanoarrays can offer several important advantages, including higher specific area and more active sites for

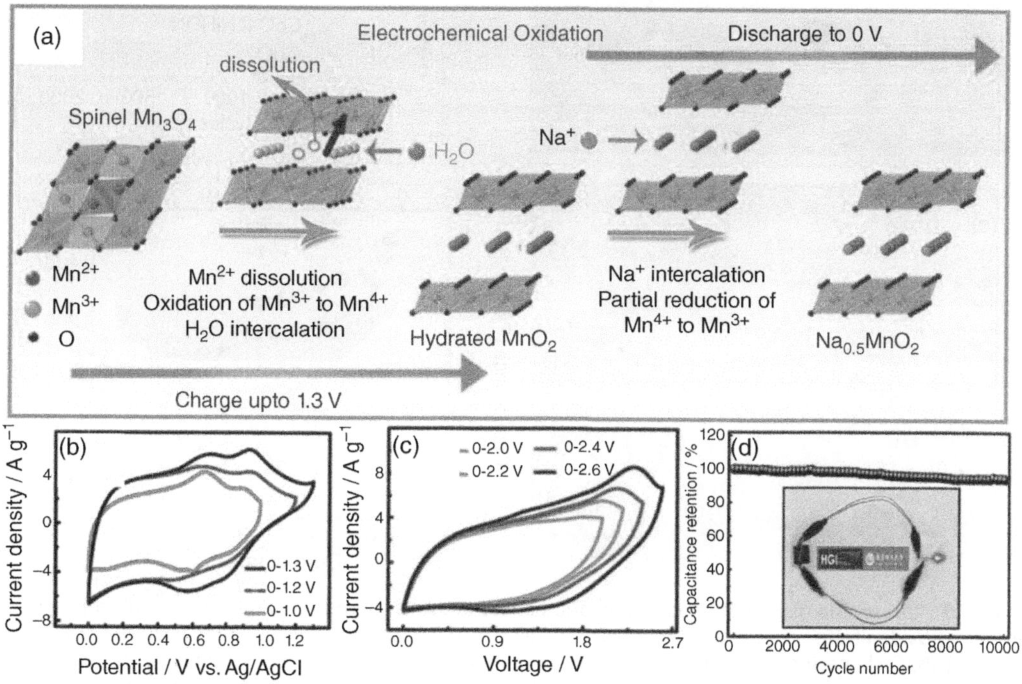

Figure 7.8 (a) Structural evolution process from spinel Mn_3O_4 to birnessite-$Na_{0.5}MnO_2$ during the electrochemical oxidation. (b) Typical CV curves of the birnessite-$Na_{0.5}MnO_2$ nanowall arrays in different potential windows. (c) CV curves of the birnessite-$Na_{0.5}MnO_2$//Fe_3O_4@C asymmetric supercapacitor in different voltage windows at a scan rate of 50 mV s^{-1}. (d) Cycle performance of the flexible supercapacitor [22b]. *Source:* Reproduced with permission, © 2017, Wiley-VCH.

electrolyte ions, prevention of aggregation, short electron pathway along the axle and alleviation of the structural strain from the void space. Thus, tremendous efforts have been devoted to the construction of 3D coaxial branched and core-shell nanostructures in the flexible supercapacitors for desirable functions.

In most cases, these hybrid nanostructure arrays are fabricated by multi-step approaches which are often designed by interfacial reactions between initially formed base-materials and the later introduced materials. As a result, a number of subtly designed coaxial branched and core-shell nanostructures have been developed [8a, 8b, 17, 27–28, 48, 72a, 74]. Some typical examples have been reported by Fan's group. For instance, in the early years, a hybrid nanostructure of porous CoO nanowires@ultra-thin nickel hydroxidenitrate nanoflakes on a nickel foam substrate was fabricated by introducing the CoO nanowire arrays into the $Ni(NO_3)_2$ aqueous solution (Figure 7.9a) [74c]. As an electrode for supercapacitors, the hybrid core-branch structure exhibits good electrochemical stability and high-rate performance compared to pure CoO nanowires and

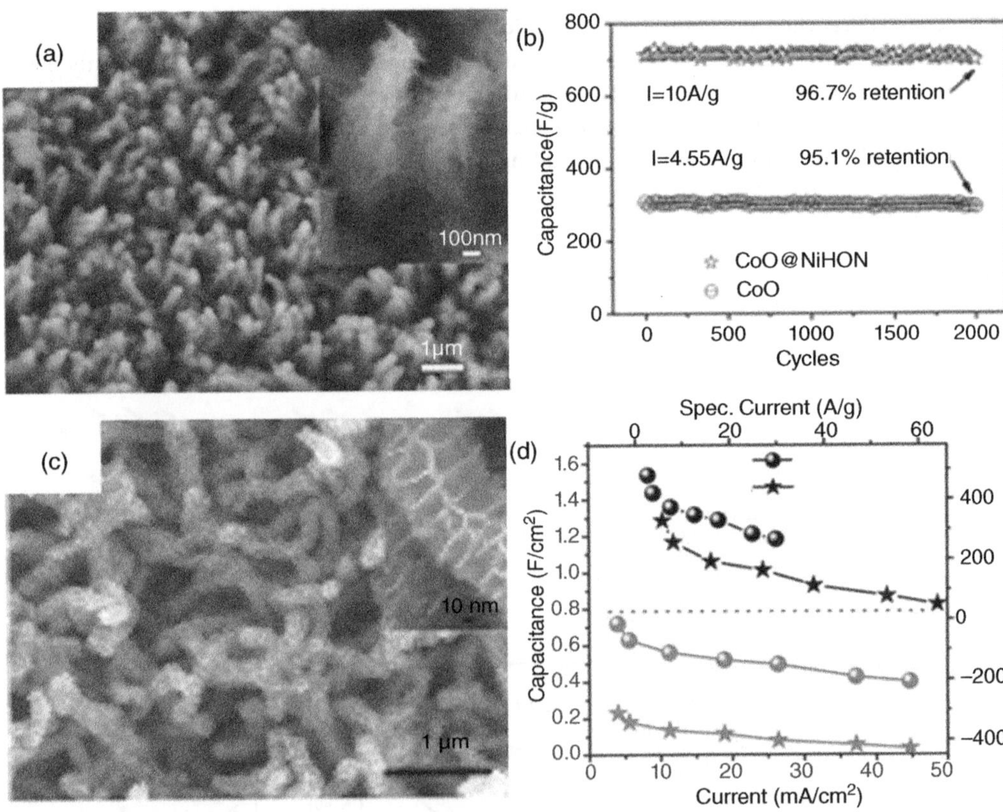

Figure 7.9 (a, c, e, and g) FESEM images of hybrid nanostructure of porous CoO nanowires@ultra-thin nickel hydroxidenitrate nanoflakes [74c], $Co_3O_4@MnO_2$ [74a], 3D carbon/$CoNi_3O_4$ [74d], and α-Fe_2O_3/PPy [74b]. (b) Cycle performance of pure CoO electrode and the hybrid CoO nanowires@ultra-thin nickel hydroxidenitrate electrode [74c]. *Source:* Reproduced with permission, © 2011, The Royal Society of Chemistry. (d) Rate performance $Co_3O_4@MnO_2$ and pure Co_3O_4 electrodes [74a]. *Source:* Reproduced with permission, © 2011, Wiley-VCH. (f) Cycle performance of 3D carbon/$CoNi_3O_4$ electrode and the hybrid electrode [74d]. *Source:* Reproduced with permission, © 2014, Wiley-VCH. (h) Ragone plot of the asymmetric supercapacitor based on α-Fe_2O_3/PPy anode and MnO_2 cathode [74b]. *Source:* Reproduced with permission, © 2017, Wiley-VCH.

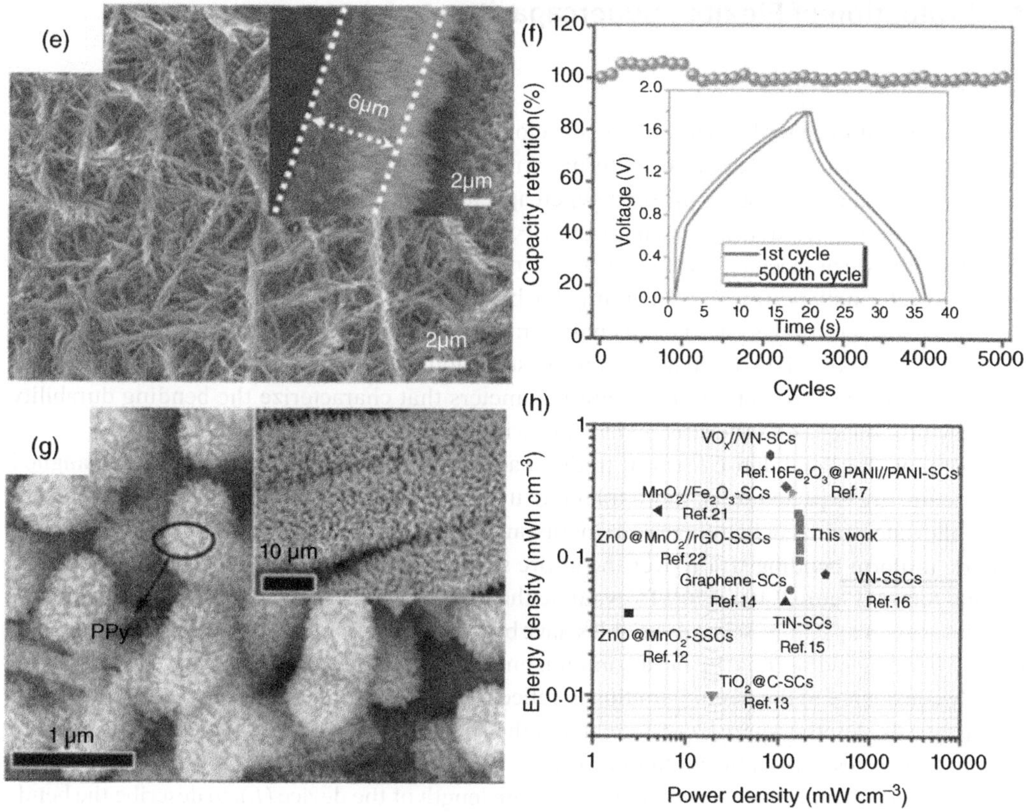

Figure 7.9 (Continued)

the film made by physically mixing these two components (~307 F g^{-1} for CoO, 545 F g^{-1} for the mixture, and 798 F g^{-1} for the hybrid system) (Figure 7.9b) [74a]. Another example is the hybrid array consisting of Co_3O_4 porous nanowires with MnO_2 ultrathin nanosheets as the shell, which are both good pseudocapacitive metal oxides (Figure 7.9c). More interesting, the nanostructured core and shell can have redox reactions with anions and cations from the electrolyte, respectively, synergistically contributed to the electrochemical charge storage (Figure 7.9d). Zhu et al. reported a 3D carbon/$CoNi_3O_4$ hybrid nanostructured array on flexible stainless-steel substrate through glucose-coating and calcination processes (Figure 7.9e) [74d]. Due to the interconnected three-dimensionally arrayed architecture and mesoporous nature, the carbon/$CoNi_3O_4$ electrode exhibits a large specific capacitance of 1299 F g^{-1} and a superior rate performance. An asymmetric supercapacitor based on the 3D carbon/$CoNi_3O_4$ cathode and active carbon anode showed a superior cycling stability with the capacity retention of 99% after 5000 cycles (Figure 7.9f). Hu's group constructed a hierarchical tectorum-like α-Fe_2O_3/PPy nanoarrays on carbon cloth for flexible solid-state asymmetric supercapacitors (Figure 7.9g), which exhibits a high areal capacitance of 382.4 mF cm^{-2} at a current density of 0.5 mA cm^{-2} and good reversibility with capacitance retention of 97.2% after 5000 cycles [74b]. The maximum energy density of the asymmetric supercapacitor based on α-Fe_2O_3/PPy anode and MnO_2 cathode is 0.22 mWh cm^{-3} at a power density of 165.5 mWh cm^{-3} (Figure 7.9h), which is superior to many supercapacitors compared with many other reported values. Following these concepts, many other excellent works have been reported.

7.4 Evaluation of Flexible Supercapacitors

7.4.1 Bending Deformation

Since extensive efforts have been devoted to developing flexible electrodes for flexible supercapacitors, much progress has been achieved in terms of material design and flexible device construction. Except for the traditional tests applied on supercapacitors, the evaluation methods for their flexibility become an issue. Information about the electrochemical performance of electronic devices under the influence of mechanical stresses or strains is collected by techniques during three typical mechanical deformation processes, including the bending, stretching, and twisting processes [75]. As for the flexible electrodes based on 3D nanoarrays, the bending ability is a basic demonstration of flexibility. However, the stretchability and twisting ability may be not available for these electrodes. Therefore, in this section, important parameters that characterize the bending durability are mainly introduced. And the stretchable and twistable 3D nanoarrays are highlighted.

Compared with coated dense films in traditional supercapacitors, 3D nanoarrays with aligned nanostructure are expected to be more tolerable under bending states, attributed to the efficient space to alleviate strain. For better measurements, an ideal automatic apparatus is usually employed to impose uniform bending strain over the entire sample and control the bending strains, strain rates, and forces (Figure 7.10a) [76]. To evaluate the tolerance to bending to a certain curvature of the flexible supercapacitors, several variables have been used to describe the bending states. Among them, the bending radius, r, is the most common one. It is noted that, the thickness of the substrate is usually considered to the bending radius to reflect the actual strain that is applied to the film of active materials. Another common parameter is the "bending angle (θ)," which is usually used to compare the devices with same configurations.

Recently, Li et al. proposed another parameter, the length of the device (L), to describe the bendability of the flexible energy storage devices, which is demonstrated to be more precise compared with the end-to-end distance along the bending direction (*LEE*). As shown in Figure 7.10b, these three parameters are schematically illustrated. According to the models, they found that when R and θ are fixed, L can be different due to the stressed region is localized. Similarly, when R and L are fixed, θ can be changed, resulting in different stressed areas. PPy-based symmetric supercapacitors with different parameters have been constructed to demonstrate the results. For instance, as shown in Figure 7.10c, the capacitance retentions after 100 bending cycles decrease with smaller bending radii when the flexible supercapacitor under different bending radius is fixed in length (7 cm) and bending angle (90°). In Figure 7.10d, decreased capacitance retentions are observed for larger θ with fixed R and L, as well as smaller L with fixed R and θ. Moreover, supercapacitors based on conductive polymers are found to have a higher tolerance to mechanical bending compared with the rigid inorganic electrode materials owing to the intrinsic properties of the materials.

7.4.2 Stretching Deformation

The stretchable abilities of the flexible electrodes require highly stretchable current collectors. For applications, stretchable supercapacitors are expected to work normally when being stretched and fully recover to the initial state after being released. In the stretching-mode measurements, the electrochemical performance for the flexible supercapacitors under test is simultaneously recorded during the tensile strain processes. Similar with those reported stretchable lithium-ion batteries, the stretchability of flexible supercapacitors also depends on the configuration of stretchable substrates and structurally stretchable device configuration rather than the nanostructure of the active

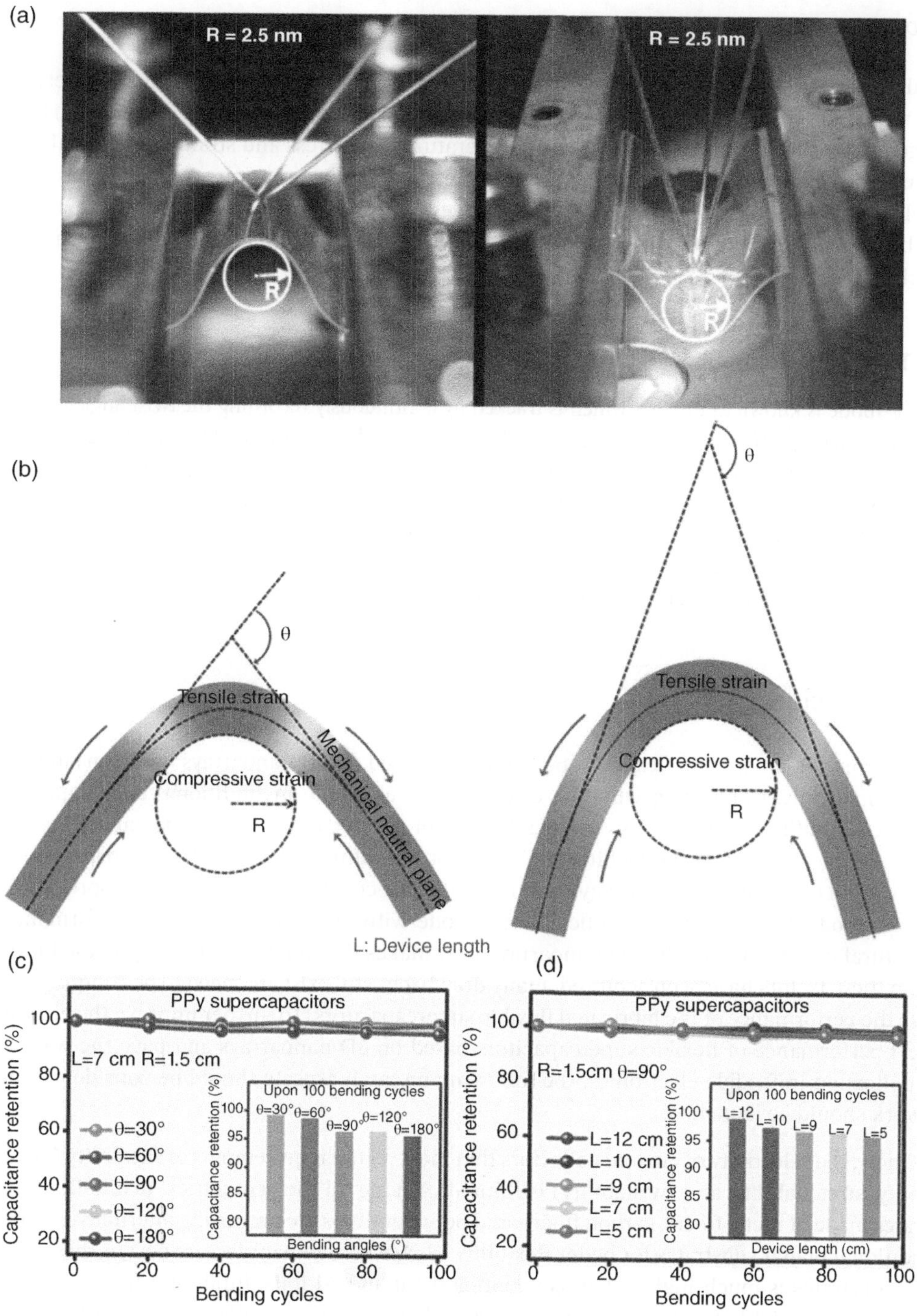

Figure 7.10 (a) The experimental apparatus for different bending states: convex (bottom left) and concave (bottom right) [76]. *Source:* Reproduced with permission, © 2009, AIP Publishing. (b) Schematics of structures and the three key parameters (*R*, *θ*, and *L*) that are commonly used to demonstrate the bending state of flexible supercapacitors. These two diagrams demonstrate that when the device length (*L*) and the bending radius (*R*) are fixed, the bending angle (*θ*) can be varied. (c) PPy supercapacitors under different bending angles at a fixed length (7 cm) and bending radius (1.5 cm). (d) PPy supercapacitors with different lengths at a certain bending angle (90°) and bending radius (1.5 cm). Insets show the capacitance retention of the devices upon 100 bending cycles [77]. *Source:* Reproduced with permission, © 2019, Elsevier Inc.

materials [78]. Similar with the tensile test standards of plastic films from the American Society for Testing and Materials (ASTM) and the International Organization for Standardization (ISO) [79], device geometry is a key factor. However, in the literatures, the stress and strain in some publications are calculated based on the initial sample geometry, while the others are calculated based on the evolving geometry [75]. In addition, the recoverability of the stretchable supercapacitors should be paid more attention since the residual strain of the device may significantly influence the mechanical stability [77].

7.4.3 Twisting Deformation

A twisting-mode is known as torsion, which is tracked by continuously recording the twist angle (θ) and the applied torque (T). The twistable abilities of the flexible electrodes require highly twistable current collectors, in order to undertake the stress and strain simultaneously. Till now, few descriptions of mechanical tests of flexible supercapacitors based on 3D nanoarrays in twisted conditions have appeared in the literature, due to the high risk of detachment from the substrates. In most cases, simple twisting-mode experiments were conducted on the flexible supercapacitors by measuring the CV and charge–discharge curves or powering a light-emitting diode light during this process [19a, 61b].

7.5 Conclusion

In this chapter, the progress in flexible supercapacitors based on 3D nanoarrays has been summarized systematically. Comparing with the coated dense electrode in traditional supercapacitor devices, free-standing 3D nanoarrays can exhibit higher specific capacitance, rate capability, and cycling stability, ascribing to the facilitated abilities of ion diffusion and electron transportation. The fabrication of flexible 3D nanoarrays requires a flexible current collector and appropriate fabrication methods. Furthermore, for a flexible electrode with high electrochemical performance, the structural engineering of the 3D nanoarrays also makes difference. Although great achievements on these factors have been acquired, many drawbacks still exist and need to be conquered to improve the performance of the fabricated flexible supercapacitors. To further improve the electrochemical performance of flexible supercapacitors based on 3D nanoarrays and pave the way for their application in flexible electronics, the following research aspects should be considered and more work should be launched:

i) Till now, the selectivity of current collectors that meet all the requirements of flexibility (bendability, stretchability, and twistability) is limited. Among all the geometries, textile-like substrates made of metal fibers, carbon fibers, and polymer fibers are promising alternatives to the traditional planar substrates for better flexibility. However, the manufacture cost of these flexible substrates is much higher than conventional Cu and Al foils, inhibiting the large-scale production of highly flexible electronic devices. Therefore, it is a critical issue in this area to develop more flexible and cost-effective substrates.
ii) Directly growing free-standing 3D nanoarrays on current collectors avoids the utilization of binder and conductive additives. In this light, it is beneficial for the researchers to study the academic issues on electrode/electrolyte interface and achieve boosted electrochemical performance based on the weight of active materials. However, it is a common problem that loading mass of 3D nanoarrays is usually less than 4 mg cm^{-2}, which is much less than the value in the commercial supercapacitors (10 mg cm^{-2}). Therefore, the gravimetric and volumetric power

density and energy density are significantly limited based on the total mass of the device, which may be another obstacle for the application of flexible supercapacitors base on 3D nanoarrays.

iii) Till now, the specific evaluation methods and parameters about the "flexibility" adopted in the literature are quite diversified. In many articles, the specific parameters of the flexible supercapacitors are even unstated during the evaluation of flexibility, which makes them difficult to follow and make a fair comparison of the performance benchmarks. Therefore, more uniform evaluation standards on flexibility of flexible supercapacitors are necessary to pave the way for the practical application of flexible electronics.

Overall, the innovative flexible supercapacitors based on 3D nanoarrays have endowed a new research field in the energy storage technology, which hold great potential in many future applications. Nowadays, since tremendous efforts are being devoted to the scientific and technological issues in this area, more encouraging achievements are expected to make a contribution to the large-scale application of flexible electronics.

Acknowledgments

We acknowledge the funding support from International S&T Cooperation Program of China (No. 2018YFE0181300) and National Key Research and Development Program of China Grant No. 2020YFB2007400.

References

1 Hamedi, M., Forchheimer, R., and Inganas, O. (2007). *Nat. Mater.* 6: 357–362.

2 a) Hu, L. and Cui, Y. (2012). *Energy Environ. Sci.* 5: 6423–6435. b) Zhou, G., Li, F., and Cheng, H.-M. (2014). *Energy Environ. Sci.* 7: 1307–1338.

3 Miller, J.R. and Simon, P. (2008). *Science* 321: 651–652.

4 Simon, P., Gogotsi, Y., and Dunn, B. (2014). *Science* 343: 1210–1211.

5 a) Choi, N.-S., Chen, Z., Freunberger, S.A. et al. (2012). *Angew. Chem. Int. Ed.* 51: 9994–10024. b) Lu, Q., Chen, J.G., and Xiao, J.Q. (2013). *Angew. Chem. Int. Ed.* 52: 1882–1889. c) Simon, P. and Gogotsi, Y. (2008). *Nat. Mater.* 7: 845–854.

6 Jiang, J., Li, Y., Liu, J. et al. (2012). *Adv. Mater.* 24: 5166–5180.

7 a) Yan, X., Tai, Z., Chen, J. et al. (2011). *Nanoscale* 3: 212–216. b) Wang, K., Zhao, P., Zhou, X. et al. (2011). *J. Mater. Chem.* c) Hu, L., Chen, W., Xie, X. et al. (2011). *ACS Nano* 5: 8904–8913.

8 a) Xia, X., Tu, J., Zhang, Y. et al. (2012). *ACS Nano* 6: 5531–5538. b) Liu, J., Jiang, J., Bosman, M. et al. (2012). *J. Mater. Chem.* 22: 2419–2426. c) Lu, X., Wang, G., Zhai, T. et al. (2012). *Nano lett.* 12: 1690–1696.

9 a) Jung, S.M., Jung, H.Y., Dresselhaus, M.S. et al. (2012). *Sci. Rep.*: 2. b) Wu, X.-L., Wen, T., Guo, H.-L. et al. (2013). *ACS Nano* 7: 3589–3597.

10 a) Li, Z., Zhang, L., Amirkhiz, B.S. et al. (2012). *Adv. Energy Mater.* 2: 431–437. b) Hu, L., Pasta, M., Mantia, F.L. et al. (2010). *Nano lett.* 10: 708–714. c) Che, G., Lakshmi, B.B., Fisher, E.R. et al. (1998). *Nature* 393: 346–349.

11 a) Deng, B., Hsu, P.-C., Chen, G. et al. (2015). *Nano lett.* 15: 4206–4213. b) Jin, J., Lee, D., Im, H.-G. et al. (2016). *Adv. Mater.* 28: 5169–5175. c) Wang, Y., Bai, S., Cheng, L. et al. (2016). *Adv. Mater.* 28: 4532–4540. d) Zhang, Y., Zhang, L., Cui, K. et al. (2018). *Adv. Mater.* 30: 1801588.

12 Kamyshny, A. and Magdassi, S. (2019). *Chem. Soc. Rev.* **48**: 1712–1740.

13 a) Kim, D.-H., Park, M.-R., Lee, H.-J., and Lee, G.-H. (2006). *Appl. Surf. Sci.* 253: 409–411. b) Kim, H., Horwitz, J.S., Kushto, G.P. et al. (2001). *Appl. Phys. Lett.* 79: 284–286.

14 Hecht, D.S., Thomas, D., Hu, L. et al. (2009). *J. Soc. Inf. Disp.* 17: 941–946.

15 a) Lee, S.-K., Jang, H.Y., Jang, S. et al. (2012). *Nano Lett.* 12: 3472–3476. b) Wee, B.-H. and Hong, J.-D. (2013). *Adv. Funct. Mater.* 23: 4657–4666.

16 a) Lee, J.-Y., Connor, S.T., Cui, Y., and Peumans, P. (2008). *Nano Lett.* 8: 689–692. b) Rathmell, A.R. and Wiley, B.J. (2011). *Adv. Mater.* 23: 4798–4803.

17 Xu, J., Wang, Q., Wang, X. et al. (2013). *ACS Nano* 7: 5453–5462.

18 a) Xu, J., Wu, H., Xu, C. et al. (2013). *Chemistry* 19: 6451–6458. b) Yu, G., Hu, L., Vosgueritchian, M. et al. (2011). *Nano Lett.* 11: 2905–2911. c) Yu, G., Hu, L., Liu, N. et al. (2011). *Nano Lett.* 11: 4438–4442.

19 a) Xu, J., Wang, Q., Wang, X. et al. (2013). *ACS Nano* 7: 5453–5462. b) Liu, B., Zhang, J., Wang, X. et al. (2012). *Nano Lett.* 12: 3005–3011. c) Wang, D., Song, P., Liu, C. et al. (2008). *Nanotechnology* 19: 075609. d) Lin, X., Shen, X., Zheng, Q. et al. (2012). *ACS Nano* 6: 10708–10719.

20 a) Chen, B.R., Sun, W., Kitchaev, D.A. et al. (2018). *Nat. Commun.* 9: 2553. b) Guan, C., Liu, X., Ren, W. et al. (2017). *Adv. Energy Mater.* 7: 1602391.

21 Li, Y., Xu, J., Feng, T. et al. (2017). *Adv. Funct. Mater.* 27: 1606728.

22 a) Xia, H., Feng, J., Wang, H. et al. (2010). *J. Power Sources* 195: 4410–4413. b) Jabeen, N., Hussain, A., Xia, Q. et al. (2017). *Adv. Mater.*: 1700804.

23 Huang, J., Xu, P., Cao, D. et al. (2014). *J. Power Sources* 246: 371–376.

24 Xu, J., Wu, H., Lu, L. et al. (2014). *Adv. Funct. Mater.* 24: 1840–1846.

25 Gao, L., Wang, X., Xie, Z. et al. (2013). *Chem. A* 1: 7167–7173.

26 Zhai, T., Wan, L., Sun, S. et al. (2017). *Adv. Mater.* 29: 1604167.

27 Wu, X., Han, Z., Zheng, X. et al. (2017). *Nano Energy* 31: 410–417.

28 Wang, Q., Xu, J., Wang, X. et al. (2014). *ChemElectroChem* 1: 559–564.

29 a) Hui, K.N., Hui, K.S., Tang, Z. et al. (2016). *J. Power Sources* 330: 195–203. b) Liu, G., Wang, B., Liu, T. et al. (2018). *Chem. A* 6: 1822–1831.

30 Young, M.J., Holder, A.M., George, S.M., and Musgrave, C.B. (2015). *Chem. Mater.* 27: 1172–1180.

31 a) Jabeen, N., Xia, Q., Savilov, S.V. et al. (2016). *ACS Appl. Mater. Interfaces* 8: 33732–33740. b) Peng, L., Peng, X., Liu, B. et al. (2013). *Nano Lett.* 13: 2151–2157.

32 Zhu, S., Li, L., Liu, J. et al. (2018). *ACS Nano* 12: 1033–1042.

33 Zhu, S., Wang, T., Liu, X. et al. (2019). *ACS Appl. Energy Mater.* 2: 1051–1059.

34 Lin, J., Zhang, C., Yan, Z. et al. (2012). *Nano Lett.* 13: 72–78.

35 a) Deng, S., Zhong, Y., Zeng, Y. et al. (2017). *Adv. Mater.* 29: 1700748. b) Deng, S., Chao, D., Zhong, Y. et al. (2018). *Energy Stor. Mater.* 12: 137–144.

36 a) Kaiser, N. (2002). *Appl. Opt.* 41: 3053–3060. b) Xia, Q., Sun, S., Xu, J. et al. (2018). *Small* 14: e1804149–e1804149. c) Sun, S., Xia, Q., Liu, J. et al. (2019). *J. Mater.* 5: 229–236.

37 Lorenz, W.J. and Staikov, G. (1995). *Surf. Sci.* 335: 32–43.

38 Liu, R. and Lee, S.B. (2008). *J. Am. Chem. Soc.* 130: 2942–2943.

39 Liu, R., Duay, J., Lane, T., and Bok Lee, S. (2010). *Phys. Chem. Chem. Phys.* 12: 4309–4316.

40 Yuan, C., Yang, L., Hou, L. et al. (2012). *Energy Environ. Sci.* 5: 7883–7887.

41 Borissov, D., Isik-Uppenkamp, S., Rohwerder, M., and Phys, J. (2009). *Chem. C* 113: 3133–3138.

42 Xie, Y., Xia, C., Du, H., and Wang, W. (2015). *J. Power Sources* 286: 561–570.

43 Xie, Y., Du, H., and Xia, C. (2015). *Microporous Mesoporous Mater.* 204: 163–172.

44 Du, H., Xie, Y., Xia, C. et al. (2014). *Mater. Lett.* 132: 417–420.

45 Tench, D. and Warren, L.F. (1983). *J. Electrochem. Soc.* 130: 869–872.

46 Broughton, J.N. and Brett, M.J. (2005). *Electrochim. Acta* 50: 4814–4819.

47 Chou, S., Cheng, F., and Chen, J. (2006). *J. Power Sources* 162: 727–734.

48 Yuan, L., Lu, X.-H., Xiao, X. et al. (2012). *ACS Nano* 6: 656–661.

49 a) Che, G., Lakshmi, B.B., Fisher, E.R., and Martin, C.R. (1998). *Nature* 393: 346. b) Wei, Q., Fu, Y., Zhang, G. et al. (2019). *Nano Energy* 55: 234–259.

50 Shaijumon, M.M., Ou, F.S., Ci, L., and Ajayan, P.M. (2008). *ChemComm*: 2373–2375.

51 a) Xia, X., Zhang, Y., Fan, Z. et al. (2015). *Adv. Energy Mater.* 5: 1401709. b) Chao, D., Xia, X., Zhu, C. et al. (2014). *Nanoscale* 6: 5691–5697.

52 Zhang, G., Song, Y., Zhang, H. et al. (2016). *Adv. Funct. Mater.* 26: 3012–3020.

53 a) Qu, Q., Zhang, P., Wang, B. et al. (2009). *Chem. C* 113: 14020–14027. b) Brousse, T., Toupin, M., Dugas, R. et al. (2006). *J. Electrochem. Soc.* 153: A2171–A2180.

54 a) Xia, X.-h., Tu, J.-p., Mai, Y.-j. et al. (2011). *J. Mater. Chem.* 21: 9319–9325. b) Meher, S.K., Rao, G.R., and Phys, J. (2011). *Chem. C* 115: 15646–15654.

55 a) Liu, T., Pell, W.G., and Conway, B.E. (1997). *Electrochim. Acta* 42: 3541–3552. b) Park, J.H., Ko, J.M., and Ok Park, O. (2003). *J. Electrochem. Soc.* 150: A864–A867.

56 a) Wang, Q., Wang, X., Liu, B. et al. (2013). *Chem. A* 1: 2468–2473. b) Zhang, G. and Lou, X.W. (2013). *Adv. Mater.* 25: 976–979.

57 Wang, X., Liu, B., Liu, R. et al. (2014). *Angew. Chem. Int. Ed.* 53: 1849–1853.

58 a) Li, M., Sun, G., Yin, P. et al. (2013). *ACS Appl. Mater. Interfaces* 5: 11462–11470. b) Boukhalfa, S., Evanoff, K., and Yushin, G. (2012). *Energy Environ. Sci.* 5: 6872–6879.

59 a) Liu, C., Xie, Z., Wang, W. et al. (2014). *J. Electrochem. Soc.* 161: A1051–A1057. b) Peng, X., Huo, K., Fu, J. et al. (2015). *ChemElectroChem* 2: 512–517.

60 Xiao, X., Peng, X., Jin, H. et al. (2013). *Adv. Mater.* 25: 5091–5097.

61 a) Raj, C.C. and Prasanth, R. (2018). *J. Electrochem. Soc.* 165: E345–E358. b) Lu, X., Yu, M., Wang, G. et al. (2013). *Adv. Mater.* 25: 267–272.

62 Li, T., Yu, H., Zhi, L. et al. (2017). *Chem. C* 121: 18982–18991.

63 a) Liu, J., Zheng, M., Shi, X. et al. (2016). *Adv. Funct. Mater.* 26: 919–930. b) Nguyen, T., Fátima Montemor, M., and Mater, J. (2018). *Chem. A* 6: 2612–2624.

64 a) Wu, H., Xu, C., Xu, J. et al. (2013). *Nanotechnology* 24: 455401. b) Pan, X., Zhao, Y., Ren, G., and Fan, Z. (2013). *Chem. Commun.* 49: 3943–3945.

65 Acerce, M., Voiry, D., and Chhowalla, M. (2015). *Nat. Nanotechnol.* 10: 313.

66 a) Wang, X., Liu, B., Wang, Q. et al. (2013). *Adv. Mater.* 25: 1479–1486. b) Yu, N., Zhu, M.-Q., Chen, D., and Mater, J. (2015). *Chem. A* 3: 7910–7918. c) Lin, J., Wang, H., Yan, Y. et al. (2018). *Chem. A* 6: 19151–19158.

67 a) Elshahawy, A.M., Guan, C., Li, X. et al. (2017). *Nano Energy* 39: 162–171. b) Liang, H., Xia, C., Jiang, Q. et al. (2017). *Nano Energy* 35: 331–340.

68 a) Xia, X., Zhang, Y., Chao, D. et al. (2015). *Energy Environ. Sci.* 8: 1559–1568. b) Xia, X., Zhan, J., Zhong, Y. et al. (2017). *Small* 13: 1602742.

69 Lu, X., Wang, G., Zhai, T. et al. (2012). *Nano Lett.* 12: 5376–5381.

70 a) Paraknowitsch, J.P. and Thomas, A. (2013). *Energy Environ. Sci.* 6: 2839–2855. b) Han, J., Zhang, L.L., Lee, S. et al. (2013). *ACS Nano* 7: 19–26. c) Stoller, M.D., Park, S., Zhu, Y. et al. (2008). *Nano Lett.* 8: 3498–3502. d) Hulicova-Jurcakova, D., Seredych, M., Lu, G.Q., and Bandosz, T.J. (2009). *Adv. Funct. Mater.* 19: 438–447. e) Chen, L.-F., Huang, Z.-H., Liang, H.-W. et al. (2014). *Adv. Funct. Mater.* 24: 5104–5111. f) Lin, T., Chen, I.-W., Liu, F. et al. (2015). *Science* 350: 1508–1513.

71 Andreas, H.A. and Conway, B.E. (2006). *Electrochim. Acta* 51: 6510–6520.

72 a) Zhang, Z., Wang, L., Li, Y. et al. (2017). *Adv. Energy Mater.* 7: 1601814. b) Choi, W.H., Choi, M.J., and Bang, J.H. (2015). *ACS Appl. Mater. Interfaces* 7: 19370–19381. c) Zhou, D.-D., Li, W.-Y., Dong, X.-L. et al. (2013). *Chem. A* 1: 8488–8496.

73 Mai, L., Li, H., Zhao, Y. et al. (2013). *Sci. Rep.* 3: 1718.

74 a) Liu, J., Jiang, J., Cheng, C. et al. (2011). *Adv. Mater.* 23: 2076–2081. b) Wang, L., Yang, H., Liu, X. et al. (2017). *Angew. Chem. Int. Ed.* 56: 1105–1110. c) Guan, C., Liu, J., Cheng, C. et al. (2011). *Energy Environ. Sci.* 4: 4496–4499. d) Zhu, J., Jiang, J., Sun, Z. et al. (2014). *Small* 10: 2937–2945. e) Pang, H., Li, X., Zhao, Q. et al. (2017). *Nano Energy* 35: 138–145. f) Xia, X., Chao, D., Fan, Z. et al. (2014). *Nano Lett.* 14: 1651–1658. g) Guan, C., Zeng, Z., Li, X. et al. (2014). *Small* 10: 300–307. h) Liang, M., Zhao, M., Wang, H. et al. (2018). *Chem. A* 6: 2482–2493. i) Liang, H., Lin, J., Jia, H. et al. (2018). *J. Power Sources* 378: 248–254.

75 Harris, K.D., Elias, A.L., and Chung, H.-J. (2016). *J. Mater. Sci.* 51: 2771–2805.

76 Jedaa, A. and Halik, M. (2009). *Appl. Phys. Lett.* 95: 103309.

77 Li, H., Tang, Z., Liu, Z., and Zhi, C. (2019). *Joule* 3: 613–619.

78 a) Chen, D., Lou, Z., Jiang, K., and Shen, G. (2018). *Adv. Funct. Mater.* 28: 1805596. b) Wang, K., Meng, Q., Zhang, Y. et al. (2013). *Adv. Mater.* 25: 1494–1498. c) Ren, J., Li, L., Chen, C. et al. (2013). *Adv. Mater.* 25: 1155–1159. d) Pu, X., Li, L., Liu, M. et al. (2016). *Adv. Mater.* 28: 98–105. e) Bae, J., Song, M.K., Park, Y.J. et al. (2011). *Angew. Chem. Int. Ed.* 50: 1683–1687.

79 ASTM. D882-12 (2012). *Standard test method for tensile properties of thin plastic sheeting*. ASTM Compass.

8

Metal Oxides Nanoarray Electrodes for Flexible Supercapacitors

Ting Meng and Cao Guan

Frontiers Science Center for Flexible Electronics, Institute of Flexible Electronics, Northwestern Polytechnical University, Xi'an, 710072, P. R. China

8.1 Introduction

With the fast development of electric vehicles and flexible electronics, it is imminent to realize lightweight, low-cost, and highly-efficient energy storage systems with mass production properties. As considerable alternative candidates to secondary batteries, supercapacitors, also known as electrochemical capacitors, have high-power density (>10 kW kg^{-1}), fast charging and discharging capability as well as excellent long-term cycle stability [1–4]. However, the use of commercial supercapacitors with intrinsically low energy density (<10 Wh kg^{-1}) limits their widespread application [5]. Therefore, many researchers are on the verge of exploring supercapacitors with superior performance.

Generally, the state-of-the-art supercapacitors can be divided into three main categories depending on storage standards viz.; electric double layer capacitors (EDLCs), pseudocapacitors and their combination named hybrid capacitors [6, 7]. Compared with the prevailing secondary batteries, all of them furnish much higher power density, mainly because the charge storage is performed only on the surface or in the near-surface area of the active materials, which avoids the diffusion of ions between the solid bulk phases [8]. In EDLCs, charges are stored by electrostatic interaction and occur directly on the electric double layer of the electrode material (without any charge transfer on the interface between the electrode and electrolyte), which endows them more electrochemically reversible and cyclically stable [9–11]. In contrast to the EDLCs, pseudocapacitors possess much higher capacitance and energy density owing to the Faradaic process that involves electro-adsorption, redox reactions and intercalation and desorption mechanism [12, 13]. With electrochemical charge transfer, active materials gradually crack and degrade during long-term reactions accompanied by the poor cycle life. The hybrid capacitors with more improved characteristics are proposed by combining these two energy storage mechanisms [12]. In this system, one-half behaves as a pseudocapacitor to attain higher energy density and specific capacitance, while the other part plays as EDLCs to provide higher power density [6]. The highly efficient supercapacitors devices need independent but mutually constrained components in a small space, including a separator, current collector, electrolyte, and electrodes. Thereinto, the electrodes as the core units have received a plethora of research. Electrode materials for EDLCs are made of nano-structured

Flexible Supercapacitors: Materials and Applications, First Edition. Edited by Guozhen Shen, Zheng Lou and Di Chen.

carbon-based materials like carbon black, carbon aerogel, carbon nanotubes, and graphene, etc. [14] They have the advantages of superior conductivity, sufficient specific surface, as well as good mechanical and chemical stability [15]. Meanwhile, the most widely investigated electrode materials with reversible redox behavior for pseudocapacitors are metal oxides/sulfides/hydroxides and conducting polymers [16].

It is well known that flexible functions are considered to be vital features for the next generation of electronic production. Therefore, flexible supercapacitors with high flexibility integrated on a soft material substrate are required to catch up with the fast development of the demand for portable electronic devices (mobile phones, Bluetooth earphones, and smart watches). As an important kind of energy storage systems, they need not only excellent electrochemical performance, but also stable mechanical integrity under various deformation conditions (folding, bending, stretching, and twisting) [12, 17]. In general, the design of electrode materials for flexible supercapacitors is of great importance, includes two strategies: (i) manufacture of self-supporting and porous conductive materials as flexible substrates; and (ii) optional coating/growing Faraday active materials on the flexible substrates to further increase the specific capacitance. Metal oxides especially in the form of nanostructures [18], have been explored extensively as electrodes in flexible supercapacitors systems on account of certain intriguing characteristics such as variable chemical valence and large-scale production. In particular, transition metal oxides with the advantages of abundant reserves and environment friendly have become research hotspots. Although there are significant benefits of downsizing the pseudocapacitive materials to a nano-size level, some inherent shortcomings still block their practical application [2, 19]. Firstly, most widely studied metal oxides are semiconductors or even insulators, with limited conductivity, suffering from slow kinetics and safety issues. Secondly, the metal oxides inescapably undergo uncontrollable volume changes during the deep cycling processes which will crush and destroy the micro/nano-structure of the electrodes along with poor cyclic stability. Furthermore, the feasible application of metal oxides in flexible devices is hindered due to weak mechanical properties (brittleness and stiffness) [15]. In order to settle these points, more strategies for developing novel electrode materials and functional electrode architectures are urgently required.

It is generally accepted that nanoscale electrode materials can expose larger surfaces and provide efficient ion transport channels to promote thorough penetration of electrolytes. However, such an avenue will inevitably lead to raised electrochemical impedance in the solid electrolyte, and the aggregation of active materials with smaller size reduces the accessible surface area involved in the redox reactions. In this sense, vertically grown nanoarrays with ordered architecture have been affirmed as one of the promising nanostructures as electrodes for supercapacitors which exhibits fascinating advantages. For instance, aligned nanoarrays together with the open space tend to diffuse the electrolyte into the inner side of the electrode matrix, thereby providing more active sites and efficiently reducing the electrolyte diffusion resistance compared to powder materials [2, 20]. In addition, the nanoarray electrodes directly connected to the conductive substrates (current collectors) avoids the use of any binder or additives, resulting in good rate capability and cycling stability [8, 21]. On account of the above unique features, the ordered hybrid nanoarrays formed by combining different kinds of metal oxide materials can significantly improve the electrochemical performance of flexible supercapacitors.

Till now, diverse satisfactory electrode materials for energy storage systems have been realized in pioneering works. This review summarizes the recent process of nanoarray electrodes for flexible supercapacitors. Figure 8.1 represents the main structure of metal oxide nanoarrays for supercapacitors. At first, we focus on synthetic strategies of self-supported metal oxide array materials including solution-based method, electrochemical assisted electrodeposition and chemical vapor

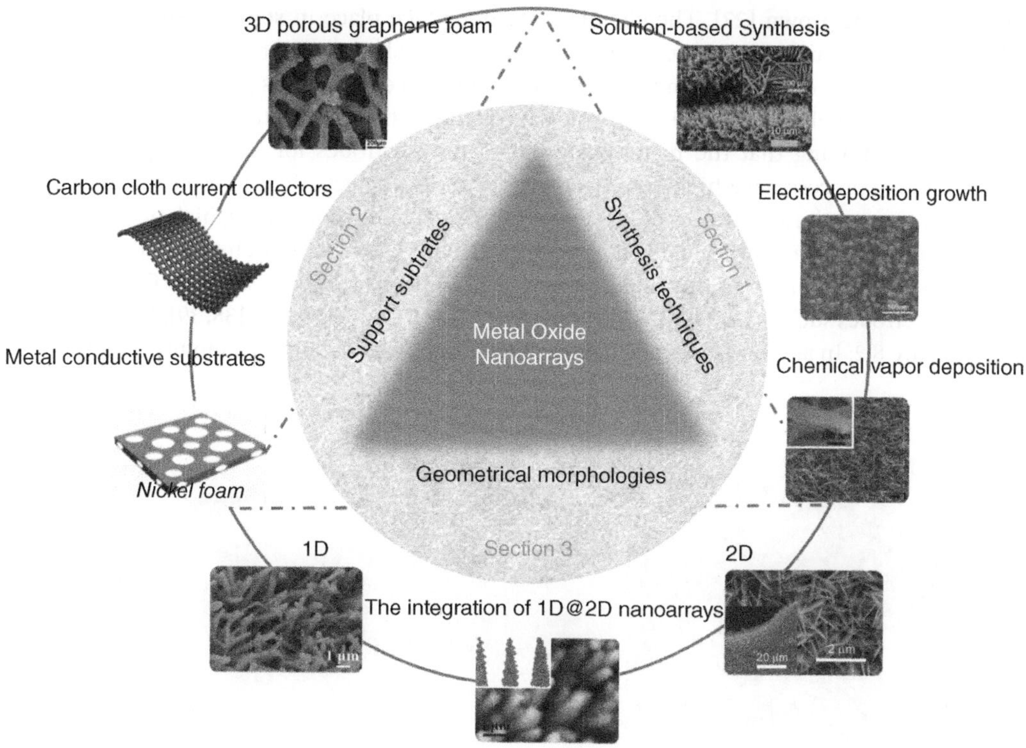

Figure 8.1 Schematic diagram of the metal oxide nanoarrays for supercapacitors.

deposition (CVD). Secondly, the original flexible support substrate for loading nanoarrays will be summarized. In the following sections, we highlight the geometry of nanostructured arrays such as one-dimension (1D) nanowires, nanorods, nanotubes, nanospikes, two-dimension (2D) nanoflakes/nanowalls and the integration of 1D@2D nanoarrays. Finally, the conclusions and an outlook on this research field will be given as well, pointing out several future evolution trends such as flexible supercapacitors.

8.2 Synthesis Techniques of Metal Oxide Nanoarrays

A variety of synthesis strategies have been developed to output well-organized metal-oxide nanoarrays in the past several years. This section mainly focused on the solution-based synthesis, electrodeposition growth, and CVD growth. Each of the above methods has its own distinctive advantages. In most cases, hybrid nanostructure arrays are usually fabricated by multi-step approaches assisted by the interfacial reaction between the different oxide nanostructures arrays [22, 23].

8.2.1 Solution-based Route

Hydrothermal technology is a simple and versatile synthetic way that can obtain array compounds in aqueous solvents at appropriate temperature and pressure, while the solvothermal technique utilizes a non-aqueous precursor solvent [24–26]. Typically, self-supported electrodes are prepared by submerging the conductive substrate in a dilute salt solution containing metal ions of the desired oxide to be grown [27]. The chemical reactions take place in stainless steel autoclaves, where the working temperature, pressure, and reaction time all play vital roles in the crystal structure and morphology of the as-formed oxides. The formation of crystals is generally accompanied by three stages, namely the formation of supersaturated solution, nucleation, and crystal growth. It is worth mentioning that the metal oxide nanoarray electrodes for flexible supercapacitors obtained by this process have high-quality crystals with less defects than that obtained by other methods. Moreover, the advantages of low toxicity, relatively mild conditions, high product purity as well as low-cost raw materials make it more attractive [28]. Many research groups [29–31] have reported that a broad range of array materials has been achieved by this method for flexible supercapacitor electrodes such as Co_3O_4 [32–34], $NiCo_2O_4$ [35–37], CoO [31], NiO [38, 39], MnO_2 [40–42] etc., see Figure 8.2. In addition, for a given oxide, the morphology including nanotubes, nanowires,

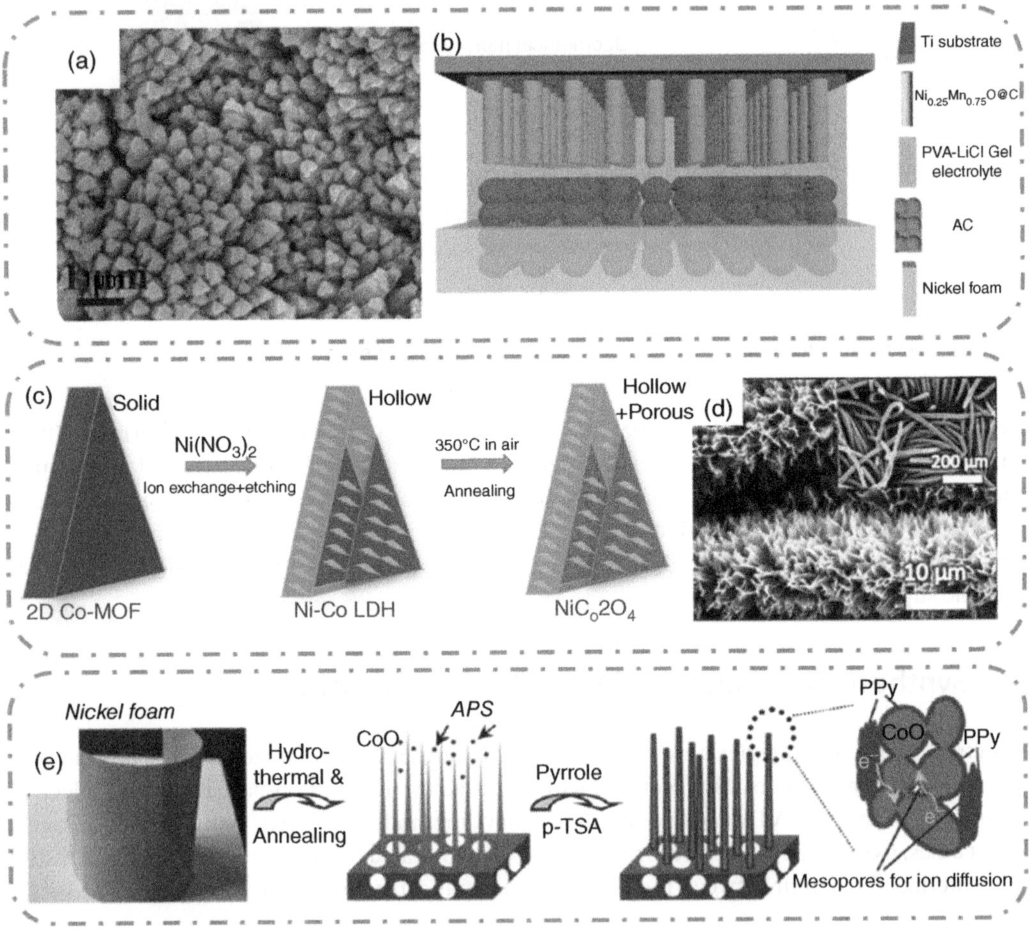

Figure 8.2 Solution-based method to synthesize metal oxide arrays. (a) SEM image of $Ni_{0.25}Mn^{0.75}O$ nanoprism arrays and (b) Schematic illustration of the $Ni_{0.25}Mn_{0.75}O$@C//activated carbon (AC) quasi-solid-state device. *Source:* Reproduced with permission [29]. © 2017, Wiley-VCH. (c) Schematic illustration of the formation of hollow $NiCo_2O_4$ nanowall array and (d) SEM images of $NiCo_2O_4$ hollow nanostructure. *Source:* Reproduced with permission [30]. © 2017, Wiley-VCH. (e) The synthetic diagram of the 3D CoO@ Polypyrrole nanowire electrode. *Source:* Reproduced with permission [31]. © 2013, American Chemical Society.

Figure 8.3 $NiCo_2O_4$ nanoarrays with different structures synthesized by solution-based method. (a) $NiCo_2O_4$@MnO_2 core-shell nanowire arrays. *Source:* Reproduced with permission [40]. © 2014, The Royal Society of Chemistry. (b) $NiCo_2O_4$ nanograss arrays. *Source:* Reproduced with permission [43]. © 2014, The Royal Society of Chemistry. (c) $NiCo_2O_4$@Ni_3S_2 core-shell mesoporous nanothorn arrays. *Source:* Reproduced with permission [44]. © 2014, The Royal Society of Chemistry. (d) 3D hierarchical porous $NiCo_2O_4$ nanosheet-derived nanowire cluster array. *Source:* Reproduced with permission [45]. © 2013, The Royal Society of Chemistry. (e) Mesoporous $NiCo_2O_4$ nanosheet and nanocloth arrays. *Source:* Reproduced with permission [46]. © 2013, The Royal Society of Chemistry. (f) 3D hierarchical multi-villous $NiCo_2O_4$ nanocyclobenzene arrays. *Source:* Reproduced with permission [47]. © 2014, The Royal Society of Chemistry.

and nanosheets can be obtained by controlling the type and concentration of the precursor used, temperature and duration of the reaction [27]. Furthermore, complex composites and core-shell/hierarchical structures can also be yielded by continuous solution deposition reactions. Take $NiCo_2O_4$ nanoarray as an example, as shown in Figure 8.3 [40, 43–47].

In order to curtail the synthesis time, the microwave-assisted solution-based method is commonly used in synthesizing metal oxides. This is a rapid synthetic technique which can dramatically reduce reaction time from days and hours to minutes and seconds by controlling of the reaction kinetics [48]. Additionally, microwave-assisted method can repress side reactions and provide fast crystallization kinetics [49]. For example, Zhang et al. [50] formed hierarchical $CoWO_4/Co_3O_4$ nanowire arrays using a two-step microwave-assisted method, in which highly interconnected $CoWO_4$ nanoparticles blanket the Co_3O_4 nanowire arrays. The experiments demonstrated that this method settles the matter of poor dispersion and easy accumulation of nano sized $CoWO_4$ and inhibits the volume change of nanomaterials during energy storage. The hierarchical hybrid electrode displayed high specific capacity of $1728\,F\,g^{-1}$ at the current density of $1\,A\,g^{-1}$($2.7\,mA\,cm^{-2}$) with an excellent cycling stability and rate capability. Meanwhile, the assembled asymmetric supercapacitor device yielded high energy density of $57.8\,Wh\,kg^{-1}$ at a power density of $750\,W\,kg^{-1}$, almost nearing the lower limit of lithium-ion cells. Thus, the unique nanomaterial electrodes synthesized by microwave-assisted approach is both economical and efficient, revealing the potential of mass production and promising prospects for applications in supercapacitors.

8.2.2 Electrodeposition Growth

Electrodeposition with templates is another common synthesis technique for preparing metal oxide nanoarrays, which involves one synthesis step throughout entire course of the experiment. The principle of this method is based on the electrochemical redox process, mainly including the following steps: (i) nucleation on substates or on seed surfaces, (ii) directional growth according to their intrinsic structures of metal oxide arrays [28, 51]. With proper selection of the metal salt solution, the metal oxide is deposited on the cathode when the current passes. Electrodeposition with simple processing conditions and not much toxic chemicals has been widely used for the growth of oxide nanoarrays coatings on various conductive substrates [9, 52, 53]. Furthermore, this strategy also can eliminate ohmic resistance between the current collector and the electrode active material. However, it still faces some challenges such as the limited template area and the introduced impurities, coupled with large-scale preparation with high quality is still difficult to come true [54, 55]. Therefore, it is of great significance to design a feasible and convenient template-free electrochemical method to prepare large-scale metal oxide-based nanoarrays. Lu et al. [56] have successfully yielded large area MnO_x nanorod arrays on F-doped SnO_2 coated glass for supercapacitor applications. For the first time, as- prepared MnOx nanorods with diameters of 70–100 nm and length of ~1.5 nm were prepared by anodic electrodeposition without any template, as shown Figure 8.4a and b. Importantly, around 50% of the initial capacitance can be retained for the as-obtained MnO_x nanoarray electrodes (232.4 $F\,g^{-1}$) even when the current rate increased as much as 10 times to 30 $A\,g^{-1}$ (Figure 8.4c), meaning a remarkable rate capability. Furthermore, such presented method was extended to grow MnO_x arrays on Ti foil. The as-prepared electrode is flexible and it can even be folded (Figure 8.4d and e) with high specific capacitance of 583.6 $F\,g^{-1}$ at a current rate of 10 $mV\,s^{-1}$ (Figure 8.4f), exhibiting their great potential for commercial application in flexible supercapacitors.

8.2.3 Chemical Vapor Deposition

CVD is a method in which materials react on the surface of a special substrate, or in which vaporized precursors decompose to form typical one-dimensional (1D) solid material, as shown in Figure 8.5 [57–59]. The precursor material is prepared in an inert gas, and the reaction takes place in the vacuum chamber. CVD technique is usually used where the porosity is very important, and the as-prepared structures have uniform/even morphology [60–62]. Apart from this, the CVD synthesis is capable of producing nanoarrays with high crystallinity and well-defined interfaces [63, 64]. For instance, Zhong et al. [65] synthesized a dense array of vertically aligned $ZnO\text{-}ZnGa_2O_4$ core-shell nanowires on an a-plane sapphire substrate by a simple two-step CVD method in which the ZnO cores and $ZnGa_2O_4$ shells of the nanowires are of single crystal quality and have aligned crystallographic orientations. Similarly, FeOCl nanosheet arrays with a height of around 1 μm and a thickness of 50–60 nm were deposited on F-doped tin oxide glass substrates through a CVD method and further converted to porous $\alpha\text{-}Fe_2O_3$ nanosheet arrays by high-temperature annealing [66]. However, this deposition method has not been used extensively due to the complicated procedures, intricate operation equipment, unsatisfactory yield and high temperature. Therefore, Zhou et al. [61] successfully fabricate an interesting, branched hybrid structure composed of SnO_2 nanowire stem and $\alpha\text{-}Fe_2O_3$ nanorod branches directly on stainless steel foil combining CVD and hydrothermal methods.

All the above-mentioned methods with their inherent merits and drawbacks play different roles in the synthesis of metal oxide nanoarrays and their hybrid composites for flexible supercapacitors.

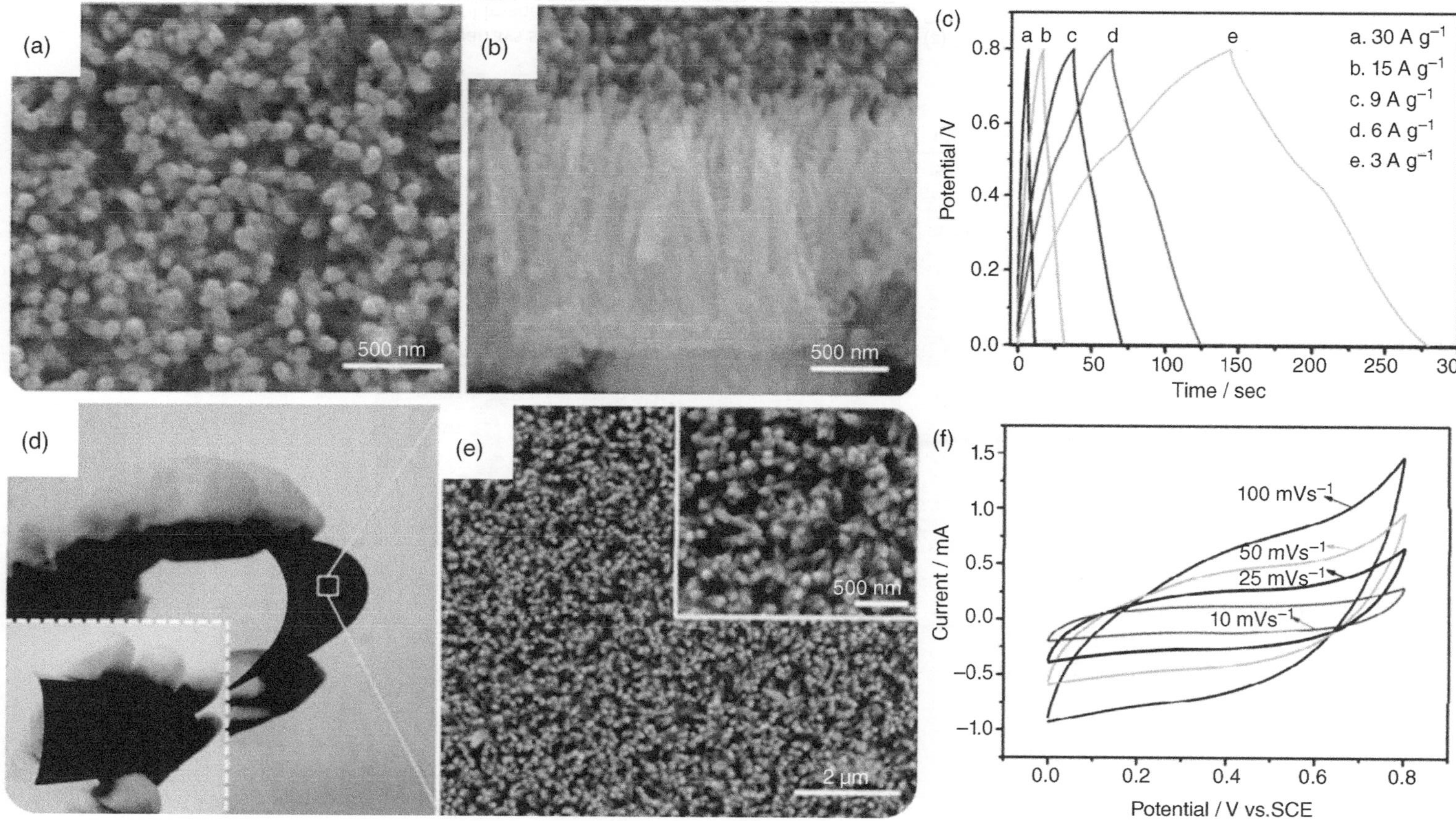

Figure 8.4 Electrodeposition to prepare metal oxide arrays. (a–b) SEM images, (c) Galvanostatic current charge/discharge curves at different current densities for the MnO_x nanorod arrays yielded on F-doped SnO_2 coated glass, (d) Optical images, (e) SEM image and (f) Cyclic voltammetry curves of MnO_x grown on a Ti substrate. *Source:* Reproduced with permission [56]. © 2011, The Royal Society of Chemistry.

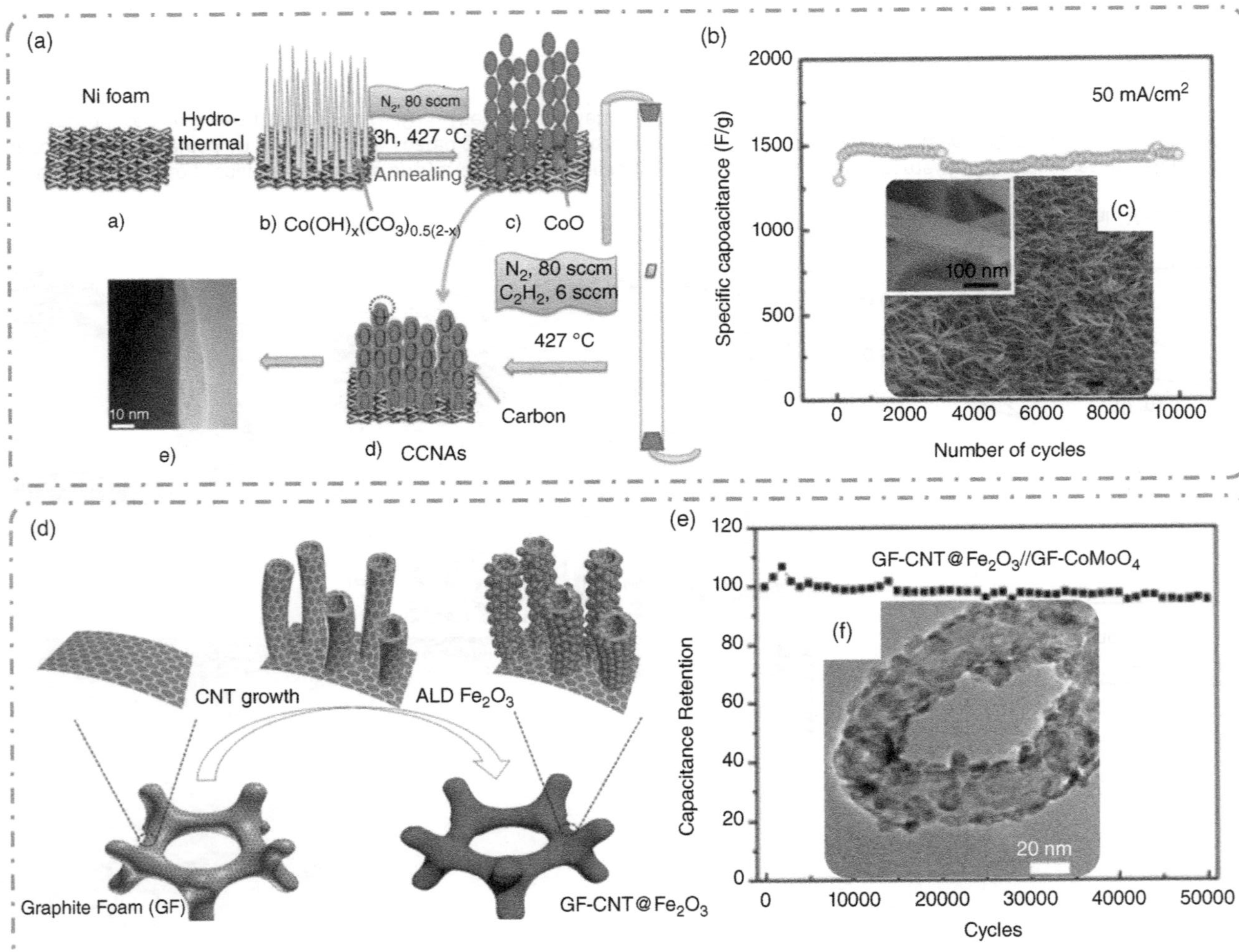

Figure 8.5 Metal oxide-based nanoarrays synthesized by chemical vapor deposition. (a) The growth process of carbon-CoO nanostructure arrays; (b) Cycling records and (c) SEM image of the optimized carbon-CoO nanoarray. *Source:* Reproduced with permission [57]. © 2014, The Royal Society of Chemistry. (d) The fabrication precedure of graphite foam-carbon natube@iron oxide (GF-CNT@Fe_2O_3); (e) Long-time cycling ability with a current of $7 A g^{-1}$ of GF-CNT@Fe_2O_3//GF-$CoMoO_4$ full cell and (f) TEM image. *Source:* Reproduced with permission [58]. © 2015, American Chemical Society.

Accordingly, these various methods may provide useful guidelines for preparing novel hierarchical array architectures.

8.3 The Flexible Support Substrate for Loading Nanoarrays

In practical applications, the energy density and power density of metal oxide electrodes are two crucial issues. These could be obtained according to the following formula:

$$E = 0.5CV^2$$

$$P = V^2 / 4R$$

where C is the specific capacitance, V is the cell voltage, and R represents the equivalent series resistance (ESR). Obviously, the ESR plays a significant role in the power density and is largely related to the electrode system in supercapacitors. One promising approach is to fundamentally increase the electrical conductivity of metal oxides by fabricating nanoarrays, aiming to enlarge the specific surface area and shorten the electron conduction length [67]. Another way is to load electrode materials onto the conductive and flexible array supports. The array nanostructures with large surface areas can not only accommodate more active materials, thereby increasing the energy density per unit area, but also furnish well-defined pathways to access the electrolyte [68, 69]. In addition, mechanical stability is another feature of nanostructured current collectors for efficient supercapacitor electrodes.

8.3.1 3D Porous Graphene Foam

Recently, graphene has received increasing attention featured by excellent electrical, high specific surface area (up to $2675\,m^2\,g^{-1}$) as well as mechanical properties and chemical stability [70, 71]. However, graphene-based nanocomposites suffer from undesired aggregation, leading to poor ionic accessibility and limited improvement in electrochemical performance [70]. Therefore, CVD-derived three dimensions (3D) graphene foam as superior backbone has been exploited to construct integrated and bind-free supercapacitor electrodes [72–74]. Such graphene foam owns obvious advantages of good electrical conductivity (~$1000\,S\,m^{-1}$), high porosity (~99.7%) and large specific surface area (~$850\,m^2\,h^{-1}$). For example, Chao et al. [75] reported lightweight V_2O_5 nanoarray electrodes via directly depositing V_2O_5 nanobelt array on 3D graphene foam which acts facile scaffolds and efficient current collector (Figure 8.6a and b). Dong et al. [76] also produced a 3D graphene/Co_3O_4 hybrid composite through the growth of CVD graphene foam followed by in-situ hydrothermal synthesis of Co_3O_4 nanowires array (Figure 8.6c and d). The monolithic composite electrode provided a higher specific capacitance of $1100\,F\,g^{-1}$ at $10\,A\,g^{-1}$ than that of many previously reported graphene-based/Co_3O_4 electrodes and outstanding cycling stability because of the synergistic integration between the graphene (excellent mechanical strength) and Co_3O_4 nanowire (intriguing electrochemical properties), as shown in Figure 8.6e and f.

8.3.2 Carbon Cloth Current Collectors

The developments of ultrathin and flexible power sources have attracted much attention driven by the increased demand for modern electronic systems [77–79]. In this respect, the design of soft and

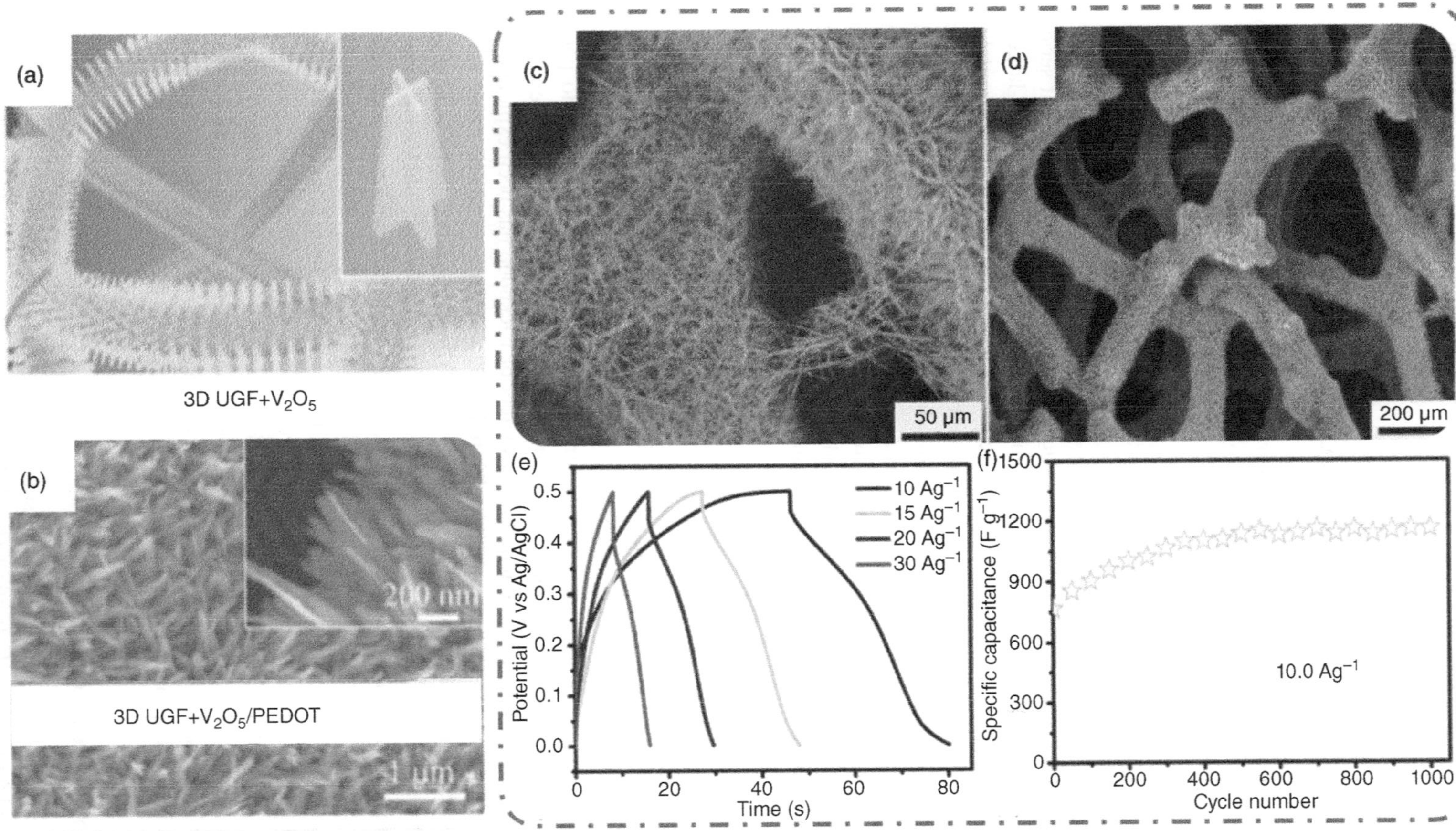

Figure 8.6 3D graphene foam used as facile scaffolds and efficient current collector. (a) V_2O_5 nanobelt array directly on 3D ultrathin graphite foam (3D UCF+V_2O_5) and (b) 3D UCF+V_2O_5 coated with polymer poly(3,4-ethylenedioxythiophene) (3D UGF−V_2O_5/PEDOT). *Source:* Reproduced with permission [75]. © 2014, Wiley-VCH. (c and d) SEM images of graphene/Co_3O_4 nanowire composite; (e) Galvanostatic current charge/discharge curves at the different current densities and (f) Cycling stability of graphene/Co_3O_4 nanoarray composite. *Source:* Reproduced with permission [76]. © 2012, American Chemical Society.

elastic current collectors is the prerequisite for flexible supercapacitors. Generally, such flexible current collectors can be obtained via pre-coating various frames with certain conductive materials. The flexible substrates at present are mainly stretchable polymers (polydimethylsiloxane, polyurethane et al.) textiles, cellulose papers, and carbon cloth. Therein, carbon cloth not only has electronical conductivity and mechanical flexibility, but also has light weight, high porosity, and chemical stability [80–82]. More importantly, it can be easily rolled, packaged, and is compatible with various deposition techniques and electrode materials, thus further meeting the needs of flexible systems. In a related report, Guan et al. [30] designed hollow $NiCo_2O_4$ nano-wall arrays on the elastic carbon cloth substrate. The as-fabricated array electrodes display prominent electrochemical performance including remarkable rate capability ($1055.3\,F\,g^{-1}$ at a small current of $2.5\,mA\,cm^{-2}$; $483.3\,F\,g^{-1}$ at the current density of $60\,mA\,cm^{-2}$) and long cycle time, which are attributed to abundant reaction sites and convenient ion diffusion path. Similarly, a nanocomposite electrode of $NiMoO_4$/polyaniline (PANI) core-shell arrays have been synthesized on carbon cloth by a facile two-step solution-based method for asymmetric pseudocapacitors (Figure 8.7a–c) [80]. The hybrid nanowire electrode yielded a favorable specific capacitance of $1340\,F\,g^{-1}$ at $1\,mA\,cm^{-2}$, and superior capacitance retention of 96.7% after charge/discharge 2000 cycles with an average coulombic efficiency of over 99.5%, see in Figure 8.7d. Moreover, such device delivered excellent capacitive properties and bending dynamics, demonstrating promising supercapacitor electrodes for practical applications in flexible electronics. Yang et al [83] also designed and fabricated unique hybrid architecture of hydrogenated single-crystal ZnO@amorphous ZnO-doped MnO_2 nanocables (HZM) on carbon cloth used as supercapacitor electrodes, as shown in Figure 8.7e. The areal capacitance of the HZM nanocable electrodes achieved $138.7\ mF\ cm^{-2}$ at a current density of $1\,mA\,cm^{-2}$. Furthermore, highly flexible all-solid-state supercapacitors were assembled with as-prepared HZM electrodes and polyvinyl alcohol/LiCl electrolyte. The working devices not only show superior electrochemical stability (over 87.5% capacitance retention after 10 000 times), but also can be bent and twisted without affecting the electrochemical performance much (Figure 8.7f), exhibiting excellent flexibility and stability.

8.3.3 Metal Conductive Substrates

Metal conductive substrates referring to any type of metal foam/foil or even wire are used as a backbone to support active materials such as carbon-based materials, conjugated polymers, and metal oxides to construct advanced supercapacitor electrodes. So far, various metal conductive substrates have been reported, such as Ni/Cu foam, Ti/Cu wire as well as stainless-steel foil. These metal current collector electrodes enable efficient transport of charges produced by the active materials along the length of the supercapacitor [84]. Typically, Liu et al. [38] prepared MnO_2-NiO nanoflake-assembled tubular array on stainless steel substrates via programmed 3D interfacial reactions (Figure 8.8a and b). A piece of $2\,cm^2$ as-achieved hybrid tubular array was directly used as the working electrode. The constituent tubular structure array was aligned on the current collector (without conductive agents or binder additives), which ensures large electrochemically active area and offers improved electron transfer pathways, thereby contributing to good rate capability and excellent long- term cyclability (retains 87.5% of the initial capacitance after 1500 cycles, displayed in Figure 8.8c).

Ti wire is one of the most commonly used fiber-shaped electrode substrates for flexible supercapacitors, dye-sensitized solar cells, and perovskite solar cells, and so on [85]. Huo et al. [71] reported that the hybrid TiO_2/reduced graphene oxide (TiO_2/rGO) nanostructure was fabricated on Ti wire used as a Ti source and scaffold to synthesize array composites and further form fiber-shaped electrodes. The TiO_2 nanoarrays were conductive to directional transfer electrons and penetrate

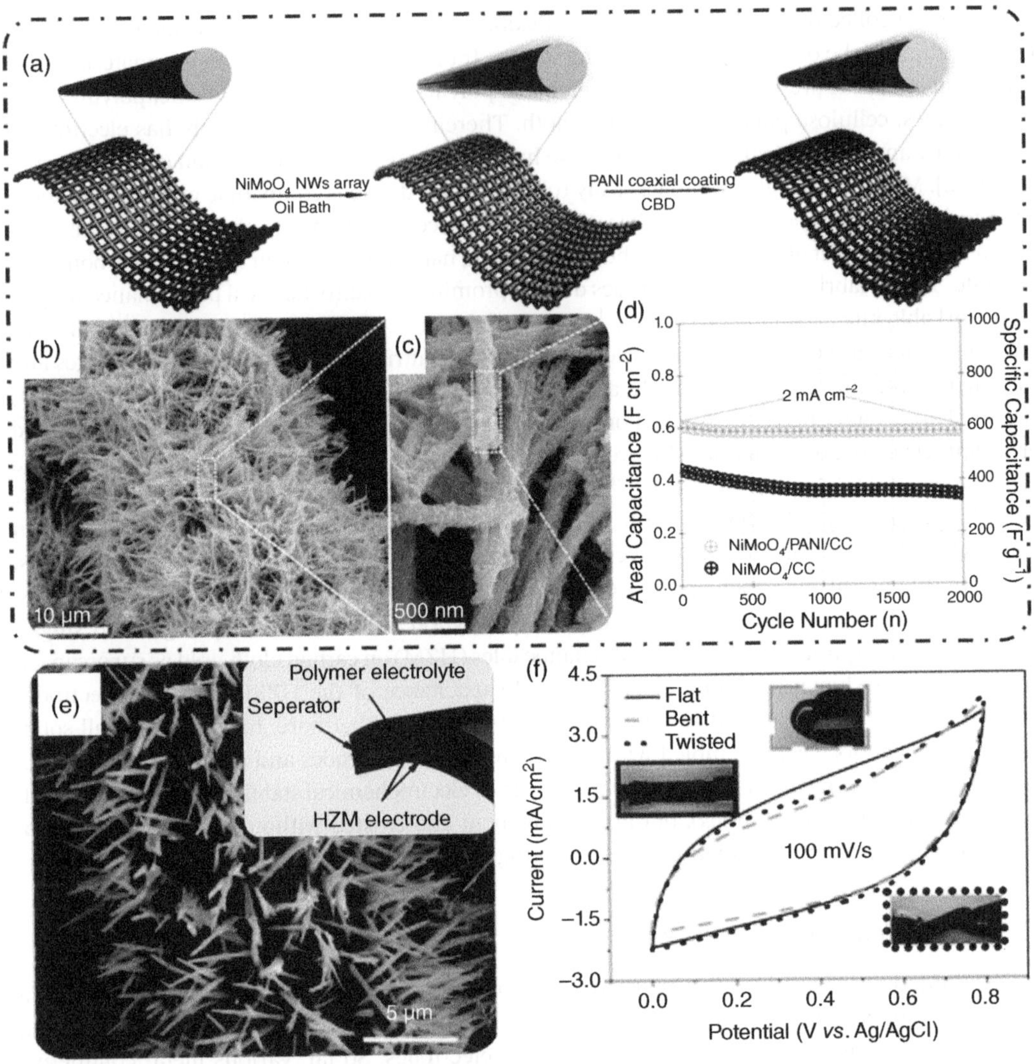

Figure 8.7 Flexible carbon cloth for current collectors. (a) Schematic illustrations of the synthesis processes of $NiMoO_4$/polyaniline (PANI) nanowire array electrode; (b and c) SEM images of the $NiMoO_4$/PANI core-shell nanoarrays growth on carbon cloth and (d) Cycling stability of nanoarray electrodes. *Source:* Reproduced with permission [80]. © 2015, Elsevier Ltd. (e) SEM image of hydrogenated single-crystal ZnO@ amorphous ZnO-doped MnO_2 nanocables (HZM) on carbon cloth, inset: schematic diagram of a solid-state supercapacitor assembled with two HZM electrodes and (f) CV curves for the flexible HZM device under flat, bent, and twisted conditions. *Source:* Reproduced with permission [83]. © 2013, American Chemical Society.

electrolyte, while the rGO nanosheets were beneficial to elevate the conductivity and active areas. Therefore, such TiO_2/rGO electrodes delivered a superior capacitive property in the optimum potential window (−0.4–0.6 V) due to the synergistic effect between the components. Furthermore, solid-state cable-type supercapacitors matched with two same TiO_2/rGO electrodes shown a high energy density of 12.2 mWh cm^{-1} at power density of 2988 mW cm^{-1} as well as a noticeable bending property at different state. Additionally, considering the advantages of high storage abundance and

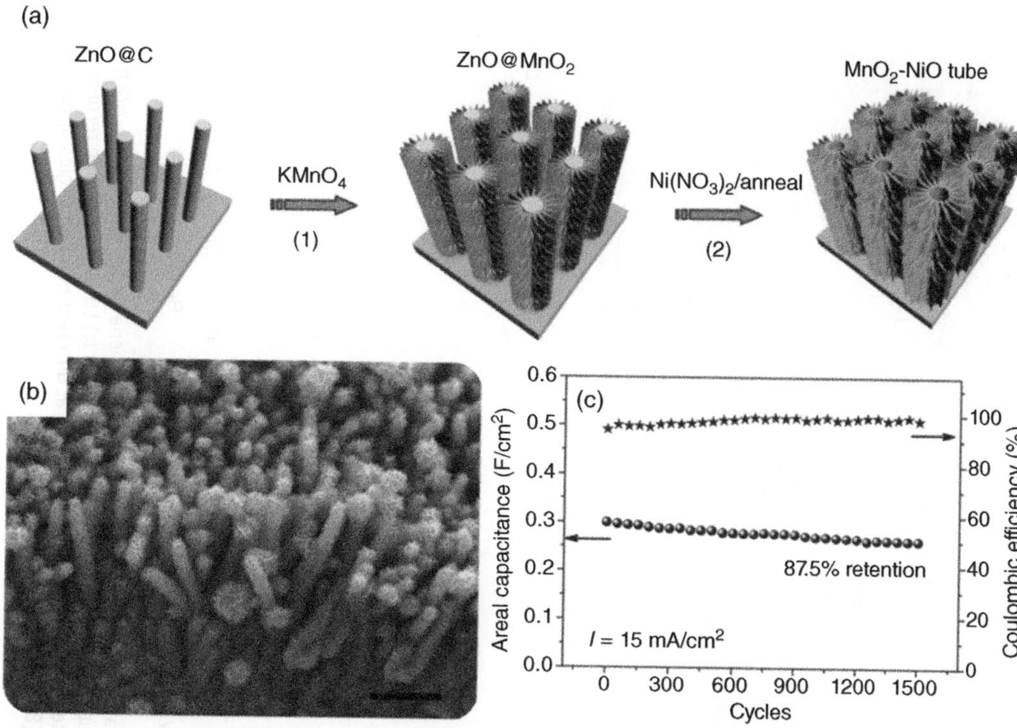

Figure 8.8 (a) Schematic illustration of fabrication procedure of 3D MnO_2-NiO tubular arrays; (b) Cross-sectional SEM image and (c) Cycling records of the array electrodes. *Source:* Reproduced with permission [38]. © 2012, The Royal Society of Chemistry.

high stability of Cu metal, the Cu wire as current collector is becoming one of the promising choices. For instance, the vertically-aligned CuO nanotube arrays/$Co(OH)_2$ nanosheets/rGo (CuO/$Co(OH)_2$/rGO) hierarchical nanoarchitectures were prepared on a thin Cu wire substrate (see in Figure 8.9a and b) [86]. The obtained porous electrode illustrated significantly improved charge storage ($22.3\,mC\,cm^{-1}$), superior rate capability as well as high electrochemical stability with 139% retention after 10000 deep times (Figure 8.9c). Most importantly, flexible all-solid asymmetric supercapacitor device assembled by CuO/$Co(OH)_2$/rGO and active carbon electrodes manifested superior energy density $3.33\,\mu Wh\,cm^{-2}$ and power density ($4.62\,mW\,cm^{-2}$) as well as outstanding flexibility and stability with the retention of 109% at 180° bending angle, as shown in Figure 8.9d and e, unveiling its potential as energy storage in portable and wearable electronics.

Metal foams like Ni foam and Cu foam are conductive to provide highly porous and conducting substrates employed to deposit active materials. Compared with 3D graphene foam, metal foams own higher electrical conductivity and better mechanical properties but with lower specific surface area. For instance, Liu et al. [87] obtained a unique hierarchical cyclic voltammetry oxidation (CVO) Cu@CuO nanorod arrays growing on the 3D conductive Cu foams (Figure 8.10a and b). Subsequently, the as-formed hierarchical CVO Cu@CuO nanoarrays exhibited a remarkable area capacitance of $1.674\,F\,cm^{-2}$ at a current density of $2\,mA\,cm^{-2}$, outstanding cycling stability (96.45% over 4000 cycles) in a three-electrode cell with 6 M KOH solution as the electrolyte (Figure 8.10c). The excellent electrochemical properties can be ascribed to the fine interfacial contact between the nanorod backbones and the Cu foam substrates, prominent conductivity, as well as good mechanical integrity.

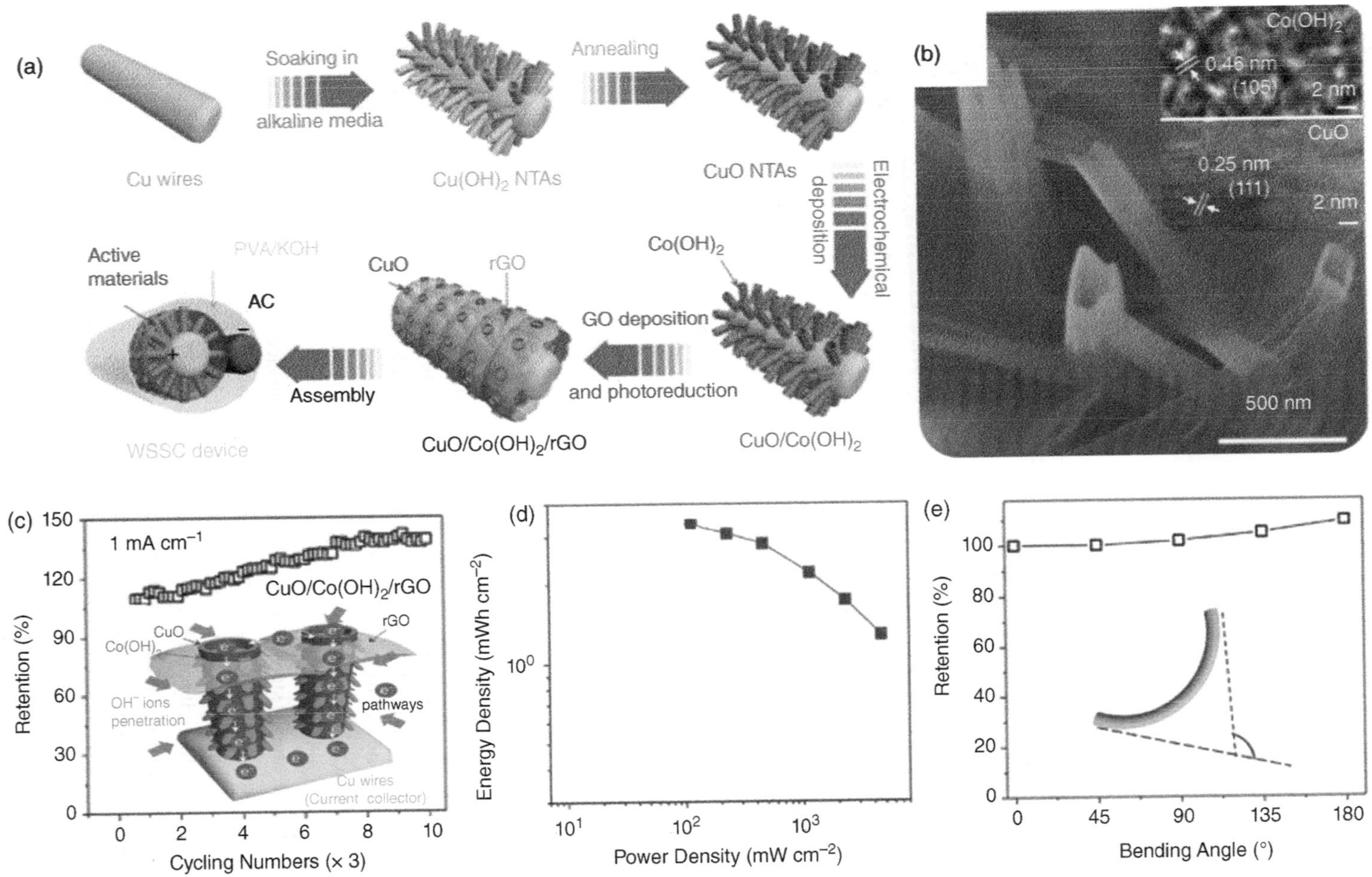

Figure 8.9 (a) Schematic illustration of fabrication process of CuO nanotube arrays/$Co(OH)_2$ nanosheets/reduced graphene oxide (CuO/$Co(OH)_2$/rGO) composite prepared on a thin Cu wire substrate; (b) SEM image, inset: high resolution TEM images of CuO/$Co(OH)_2$/rGO; (c) Long-time cycling records of the array electrodes, inset: schematic of the electrochemical processes; (d) Ragone plot; (e) Capacitance retention at different bending angles of flexible all-solid asymmetric supercapacitor device. *Source:* Reproduced with permission [86]. © 2019, Elsevier BV.

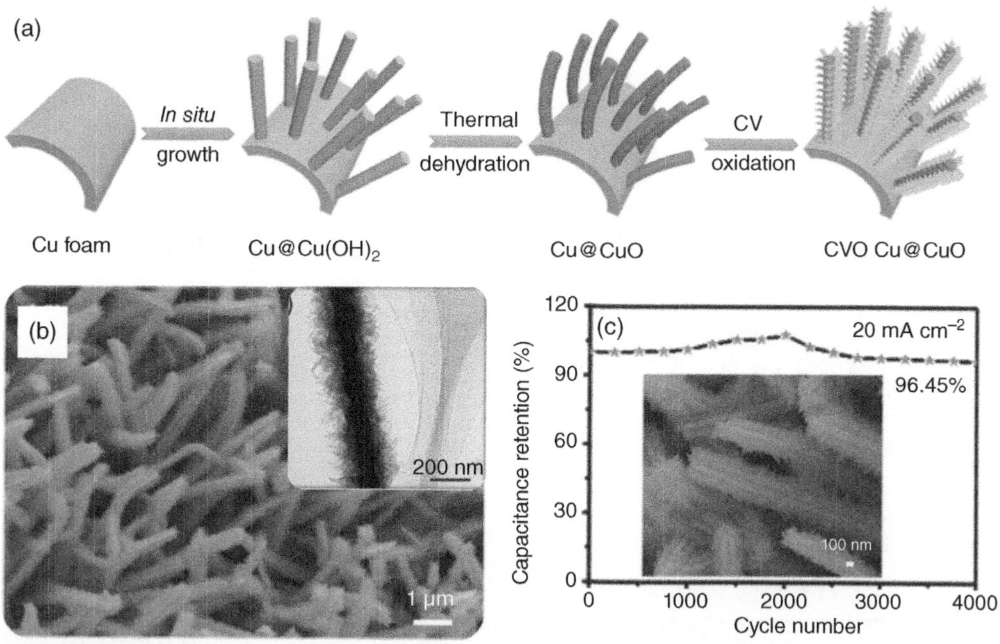

Figure 8.10 (a) Schematic preparation illustration of cyclic voltammetry oxidation (CVO) Cu@CuO nanorod arrays growing on the 3D conductive Cu foams; (b) SEM image of CVO Cu@CuO arrays, inset: TEM image and (c) Cycling performance of the array electrode, inset: SEM image of CVO Cu@CuO electrode after 4000 cycles. *Source:* Reproduced with permission [87]. © 2018, The Royal Society of Chemistry.

Meanwhile, core-shell $Ni(OH)_2$@CuO electrodes grown on 3D Cu foams were synthesized by a facile thermal oxidation and chemical bath deposition method [88]. The electrochemical experiments displayed that the $Ni(OH)_2$@CuO hierarchical nanoarrays afforded high areal capacitance of 1.625 F cm^{-2} at 3 mA cm^{-2} with excellent cycling stability (retaining 96.4% after 5000 cycles) and superior rate capacity of 1.285 F cm^{-2} at a current density of 30 mA cm^{-2}. Moreover, the self-assembled $Ni(OH)_2$@CuO//AC device presented a high energy density of 58.59 Wh kg^{-1} at a power density of 686.45 W kg^{-1} and superior cycling stability (92.1% retention after 10 000 deep cycles).

Likewise, Ni is frequently used due to its high conductivity, structural robustness, and low cost. In addition, the Ni foam with 3D network structures serves both the support of active nanoarrays and the current collector of the electrode, which makes the electrodes have increased surface area and ideal electron paths as well as eliminates the use of ancillary conductive material and binder [89–91]. An excellent example is the fabrication of a highly ordered 3D Co_3O_4@MnO_2 hierarchical porous nanoneedle array on Ni foam through a stepwise hydrothermal method (Figure 8.11a and b) [92]. As-obtained core-shell nanoneedle arrays delivered favorable electrochemical performance in terms of high specific capacitances of 932.8 F g^{-1} at a scan rate of 10 mV s^{-1}, high energy density of 66.2 Wh Kg^{-1} at a power density of 0.25 kW kg^{-1} and long-term cycling stability, which are better than that of the individual components of Co_3O_4 nanoneedles and MnO_2 nanosheets (Figure 8.11c). Significantly, the synergist effect between Co_3O_4 and MnO_2 inhibited the anodic dissolution of MnO_2, leading to the improvement of the electrochemical reversibility and stability of the core-shell electrode. Another typical example is the fabrication of binary metal oxides $NiCo_2O_4$@$NiWO_4$ nanowire arrays as binder-free electrodes for supercapacitors [93]. The optimized hierarchical electrode (eight hours) demonstrated a large specific capacitance of 1384 F g^{-1} at a current density of

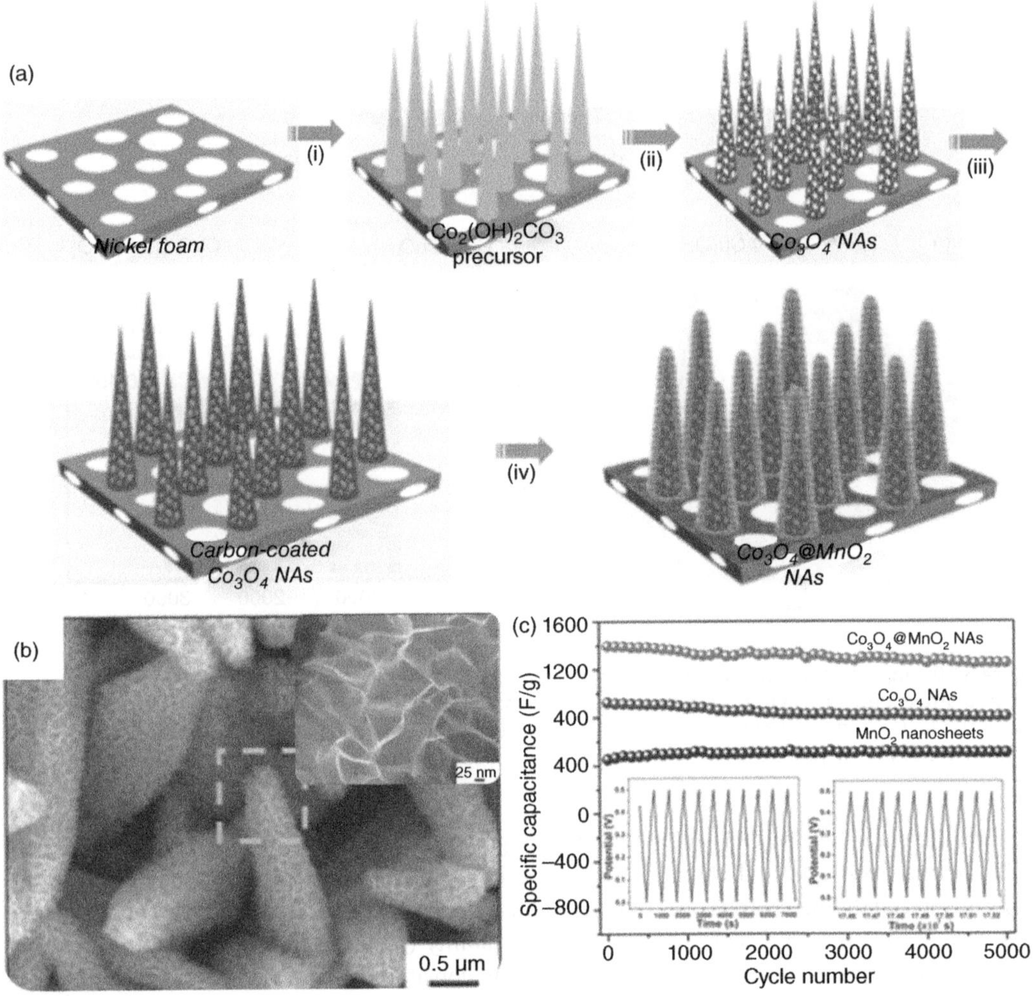

Figure 8.11 (a) Schematic illustration of the generation process of Co_3O_4@MnO_2 hierarchical porous nanoneedle array on Ni foam; (b) SEM image of Co_3O_4@MnO_2 arrays and (c) Cycling performance of the nanoarray electrodes, inset: charge/discharge curves [92]. *Source:* Reproduced with permission. © 2014, Wiley-VCH.

$1\,A\,g^{-1}$, desirable cycling stability (87.6% retention over 6000 cycles at a current rate of $5\,A\,g^{-1}$), and intriguing rate capability. Teng et al. [94] resoundingly fabricated hierarchical transition metal oxides nanoarray material ($MgCo_2O_4$@MnO_2) by a facile solvothermal reaction and a simple calcination afterwards. As assembled core-shell electrode displayed the enhancement of electrochemical properties (considerable capacitive property and the remarkable cycling performance), which is incited by the abundant electroactive sites at the electrolyte-electrode interface.

8.4 The Geometry of Nanostructured Arrays

According to geometrical/dimensional characteristics, nanostructured array electrodes can be generally classified into 1D, 2D and the integration of 1D@2D nanoarrays. Generally, 1D nanoarrays such as nanowires, nanotubes, nanopillars, nanobelts/nanoribbons are fascinating because of their dimensionality dependence on its functional feature. The 2D nanostructured array (nanowall/nanoflake matrix) refer to materials that have a thickness of several atomic layers in a certain dimension while other dimensions exceed the nanometer size range.

8.4.1 The 1D Nanostructured Arrays

1D nanostructures have been attractive as components for flexible energy storage systems due to their unique electronic properties. Well-aligned array of 1D nanostructures is a novel and promising architecture that exhibits superior geometric and morphologic characteristics. For example, the longitudinal axis of 1D nanostructures can provide efficient transport pathway for both electrons and ions [95]. Using 1D arrays growth directly on the conductive substrates as electrodes either by a "bottom-up" or "top-down" method guarantees efficient transport of electrons as well as robust contact between the current collector and active material [14]. Thus, conductive electrons can be quickly transferred from the active redox sites to the working electrode along a superhighway instead of randomly entering the disordered nanocrystalline network by moving through interparticle contact areas which results in poor electrochemical kinetics and actives utilization [96].

Nanowires-based arrays are one of the most common nanoarrays due to their high aspect ratio can furnish a large surface area to store charges and a direct/rapid pathway to transport charges. Gao et al. [89] prepared Co_3O_4 nanowire arrays freely standing on Ni foam by template-free growth followed by thermal treatment at 300 °C for two hours in air. The obtained nanowires densely covered the Ni foam substrate and had relatively uniform size (the diameter around 250 nm and the length up to around 15 μm, Figure 8.12a), displaying a specific capacitance of 746 $F g^{-1}$ at a current density of 5 $mA cm^{-2}$. In addition, the nanowire electrodes delivered relatively good electrochemical stability (specific capacitance decreased to 86% after charge/discharge 500 cycles, Figure 8.12b). Ni foam supported hierarchical mesoporous Zn-Ni-Co ternary oxide (ZNCO) nanowire arrays were synthesized for supercapacitive investigation for the first time, see in Figure 8.12c and d [97]. Such hierarchical ZNCO nanoarrays provided a superior specific capacitance of 2481.8 $F g^{-1}$ at 1 A g^{-1} and wonderful rate capability (capacitance retention of 91.9% even at 5 A g^{-1}). An asymmetric supercapacitor using the ZNCO electrode materials as positive electrode and AC materials as negative electrode (ZNCO//AC) exhibited a high energy density of 35.6 Wh kg^{-1} at a power density of 187.6 W kg^{-1} and an attractive capacitance retention (71.2%, 6000 cycles) as well as 100% coulombic efficiency, Figure 8.12e. Moreover, Xia et al. [98] fabricated metal oxide core-shell nanoarrays on various conductive substrates such as transparent conducting glass and Ni foil/foam. Supercapacitor electrodes based on the Co_3O_4/NiO nanowire arrays grown on Ni foam shown superior electrochemical performance owing to the enough surface areas, well conductivity, and strong mechanical stability.

Compared with nanowire arrays, nanotube-based arrays seem to be more promising because they supply additional degrees of freedom, not only the length and diameter can be changed but also the thickness. Simultaneously, the thin nanotube walls can enlarge surface area and enable the electrolyte to permeate the entire internal and external surfaces [51]. Long before, Kim et al. [99] fabricated oriented NiO-TiO_2 nanotube arrays as pseudocapacitor electrodes via electrochemically anodizing Ni-Ti alloy foils and then thermal annealing. The faster charge/discharge kinetics of nanoarray electrodes even at high scan rates were attributed to the more ordered nanotube architecture, which promotes both electron and ion transport during the redox process. After that, a novel nanoarchitecture of MnO_2/Mn/MnO_2 sandwich-like nanotube arrays was designed by Tong and coworkers (Figure 8.13a) [100]. The well-ordered nanotube electrodes allowed efficient use of MnO_2 for charge storage and promoted the transport of electrons and ions, thus

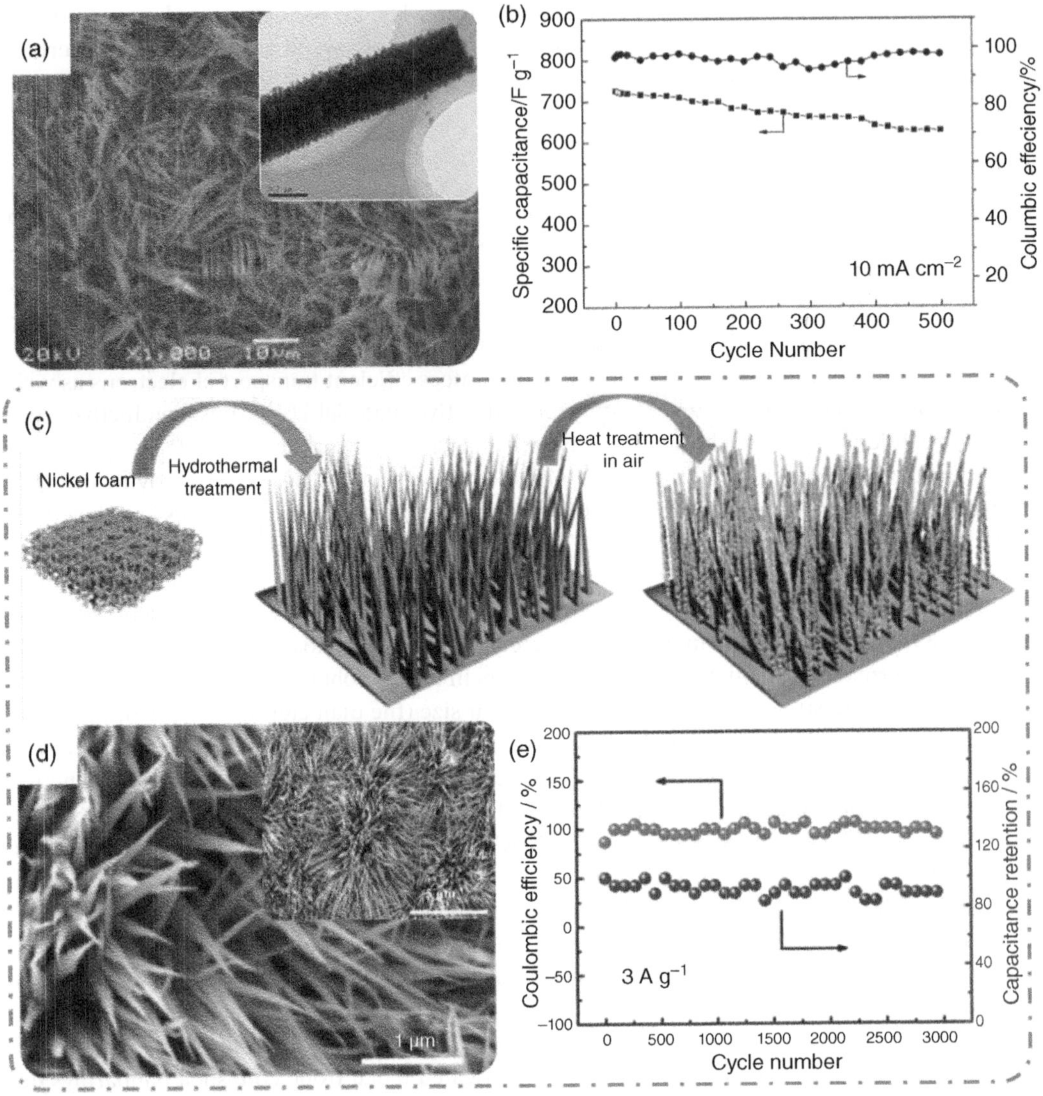

Figure 8.12 1D nanowire-based arrays. (a) SEM image of the Co_3O_4 nanowire arrays, inset: TEM image and (b) Cycling performance of the nanoarray electrodes. *Source:* Reproduced with permission [89]. © 2010, Elsevier BV. (c) Scheme of the formation process of the mesoporous Zn-Ni-Co ternary oxide (ZNCO) nanowire arrays; (d) the SEM image of ZNCO nanowire arrays, inset: TEM image and (e) Cycling records of the ZNCO//AC asymmetric supercapacitor device. *Source:* Reproduced with permission [97]. © 2015, American Chemical Society.

making the unique composite electrodes have excellent properties (Figure 8.13b and d), such as large specific capacitance, excellent rate capability, and remarkable long-term cycle stability. Li et al. [101] reported aligned TiO_2 nanotube arrays as a high-surface-area framework to construct nanostructured electrode on planar silicon substrate for on-chip supercapacitors for the first time. The MnO_2 decorated TiO_2 nanotube arrays exhibited maximum specific capacitance 20.6 mF cm^{-2} at a current density of 0.05 mA cm^{-2} and excellent cycling stability (82.1% capacitance retention after 3000 cycles).

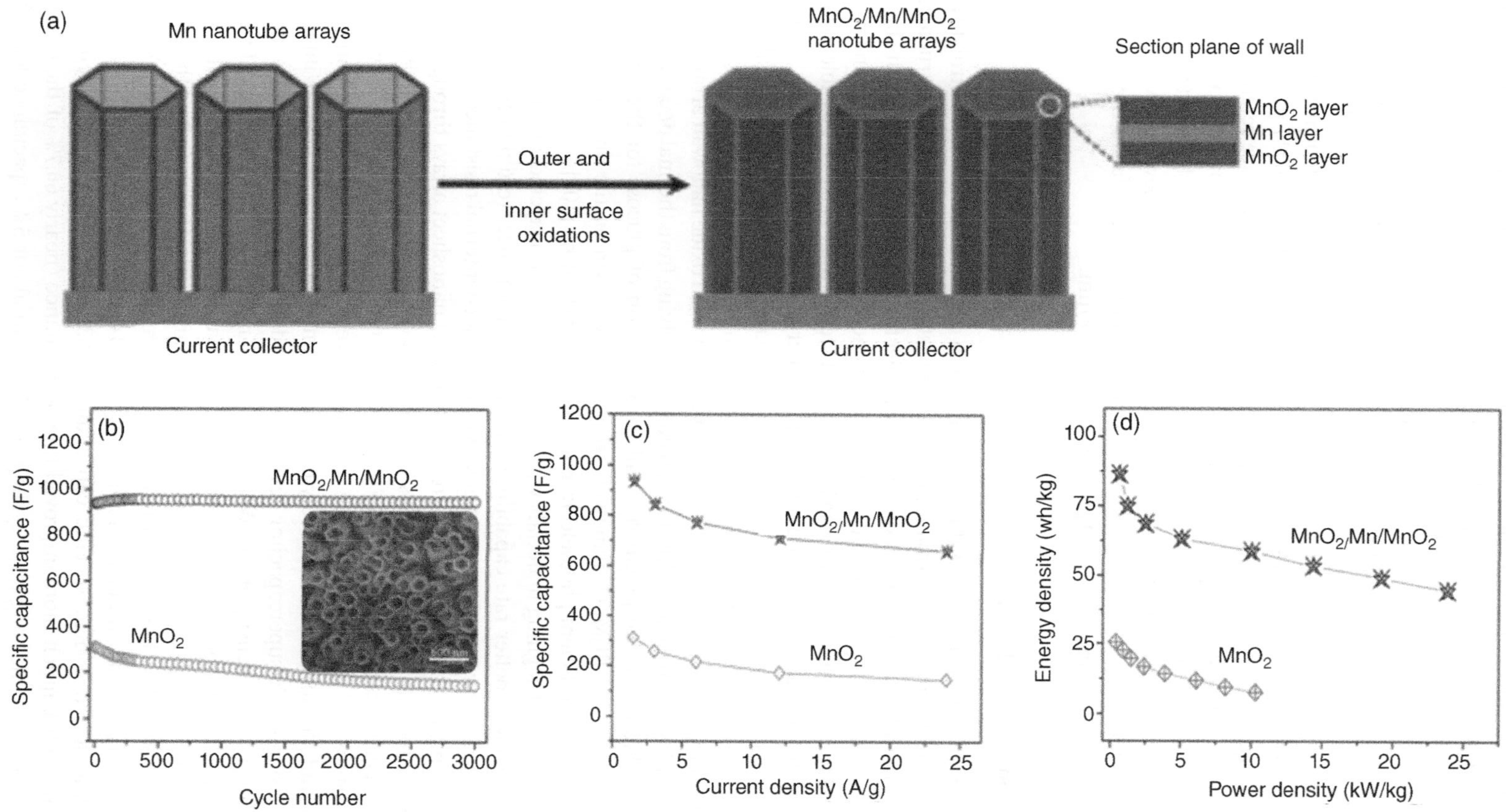

Figure 8.13 1D nanotube-based arrays. (a) Schematic illustration of the synthesis of MnO_2/Mn/MnO_2 sandwich-like nanotube arrays; (b) Cycling performance of the nanoarray electrodes, inset: SEM image; (c) Rate behaviors and (d) Ragone plots. *Source:* Reproduced with permission [100]. © 2012, American Chemical Society.

Although the moderate aspect ratio of nanopillars limits the increase in surface area compared to nanowires, the chance of a single nanopillar collapsing into its adjacent structure is significantly reduced, which makes them easier to enter the electrolyte and effectively increases the ionic conductivity. Thomas and coworkers described the development of MnO_2 decorated, highly ordered nanopillar array electrodes for supercapacitors [102]. Such array-based nanopillars have diameter of ~140 nm and height of ~250 nm, and center-to-center distance of ~200 nm. Compared with planar electrodes, this novel nanostructure design increased the surface area by nearly 4.5 times, manifesting significant enhancements in specific capacitance (reached up to 603 $F g^{-1}$ at a scan rate of 5 $mV s^{-1}$), charge/discharge capability and cycle stability (93% capacitance retention even after 5000 cycles). Besides, nanobelts offer larger surface areas than nanowires owing to the higher surface-to-volume rations. Zheng and coworkers [103] explored an atmospheric and rapid flame synthesis technique to controllably grow single α-MoO_3 nanobelt arrays. The newly grown α-MoO_3 nanobelts are single-crystalline, rectangular, closely packed, and perpendicular to the substrate with a typical thickness of 200 nm, width of 4 μm and length of 30 μm.

8.4.2 The 2D Nanostructured Arrays

Given the importance of properties such as diffusion ion length or contact area with electrolyte for utilization of active material in the performance of the energy storage devices, 2D nanostructured metal oxides have naturally attracted enormous interest in electrode design. Meanwhile, it shows promising signs thanks to the high mechanical strength and chemical stability. The size and morphology of metal oxide nanostructures can also be easily adjusted, which makes it possible to systematically investigate the structure-electrochemical property relationship. Among various nanostructured strategies, 2D nanoarray-based metal oxides (nanowall/nanosheet/nanoflake) are always prospective owing to efficient charge transport without obviously compromising surface area. Typically, Jiang et al. [104] prepared carbon coated Co-Fe mixed oxide nanowall arrays grown directly on a flexible alloy substrate for lithium-ion batteries, evolving from thermal decomposition of Co-Fe layered double hydroxide precursors and carbonization of glucose for the first time. Moreover, a vertically aligned Bi_2MoO_6 nanowall array directly grown on the tin-doped indium oxide (ITO) glass for highly efficient photoelectrochemical water splitting [105].

Inspired by this concept, designing materials with long range 2D ultra layered structures can be an efficient road to achieve better rate capabilities and high capacitance, resulting from a large specific surface area, fast electron and ion transfer as well as reduced internal resistance. Therefore, Han et al. [106] presented a core-shell ultralayered Co_3O_4@NiO nanosheet arrays through a two-step low temperature hydrothermal method. As-prepared nanosheet array electrode delivered a notably enhanced specific capacitance of 715 $F g^{-1}$ at the current density of 0.5 $A g^{-1}$, which is attributed to the porous and ultrathin structure does not prevent contact between the core-shell nanosheet arrays and the ions in the electrolyte whilst maintaining their structural integrity. Noticeably, a new asymmetric supercapacitor device assembled by Co_3O_4@NiO nanosheet arrays and AC yielded a large energy density of 30.2 $W h kg^{-1}$ at a power density of 201.5 $W h kg^{-1}$. In addition, it also exhibited a low charge transfer resistance and a remarkable cycling stability with about 2% capacitance increase after 6000 cycles. In another study, Zhang et al. [107] described the grown of an ordered core-shell $CoMoO_4$@$NiMoO_4$ nanosheet (the thickness is approximately 100 nm) array composites on Ni foam substrate, as shown in Figure 8.14a. As expected, such hierarchical array electrodes manifested a high specific capacitance (1639.8 $F g^{-1}$ at a current density of 10 $mA cm^{-2}$, Figure 8.14b), much more improved rate performance (nearly 66.7% of the original capacitance at 60 $mA cm^{-2}$) and long-term cycling stability with about 5% capacitance loss after

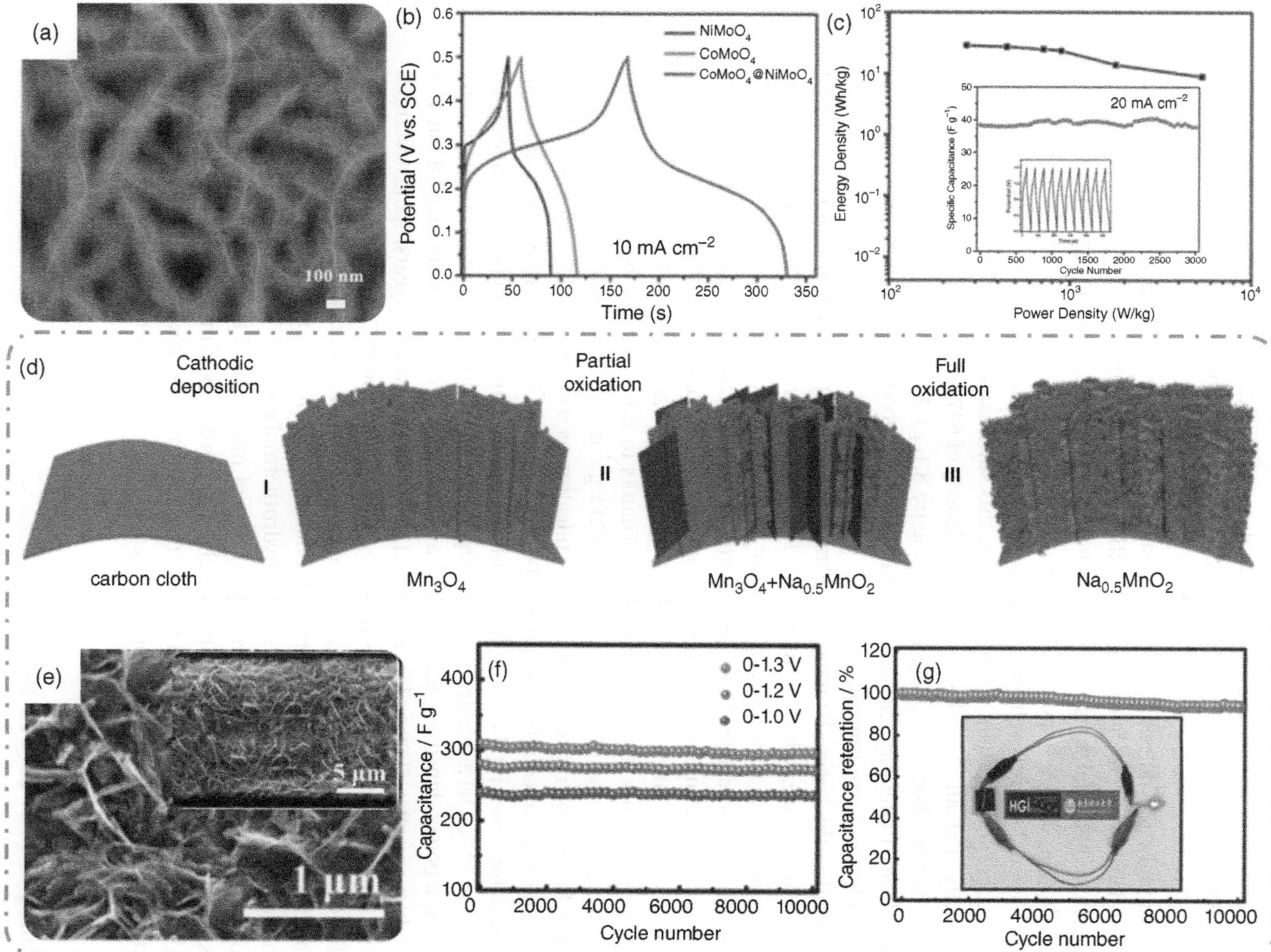

Figure 8.14 2D metal oxide nanoarrays. (a) SEM image; (b) Galvanostatic charge/discharge curves of core-shell $CoMoO_4$@$NiMoO_4$ nanosheet and (c) Ragone plot of the $CoMoO_4$@$NiMoO_4$/Ni foam//AC asymmetric supercapacitors, inset: cycling performance. *Source:* Reproduced with permission [107]. © 2016, The Royal Society of Chemistry. (d) Schematic illustration of the synthesis process of the $Na_{0.5}MnO_2$ nanosheet assembled nanowall arrays; (e) the SEM image $Na_{0.5}MnO_2$ nanosheet; (f) Cycle performances in different potential windows and (g) Cycling records of the $Na_{0.5}MnO_2$//Fe_3O_4@C asymmetric supercapacitor, inset: the optical image of the array device lighting an 2.2 V LED. *Source:* Reproduced with permission [[108]]. © 2017, Wiley-VCH.

3000 cycles, attributing to the larger specific surface area and more active sites during the electrochemical redox process. Additionally, an asymmetric supercapacitor was constructed ($CoMoO_4$@$NiMoO_4$/Ni foam//AC), which shown a high energy density of 28.7 W h kg^{-1} at a power density of 267 W kg^{-1} and outstanding cycling stability with 99% specific capacitance retention after 3000 times (Figure 8.14c). Recently, a hierarchical architecture of Birnessite $Na_{0.5}MnO_2$ nanosheet assembled nanowall arrays was synthesized on flexible carbon cloth via electrochemical oxidation (Figure 8.14d) [108]. Each nanowall was composed of numerous tiny nanosheets with a highly porous structure (Figure 8.14e), which is beneficial to increase the surface area of fast faradaic reactions. Interestingly, the potential window of such nanoarray electrode can be extended to 0–1.3 V accompanied by extremely increased specific capacitance up to 366 F g^{-1} (Figure 8.14f). Based on the separate potential windows and matchable capacitance, an asymmetric supercapacitor integrating the $Na_{0.5}MnO_2$ nanowall arrays cathode and the carbon-coated Fe_3O_4 nanorod arrays anode achieved 2.6 V cell voltage theoretically. Apart from the large specific capacitance, high working voltage, and excellent cycling stability, this supercapacitor system afforded a maximum energy density of about 81 Wh kg^{-1} at a power density of 647 W kg^{-1}, and successfully driven a 2.2 V red light emitting diode (LED) lightening (Figure 8.14g). Meanwhile, the electrochemical behaviors of the device under different bending states unveiled good flexibility and potential application in flexible electronic equipment.

Due to the ultrathin thickness, 2D nanoflake array have a high packing density that leads to a high volumetric capacitance, which is important for manufacturing thin-film supercapacitors. However, the re-stacking of nanoflakes is a major challenge for the potential utilization of 2D nanomaterials in energy storage applications. To settle this issue, Zhang and coworkers reported a novel approach to form hierarchical composite electrodes of NiO nanoflake (the length of 1–2 μm with ultra-thin thickness) on a 3D foothill-like graphene architecture atop Ni foams for high-performance supercapacitors (Figure 8.15a) [109]. Thanks to this unique structure, as-prepared electrode provided a high specific capacitance of 1421 F g^{-1} even at a current density of 5 A g^{-1} without needing binders or metal-based current collectors as well as excellent cycling stability (81% of the initial capacitance retention after 5000 cycles, Figure 8.15b). Furthermore, an asymmetric supercapacitor was setup based on hierarchical NiO-3D graphene composites (anode) and AC (cathode) with 1 M NaOH as electrolyte. As-designed device afforded a superior energy density of 138 Wh kg^{-1} at a power density of 5.25 kW kg^{-1} and excellent cycling performance (about 15% capacitance reductions after 5000 cycles, Figure 8.15c). Over the past few years, nanostructured WO_3 has attracted great attention due to its high electronic conductivity ($10–10^{-6}$ S cm^{-1}), facilitating ions diffusion/transport for supercapacitor electrochemical systems. For example, Wang et al. [110] obtained three types of hexagonal WO_3 nanoflake arrays including single crystal, polycrystal, and hierarchical by using an easy/simple hydrothermal method. The width and thickness of a single polycrystal WO_3 nanoflake were kept at tens of microns and less than 100 nm, respectively. Impressively, the crystal WO_3 nanoarrays yielded a high-rate performance (272 F g^{-1}) at high scan rate of 10 A g^{-1} which is derived from fast electron transmitting channel, while the other two nanoflake arrays with broadened ion storage spaces performed relatively high specific capacitances at low scan rate.

8.4.3 The Integration of 1D@2D Nanoarrays

Constructing advanced materials for electrochemical energy storage devices is a perplexing challenge faced by many researchers. Currently, despite some encouraging progress having been achieved, the development based on metal oxide nanoarrays for supercapacitors is still a huge

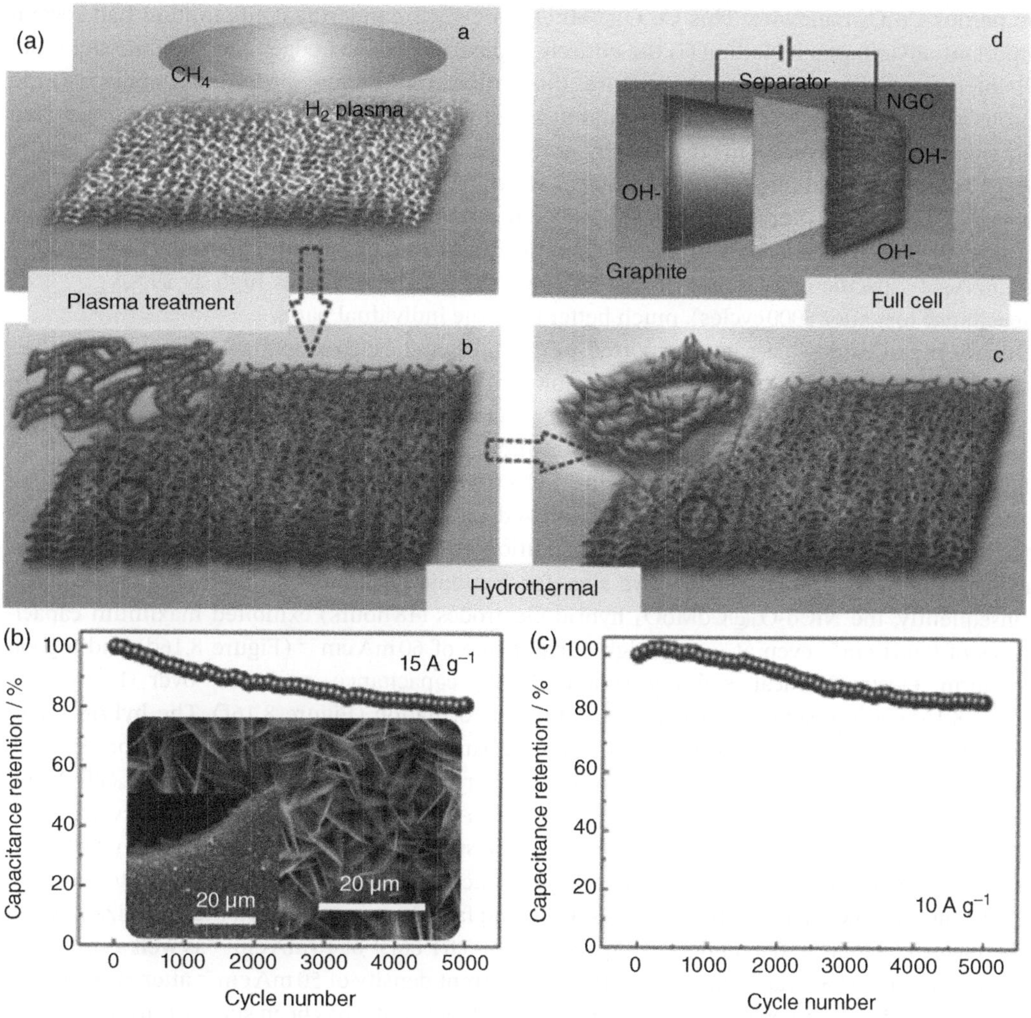

Figure 8.15 (a) Schematic illustration of the synthesis process of NiO nanoflake on a 3D foothill-like graphene architecture atop Ni foams; (b) Cycling performance of the nanoarray electrodes, inset: SEM image; (c) Long-time cycling stability of NiO-3D foothill-like graphene//AC asymmetric supercapacitor. *Source:* Reproduced with permission [109]. © 2014, Wiley-VCH.

space. It is well-known that 1D nanostructure can provide efficient transport pathways for both electrons and ions in the supercapacitor electrode. While 2D nanomaterials have brought synergic advantages of controllable electrical properties and high specific surface areas as well as superior electrochemical activities. As a result, the integration of hybrid systems of different dimension nanoarrays would synergistically enhance the intrinsic properties of each component such as electrical/ionic conductivity, electrochemical reactivity, and mechanical stability. For instance, 3D ordered amorphous carbon (originated from glucose) layers as a sacrificial reactive template to anchor MnO_2 nanosheet onto Co_3O_4 nanowire scaffolds pre-grown on a flexible stainless-steel foil, leading to a well-designed hybrid architecture for supercapacitor applications [111]. In individual hybrid nanostructure, thin nanosheets (only 4.3 nm) branching uniformly covered the surface of the porous Co_3O_4 nanowire. This Co_3O_4@MnO_2 core-shell nanoarray configuration had several important advantages, including (i) the entirely exposed nanosheet edges can facilitate the rapid Li^+ intercalation into the layered structure, thus enhancing the redox reaction kinetics, (ii) the porous structure of Co_3O_4 nanowires would further increase the electrolyte-material contact area and enhance ion diffusion, (iii) the directly grown array on a stainless steel substrate can ensure robust mechanical stability (good mechanical adhesion and electrical connection to the current collector). Thence, the purposely built smart electrode made of a Co_3O_4@MnO_2 core-shell nanowire array displayed excellent electrochemical performance (a high specific capacitance of 480 $F g^{-1}$ at 2.67 $A g^{-1}$, specific capacitance of 267 $F g^{-1}$ even at a current rate as high as 29.8 $A g^{-1}$, 2.7% capacitance loss after 5000 cycles), much better than the individual parts.

Binary metal oxides have been reported to exhibit better electrochemical performances than single component oxides due to their achievable oxidation states and high electrical conductivity. In this regard, Cai et al. [112] reported a construction of unique $NiCo_2O_4$ nanowire@$CoMoO_4$ nanoplate core-shell arrays on Ni foam for high areal capacitance supercapacitors, see in Figure 8.16a–c. In such designed hierarchical electrodes, the open space between the core-shell nanowire/nanoplate array can be efficiently utilized, ensuring a prompt diffusion of the electrolyte ions to the surface of the active material. In addition, this hybrid arrays greatly increased the surface area, providing more electroactive sites for Faradaic reactions and high area capacitance. Consequently, the $NiCo_2O_4$@$CoMoO_4$ hybrid electrodes (18 hours) exhibited maximum capacitances of 9.66 $F cm^{-2}$ even at a very high current rate of 60 $mA cm^{-2}$ (Figure 8.16d) and a good long-term electrochemical stability (about 74.1% capacitance retention over 1000 cycles, Figure 8.16e) accompanied by slightly structural deformation (Figure 8.16f). The hybrid nanostructure arrays possess all advantages of ordered single-phase arrays, and further bring along some unique merits of hybrid composites. In another work, the hierarchical $NiCo_2O_4$@$NiMoO_4$ core-shell hybrid nanowire/nanosheet arrays for high-performance pseudocapacitors were fabricated directly on Ni foam [113]. The as-prepared free standing $NiCo_2O_4$ nanowires are nearly uniformly aligned on the Ni foam with an average diameter of ca. 60 nm and length up to ca. 3 μm. Benefiting from the facile electrolyte penetration and fast electron transport, the optimized hybrid array electrodes delivered a high areal capacitance of 5.8 $F cm^{-2}$ at 10 $mA cm^{-2}$ as well as excellent cycling stability (81.8% capacitance retention at a current density of 50 $mA cm^{-2}$ after 5000 cycles). The asymmetric supercapacitor device ($NiCo_2O_4$@$NiMoO_4$//AC) has been successfully assembled, which can provide a high energy density of 21.7 $Wh kg^{-1}$ at a power density of 157 $W kg^{-1}$. Following a similar concept, 3D $CuCo_2O_4$@$NiCo_2O_4$ core-shell nanostructures also have been synthesized, in which the porous $CuCo_2O_4$ nanowires as the scaffold were well covered by $NiCo_2O_4$ nanosheets [37]. Such hierarchical electrode exhibited good electrochemical performances such as large specific capacitance of 2029 $F g^{-1}$ at the current density of 10 $mA cm^{-2}$ and notable rate capacitance (1551 $F g^{-1}$ at a current rate of 30 $mA cm^{-2}$) as well as remarkable cycling stability (the 76% capacitance retention after 4500 cycles).

8.5 Conclusions and Prospects

In brief, nanostructured metal oxide arrays have exhibited significant advantages in fabricating current cutting-edge flexible supercapacitors. By adjusting the composition and nanostructure of the array electrode, metal oxides have displayed enhanced conductivity, abundant surface specific area and sufficient electrochemical actives sites, thereby further leading to the improvement of electrochemical properties. For example, when combined with highly porous structural morphologies,

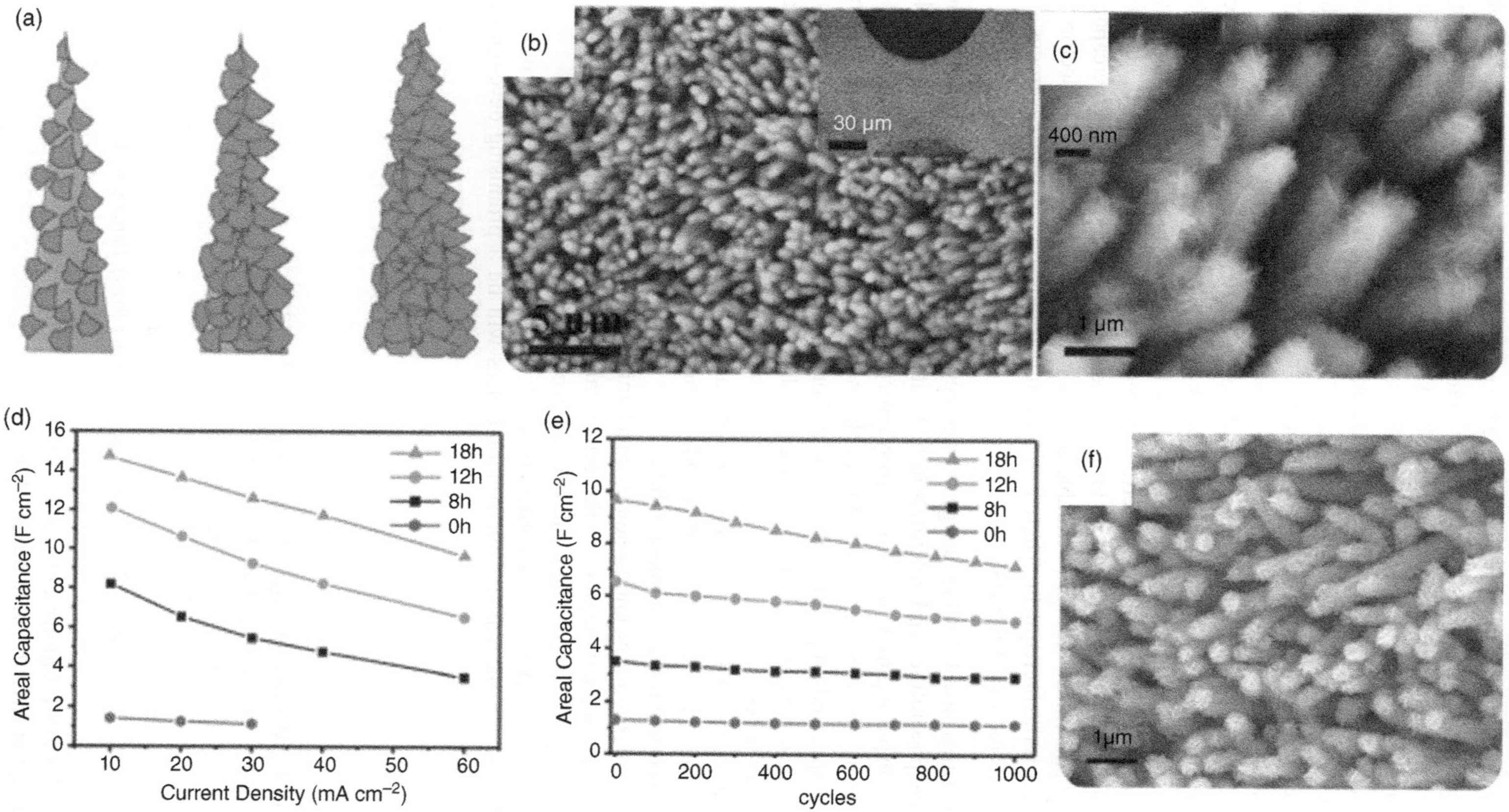

Figure 8.16 (a) Schematic illustration of the possible synthesis mechanism of $NiCo_2O_4$ nanowire@$CoMoO_4$ nanoplate core-shell arrays on Ni foam; (b and c) SEM images; (d) The areal capacitance as a function of current density plot; (e) Cycling records of the hybrid nanoarray electrodes and (f) SEM image of the $NiCo_2O_4$ nanowire@$CoMoO_4$ nanoplate hybrid electrodes after 1000 cycles. *Source:* Reproduced with permission [112]. © 2014, The Royal Society of Chemistry.

nanostructure arrays represent an attractive architecture for rapid reaction kinetics thanks to their significantly enhanced surface area. Nevertheless, several inherent obstacles still remain on the way to the development of advanced supercapacitors, mainly: (i) poor electrical conductivity, and (ii) unavoidable volumetric expansion. In this article, we first discuss the approaches toward metal oxide arrays from three aspects, including solution-based method, electrochemical assisted electrodeposition and CVD. The common supporting substrates (3D porous graphene foam, carbon cloth current collectors and metal conductive substrates) of nanoarray electrodes, along with geometry/dimension design strategies in advanced supercapacitor devices are also introduced.

So far, despite the current research achievements on the design and fabrication of metal oxide-based array electrodes for flexible supercapacitors, these are still at the investigative stage and the nanoarray electrodes cannot satisfy many requirements of future functional power sources (e.g. high energy and power densities, operational safety, long lifetime, low cost, in addition to mechanical flexibility, light weight, thin thickness, etc.). From the perspective of the flexible supercapacitor devices, the use of substrates with soft, bendable, and elastic properties as current collectors to load metal oxide arrays is the prerequisite. Therefore, the future trends of nanoarray electrodes are expected to focus on the following aspects:

1) It can be expected that novel flexible energy storage devices will be further developed. And there are still huge possibilities and potentials in fabricating stretchable or arbitrarily variable supercapacitors by exploring more novel nanoarray structures and materials.
2) There is an urgent need to develop some controllable and cost-effective manufacturing methods to simplify complicated preparation technologies. For example, the advance in 3D printing techniques offers new possibilities for improving the electrochemical performance of metal oxide nanoarray electrodes.
3) The design of scaffold (used as current collectors) for loading nanoarrays should have enough flexibility because they are not limited by morphology and composition as long as being highly conductive and electrochemical/chemical stability.
4) In order to better understand the energy storage mechanism of metal oxide nanoarrays, multiple in-situ characterization techniques and theoretical calculations need to be applied simultaneously to achieve superior electrochemical performance.
5) Integrating flexible supercapacitors with other energy storage devices to form a series of novel functional electronic and mechanical micro/nanodevices will be a promising development direction in the future.

References

1 Choudhary, N., Li, C., Moore, J. et al. (2017). *Adv. Mater.* 29: 1605336.
2 Guo, D., Lai, L., Cao, A. et al. (2015). *RSC Adv.* 5: 55856.
3 Ho, K.-C. and Lin, L.-Y. (2019). *J. Mater. Chem. A* 7: 3516.
4 Liu, L., Niu, Z., and Chen, J. (2016). *Chem. Soc. Rev.* 45: 4340.
5 Yan, J., Li, S., Lan, B. et al. (2019). *Adv. Funct. Mater.* 30: 1902564.
6 Muzaffar, A., Ahamed, M.B., Deshmukh, K., and Thirumalai, J. (2019). *Renew. Sust. Energ. Rev.* 101: 123.
7 Chen, T. and Dai, L. (2014). *J. Mater. Chem. A* 2: 10756.
8 Jiang, J., Li, Y., Liu, J. et al. (2012). *Adv. Mater.* 24: 5166.
9 Poonam, K., Sharma, A., Arora, S.K., and Tripathi, J. (2019). *Energy Storage* 21: 801.

10 Liu, S., Sun, S., and You, X.Z. (2014). *Nanoscale* 6: 2037.
11 Liu, T., Zhang, F., Song, Y., and Li, Y. (2017). *J. Mater. Chem. A* 5: 17705.
12 Liang, J., Jiang, C., and Wu, W. (2019). *Nanoscale* 11: 7041.
13 Wang, F., Wu, X., Yuan, X. et al. (2017). *Chem. Soc. Rev.* 46: 6816.
14 Yu, Z., Tetard, L., Zhai, L., and Thomas, J. (2015). *Energy Environ. Sci.* 8: 702.
15 An, T. and Cheng, W. (2018). *J. Mater. Chem. A* 6: 15478.
16 Bhoyate, S., Kahol, P.K., and Gupta, R.K. (2019). *Nano* 5: 1.
17 Peng, X., Peng, L., Wu, C., and Xie, Y. (2014). *Chem. Soc. Rev.* 43: 3303.
18 Wang, Y. and Xia, Y. (2013). *Adv. Mater.* 25: 5336.
19 An, C., Zhang, Y., Guo, H., and Wang, Y. (2019). *Nanoscale Adv.* 1: 4644.
20 Yang, Q., Lu, Z., Liu, J. et al. (2013). *Prog. Nat. Sci. Mater.* 23: 351.
21 Wang, K., Wu, H., Meng, Y., and Wei, Z. (2014). *Small* 10: 14.
22 Zeng, W., Zheng, F., Li, R. et al. (2012). *Nanoscale* 4: 2760.
23 Ma, Z., Shao, G., Fan, Y. et al. (2017). *ACS Sustain. Chem. Eng.* 5: 4856.
24 Devan, R.S., Patil, R.A., Lin, J.-H., and Ma, Y.-R. (2012). *Adv. Funct. Mater.* 22: 3326.
25 Zhao, D., Dai, M., Tong, Y. et al. (2019). *CrystEngComm* 21: 5789.
26 Fang, X., Wu, L., and Hu, L. (2011). *Adv. Mater.* 23: 585.
27 Ellis, B.L., Knauth, P., and Djenizian, T. (2014). *Adv. Mater.* 26: 3368.
28 Zhang, Y., Li, L., Su, H. et al. (2015). *J. Mater. Chem. A* 3: 43.
29 Zuo, W., Xie, C., Xu, P. et al. (2017). *Adv. Mater.* 29: 1703463.
30 Guan, C., Liu, X., Ren, W. et al. (2017). *Adv. Energy Mater.* 7: 1602391.
31 Zhou, C., Zhang, Y., Li, Y., and Liu, J. (2013). *Nano Lett.* 13: 2078.
32 Yuan, C., Yang, L., Hou, L. et al. (2012). *Adv. Funct. Mater.* 22: 2560.
33 Bai, X., Liu, Q., Liu, J. et al. (2017). *Chem. Eng. J.* 315: 35.
34 Xia, X.-h., Tu, J.-p., Zhang, Y.-q. et al. (2012). *RSC Adv.* 2: 1835.
35 Wu, C., Cai, J., Zhu, Y., and Zhang, K. (2016). *RSC Adv.* 6: 63905.
36 Wu, J., Mi, R., Li, S. et al. (2015). *RSC Adv.* 5: 25304.
37 Zhang, K., Zeng, W., Zhang, G. et al. (2015). *RSC Adv.* 5: 69636.
38 Liu, J., Jiang, J., Bosman, M., and Fan, H.J. (2012). *J. Mater. Chem.* 22: 2419.
39 Huang, M., Li, F., Ji, J.Y. et al. (2014). *CrystEngComm* 16: 2878.
40 Xu, K., Li, W., Liu, Q. et al. (2014). *J. Mater. Chem. A* 2: 4795.
41 Yu, D., Zhang, Z., Meng, Y.n. et al. (2018). *Inorg. Chem. Front.* 5: 597.
42 Tang, C., Mao, Y., Xie, J. et al. (2018). *Inorg. Chem. Front.* 5: 1707.
43 Wang, Z., Zhang, Y., Li, Y., and Fu, H. (2014). *RSC Adv.* 4: 20234.
44 Wang, J., Wang, S., Huang, Z., and Yu, Y. (2014). *J. Mater. Chem. A* 2: 17595.
45 Chen, Y., Qu, B., Hu, L. et al. (2013). *Nanoscale* 5: 9812.
46 Wang, W., Chu, Q., Zhang, Y. et al. (2015). *New J. Chem.* 39: 6491.
47 Cheng, J., Lu, Y., Qiu, K. et al. (2014). *CrystEngComm* 16: 9735.
48 Tian, B., Liu, X., Yang, H. et al. (2003). *Adv. Mater.* 15: 1370.
49 Bilecka, I. and Niederberger, M. (2010). *Nanoscale* 2: 1358.
50 Zhang, M., Fan, H., Zhao, N. et al. (2018). *Chem. Eng. J.* 347: 291.
51 Qiu, Y., Pan, Z., Chen, H. et al. (2019). *Sci. Bull.* 64: 1348.
52 Lu, X.-F., Wu, D.-J., Li, R.-Z. et al. (2014). *J. Mater. Chem. A* 2: 4706.
53 Yang, D., Wang, Y., Wang, Q. et al. (2018). *Mater. Lett.* 213: 222.
54 Zhang, Z., Xu, Z., Yao, Z. et al. (2019). *J. Alloys Compd.* 805: 396.
55 Zhang, B., Sun, S., Shi, N. et al. (2020). *J. Alloys Compd.* 820: 153066.
56 Lu, X., Zheng, D., Zhai, T. et al. (2011). *Energy Environ. Sci.* 4: 2915.

57 Wang, H., Qing, C., Guo, J. et al. (2014). *J. Mater. Chem. A* 2: 11776.
58 Guan, C., Liu, J., Wang, Y. et al. (2015). *ACS Nano* 9: 5198.
59 Zhang, G., Xiao, X., Li, B. et al. (2017). *J. Mater. Chem. A* 5: 8155.
60 Chavez-Valdez, A., Shaffer, M.S.P., and Boccaccini, A.R. (2013). *J. Phys. Chem. B* 117: 1502.
61 Zhou, W., Cheng, C., Liu, J. et al. (2011). *Adv. Funct. Mater.* 21: 2439.
62 Guan, C., Wang, Y., Hu, Y. et al. (2015). *J. Mater. Chem. A* 3: 23283.
63 Cheng, C., Liu, B., Yang, H. et al. (2009). *ACS Nano* 3: 3069.
64 Bi, T., Du, J., Fang, H. et al. (2019). *Mater. Res. Express* 6: 115630.
65 Zhong, M., Li, Y., Yamada, I., and Delaunay, J.J. (2012). *Nanoscale* 4: 1509.
66 Wang, C.W., Yang, S., Jiang, H.B., and Yang, H. (2015). *Chem. Eur. J.* 21: 18024.
67 Jiang, Y. and Liu, J. (2019). *Energy Environ. Mater.* 2: 30.
68 Qiu, H., An, S., Sun, X. et al. (2020). *Chem. Eng. J.* 380: 122490.
69 Bandyopadhyay, P., Saeed, G., Kim, N.H., and Lee, J.H. (2020). *Chem. Eng. J.* 384: 123357.
70 Chen, W., Li, S., Chen, C., and Yan, L. (2011). *Adv. Mater.* 23: 5679.
71 Huo, J., Xue, Y., Cheng, C., and Guo, S. (2019). *Ionics* 26: 971.
72 Ji, H., Zhang, L., Pettes, M.T. et al. (2012). *Nano Lett.* 12: 2446.
73 Luo, J., Liu, J., Zeng, Z. et al. (2013). *Nano Lett.* 13: 6136.
74 Cao, X., Shi, Y., Shi, W. et al. (2011). *Small* 7: 3163.
75 Chao, D., Xia, X., Liu, J. et al. (2014). *Adv. Mater.* 26: 5794.
76 Dong, X.-C., Xu, H., Wang, X.-W. et al. (2012). *ACS Nano* 6: 3206.
77 Panda, P.K., Grigoriev, A., Mishra, Y.K., and Ahuja, R. (2020). *Nanoscale Adv.* 2: 70.
78 Salunkhe, R.R., Kaneti, Y.V., and Yamauchi, Y. (2017). *ACS Nano* 11: 5293.
79 Chen, W., Xia, C., and Alshareef, H.N. (2014). *ACS Nano* 8: 9531.
80 Chen, Y., Liu, B., Liu, Q. et al. (2015). *Electrochim. Acta* 178: 429.
81 Gao, L., Wang, X., Xie, Z. et al. (2013). *J. Mater. Chem. A* 1: 7167.
82 Jiao, Y., Liu, Y., Yin, B. et al. (2014). *Nano Energy* 10: 90.
83 Yang, P., Xiao, X., Li, Y. et al. (2013). *ACS Nano* 7: 2617.
84 Varma, S.J., Kumar, K.S., Seal, S. et al. (2018). *Adv. Sci.* 5: 1800340.
85 Zhang, Z., Chen, X., Chen, P. et al. (2014). *Adv. Mater.* 26: 466.
86 Lan, W., Zhang, X., Zhai, A. et al. (2019). *Chem. Eng. J.* 374: 181.
87 Liu, Y., Cao, X., Jiang, D. et al. (2018). *J. Mater. Chem. A* 6: 10474.
88 Xu, L.n., Zhang, H., Li, J. et al. (2019). *ChemElectroChem* 6: 5462.
89 Gao, Y., Chen, S., Cao, D. et al. (2010). *J. Power Sources* 195: 1757.
90 Yu, L., Zhang, G., Yuan, C., and Lou, X.W. (2013). *Chem. Commun.* 49: 137.
91 Cai, D., Wang, D., Liu, B. et al. (2014). *ACS Appl. Mater. Interfaces* 6: 5050.
92 Kong, D., Luo, J., Wang, Y. et al. (2014). *Adv. Funct. Mater.* 24: 3815.
93 Chen, S., Yang, G., Jia, Y., and Zheng, H. (2017). *J. Mater. Chem. A* 5: 1028.
94 Teng, Y., Yu, D., Li, Y. et al. (2020). *J. Electrochem. Soc.* 167: 020510.
95 Wang, H., Gao, Q., and Jiang, L. (2011). *Small* 7: 2454.
96 Jennings, J.R., Ghicov, A., Peter, L.M. et al. (2008). *J. Am. Chem. Soc.* 130: 13364.
97 Wu, C., Cai, J., Zhang, Q. et al. (2015). *ACS Appl. Mater. Interfaces* 7: 26512.
98 Xia, X., Tu, J., Zhang, Y. et al. (2012). *ACS Nano* 6: 5531.
99 Kim, J.H., Zhu, K., Yan, Y. et al. (2010). *Nano Lett.* 10: 4099.
100 Li, Q., Wang, Z.L., Li, G.R. et al. (2012). *Nano Lett.* 12: 3803.
101 Li, G., Li, J., Li, T. et al. (2019). *Sources* 425: 39.
102 Yu, Z., Duong, B., Abbitt, D., and Thomas, J. (2013). *Adv. Mater.* 25: 3302.
103 Cai, L., Rao, P.M., and Zheng, X. (2011). *Nano Lett.* 11: 872.

104 Jiang, J., Zhu, J., Ding, R. et al. (2011). *J. Mater. Chem.* 21: 15969.
105 Wu, M., Wang, Y., Xu, Y. et al. (2017). *ACS Appl. Mater. Interfaces* 9: 23647.
106 Han, D., Shen, Y., Pan, Y. et al. (2018). *Sustain. Energy Fuels* 2: 2115.
107 Zhang, Z., Zhang, H., Zhang, X. et al. (2016). *J. Mater. Chem. A* 4: 18578.
108 Jabeen, N., Hussain, A., Xia, Q. et al. (2017). *Adv. Mater.* 29: 1700804.
109 Wang, C., Xu, J., Yuen, M.-F. et al. (2014). *Adv. Funct. Mater.* 24: 6372.
110 Zheng, F., Wang, J., Liu, W. et al. (2020). *Electrochim. Acta* 334: 135641.
111 Liu, J., Jiang, J., Cheng, C. et al. (2011). *Adv. Mater.* 23: 2076.
112 Cai, D., Liu, B., Wang, D. et al. (2014). *J. Mater. Chem. A* 2: 4954.
113 Cheng, D., Yang, Y., Xie, J. et al. (2015). *J. Mater. Chem. A* 3: 14348.

9

Printed Flexible Supercapacitors

Yizhou Zhang[1,2] *and Wen-Yong Lai*[1,3]

[1]*State Key Laboratory of Organic Electronics and Information Displays (SKLOEI), Institute of Advanced Materials (IAM), Nanjing University of Posts & Telecommunications, 9 Wenyuan Road, Nanjing, 210023, Jiangsu, China*
[2]*Institute of Advanced Materials and Flexible Electronics (IAMFE), School of Chemistry and Materials Science, Nanjing University of Information Science & Technology, 219 Ningliu Road, 210044, Nanjing, Jiangsu, China*
[3]*Frontiers Science Center for Flexible Electronics (FSCFE), MIIT Key Laboratory of Flexible Electronics (KLoFE), Northwestern Polytechnical University, 127 West Youyi Road, Xi'an, 710072, Shaanxi, China*

List of Abbreviations

supercapacitor	(SC)
micro-supercapacitor	(MSC)
Asymmetric SC	(ASC)
polyethylene terephthalate	(PET)
2D metal carbides and nitrides	(MXenes)
metal-organic frameworks	(MOFs)
activated carbon	(AC)
graphene and carbon nanotube	(CNT)
electric double-layer capacitor	(EDLC)
polythiophene	(PTH)
polypyrrole	(PPy)
polyaniline	(PANI)
covalent organic framework	(COF)
polyvinyl alcohol	(PVA)
peracetic acid	(PAA)
polyethylene Oxide	(PEO)
poly(methylmethacrylate)	(PMMA)
Poly(vinylidene fluoride-hexafluoropropylene)	(PVDF-HFP)
propylene carbonate	(PC)
dimethyl formamide	(DMF)
ethylene carbonate	(EC)
P-phenylenediamine	(PPD)
drop-on-demand	(DOD)
cellulose nanofibril	(CNF)

Flexible Supercapacitors: Materials and Applications, First Edition. Edited by Guozhen Shen, Zheng Lou and Di Chen.

stereolithography	(SLA)
selective laser sintering	(SLS)
fused deposition modeling	(FDM)
direct ink writing	(DIW)

9.1 Overview of Printed Flexible Supercapacitor

Flexible supercapacitors (SCs) have been receiving heated attention recently, as shown in previous chapters. One typical application of flexible SCs is as energy storage units for flexible sensing circuits. In the context of the Internet of Things (IoT) where daily objects are interconnected through ubiquitous, low-cost, thin, and flexible sensors, miniaturized sensing circuits are in urgent need of health and environmental monitoring, industrial process tracking, and transport surveillance [1–3]. To harvest energy from ambient environments and store it in flexible energy storage devices provides a promising approach to power these circuits. Toward this end, SCs stand out for their long cycle lifetime, high power output, and simple structure. Moreover, SCs can be made from non-toxic materials that are safer and more stable, enabling safe usage in a wide range of conditions and easily disposed of [4, 5]. Particularly, micro-supercapacitors (MSCs) are attracting significant attention, especially for powering flexible electronic systems, due to ease of integration on chips or flexible substrates [6, 7]. Despite their relatively low energy density, MSCs are indispensable for applications that require high power and long cycle life, complementary to the more matured microbatteries.

The rise of flexible SCs presents new challenges in terms of device fabrication and has sparked significant interest toward the field of printed electronics, which means applying printing technologies from the graphic industry (e.g. screen, transfer, inkjet printing) to the fabrication of electronic devices. Printed electronics provides a whole range of low-cost, time-saving, versatile, and environmental-friendly technologies to fabricate SCs with desirable structures and integration possibilities, thus unleashing the full potential of SCs for future electronics.

Printed electronics falls within additive manufacturing, which features upscale production, design flexibility, and low materials consumption compared to traditional subtractive manufacturing [8, 9]. Depending on whether a template is used, printing techniques can be divided into digital printing, such as inkjet printing and 3D printing, and non-digital printing, such as screen and transfer printing [10]. These different printing techniques differ in ink fluidic properties, resolution, and scalability. So far, inkjet, screen, 3D, and transfer printing are the main methods used for printing flexible SCs.

The printing of SCs generally starts with dispersing active materials into a suspension or a viscous paste to meet the requirements of various printing and post-treatment methods. Therefore, materials chemistries and their solution processing remain the core issues for the successful printing of SCs. For printed MSCs, a typical footprint area is less than 1 cm^2, with the feature sizes in the micrometer range [11]. The challenge is to obtain well-defined electrodes with μm-level precision and good adhesion to the substrate without shorting between the positive and negative electrodes. The general process of printing SCs includes ink preparation, deposition, and solidification. Correspondingly, the critical issues involved in printing SCs mainly include the formulation of well-dispersed and printable inks and their stable and on-demand deposition on substrates; solidification of inks into desired 2D/3D structures, meanwhile ensuring good adhesion between electrodes and substrates; tuning of the wetting behavior of inks on substrates without compromising feature resolution nor causing shorts.

Besides the general advantages such as low cost, large area, design flexibility and easy integration, printing offers some unique advantages for the fabrication of SCs: (i) The 3D electrode with tunable architecture and the porous structure can be easily obtained using 3D printing, resulting in increased exposed active surface area and ionic/electronic transport rate; increased height/footprint ratio of the interdigitated electrodes can be easily achieved using screen printing and 3D printing, both can maximize the specific energy/power density of SCs. (ii) The design flexibility enables easy optimization of SC performances, for instance, to increase energy density by making asymmetric device configurations with an optimized weight ratio between the two electrodes. (iii) Materials that are difficult to process using traditional manufacturing techniques (e.g. ceramics) can be patterned into electrodes through printing. (iv) Through printing, prototypes are much easier to be manufactured and tested, both in terms of materials selection and device configuration design; thus research process can be greatly accelerated without wasting large quantities of materials, some of which are expensive to obtain. In this sense, printing is not only a fabrication technique but also an enabling tool for new applications as well as for industrialization.

In this chapter, we first introduce the structure of SCs, especially the novel structures facilitated by printing; then the printable materials involved in the fabrication of printed SCs are discussed, followed by the recent progress of printed SCs in terms of different printing methods (mainly inkjet, transfer, screen, and 3D printing, Figure 9.1); next, come the applications of printed SCs in integrated printed systems; finally, we summarize the chapter and present our perspectives on the challenges and future research directions of printing SCs.

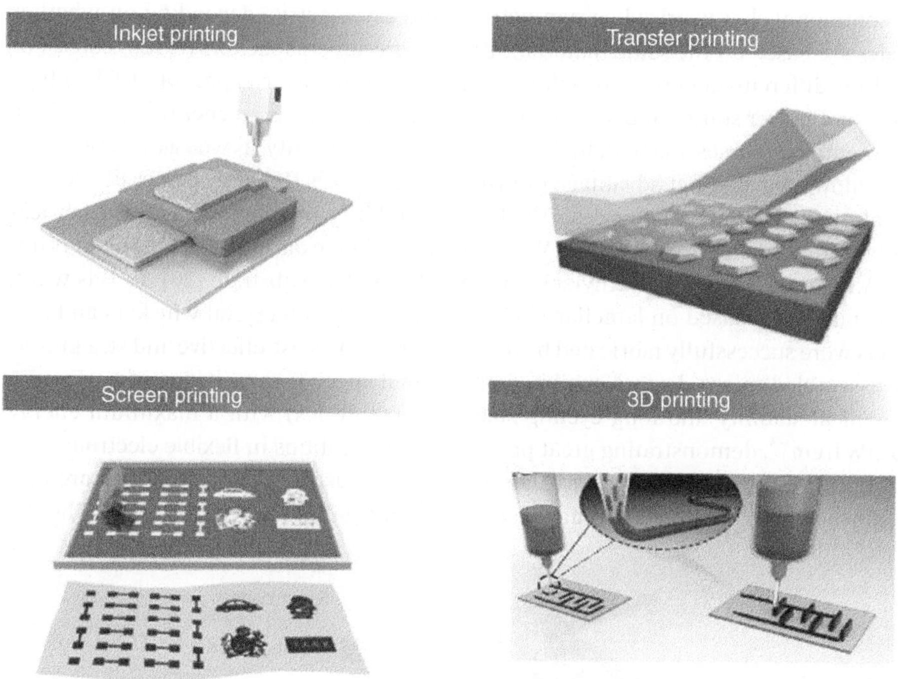

Figure 9.1 The main printing methods for flexible supercapacitors: inkjet [12], transfer [13], screen [14] and 3D printing [15]. *Source:* Reproduced with permission. Reproduced with permission. © 2018, Elsevier; Reproduced with permission. © 2006, Nature Publishing Group; Reproduced with permission. © 2020, WILEY-VCH; Reproduced with permission. © 2019, American Chemical Society.

9.2 Devices Structure of Printed SCs

Most printed SCs adopt either sandwich-structured or planar interdigitated configurations, as shown in Figure 9.2, with a preference for the latter. The small interspacing between the interdigitated electrode arrays can help prevent electrical short-circuiting and decrease the ionic transport resistance [16]. Meanwhile, the planar configuration can avert displacement between different layers as often observed in the sandwiched SCs and therefore is better suited for flexible applications. It is worth noting that although the interdigital patterns are used in most of the MSC literature, other designs such as spiral-shaped electrodes have been shown to lead to larger energy/power densities [17]. More advanced electrode designs are expected to emerge, assisted by the design freedom offered by printing. Lastly, sandwiched configuration is often impractical in the context of printed SCs because it is difficult to deposit an electrode layer on top of the gel or liquid electrolyte. By contrast, planar MSCs can be easily integrated into on-chip microelectronic devices due to the in-plane configuration and the small sizes. Compared to traditional microfabrication methods such as photolithography and vacuum deposition, printing methods provide a series of feasible and effective manufacturing techniques for low-cost and mass production of MSCs with interdigitated electrode configuration. As a typical example, inkjet printing was used to construct hierarchical nanocoral-structured interdigitated electrodes for flexible MSCs [18]. The versatile printing process enables the deposition of precisely designed and artistically pleasing electrode patterns. Furthermore, the areal specific capacitance of the MSCs was tunable by changing the interspace between the adjacent two electrodes [19].

SCs are also commonly characterized as symmetrical and asymmetric, depending on whether the two electrodes are based on the same material. Asymmetric supercapacitors (ASCs) consist of electrodes based on different materials, usually an electric double-layer capacitor (EDLC)-type electrode (anode) as a power source and a Faradaic electrode (cathode) as an energy source. Thus, ASCs have the potential of possessing both high energy and power density, as well as a large operational voltage window. One distinct advantage of printing methods is that they can easily deposit different materials with different mass loading/thicknesses, holding great potential for the fabrication of ASC with optimized electrode weights. We reported the first example of using inkjet printing to fabricate ASCs on a flexible polyethylene terephthalate (PET) substrate [20]. In this work, flexible all-solid-state ASCs based on lamellar $K_2Co_3(P_2O_7)_2{\cdot}2H_2O$ nanocrystal whiskers and graphene nanosheets were successfully fabricated by inkjet printing in a cost-effective and straightforward way. The assembled micro-devices exhibited an optimal specific capacitance of $6.0\,F\,cm^{-3}$, good rate/mechanical stability and long cycling stability (5000 cycles) with a maximum energy density of $0.96\,mW\,h\,cm^{-3}$, demonstrating great promise for applications in flexible electronics.

To maximize areal energy density, thicker (high height/width aspect ratio) electrodes are preferred, which can be easily obtained through using 3D printing and screen printing. The 3D

Figure 9.2 Common device configurations for printed SCs: sandwiched (left) and planar with interdigitated electrodes.

structure will expose electrodes' surfaces to the electrolyte in three dimensions, facilitating charge storage processes and leading to both larger effective surface area and enhanced active mass utilization. Moreover, the current collector with a 3D porous structure can also lead to lower resistance. Although thick electrodes may reduce power density, the largely increased energy density is usually more critical for applying SCs. Through 3D printing, thick yet porous electrodes, which can simultaneously boost the energy and power density, can be facilely obtained.

To evaluate the performance of printed SCs, the gravimetric properties are often inappropriate, particularly for MSCs [1]. The volumetric properties of electrodes can also be misleading, especially if they are obtained from very thin electrodes. The outstanding volumetric values are likely to significantly decrease if the electrode thickness is largely increased, which is often the case for printed SCs. Therefore, the areal properties normalized to the footprint area are often preferred.

9.3 Printable Materials for SCs

In the context of printing SCs, material parameters, including processability, electrochemical performance, and long-term stability, need to be considered collectively with the intended printing methods in mind. In this section, we will discuss printable electrodes, current collectors, electrolytes, and flexible substrates.

9.3.1 Electrodes Materials

The electrode is arguably the most critical factor in determining the electrochemical performance of SCs. Therefore, the development of printable electrode materials with high performance at low cost is the focus of most research on printed SCs. As discussed in previous chapters, ideal SC electrode materials should possess high specific surface area, tunable hierarchical porosity, abundant electroactive sites, high stability, and high electronic conductivity [21]. To date, electrode materials for printed SCs are mainly carbon-based materials, conducting polymers, metal oxides, 2D metal carbides and nitrides (MXenes), metal-organic frameworks (MOFs), etc. [22] In practice, electrode materials usually need to be combined with current collectors/conductive additives to facilitate charge transfer. Metal nanomaterials are the most widely used current collectors for printed SCs due to their high electrical conductivity and ease of being formulated into inks. One advantage of the printing method is the ease to fabricate 3D current collectors with hierarchical structures, which can improve capacitance, rate capability, flexibility and cyclic stability.

9.3.1.1 Carbon-Based Materials

Carbon-based materials possess large specific surface area, good conductivity, low cost, and good wettability to electrolytes, making them ideal EDLC electrode materials. Common ones include activated carbon (AC), graphene and carbon nanotubes (CNTs) [23].

Activated carbon materials can be easily obtained from various kinds of biomass; they are low cost and stable, making them widely applied in commercial SCs. ACs tend to have a hierarchically porous structure with wide pore-size distribution, including micropores and mesopores. Since the specific surface area, pore-size distribution, pore density, and surface functional groups all affect the electrochemical performance of ACs, the optimization of AC electrode materials can be achieved through tuning of pore-size distribution and surface modification. For printed SCs, AC inks/slurries should be developed with specific formulation and rheological properties. In one example, an inkjet ink was successfully prepared by mixing activated carbon powders with polytetrafluoroethylene (PTFE) polymer binders in ethylene glycol stabilized with a surfactant [24]. MSCs were then fabricated by inkjet printing the activated carbon ink on patterned gold current collectors; the device showed an excellent capacitive behavior over a wide potential range of 2.5 V and a cell capacitance of 2.1 mF cm$^{-2.}$ In another example, an aqueous screen-printable ink containing 0.2 g ml^{-1} of activated carbon was developed, SC fabricated from this ink demonstrated capacitance as high as 0.51 F cm^{-2} at 10 mV s^{-1} [25].

Graphene has aroused great attention since Novoselov and Geim isolated it from graphite in 2004 [26]. It possesses many excellent properties that make them ideal SC electrode materials, including a large theoretical specific surface area, high electrical conductivity, excellent thermal and chemical stability, superior mechanical flexibility, and wide potential window [27]. Among the various methods developed to prepare graphene, solution-based chemical exfoliation is the most common mass production technique of printable graphene inks. For example, stable graphene inks with high concentration and compatibility for inkjet printing were prepared through a simple and effective polymer stabilization and solvent exchange technique. The fully printed MSCs using this graphene ink exhibited an areal capacitance of around 0.59 mF cm^{-2} [28]. CNTs are another group of carbon allotropes that can be regarded as curled graphene nanosheets. Because of their large specific surface area, high electrical conductivity, hierarchical pore structure, and chemical stability, CNTs inks have also been formulated for printed SCs application [29].

9.3.1.2 Metal Oxides

Metal oxides are the most widely explored pseudocapacitive electrode materials for SCs [30]. Common metal oxides include NiO, CoO, $Co(OH)_2$, MnO_2, $NiCo_2O_4$, and V_2O_5 [31]. However, metal oxides' structure tends to collapse after long cycling because of their redox reaction with electrolytes. Additionally, the poor electrical conductivity of metal oxides can also decrease the electrochemical performance of SCs. Thus, binder and conductive additives are often needed to prepare metal oxides inks.

Conducting Polymers. Conductive polymers are another kind of widely explored pseudocapacitive electrode materials for SCs; they usually have higher electrical conductivity than metal oxides; thus, they can be directly used as electrodes for SCs without extra conductive additives [32]. Moreover, their solution processability makes them compatible with printing methods. The commonly used conductive polymers are polythiophene (PTH), polypyrrole (PPy), and polyaniline (PANI), among others, all of which have been used for various printing applications.

9.3.1.3 2D Transition Metal Carbides, Nitrides, and Carbonitrides (MXenes)

MXenes are a new member of the 2D materials family that dates back to 2011, and has since evolved into a worldwide joint effort dealing with a material family of almost 30 members. MXenes possess outstanding electrochemical, electronic, optical, and mechanical properties and are promising materials in many applications, including energy storage, electronics, optoelectronics, biomedicine, sensors, and catalysis [33]. MXenes possess high electrical conductivity and structures that allow easy ion intercalation between layers, leading to excellent performance as SC electrodes. Compared to other 2D materials, MXenes possess excellent dispersion quality, negative surface charge, excellent hydrophilicity, and solution-based synthesis, making them particularly suitable as inks for printing applications. As a result, we have seen a booming development of MXene-based printed SCs. MXene is a perfect example that illustrates printing is more than merely a device fabrication method, but instead a versatile approach to broaden the application range of materials, and as such, the performance of printed devices are influenced by the whole synthesis, modification, and application process cycle (Figure 9.3).

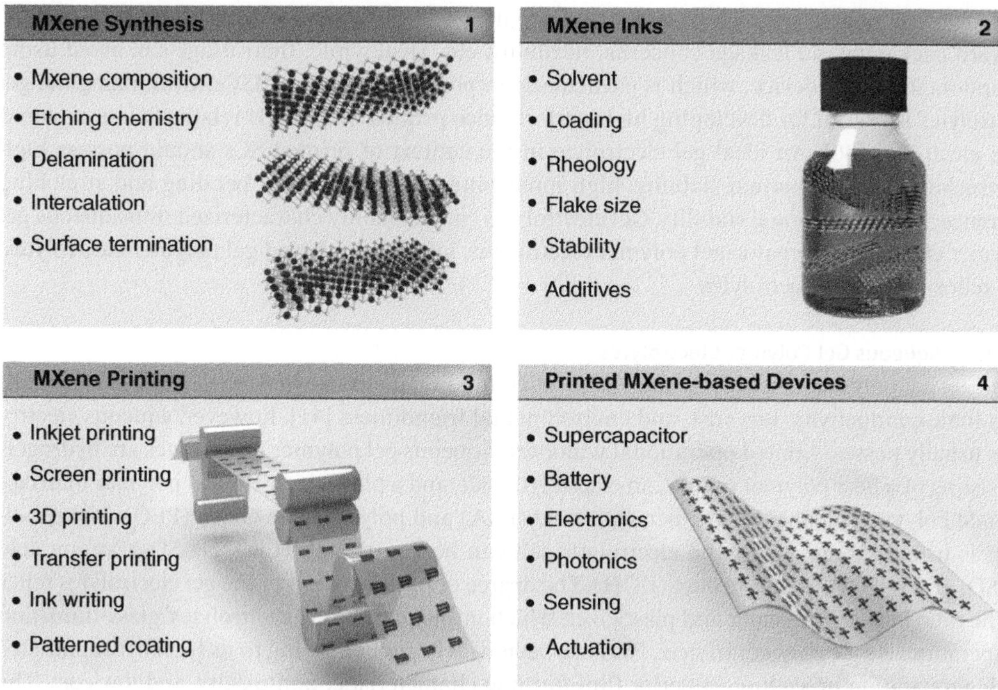

Figure 9.3 MXene is chosen here as an example to show the critical elements involved in the realization of printed devices. Synthesis, ink formulations, printing process, the design and fabrication of devices all have to be collectively optimized. *Source:* Reproduced with permission [33]. © 2020, Wiley-VCH.

9.3.1.4 Metal-Organic Frameworks (MOFs)

MOFs are porous coordination polymers, promising as SC electrode materials because of their high porosity, controlled pore size, diverse structures, chemical stability, and tunable chemical compositions [34–36]. It has been observed that MOFs can have a specific surface area as high as $46\,000\,m^2\,g^{-1}$, comparatively uniform pore sizes, and large pore volumes, showing great promise in the field of energy storage. More recently, covalent organic frameworks (COFs) have also shown promise as SC electrodes. COFs are an interesting class of covalent porous crystalline polymers that enable the elaborate integration of organic building blocks into an ordered configuration with atomic precision [37]. The 2D or 3D crystalline structures of COFs are well-defined and predictable because of the strong covalent bonds formed between the building units through various synthetic organic reactions [38]. The strong covalent bonds typically include B–O, C–N, B–N, and B–O–Si. COFs also have controllable pore sizes, high surface area, and superior flexibility in molecular designs, showing great potential in energy storage. Due to the organic and soluble nature, both MOFs and COFs are promising materials for printed SC application.

9.3.2 Electrolytes

Electrolytes have a significant influence on the electrochemical properties of devices, especially in terms of voltage window, rate capability, and cycling stability [39]. Parameters to characterize electrolytes mainly include ionic conductivity, electrochemical window, operating temperature range, stability, and safety. So far, printed SCs mostly use gel electrolytes. The advantages of gel electrolytes include relatively high ionic conductivity, facile fabrication process, cheap and straightforward packaging, no leakage concerns, flexibility, etc. Meanwhile, their usage can avoid using separators in the SC device, which is particularly important for planar MSCs. Optimizing the gel electrolytes is crucial for developing high-performance printed SCs, but is relatively less explored than electrodes [40]. An ideal gel electrolyte in the context of printed SCs should possess high electrochemical and thermal stability, high ionic conductivity, excellent bending and stretching tolerance, and dimensional stability. Gel electrolytes can be mainly characterized into aqueous gel polymer electrolytes, organic gel polymer electrolytes, ionic liquid-based gel polymer electrolytes, and redox-active gel electrolytes.

9.3.2.1 Aqueous Gel Polymer Electrolytes

Aqueous gel polymer electrolytes are the most common gel electrolytes for SCs because of their high ionic conductivity, low cost, and environmental friendliness [41]. however, aqueous electrolytes usually possess limited operational windows. Aqueous gel polymer electrolytes are hydrogels consisting of a host polymer matrix, an electrolytic salt, and a plasticizer. Typical polymer matrixes include polyvinyl alcohol (PVA), peracetic acid (PAA) and polyethylene Oxide (PEO). The plasticizer is usually water, while the electrolytic salt can be neutral salt (LiCl, Na_2SO_4), strong acid (H_2SO_4 and H_3PO_4), or strong base (KOH). The degree of plasticization of the gel electrolytes relies on the ratio between polymer and plasticizer, which influences the gel electrolytes' glass-transition temperature. As for polymer matrix, PVA has been widely applied owing to its high hydrophilicity, facile preparation procedure, superior film-forming characteristic, nontoxicity, and low cost. The choice of electrolyte salt depends on the electrode materials. It is worth noting that acid or alkaline solutions can cause damage to many dispensing nozzles; thus, care should be taken when printing aqueous electrolytes.

9.3.2.2 Organic Gel Polymer Electrolytes

Despite the excellent attributes of the aqueous gel polymer electrolytes, their narrow operating window limits their extensive applications. This problem can be overcome by using organic gel polymer electrolytes. The preparation methods for organic gel polymer electrolytes significantly affect the ionic conductivity and mechanical properties of the system. Generally, organic gel polymer electrolytes are prepared by physically blending a polymer with high molecular weight, e.g. poly(methylmethacrylate) (PMMA) and Poly(vinylidene fluoride-hexafluoropropylene) (PVDF-HFP), with a conducting salt in a non-aqueous solvent system followed by gelation. Organic solvents, such as propylene carbonate (PC), dimethyl formamide (DMF), ethylene carbonate (EC), and their mixtures have been commonly employed as plasticizers to broaden the operating voltage range [42]. The cell voltage could be enhanced to 2.5–3 V, which are significantly larger than those using aqueous gel polymer electrolytes, which is beneficial for improving the energy density of SCs.

9.3.2.3 Ionic Liquid-Based Gel Polymer Electrolytes

Compared with aqueous and organic electrolytes, ionic liquid-based gel electrolytes show some extra advantages, such as nonvolatility, high ionic conductivities, nonflammability, wide operating voltage windows (up to 3.5 V) as well as superior mechanical compliance. Thus, they have shown great promise for application in flexible and stretchable SCs. In the context of printed SCs, they have additional advantages: superior stability and no danger of corrosion to the printing apparatus. Similar to aqueous gel polymer electrolytes and organic gel polymer electrolytes, a variety of polymer hosts have been studied for ionic liquid gel electrolytes, such as PVA, PVDF–HFP and PEO [43].

9.3.2.4 Redox-Active Gel Electrolytes

Redox-active species can be added to the gel electrolyte to improve the capacitance and the energy density of SCs. Redox additives can introduce additional pseudocapacitance via reversible faradaic reactions and fast electron transfer at the electrode-electrolyte interface, leading to improvement of the specific capacitance [44]. A large number of redox couples such as $K_3Fe(CN)_6$ [45], iodides (KI) [46], and Na_2MoO_4 [47], and organic redox mediators such as P-phenylenediamine (PPD) [48], hydroquinone [49], etc., have been used to prepare redox-active gel electrolytes.

9.3.3 Flexible Substrates

The flexible substrate is another essential component for printed flexible SCs. The parameters to evaluate flexible substrates in the context of printed SCs include mechanical flexibility, wettability, thinness, surface smoothness, density, heat resistance, cost, and even transparency in some applications [50].

Wettability deserves special attention for printed SCs because of its significant influence on the morphology of printed films. For example, it is challenging to print aqueous ink on hydrophobic surfaces because the aqueous solution is inclined to contract, thus unable to effectively spread on the hydrophobic substrate, leading to non-uniform film morphology [51]. To solve this problem, pre-treatments of the substrate, such as plasma treatment or chemical modification, are usually performed before printing. Furthermore, as heating is typically required during the printing processes, heat resistance is another crucial parameter for evaluating substrates, particularly for polymer-based plastic substrates. Additionally, mechanical flexibility is usually preferred for printed SCs, especially when the SCs are integrated into flexible circuits through a roll-to-roll printing process. Common flexible substrates for SCs include metal foil, PET, polyimide (PI), polydimethylsiloxane (PDMS), paper, etc. Each kind of substrates possesses unique attributes and advantages as well as shortcomings. Here we will briefly introduce these common substrates for printed SCs.

9.3.3.1 Metal Substrates

Metal foils, such as gold (Au), copper (Cu), aluminum (Al), and nickel (Ni) have been widely utilized as flexible substrates for SCs because of their superior electrical conductivity and good mechanical ductility [52]. Metal foils can simultaneously act as the current collector due to the excellent conductivity. Metal foils are usually much thicker and heavier than the active materials coated on their surface, which may decrease the gravimetric specific capacitance of the SC devices. Moreover, it is difficult to effectively transport charges from the thick electrode materials to the surface of the conducting substrates, which can decrease the rate capability and capacitance of SCs based on thick electrodes. 3D structured metal mesh or foams have been developed as substrates for flexible SCs to increase the charge transport properties and the mass loading of the active materials [52, 53].

9.3.3.2 Synthetic Polymer-Based Substrates

Synthetic polymer-based substrates are extensively used in printed SCs because of their low cost, thinness, lightweight, and excellent mechanical deformability [54]. However, different from metal foils, common plastic substrates are non-conductive; SCs based on them need current collectors. PET film is presently the most commonly used plastic substrate owing to its easy availability, moisture barrier property, the high optical transparency of above 90%, and relatively low cost [55]. Nevertheless, the heat resistance of PET is comparatively low, and therefore, the temperature in the printing process must be controlled below around 130 °C. Polyethylene naphthalate (PEN) has better heat resistance while possessing higher cost and lower transparency [56]. PI has the best heat resistance with the highest cost and lowest transparency [57].

9.4 Fabrication of Flexible SCs Using Various Printing Methods

9.4.1 Inkjet Printing

Inkjet printing is a digital high precision printing technique widely used in people's daily lives. Therefore, it is no surprise that inkjet is a popular technique to deposit materials onto various flexible substrates such as paper and polymer films [58]. Most inkjet printers for the fabrication of electronic devices are based on the drop-on-demand (DOD) mode, which can be divided into thermal and piezoelectric types based on the mechanism through which inks are jetted out of nozzles. In the thermal inkjet process, a resistive element is heated and then creates a thin film, which in turn forms an expanding bubble that injects droplets through the nozzle onto the substrate. By contrast, during the piezoelectric process, an electrical field is applied to the piezoelectric materials. It creates a distortion, which then generates pressure to the chamber, pushing ink through the nozzle onto the substrate.

The key issues that concern researchers in inkjet printing are mainly droplet formation and morphology optimization of printed patterns. Even tiny printing instabilities, such as satellite drops or tails, could decrease both the shape fidelity and resolution of the printed patterns. Therefore, inks need to be carefully optimized to ensure smooth jetting. The inverse Ohnesorge number Z, which is defined as

$$Z = \frac{(\alpha\rho\gamma)^{1/2}}{\eta}$$

is often used to describe the fluidic properties of inks and thus the ink printability, where α is the nozzle diameter, ρ is the density, γ is the surface tension, η is the viscosity of the fluid. It is suggested that the optimal Z value is in the range of 4–14, considering the characteristics such as single droplet formability, position accuracy, and maximum allowable jetting frequency [59]. However, there are also many reports of inks with Z values outside of this range that print smoothly [60–62].

Ink formulation engineering is an efficient way to control the droplet formation and, thus, the morphology of the printed patterns. The final resolution and morphology of the printed patterns are also influenced by the printing parameters, interface engineering, and post-treatments [63]. Inkjet printing has the following advantages: (i) Low viscosity inks such as water-based inks can be inkjet-printed. (ii) As a non-contact printing technique, the print head does not need to contact the substrate and thus avoid cross-contamination. As a digital technique, inkjet printing has high design flexibility. (iii) Only a small amount of ink is required: the typical volume to fill an inkjet printer cartridge is in the range of a few milliliters, and therefore little waste is produced, which helps save expensive materials and reduce costs. Meanwhile, inkjet printing also presents several limitations: (i) Stable ink (proper fluidic physical properties, filler sizes) is difficult to prepare to meet the strict requirements of inkjet printers. (ii) Standard inkjet printers have low printing speed and throughput.

Inkjet printing is among the first technique for printed electronic applications; it has already been incorporated into the industrial-scale production lines of printed organic light-emitting diodes and organic solar cells (OSCs). In comparison, inkjet-printed flexible SCs are mainly in the laboratory stage. As an example, we have reported a flexible and transparent SC through inkjet printing [64]. By inkjet-printing PEDOT: PSS (poly(3,4-ethylenedioxythiophene) polystyrene sulfonate)/Ag grid electrodes, flexible and transparent all-solid-state SCs with a specific areal capacitance of 4.52 $mF\,cm^{-2}$ and energy density of 0.59 $mW\,h\,cm^{-2}$ were fabricated. To show the design flexibility endowed by inkjet printing, SCs with various patterns, including logos, flowers and dragons, were fabricated. The printed SCs possessed superior flexibility, aesthetic appearance, and relatively high optical transparency, showing great promise for flexible and wearable electronic devices. The conductivity of PEDOT:PSS can be improved by adding conductive enhancers such as Ag nanoparticles and CNTs to the ink [65]. The highly conductive PEDOT: PSS-CNT/Ag composite ink served as both interdigitated electrodes and current collector, due to the synergistic effects formed between the PEDOT:PSS-CNT/Ag composite film and the paper substrate, the printed all-solid-state MSCs (Figure 9.4a) delivered high volumetric specific capacitance of 23.6 $F\,cm^{-3}$.

Carbon materials represented by graphene are popular electrode materials for inkjet-printed SCs [70]. For instance, Li et al. prepared MSCs by inkjet printing liquid exfoliated graphene inks into interdigitated electrodes on top of printed Ag current collectors [28]. The high-concentration graphene ink was prepared by a distillation-assisted solvent exchange and polymer stabilization process. The printed MSC showed an areal specific capacitance of 0.19 $mF\,cm^{-2}$ at 100 $mV\,s^{-1}$, owing to the high conductivity and large surface area of graphene. The device performance can be further enhanced by printing more layers of graphene, i.e. the areal specific capacitance could reach 0.59 $mF\,cm^{-2}$ with 8 layers of graphene. Li et al. reported a scalable fabrication of graphene-based flexible SCs on various substrates using inkjet printing [66]. The fully printed SCs exhibited a high areal capacitance of ~0.7 $mF\,cm^{-2}$. Furthermore, the printing method enables the effective connection of SCs in parallel and/or in series at multiple scales (Figure 9.4b–d). Remarkably, large-scale SC arrays comprising more than 100 devices were prepared on Kapton, which can be charged up to 10 V at a scan tare of 1000 $mV\,s^{-1}$. The devices exhibited a high energy density of 1 $mWh\,cm^{-3}$ at the power density of 0.1 $W\,cm^{-3}$. Hersam et al. reported MSCs based on inkjet-printed interdigitated graphene electrodes as both electrodes and current collectors [67]. An all-solid-state MSC was fabricated with the volumetric capacitance of 9.3 $F\,cm^{-3}$ at a current density of 0.25 $A\,cm^{-3}$ and a volumetric energy density of 1.29 $mW\,h\,cm^{-3}$ at a power density of 278 $W\,cm^{-3}$, respectively. The MSC also showed excellent operational stability with minimal capacitance decay after 10 000 cycles. Furthermore, the graphene electrodes can be inkjet-printed at a large scale, which offers compelling prospects for practical utility (Figure 9.4e). Wang et al. inkjet-printed the 2D δ-MnO_2 nanosheets to construct all-solid-state MSCs (Figure 9.4f) [68]. A highly concentrated water-based δ-MnO_2 nanosheets ink was developed with an average lateral size of 89 nm and a thickness of 1 nm. By engineering the formulation of the δ-MnO_2 ink, the printed δ-MnO_2 patterns showed uniform morphology without the undesired "coffee-ring" effect. The MSCs exhibited a volumetric capacitance as high as 2.4 $F\,cm^{-3}$, and an energy density of 1.8×10^{-4} $Wh\,cm^{-3}$ at a power density of 0.018 $W\,cm^{-3}$. Furthermore, the fabricated MSCs showed good mechanical flexibility and excellent cycling stability with capacitance retention of 88% after 3600 charge-discharge cycles.

Lee et al. reported an interesting study of fully inkjet-printed solid-state SCs on a piece of conventional A4 paper [69]. They inkjet-printed activated carbon/ CNTs ink as electrodes, and ionic liquid/ultraviolet-cured triacrylate polymer as the electrolyte. As illustrated in Figure 9.4g and h, cellulose nanofibril (CNF)-mediated nanoporous mats were first printed on the top of an A4 paper to make high-resolution patterns. Then SWNT (Single-walled Carbon Nanotubes)/AC inks were inkjet-printed on the top of the CNF-paper substrate, followed by printing Ag NW (nanowire) inks. To improve the electrical conductivity of the electrode, a photonic sintering technique was

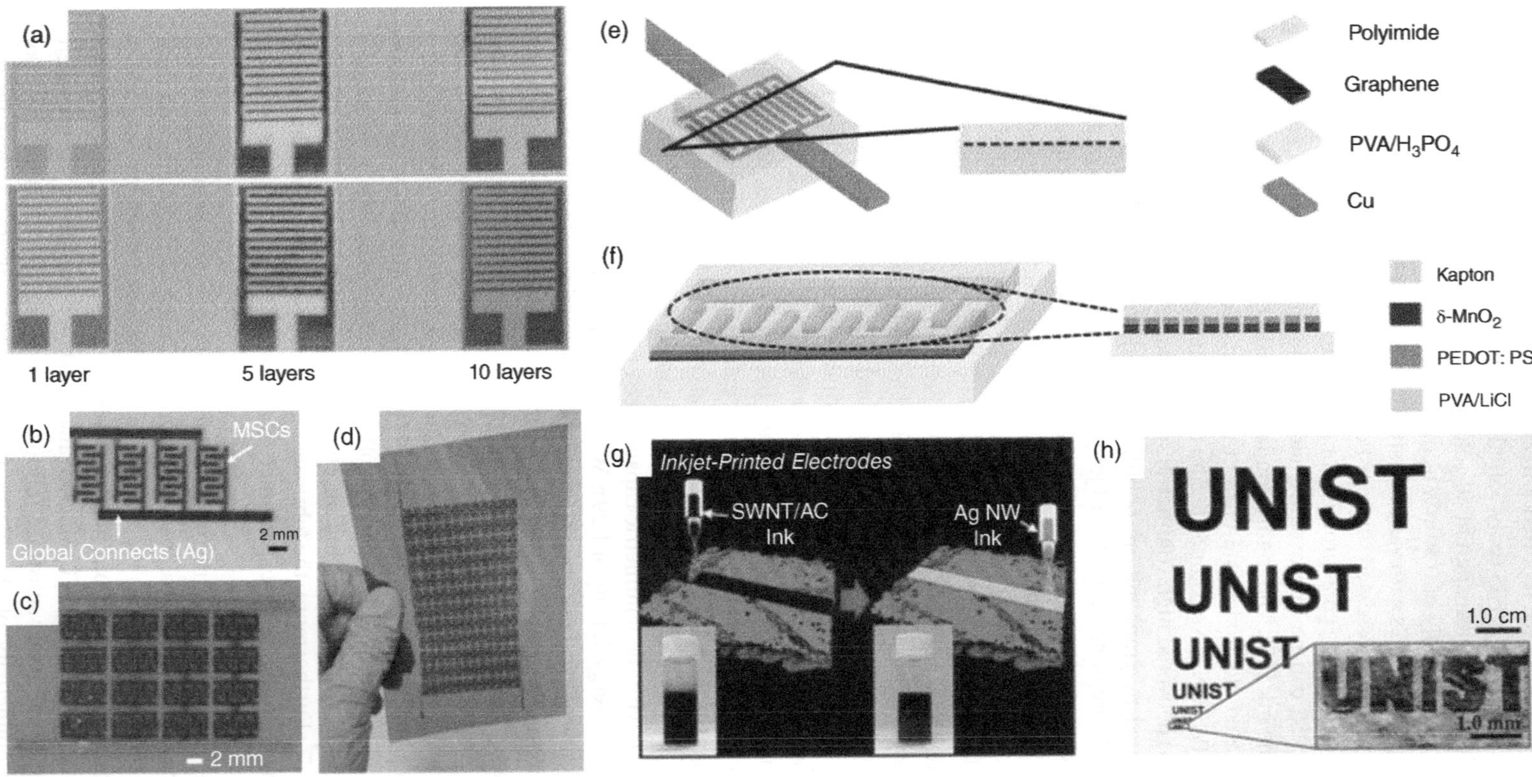

Figure 9.4 Representative works of inkjet-printed flexible SCs. (a) Optical images of SCs with the interdigitated PEDOT:PSS-CNT film before (top) and after (bottom) overprinting of the Ag film. *Source:* Reproduced with permission [65]. © 2016, The Royal Society of Chemistry. Photographs of (b) 4 parallel-connected SCs on glass, (c) 4 parallel rows of 4 series-connected SCs on Kapton. (d) Large-scale integration of fully printed SCs on Kapton. *Source:* Reproduced with permission [66]. © 2017, American Chemical Society. (e) Schematic diagram of a SC with interdigitated structure, including both a 3D representation and 2D cross-section. *Source:* Reproduced with permission [67]. © 2016, WILEY-VCH. (f) Schematic diagram of inkjet printed MSC with interdigitated electrode configuration. *Source:* Reproduced with permission [68]. © 2018, Elsevier. (g) Schematic illustration of the fabrication procedure of the inkjet-printed SC electrodes. The insets are the photographs of an electrochemically active ink (SWNT/AC in water) and an electrically conductive ink (Ag NWs in water/IPA = 1/1 (v/v)). (h) Photograph of the inkjet-printed, letter ("UNIST")-shaped electrodes of different sizes (ranging from the centimeter to micrometer scale) on A4 paper. *Source:* Reproduced with permission [69]. © 2016, The Royal Society of Chemistry.

employed for the post-treatment of electrodes. Afterward, the electrolyte ink was inkjet-printed on the top of the treated electrode and then dried at 80 °C to remove the residual water. Consecutively, a UV-curing reaction was used to solidify the gel polymer electrolyte. The fully inkjet-printed SC exhibited excellent cycling stability after more than 10 000 charge/discharge cycles.

9.4.2 Screen Printing

Screen printing is another common mass-printing method borrowed from the graphic printing industry to the manufacturing of flexible SCs. It works by squeezing ink through a stencil screen to be deposited onto the substrate. A wide variety of functional inks and substrates can be used for screen printing due to its simplicity and versatility [71, 72]. There are mainly two types of screen printing, i.e. the flat-bed and the rotary modes. The former is a stepwise process in which the patterned screen is lifted up to enable changing or removing the substrate manually. Thus, the flat-bed screen technique enables facile multilayer printing. It also shows great potential for upscale printing and printing of high aspect-ratio structures. On the other hand, the rotary technique works by printing ink through a polyester screen cylinder and a perforated metal onto the substrate. Since the screen rotates at the same speed as the substrate, the ink is continuously pressed onto the substrate by the stationary squeeze, making a full print cycle upon every rotation. Therefore, the printing speed of the rotary technique is much higher than the flat-bed one. The wet thickness of the screen-printed pattern is determined by a number of parameters, such as the concentration of functional ink, the mesh count per inch, the emulsion thickness, and the fraction of ink transferred onto the substrate. The quality of the screen-printed patterns depends on the screen mesh, the printing parameters, the substrates, and the ink viscosity.

Screen printing has the following advantages: (i) Screen printing is suitable for high aspect ratio patterns and thick films. (ii) Paste movements can be easily controlled during printing, resulting in better printing shape fidelity. (iii) The overall cost of screen-printing techniques is lower than other techniques. On the other hand, due to the high viscosity of the inks, the screen-printed films suffer from high roughness and low resolution, which limits the application of screen printing in printed electronics, especially when accurate overlays are required.

Screen printing is routinely used for printing on textiles and clothing; thus, it is no surprise that screen-printed textile SCs have received heated attention. Jost et al. demonstrated an "energy textile" which can be integrated into the garment by screen-printing porous carbon electrode materials into woven cotton and polyester fabrics in order to prepare a flexible and lightweight fabric SC [73]. The porous structure of these substrates facilitates ion transfer between the electrodes and is beneficial for SC applications. The screen-printed porous carbon electrodes achieved a high specific capacitance of $90\,F\,g^{-1}$ with an areal capacitance of $0.43\,F\,cm^{-2}$.

Wang et al. demonstrated an ASC by screen-printing MWCNTs (Multi-walled Carbon Nanotubes)-MnO_2 anodes and MWCNTs-MoO_3 cathode onto the screen-printed current collectors made of commercial carbon nanoparticles. They extended their approach to various substrates, including clothes, PET, and paper [74]. The fully printed ASC showed a wide operational potential window of 1.7 V and a high energy density of $11.04\,mW\,h\,cm^{-3}$ at a power density of $614.6\,mW\,cm^{-3}$. Furthermore, the device exhibited excellent cycling stability, with ~91.3% of initial capacitance retention after 5000 charge/discharge cycles. Xu et al. reported thin film SCs based on screen-printed Graphene/PANI hybrid inks (Figure 9.5a) [75]. Graphene served as a stable and underlying conductive network for PANI, which acted as a spacer to separate graphene nanosheets. Due to the synergistic effect between graphene and PANI, the printed SC showed a high specific capacitance of $269\,F\,g^{-1}$, an energy density of $9.3\,Wh\,kg^{-1}$, and a power density of $454\,kW\,kg^{-1}$. The devices also

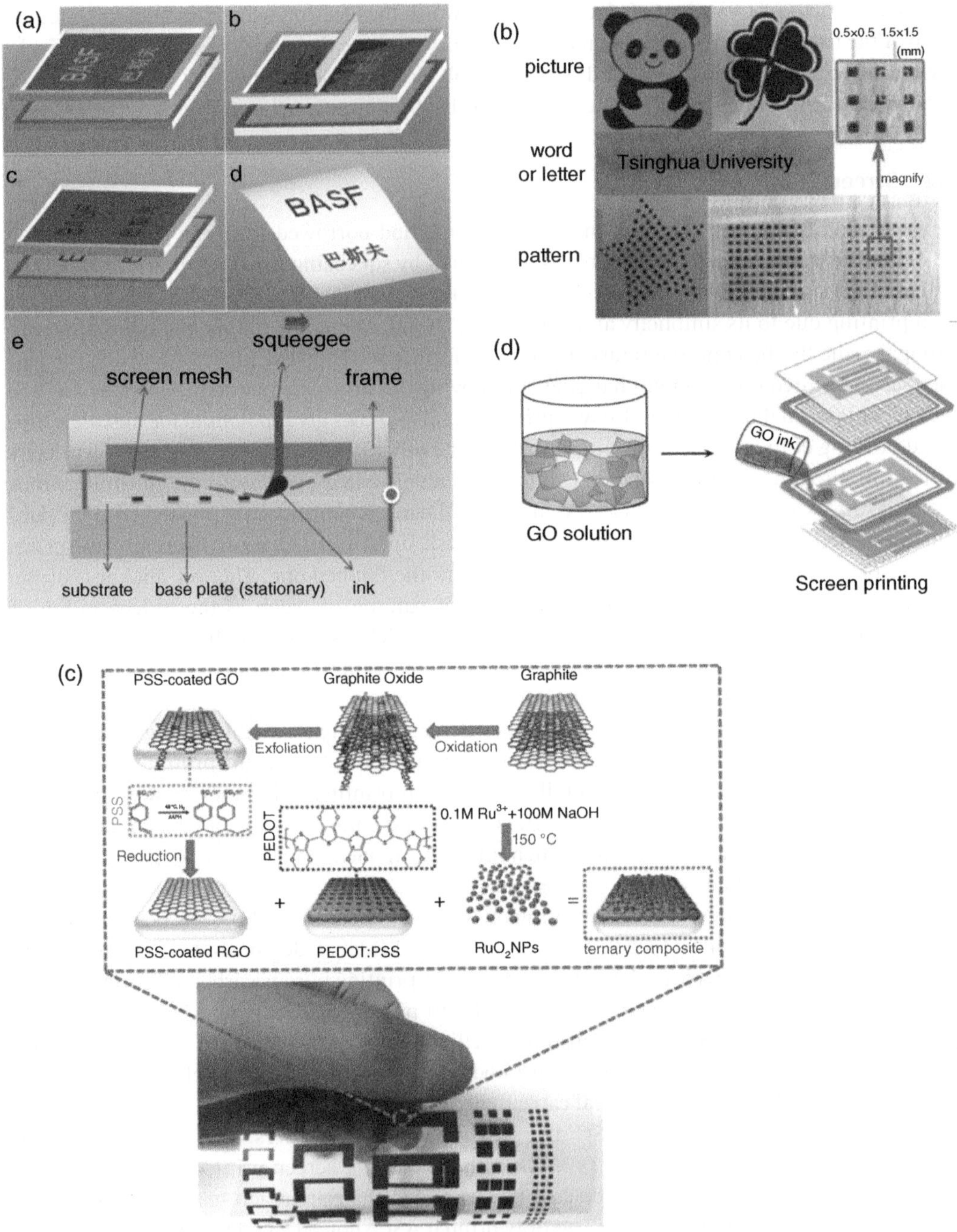

Figure 9.5 Representative screen-printed flexible SCs: (a) Schematic illustration of the screen-printing process with NGP/PANI ink. *Source:* Reproduced with permission [75]. © 2013, WILEY-VCH. (b) Photographs of the screen-printed electrodes of various designs with MnO_2 and graphene inks. *Source:* Reproduced with permission [76]. © 2013, Nature Publishing Group. (c) Overall procedure for fabricating RuO_2/PEDOT:PSS/graphene screen-printed electrode and digital camera image of screen-printed electrode. *Source:* Reproduced with permission [77]. © 2015, American Chemical Society. (d) Schematic representation of the printed SC fabrication process with GO solution. *Source:* Reproduced with permission [78]. © 2017, IOP Publishing Ltd.

showed excellent cycling stability with no degradation after over 1000 charge/discharge cycles. Shi et al. prepared ASCs by screen-printing 2D MnO_2 nanosheets and graphene as cathode and anode, respectively (Figure 9.5b) [76]. The ASCs exhibited a high energy density of 97.2 $Wh\,kg^{-1}$ with no more than 3% capacitance loss even after 10 000 cycles. Cho et al. screen-printed water-dispersible RuO_2 decorated PEDOT: PSS/graphene-based inks as electrode materials for SCs as illustrated in Figure 9.5c [77]. PEDOT: PSS, graphene, and hydrous RuO_2 NPs (nanoparticles) were mixed into an aqueous ternary electrode ink. The PSS binders in the PEDOT: PSS system could optimize the viscosity of the PEDOT: PSS-based inks to be suitable for screen-printing. The resulting RuO_2/PEDOT:PSS/graphene screen-printed electrode exhibited an impressive conductivity of 1570 $S\,cm^{-1}$, a larger specific capacitance of 820 $F\,g^{-1}$, and a better cycling performance with 81.5% retention of initial capacitance after 1000 cycles compared with PEDOT:PSS-only electrode. This work provides insights into improving the electrochemical performance of the printed electrodes and the SC devices by tuning the ink formulations. In another example, graphene oxide ink was screen-printed on textiles, followed by reduction via a rapid electrochemical method, as illustrated in Figure 9.5d [78]. The printed SC exhibited areal capacitance as high as 2.5 $mF\,cm^{-2}$, 95.6% of which was maintained under bending conditions. Moreover, the SC maintained 97% of the original capacitance even after 10 000 cycles.

9.4.3 Transfer Printing

Transfer printing means transferring material from relief features on an intermediate medium to the target substrate. It decouples the device fabrication and integration processes and enables heterogeneous integration of different classes of materials into desired layouts, thus holds great potential for flexible electronics [79]. Although transfer printing is often conducted by using a polymeric stamp in the laboratory setting, gravure and flexographic transfer printing are the two main modes for scaled production. The gravure printing system consists of an ink trough in which a gravure roller is rotated and a metal doctor blade that scrapes the excess ink from the roller surface [80]. It uses gravure rollers to transfer the designed pattern pre-engraved into the gravure roller onto the substrate under pressure. Different from gravure printing, flexographic printing is using a plate with a raised surface and fast-drying fluidic inks to print materials directly onto the substrate [80]. Differences between gravure and flexographic transfer printing are mainly the applied pressure and ink transfer pathway [22, 81]. Apart from these, the printing speed (around 1000 $m\,min^{-1}$ for gravure and 500 $m\,min^{-1}$ for flexographic), as well as the ink viscosity (100–1000 mPa s for gravure printing, 50–500 mPa s for flexographic printing), are also different [80]. To facilitate transfer printing, the adhesion between interfaces of ink/donor/substrate needs to be carefully optimized. The adhesion and fracture mechanics at the critical interfaces between the ink/donor, the mold/ink and the ink/substrate are determined by the viscosity of the functional inks and the applied pressure. The adhesion energies can also be adjusted through chemical functionalization, as illustrated in Figure 9.6a–c [13, 83, 84]. The advantages of gravure and flexographic transfer printing are high printing speed and high scalability. The main limitations are the high cost of setting up and prototyping of gravure rollers and plates, as well as high volume requirements for inks.

Transfer printing has been utilized to print various SC electrode materials such as SWCNT, graphene and MXenes. In a typical example, Zhang et al. demonstrated the production of flexible MSCs by transfer-printing MXene inks ($Ti_3C_2T_x$ and Ti_3CNT_x) [82]. As shown in Figure 9.6d and e, stamps with designed patterns were fabricated by 3D printing. Then the MXene ink was brushed onto the surface of the stamps, after which it was transferred onto paper or other substrates by

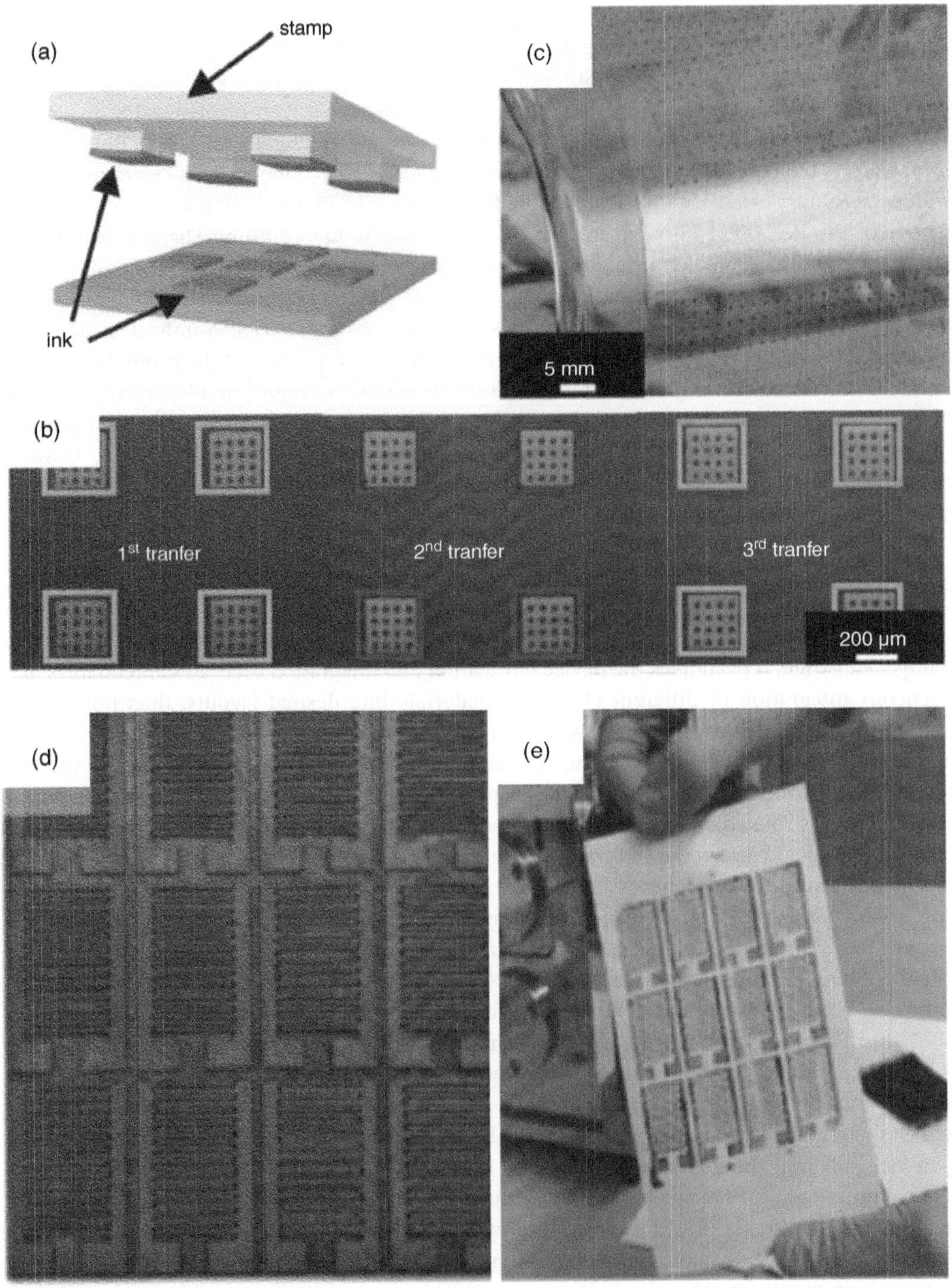

Figure 9.6 Representative transfer-printed flexible SCs. (a) Schematic illustration of the process of retrieving and printing selected sets of microstructures (b) Micrograph of a receiving substrate after printing from the donor substrate (c) Large-scale collection of structures printed onto a thin, flexible sheet of plastic, shown here wrapped onto a cylindrical glass substrate. *Source:* Reproduced with permission [79]. © 2012, WILEY-VCH. (d) As-printed pad with 12 interdigitated stamps. (e) The as-tamped I-$Ti_3C_2T_x$ produced by rolling. *Source:* Reproduced with permission [82]. © 2018, WILEY-VCH.

pressing. Finally, Ag wires were attached and PVA/H_2SO_4 gel electrolyte was drop-cast to complete the fabrication of the solid-state MXene-based MSCs. The fabricated $Ti_3C_2T_x$ MSC exhibited an areal capacitance as high as 61 mF cm^{-2} at 25 μA cm^{-2}, a high energy density of 0.76 μWh cm^{-2} and excellent cycling stability. This work demonstrates that simple stamping can produce continuous films with excellent metallic conductivity even on rough surfaces such as porous paper.

Transfer printing is a scalable, simple technique to fabricate SCs on flexible substrates. This technique is a particularly promising step toward the fabrication of flexible integrated devices.

9.4.4 3D Printing

3D printing is often used as a colloquial substitute for additive manufacturing, which can be categorized into seven groups according to the International Organization for Standardization and American Society for Testing and Materials (ISO/ASTM 52900:2015): binder jetting, directed energy deposition, material extrusion, material jetting, powder bed fusion, sheet lamination, and vat photopolymerization [85] Unlike subtractive manufacturing, additive manufacturing can produce complex 3D structure without wasting materials; thus they have shown great potential for the fabrication of energy storage devices such as SCs [86–88].

Light- and ink-based 3D printing are the leading additive manufacturing technologies in the context of printed SCs [89]. Light-based 3D printing methods mainly include stereolithography (SLA), which photopolymerizes photocurable resins into solid objects by rastering laser beams; and selective laser sintering (SLS), which works by locally heating and fusing of polymer particles in a powder bed. The light-based 3D printing methods generally show high resolution; however, they suffer from an inherent trade-off between resolution, object volume, and speed as well as limited materials selection that are mostly restricted to rigid thermoset polymers and thermoplastic polymer powders [90]. Ink-based 3D printing methods, which include fused deposition modeling (FDM), direct ink writing (DIW), and binder jetting, can deposit a wide range of materials in the form of low-viscosity fluids, continuous filaments, and viscoelastic materials. They are the most popular methods for printing SCs mainly due to the wide selection of materials. FDM works by extruding filaments usually made of a thermoplastic polymer after being heated inside the nozzle to a temperature close to its melting point. The extruded filament quickly cools down and solidifies after being deposited onto the substrate, enabling the building of the 3D structure in a layer-by-layer manner. There are commercially available conductive filaments (mostly based on graphene or CNT) that can be used as the current collector or electrode for SC; however, the low conductivity (hundreds of S/m) due to the low concentration of the active material within the filament restricts their application [91, 92]. The materials selection for FDM is somewhat limited due to the high temperature needed during the FDM process. DIW refers to the deposition of a shear-thinning (viscosity decreases as shear rate increases) ink/filament [93]. The shear-thinning behavior enables materials to be extruded through fine nozzles, as well as to possess high enough shear elastic modulus and shear yield strength in order to retain their shape. The resolution of DIW can be quite high (less than 5 μm) depending on the nozzle size [94]. DIWs are versatile manufacturing techniques that can print a wide variety of materials such as plastics, hydrogels, and nanomaterials, and thus have received the most attention for 3D printed SCs [95]. Post-treatment processes such as photopolymerization, thermal curing, freezing [96] are usually required. Binder jetting refers to the process in which special adhesives are deposited onto a thin layer of powders where the binder will bond the layer of powder and create a solid 2D pattern [97]. 3D structure can be fabricated by repeating the above procedure. Binder jetting technique alleviates the need for sacrificial structures as the powders can support themselves. A variety of powder materials, such as ceramics, plaster, and active carbon, can be used for binder jetting. Binder jetting normally produce a porous hard solid, and it is possible to infiltrate the product with a second phase to obtain a functional composite [98]. One interesting direction for binder jetting in energy storage is to make composite electrodes made of metal oxide and conductive binders.

Compared to other printing technologies, 3D printing offers the following advantages: (i) Excellent printing flexibility and geometry controllability. A wide range of materials, including powders, liquids, and metals, can be used for 3D printing. On the other hand, devices with complex 3D structures can be realized, especially the hierarchically porous structure, which is advantageous for enhancing electrochemical performances. (ii) Good control over the thickness of the printed pattern. Both thin-film and thick electrodes can be printed, especially electrodes with high aspect-ratio configurations. (iii) Cost-effectiveness and environment friendliness due to the additive manufacturing nature and simplified processes. 3D printing is particularly promising for the production of 3D MSCs due to the following reasons: (i) The thickness of interdigitated electrodes can be optimized from hundreds of nanometers to millimeters to enhance the specific areal capacitance. (ii) The symmetric and asymmetric interdigitated electrodes arrays can be directly printed on substrates promoting the integration of MSCs as energy storage units in on-chip devices. (iii) 3D printing offers a convenient technology for the encapsulation of MSCs.

Extrusion-type DIW is the most popular 3D printing method to fabricate SCs among researchers because of the simple and low-cost setup, wide material selection and well-controlled printed structures. Yang et al. reported a quasi-solid-state asymmetric MSC using DIW [99]. In a typical procedure, the cathode(V_2O_5), anode (graphene–vanadium nitride quantum dots), and electrolyte (LiCl–PVA) inks with high viscosities and shear-thinning rheological properties were 3D printed separately on the substrates, as shown in Figure 9.7(a). The 3D printed ASC exhibited high areal mass loading of 3.1 $mg\,cm^{-2}$, and a wide electrochemical potential window of 1.6 V. Furthermore, Due to the open macropores in the 3D printed electrodes, the resulted ASC achieved a maximum areal and volumetric capacitance of 207.9 $mF\,cm^{-2}$ and 5.0 $F\,cm^{-3}$, respectively, with the areal energy density and power density reaching 73.9 μWh cm^{-2} and 3.77 $mW\,cm^{-2}$, respectively. Beidaghi et al. employed DIW to print 3D electrodes with a viscoelastic, water-based, and additive-free MXene ink [15]. By increasing the number of deposited layers and thus electrode height, they were able to tune the capacitance, and an outstanding areal capacitance (1035 $mF\,cm^{-2}$) was achieved when the print run was 10. Furthermore, the maximum areal and volumetric energy and power densities of the 3D printed device were 51.7 $\mu Wh\,cm^{-2}$ and 5.7 $mW\,cm^{-2}$, respectively, significantly higher than those of previously reported MXene-based SCs. Recently, aerogels show great promise for 3D printed energy storage applications. Among them, graphene aerogel has attracted increasing attention due to its high specific surface area for ion access, tunable porosity, and pore structures as well as mechanical properties. The key for DIW of aerogels is to prepare gel-based viscoelastic functional material inks with non-Newtonian shear-thinning behaviors. As an example, Zhu et al. prepared a GO-based composite aerogels ink consisting of GO, silica fillers, catalyst, and graphene nanoplatelets which can improve electrical conductivity and lower the resistance of the composite ink without a detrimental loss of surface area [103]. In order to convert the printed structures to an aerogel, the key is to keep the composite ink wet during the printing and gelation processes. The prepared aerogels ink was extruded through a micro nozzle in an isooctane bath which was used to avoid gel collapsing. Afterward, the 3D printed porous structure was processed into aerogels by gelation, supercritical drying, and carbonization, followed by etching of silica by hydrofluoric acid. The symmetric SC printed by DIW achieved a gravimetric capacitance of 4.76 $F\,g^{-1}$ at a current density of 0.4 $A\,g^{-1}$, and 61% of the capacitance could be retained when the current density was increased by 20 times to 8 $A\,g^{-1}$. Moreover, the SC with thick

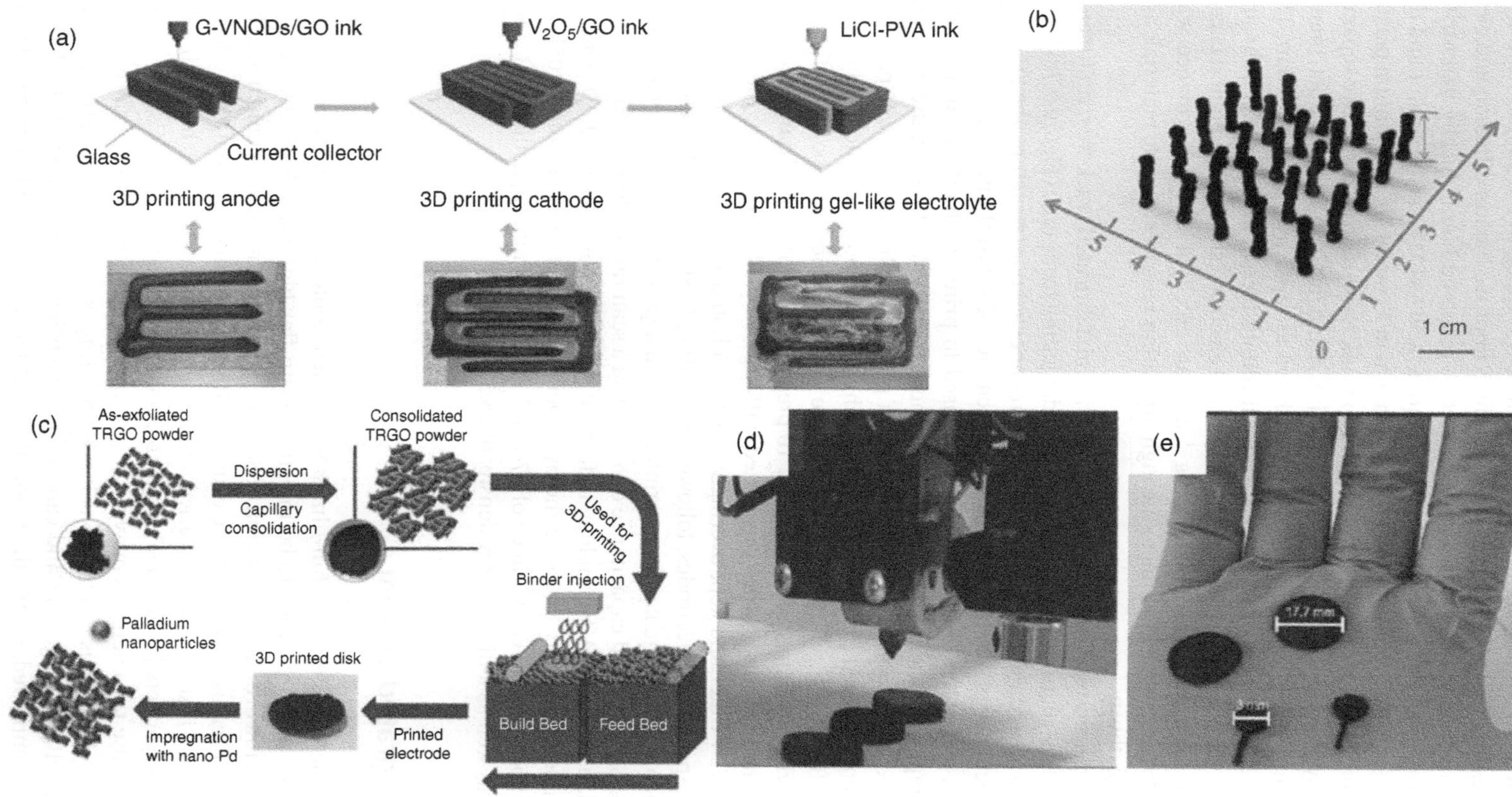

Figure 9.7 Representative 3D-printed flexible SCs. (a) Schematic illustration of 3D printing an asymmetric MSC with interdigitated electrodes. *Source:* Reproduced with permission [99]. © 2018, Wiley-VCH. (b) Micro-patterns of PAni hydrogel by 3D printing. *Source:* Reproduced with permission [100]. © 2016, Springer. (c) Schematic presentation of the binder-jetting powder-bed AM technique. *Source:* Reproduced with permission [101]. © 2017, Elsevier. (d and e) Optical images of the 3D printable graphene/PLA. *Source:* Reproduced with permission [102]. © 2017, Nature Publishing Group.

electrodes of ~1000 μm showed outstanding rate capability (87% capacitive retention from 2 to $10\,A\,g^{-1}$) due to the presence of ordered macropores that help improve the ion diffusion and electrical conductivity to enable fast charge transport. Dou et al. prepared hierarchically nanostructured PANI hydrogel by using amino trimethylene phosphonic acid (ATMP) as the gelator and the dopant to form a conducting polymer network [100]. As shown in Figure 9.7b, ammonium persulfate solution (ink A) and ATMP mixed with aniline solution (ink B) were loaded into two separate 3D printer cartridges. By alternatively printing droplets of ink A and ink B on the same spot, immediate contact and mixing of the droplets occurred, which led to rapid hydrogel formation. The interconnected 3D conductive network structure could facilitate the transport of electrons and enhance the polymer elasticity. The 3D printed PANI hydrogel electrode showed the highest capacitance of $\sim 422\,F\,g^{-1}$ at a current density of $0.2\,A\,g^{-1}$ and an excellent rate performance with $\sim 240\,F\,g^{-1}$ at a current rate of $10\,A\,g^{-1}$ owing to the high utilization ratio of electrode materials and the short ion diffusion pathway.

Binder jetting has also been used to fabricate SC. Azhari et al. fabricated crack-free, millimeter-thick, graphene-based 3D structured electrodes by the binder jetting technique (Figure 9.7c) [101]. Due to the high surface area of the electrodes and the interconnected pore structure, the SC exhibited a gravimetric capacitance of $\sim 260\,F\,g^{-1}$ and areal capacitance of $\sim 700\,mF\,cm^{-2}$, respectively, at $5\,mV\,s^{-1}$ in 1 M H_2SO_4 electrolyte. FDM has also been used to print SC. Banks et al. used this method to print a graphene-based polylactic acid filament (graphene/PLA) (Figure 9.7d and e) [102]. The 3D printed solid-state SC with a low mass of graphene exhibiting a low capacitance of 28 μF. Areir et al. demonstrated the use of the FDM technique to deposit activated carbon onto flexible fabrics [104]. By tuning printing parameters, the layer thickness can be carefully controlled, and accurate orientation has been achieved. To prepare inks with appropriate rheological properties, 2 g of activated carbon powder was mixed with 2 g of PVA binder. Then, the mixture was dissolved in 40 mL DI water and kept overnight to make a homogenous slurry. The activated carbon slurry was first printed as electrodes, followed by printing H_2SO_4/PVA gel electrolyte onto flexible electrodes. Finally, a silicone layer was printed to form a gasket to prevent any leaking or drying. The fabricated SC possessed capacitance of 68.7 mF at a scan rate of $20\,mV\,s^{-1}$.

Recently, Gao et al. used the SLA technology to fabricate the rationally designed 3D structure made of reduced graphene oxide (rGO) using the hydrogen bubbles as soft templates [105]. The electrodes achieved superior areal capacitance of $57.75\,mF\,cm^{-2}$, good rate capability (70% retention, $2\text{–}40\,mA\,cm^{-2}$), as well as a high energy density of $56.52\,mW\,cm^{-3}$.

9.5 Printed Integrated System

Printing enables scalable fabrication of thin-film SCs and their conformal lamination on various substrates; thus, printed SCs are easier to be integrated with flexible/wearable electronics than conventional bulk SCs. However, the integration of SCs can be more complicated compared to batteries [106–108]. This is because SCs have to be coupled with a power management circuit for the output voltage to be steady instead of being varied linearly with time, as well as to be amplified to meet the demand of the gadgets [109]. To fabricate integrated systems based on SCs, it is important, first of all, to increase the energy density without compromising the power performance of SCs. To print hierarchically porous electrodes using 3D printing is a promising approach in this regard. Secondly, the self-discharge rate and capacitance retention upon cycling of flexible SCs need to be significantly improved. Lastly, the fabrication method of SCs must be compatible with the processes that assemble the other parts of the circuit.

Miniaturized sensors for temperature, humidity, light, hazardous gas, and pH value are required in huge amounts in the context of the IoT; therefore, integrated sensor systems powered by SCs have aroused extensive attention recently. Printed MSC are feasible power sources for these integrated sensor systems because of their small size and planar structure. Guo et al. reported a sensor integrated with MSC arrays. The MSC was manufactured by screen printing, followed by an electrodeposition process (Figure 9.8a) [110]. The MSCs were then integrated with sensors on one piece of paper. The whole device consisted of MSCs, sensors, interconnections, and contact pads, as shown in Figure 9.8b and c. Two AMSCs (Asymmetric micro-supercapacitors) were connected in series to improve the output voltage of the SC. A ZnO-based UV sensor was used in this integrated system, which was constructed by drop-casting the suspension of ZnO nanowires on the interdigital electrodes. In another example, a foldable integrated system composed of resistive UV sensors and asymmetric MSC was fabricated [111]. The graphene/ZnO nanoparticles-based UV sensor was driven by the power stored in the AMSCs. The AMSCs used MWNTs loaded with MnO_2 nanoball and V_2O_5 wrapped MWNTs as positive and negative electrodes, respectively.

Printed SCs are also promising as the bridge between an energy harvester and functional devices to constitute a self-powered circuit. Common ambient energy sources like solar, thermal, or mechanical energy are promising for powering flexible electronics, yet they suffer from low and non-uniform power output. The fast-charging speed, high power density, and long cycle life characteristics of SCs make them ideal energy storage unit for these ambient energy sources.

Lin and co-workers have developed a scalable printing technique to construct all-solid-state planar SCs on flexible substrates, which was further integrated with solar cells [18]. The as-fabricated flexible SCs exhibited an areal capacitance up to 52.9 $mF\,cm^{-2}$, a high power density of 39.6 $W\,cm^{-3}$ and energy density of 11.1 $mWh\,cm^{-3}$. Interestingly, a self-powered T-shirt was made using commercial polycrystalline silicon solar cell panels to harvest solar energy and two thus-made SCs connected in series to collect and storage the solar energy (Figure 9.8d). In the dark condition, the LED (light-emitting diode) arrays could be lit up by the energy released by the SCs (Figure 9.8e). In another example, a simple, low-cost, and scalable strategy combining printing and electrochemical deposition has been developed to fabricate a flexible integrated system which comprises a MnO_2-PPy//V_2O_5-PANI based all-solid-state asymmetric MSC, a perovskite nanowires-based photoconductive photodetector and a wireless charging coil. The MSC could be fully charged by the power receiver and in turn drive the photodetector [112]. OSCs have also been integrated with SCs to achieve self-powered electronics. A total energy conversion and storage efficiency (ECSE) of 1.57% was achieved under one sun condition, producing maximum power of 4.1 mW. The system yielded an ECSE of 2.9% and 2.8 mW power under simulated indoor light [113].

9.6 Perspective

As the study of flexible SCs and flexible electronics in general gradually move from academic curiosity to industrial application, printing as a manufacturing tool seems more relevant than ever. In this chapter on printed flexible SCs, we try to show that printing can revolutionize the manufacturing of SCs in terms of increasing design complexity yet at a lower cost and simplifying manufacture, and facilitating integration with other flexible electronic devices. Noteworthy structural features of printed SCs mainly include complex 2D structure represented by the interdigital patterns, which can improve the ion transport behaviors thus increase the power density of SCs; and hierarchically porous 3D structures which can facilitate ion transport and

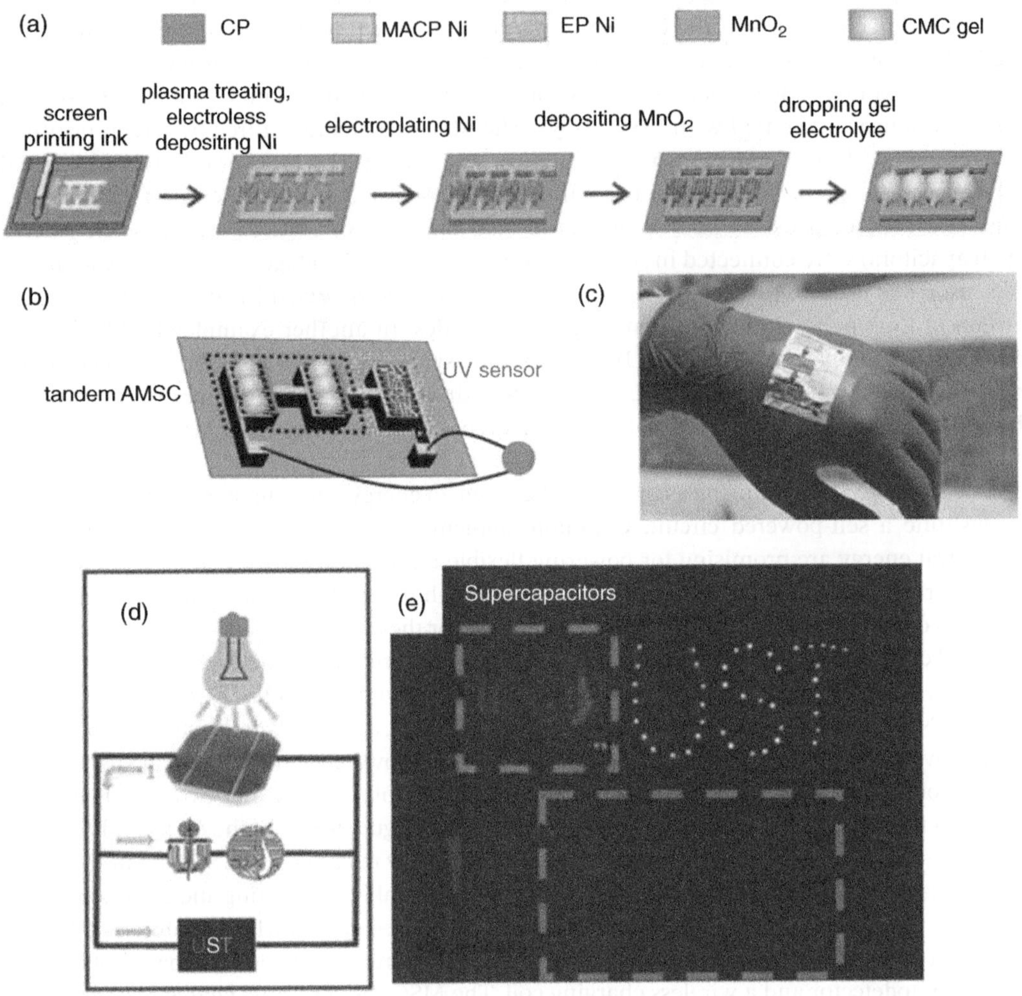

Figure 9.8 Representative printed integrated systems containing SCs. (a) Scheme of the fabrication process of a MSC on paper. (b) Integration of MSCs and UV sensor on one piece of paper. (c) Digital image of a practical integrated circuit with MSC and UV sensor. *Source:* Reproduced with permission [110]. © 2017, Wiley-VCH. (d) Schematic of a wearable self-powered system on T-shirt. (e) The LEDs on the T-shirt are lit up by the solar-charged solar cell. *Source:* Reproduced with permission [18]. © 2017, Wiley-VCH.

increase mass loading, thus simultaneously increase energy and power density. However, printing also suffers from some challenges. With these challenges also come opportunities for further research.

The layering nature of most printing processes dictates an innate accuracy limitation which brings about issues like misalignment, non-uniform solidification, and poor surface quality. These problems are fine on a macroscopic level; however, as the resolution of printing increases and the microscopic design of structure becomes feasible, these can also become serious problems. So far, the resolutions of most printing technologies are in the micrometer range, although there are some methods such as two-photon polymerization and electrodynamic jet printing, which have entered the realm of nanofabrication. This has great significance for printed SCs since many of the energy storage processes, such as ion transport and redox reactions, occur at the nanoscale. It is expected the performances of printed SCs can be significantly improved shortly, in line with the rapid developments of these advanced super-high-resolution printing methods.

Although significantly varied from method to method, the material availabilities of printing methods are generally low. Most printing techniques require materials to possess certain rheological properties that impose restrictions on the size of materials, solvents, and concentration. It is a challenging task to formulate the desired materials into printable "inks." Printing also presents requirements of materials in terms of fast solidification approaches, long-term stability and mechanical properties. For instance, it is often challenging to develop 3D printable inks with the desired fast-drying capability and mechanical toughness without compromising the conductivity or capacitive properties of the active materials. SCs, like other devices, require more than one material to complete device fabrication. Presently, multi-material printing is usually achieved either through mixing the materials into one ink before printing or printing the second material on top of/into the pre-existing layer of material. There are already some commercially available multi-material printing strategies (mainly inkjet and extrusion-type 3D printing) based on advanced printheads. It is increasingly common to simultaneously deposit different materials on either different or the same spot during one print run. Interestingly, it is also possible to create compositional variation within printed structures, which creates another dimension for optimizing the performance of printed SCs. In this chapter, we have glimpsed at the printing-endowed ability to control local microstructure and chemical composition, which makes structures like hierarchically porous electrode structure easy to obtain. Thus, through printing, new materials and structures can be tested quickly at a low cost both in terms of money and materials and the performances of printed SCs. So far, most researches have focused on the printing of electrodes. Other than electrodes, the printing of current collector, electrolyte and encapsulation largely remain to be explored. Fully printed SCs will greatly accelerate the application of SCs in a variety of novel and interesting application scenarios.

We have shown that printing technologies possess great potential in promoting the application of flexible SCs. To further advance this field, printing technologies, printable materials, and structure design should proceed hand-in-hand to make printing not only a manufacturing tool but also innovation catalysis to promote the application of flexible SCs for the next-generation electronics.

Acknowledgments

We acknowledge financial support from the National Natural Science Foundation of China (21835003, 21805136, 91833304, 21422402, 21674050), the National Key Basic Research Program of China (2017YFB0404501, 2014CB648300), the Natural Science Foundation of Jiangsu Province (BE2019120), Program for Jiangsu Specially-Appointed Professor (RK030STP15001), the Six Talent Peaks Project of Jiangsu Province (TD-XCL-009), the 333 Project of Jiangsu Province (BRA2017402), the NUPT "1311 Project" and Scientific Foundation (NY218164, NY217169), the Leading Talent of Technological Innovation of National Ten-Thousands Talents Program of China, the Excellent Scientific and Technological Innovative Teams of Jiangsu Higher Education Institutions (TJ217038), the Synergetic Innovation Center for Organic Electronics and Information Displays, the Priority Academic Program Development of Jiangsu Higher Education Institutions (PAPD, YX030003), the Open research fund from State Key Laboratory of Organic Electronics and Information Displays.

References

1 Zhang, Y.Z., Wang, Y., Cheng, T. et al. (2015). *Chem. Soc. Rev.* 44: 518.
2 Lu, X., Yu, M., Wang, G. et al. (2014). *Energy Environ. Sci.* 7: 2160.
3 Chen, D., Jiang, K., Huang, T., and Shen, G. (2020). *Adv. Mater.* 32: 1901806.
4 Dubal, D.P., Chodankar, N.R., Kim, D.-H., and Gomez-Romero, P. (2018). *Chem. Soc. Rev.* 47: 2065.
5 Peng, X., Peng, L., Wu, C., and Xie, Y. (2014). *Chem. Soc. Rev.* 43: 3303.
6 Da, Y., Liu, J., Zhou, L. et al. (2019). *Adv. Mater.* 31: 1802793.
7 Qi, D., Liu, Y., Liu, Z. et al. (2017). *Adv. Mater.* 29: 1602802.
8 Li, D., Lai, W.Y., Zhang, Y.Z., and Huang, W. (2018). *Adv. Mater.* 30: 1704738.
9 Garlapati, S.K., Divya, M., Breitung, B. et al. (2018). *Adv. Mater.* 30: 1707600.
10 Kyeremateng, N.A., Brousse, T., and Pech, D. (2017). *Nat. Nanotechnol.* 12: 7.
11 Lee, K.H., Lee, S.S., Ahn, D.B. et al. (2020). *Sci. Adv.* 6: eaaz1692.
12 Wang, Y., Mehrali, M., Zhang, Y.Z. et al. (2021). *Energy Stor. Mater.* 36: 318.
13 Meitl, M.A., Zhu, Z.T., Kumar, V. et al. (2006). *Nat. Mater.* 5: 33.
14 Liang, J., Tian, B., Li, S. et al. (2020). *Adv. Energy Mater.* 10: 2000022.
15 Orangi, J., Hamade, F., Davis, V.A., and Beidaghi, M. (2019). *ACS Nano* 14: 640.
16 Beidaghi, M. and Gogotsi, Y. (2014). *Energy Environ. Sci.* 7: 867.
17 Tian, X., Shi, M., Xu, X. et al. (2015). *Adv. Mater.* 27: 7476.
18 Lin, Y., Gao, Y., and Fan, Z. (2017). *Adv. Mater.* 29: 1701736.
19 Meng, C., Maeng, J., John, S.W., and Irazoqui, P.P. (2014). *Adv. Energy Mater.* 4: 1301269.
20 Dinh, T.M., Armstrong, K., Guay, D., and Pech, D. (2014). *J. Mater. Chem. A* 2: 7170.
21 Wang, G., Zhang, L., and Zhang, J. (2012). *Chem. Soc. Rev.* 41: 797.
22 Wang, F., Wu, X., Yuan, X. et al. (2017). *Chem. Soc. Rev.* 46: 6816.
23 Zhang, L.L. and Zhao, X. (2009). *Chem. Soc. Rev.* 38: 2520.
24 Pech, D., Brunet, M., Taberna, P.-L. et al. (2010). *Sources* 195: 1266.
25 Jost, K., Stenger, D., Perez, C.R. et al. (2013). *Energy Environ. Sci.* 6: 2698.
26 Novoselov, K.S., Geim, A.K., Morozov, S.V. et al. (2004). *Science* 306: 666.
27 Shao, Y., El-Kady, M.F., Wang, L.J. et al. (2015). *Chem. Soc. Rev.* 44: 3639.
28 Li, J., Ye, F., Vaziri, S. et al. (2013). *Adv. Mater.* 25: 3985.
29 González, A., Goikolea, E., Barrena, J.A., and Mysyk, R. (2016). *Renew. Sust. Energ. Rev.* 58: 1189.
30 Sun, J., Wu, C., Sun, X. et al. (2017). *J. Mater. Chem. A* 5: 1266.
31 Augustyn, V., Simon, P., and Dunn, B. (2014). *Energy Environ. Sci.* 7: 1597.
32 Meng, Q., Cai, K., Chen, Y., and Chen, L. (2017). *Nano Energy* 36: 268.
33 Zhang, Y.Z., Wang, Y., Jiang, Q. et al. (2020). *Adv. Mater.* 32: 1908486.
34 Kitagawa, S. (2014). *Chem. Soc. Rev.* 43: 5415.
35 Zhou, H.-C., Long, J.R., and Yaghi, O.M. (2012). *Chem. Rev.* 112: 673.
36 Furukawa, H., Cordova, K.E., O'Keeffe, M., and Yaghi, O.M. (2013). *Science* 341: 1230444.
37 Ding, S.-Y. and Wang, W. (2013). *Chem. Soc. Rev.* 42: 548.
38 El-Kaderi, H.M., Hunt, J.R., Mendoza-Cortés, J.L. et al. (2007). *Science* 316: 268.
39 Zhong, C., Deng, Y., Hu, W. et al. (2015). *Chem. Soc. Rev.* 44: 7484.
40 Cheng, X., Pan, J., Zhao, Y. et al. (2018). *Adv. Energy Mater.* 8: 1702184.
41 Choudhury, N., Sampath, S., and Shukla, A. (2009). *Energy Environ. Sci.* 2: 55.
42 Gao, H. and Lian, K. (2014). *RSC Adv.* 4: 33091.
43 Armand, M., Endres, F., MacFarlane, D.R. et al. (2011). *Materials for Sustainable Energy: A Collection of Peer-Reviewed Research and Review Articles from Nature Publishing Group*, 129. World Scientific.

44 Ma, G., Feng, E., Sun, K. et al. (2014). *Electrochim. Acta* 135: 461.
45 Ma, G., Li, J., Sun, K. et al. (2014). *Sources* 256: 281.
46 Yu, H., Wu, J., Fan, L. et al. (2011). *Electrochim. Acta* 56: 6881.
47 Senthilkumar, S., Selvan, R.K., Melo, J., and Sanjeeviraja, C. (2013). *ACS Appl. Mater. Interfaces* 5: 10541.
48 Yu, H., Wu, J., Fan, L. et al. (2012). *Sci. China Chem.* 55: 1319.
49 Roldán, S., Granda, M., Menéndez, R. et al. (2011). *J. Phys. Chem. C* 115: 17606.
50 Dubal, D.P., Kim, J.G., Kim, Y. et al. (2014). *Energ. Technol.* 2: 325.
51 Kim, H., Boulogne, F., Um, E. et al. (2016). *Phys. Rev. Lett.* 116: 124501.
52 Gupta, R.K., Candler, J., Palchoudhury, S. et al. (2015). *Sci. Rep.* 5: 15265.
53 Zhu, G., He, Z., Chen, J. et al. (2014). *Nanoscale* 6: 1079.
54 Si, W., Yan, C., Chen, Y. et al. (2013). *Energy Environ. Sci.* 6: 3218.
55 Cao, X., Zheng, B., Shi, W. et al. (2015). *Adv. Mater.* 27: 4695.
56 Kurra, N., Hota, M.K., and Alshareef, H.N. (2015). *Nano Energy* 13: 500.
57 Xu, Y., Lin, Z., Huang, X. et al. (2013). *ACS Nano* 7: 4042.
58 Sirringhaus, H., Kawase, T., Friend, R.H. et al. (2000). *Science* 290: 2123.
59 Jang, D., Kim, D., and Moon, J. (2009). *Langmuir* 25: 2629.
60 Shin, P., Sung, J., and Lee, M.H. (2011). *Microelectron. Reliab.* 51: 797.
61 de Gans, B.J., Kazancioglu, E., Meyer, W., and Schubert, U.S. (2004). *Macromol. Rapid Commun.* 25: 292.
62 Hsiao, W.-K., Hoath, S.D., Martin, G.D. et al. (2009). *Technology* 53: 050304.
63 Chang, P., Mei, H., Zhou, S. et al. (2019). *J. Mater. Chem. A* 7: 4230.
64 Cheng, T., Zhang, Y.-Z., Yi, J.-P. et al. (2016). *J. Mater. Chem. A* 4: 13754.
65 Liu, W., Lu, C., Li, H. et al. (2016). *J. Mater. Chem. A* 4: 3754.
66 Li, J., Sollami Delekta, S., Zhang, P. et al. (2017). *ACS Nano* 11: 8249.
67 Li, L., Secor, E.B., Chen, K.-S. et al. (2016). *Adv. Energy Mater.* 6: 1600909.
68 Wang, Y., Zhang, Y.-Z., Dubbink, D., and ten Elshof, J.E. (2018). *Nano Energy* 49: 481.
69 Choi, K.-H., Yoo, J., Lee, C.K., and Lee, S.-Y. (2016). *Energy Environ. Sci.* 9: 2812.
70 Zhu, D., Wang, Z., and Zhu, D. (2020). *J. Electron. Mater.* 49: 1765.
71 He, P., Cao, J., Ding, H. et al. (2019). *ACS Appl. Mater. Interfaces* 11: 32225.
72 Garnier, F., Hajlaoui, R., Yassar, A., and Srivastava, P. (1994). *Science* 265: 1684.
73 Jost, K., Perez, C.R., McDonough, J.K. et al. (2011). *Energy Environ. Sci.* 4: 5060.
74 Wang, S., Liu, N., Yang, C. et al. (2015). *RSC Adv.* 5: 85799.
75 Xu, Y., Schwab, M.G., Strudwick, A.J. et al. (2013). *Adv. Energy Mater.* 3: 1035.
76 Shi, S., Xu, C., Yang, C. et al. (2013). *Sci. Rep.* 3: 2598.
77 Cho, S., Kim, M., and Jang, J. (2015). *ACS Appl. Mater. Interfaces* 7: 10213.
78 Abdelkader, A.M., Karim, N., Valles, C. et al. (2017). *2D Mater.* 4: 035016.
79 Carlson, A., Bowen, A.M., Huang, Y. et al. (2012). *Adv. Mater.* 24: 5284.
80 Bonaccorso, F., Bartolotta, A., Coleman, J.N., and Backes, C. (2016). *Adv. Mater.* 28: 6136.
81 Liang, X., Fu, Z., and Chou, S.Y. (2007). *Nano Lett.* 7: 3840.
82 Zhang, C., Kremer, M.P., Seral-Ascaso, A. et al. (2018). *Adv. Funct. Mater.* 28: 1705506.
83 Loo, Y.-L., Willett, R.L., Baldwin, K.W., and Rogers, J.A. (2002). *J. Am. Chem. Soc.* 124: 7654.
84 Lee, J., Lee, J.Y., and Yeo, J.S. (2017). *ACS Appl. Mater. Interfaces* 9: 3251.
85 Huang, Y., Leu, M.C., Mazumder, J., and Donmez, A. (2015). *J. Manuf. Sci. Eng.* 137: 014001.
86 Zhu, C., Liu, T., Qian, F. et al. (2017). *Nano Today* 15: 107.
87 Tofail, S.A., Koumoulos, E.P., Bandyopadhyay, A. et al. (2018). *Mater. Today* 21: 22.
88 Zhang, F., Wei, M., Viswanathan, V.V. et al. (2017). *Nano Energy* 40: 418.

89 Truby, R.L. and Lewis, J.A. (2016). *Nature* 540: 371.
90 Ambrosi, A. and Pumera, M. (2016). *Chem. Soc. Rev.* 45: 2740.
91 Browne, M.P., Novotný, F., Sofer, Z.k., and Pumera, M. (2018). *ACS Appl. Mater. Interfaces* 10: 40294.
92 Kennedy, Z., Christ, J., Evans, K. et al. (2017). *Nanoscale* 9: 5458.
93 Highley, C.B., Rodell, C.B., and Burdick, J.A. (2015). *Adv. Mater.* 27: 5075.
94 Mao, M., He, J., Li, X. et al. (2017). *Micromachines* 8: 113.
95 Ruiz-Morales, J., Tarancón, A., Canales-Vázquez, J. et al. (2017). *Energy Environ. Sci.* 10: 846.
96 Zhang, Q., Zhang, F., Medarametla, S.P. et al. (2016). *Small* 12: 1702.
97 Xu, Y., Wu, X., Guo, X. et al. (2017). *Sensors (Basel)* 17: 1166.
98 Mostafaei, A., Elliott, A.M., Barnes, J.E. et al. (2021). *Prog. Materi. Sci.* 119: 100707.
99 Shen, K., Ding, J., and Yang, S. (2018). *Adv. Energy Mater.* 8: 1800408.
100 Dou, P., Liu, Z., Cao, Z. et al. (2016). *J. Mater. Sci.* 51: 4274.
101 Azhari, A., Marzbanrad, E., Yilman, D. et al. (2017). *Carbon* 119: 257.
102 Foster, C.W., Down, M.P., Zhang, Y. et al. (2017). *Sci. Rep.* 7: 42233.
103 Zhu, C., Liu, T., Qian, F. et al. (2016). *Nano Lett.* 16: 3448.
104 Areir, M., Xu, Y., Zhang, R. et al. (2017). *J. Manuf. Process.* 25: 351.
105 Xue, J., Gao, L., Hu, X. et al. (2019). *Nano-Micro Lett.* 11: 46.
106 Pu, X., Hu, W., and Wang, Z.L. (2018). *Small* 14: 1702817.
107 Bondade, R., Zhang, Y., Wei, B. et al. (2016). *IEEE Trans. Ind. Electron.* 63: 2850.
108 Chai, R. and Zhang, Y. (2015). *IEEE Trans. Power Electron.* 30: 6720.
109 Ayoubi, Y., Elsied, M., Oukaour, A. et al. (2016). *Electr. Power Syst. Res.* 134: 186.
110 Guo, R., Chen, J., Yang, B. et al. (2017). *Adv. Funct. Mater.* 27: 1702394.
111 Yun, J., Lim, Y., Lee, H. et al. (2017). *Adv. Funct. Mater.* 27: 1700135.
112 Yue, Y., Yang, Z., Liu, N. et al. (2016). *ACS Nano* 10: 11249.
113 Lechêne, B.P., Cowell, M., Pierre, A. et al. (2016). *Nano Energy* 26: 631.

10

Printing Flexible On-chip Micro-Supercapacitors

Guozhen Shen

State Key Laboratory for Superlattices and Microstructures, Institute of Semiconductors, Chinese Academy of Sciences, and Center of Materials Science and Optoelectronic Engineering, University of Chinese Academy of Sciences, Beijing, 100083, China

10.1 Introduction

With the rapid development of emerging portable and wearable electronics such as mobile phones, smart watches, wearable displays, wireless headphones, nanorobotics, healthy monitoring devices, wireless sensor networks, and virtual reality sets, there are huge requirements for portable, flexible and mobile powering sources (energy harvesters or energy storage devices) for charging these devices [1–10]. Currently, the most commonly used mobile powering sources are microbatteries. Unfortunately, the relatively slow charging speeds and limited lifetimes greatly restricted their applications in some special areas, such as next generation flexible electronic devices.

Recently occurred microsupercapacitors (MSCs), especially the planar on-chip MSCs, provided a solution for on-chip electronic devices with high volumetric performance compared with their stacked counterparts due to the planar design with electrodes closely spaced and naturally separated. Such device structures dramatically decreased the ion diffusion lengths, increased the fraction of active materials and eliminated the need for separators [11–20]. Various microfabrication, lithographic, and printing methods have been developed to pattern electrodes for planar on-chip MSCs. For example, Xu et al. fabricated reduced graphene oxide (rGO) based in-plane MSCs, which were fabricated by first spin-coating rGO film and then patterning interdigital electrodes with ultraviolet lithography [21]. As-fabricated MSCs exhibited a volumetric capacitance of 8.01 F cm^{-3} and an energy density of 6.204 Wh cm^{-3}. Bhattacharya et al. fabricated planar MSCs through inkjet printing [22]. In their work, the digitally designed interdigitated electrode patterns were first printed on paper with rGO ink. The negative electrode (activated carbon-Bi_2O_3 ink) and the positive electrode (rGO-MnO_2 ink) were then printed on the pre-printed rGO film to form the asymmetric MSCs, which delivered excellent electrochemical performance in terms of high energy density (13.28 mWh cm^{-3}), high power density (4.5 W cm^{-3}), excellent rate capability (80% retention of capacitance) and excellent stability (92.2% retention of capacitance after 20 000 charge/discharge cycles). He et al. reported the fabrication of flexible planar MSCs with crosslinked polyaniline (PANI) using laser printing lithography [23]. Patterned electrode was first printed on the flexible polyethylene terephthalate (PET) film with a commercial laser printer. Then PANI was electrodeposited on the patterned electrode. The fabricated MSCs showed high areal capacitance

Flexible Supercapacitors: Materials and Applications, First Edition. Edited by Guozhen Shen, Zheng Lou and Di Chen.

of 54 $mF\,cm^{-2}$ at the current density of 0.3 $mA\ cm^{-2}$, good capacitance retention (79.8% of the initial capacitance after 1000 cycles).

Directly printing on-chip MSCs on desired substrates with printable inks is a promising way for scalable production [24]. Compared with the conventional microfabrication or lithography methods, printing techniques allows digital and additive patterning of electrode and device structures, customization, scalable and rapid production, and so on. In this chapter, we try to summarize the recent progress of producing on-chip MSCs using printing techniques. We first focus on printing different materials on substrates, including active electrodes, electrolytes, and current collectors. Then, different printing techniques, such as inkjet printing, spray printing, screen printing, and three-dimensional (3D) printing, are illustrated in details. Finally, the challenges and future developing trends are discussed to provide readers on how to contrive this area.

10.2 Printable Materials for On-chip MSCs

Printing techniques to on-chip MSCs have two great features compared with other techniques [25, 26]. The first one is that, for printing techniques, the functional materials are added layer by layer on substrate in a very controllable way, thus providing an environmentally friendly process with quite low energy consumption and little material waste and pollution. The second feature is that the printed on-chip MSCs are independent of the substrate materials, thus the printing method enables to produce devices on almost any substrate, including flexible or stretchable substrates. Consequently, the printed MSCs have great potential for next generation flexible electronics.

Typically, on-chip MSCs are composed of substrate, active electrodes (anode and cathode), current collector, and electrolyte. To printing on-chip MSCs, current collectors were first printed on substrate (rigid, flexible, or stretchable substrate), followed with the printing of active electrodes. Solid-state or gel electrolytes were finally printed to finish the fabrication of MSCs. Considering these processes, it is very important to develop printable materials. In this section, we will provide details on different printable materials, including printable electrode materials, printable current collectors, and printable electrolytes.

10.2.1 Printable Electrode Materials

The electrochemical performance of an on-chip MSC is largely dependent on the active electrode materials. As a result, most of the state-of-the-art printing MSC works are focused on printing electrode materials. Basically, supercapacitors based on different electrode materials are classified as two types, the electric double-layer capacitors (EDLCs) and pseudocapacitors [27–40]. EDLCs are similar to the traditional capacitors, the mechanism of which is the charge, formed at the electrode/electrolyte interface, accumulated in the electric double layer. However, compared with a traditional capacitor, the specific capacitance of an EDLC mainly depends on the surface properties of the active electrodes, such as pore size distribution and specific surface area, which is usually several orders of magnitude higher than the traditional capacitor. Contrary to the EDLCs, the pseudocapacitors are based on the fast and reversible redox reactions or Faradic charge transfer reactions at the surface of active electrode materials, providing higher energy density.

Carbon materials including active carbon (AC), mesoporous carbon, carbon nanotube (CNT), and graphene are typical electrode materials for EDLCs [41–49]. Compared with other carbon materials, graphene exhibited some superior electrochemical performance due to its excellent properties. It is well known that graphene is a two-dimensional (2D) defect-free carbon monolayer.

Graphene has gained growing research interest since the first report of isolating a single layer of graphitic film in 2004 by Novoselov and Gem [50–52]. Graphene has a large theoretical specific surface area of about 2630 $m^2 g^{-1}$, high conductivity, good chemical and thermal stability, and wide potential window, which make graphene excellent electrode candidate for high performance supercapacitors. Theoretically, intrinsic EDLC of single-layer graphene is about 21 $\mu F cm^{-2}$. It means that if the entire surface area of a single layer graphene is used, a theoretical capacitance of about 550 $F g^{-1}$ could be achieved, which is superior to all the other carbon materials.

Different kinds of printing techniques have been widely applied to fabricate on-chip MSCs based on carbon materials, especially graphene. To printing graphene-based MSCs, scalable exfoliation methods to produce graphene should be applied. Besides, printable and stable graphene ink should be developed. Liu et al. reported the fabrication of ultraflexible in-plane MSCs by direct printing method [53]. In their work, graphene ink with a concentration of 0.8 $mg ml^{-1}$ was first obtained by electrochemically exfoliating graphite under mild conditions (Figure 10.1a). Studies found that the prepared graphene dispersion was stable in volatile 2-propanol (IPA) for at least four weeks. A directly printable hybrid ink composed of the electrochemically exfoliated graphene (EG) and an electrochemically active poly(3,4-ethylenedioxythiophene):poly(styrenesulfonate) (PEDOT:PSS) formulation (Clevios PH1000) (EG/PH1000 hybrid ink) was then prepared for the following printing. To get printed MSCs, mask-assisted spray printing was first performed, as shown in Figure 10.1b. EG ink was first spray-coated onto paper or PET substrate. Subsequently, a gel electrolyte of poly(vinyl alcohol) (PVA)/H_2SO_4 was drop-cast onto the printed finger electrodes. The thickness of the patterned finger electrodes can be easily tuned by adjusting the amount of EG ink and the number of spray-coating cycles, thus the capacitance of the MSCs can be controlled. Figure 10.1c showed the cyclic voltammetry curves of the MSC on a paper substrate at the scan rates of 1–100 $mV s^{-1}$. It showed that the device delivered a good areal capacitance of 800 $\mu F cm^{-2}$ at a scanning rate of 1 $mV s^{-1}$. Unfortunately, it was found that only 40% of the capacitance remained at the scanning rate of 100 $mV s^{-1}$. To improve the capacitance of the printed MSC, a EG/PH1000 hybrid ink was fabricated, which delivered greatly enhanced electrochemical performance. For the printed MSC, an areal capacitance of 1080 $\mu F cm^{-2}$ at a scan rate of 10 $mV s^{-1}$ was obtained, and there was no obvious capacitance change when the scan rate was increased to 100 $mV s^{-1}$. Besides the spray-printing, ink-jet printing was also performed in the work to show the potential of scale-up production. Figure 10.1d demonstrated the ink-jet printing of MSC arrays using conventional printer on a standard A4 paper.

MXenes, a new class of 2D metal carbides and nitrides, attracted great research interests in recent years [54–59]. They have shown significant promise for use as the electrode materials for supercapacitors. MXenes derived from the corresponding MAX phases, where A is typically Al or Ga. By etching away the A-group element, multi-layered MXene was obtained and can be expressed as $Mn_{+1}X_nT_x$, where M represents an early transition metal, X is C or N, T stands for various surface functionalities such as –OH, –O, or F [57–61]. MXene-based on-chip MSCs were also produced by the printing techniques. For example, Zhang et al. reported the produce of $Ti_3C_2T_x$ MXene based on-chip MSCs through a simple stamping printing strategy [62]. Figure 10.2a depicted the stamping strategy for the MSCs. Patterned stamp was first 3D printed assisted with computer. The MXene ink was then brushed onto the surface of the 3D printed stamp and firmly pressed onto the substrate (paper, PET, or others). After casting gel electrolyte, on-chip MSCs with desired structures were obtained. Figure 10.2b showed the photographs and scanning electron microscope (SEM) images of the stamp printed MSCs with different device configurations. As-printed MSCs exhibited areal capacitances of 61 $mF cm^{-2}$ at 25 $\mu A cm^{-2}$ and 50 $mF cm^{-2}$ at 800 $\mu A cm^{-2}$. To fulfill the mass production of on-chip MSCs, the authors also demonstrated the printing of MSCs with

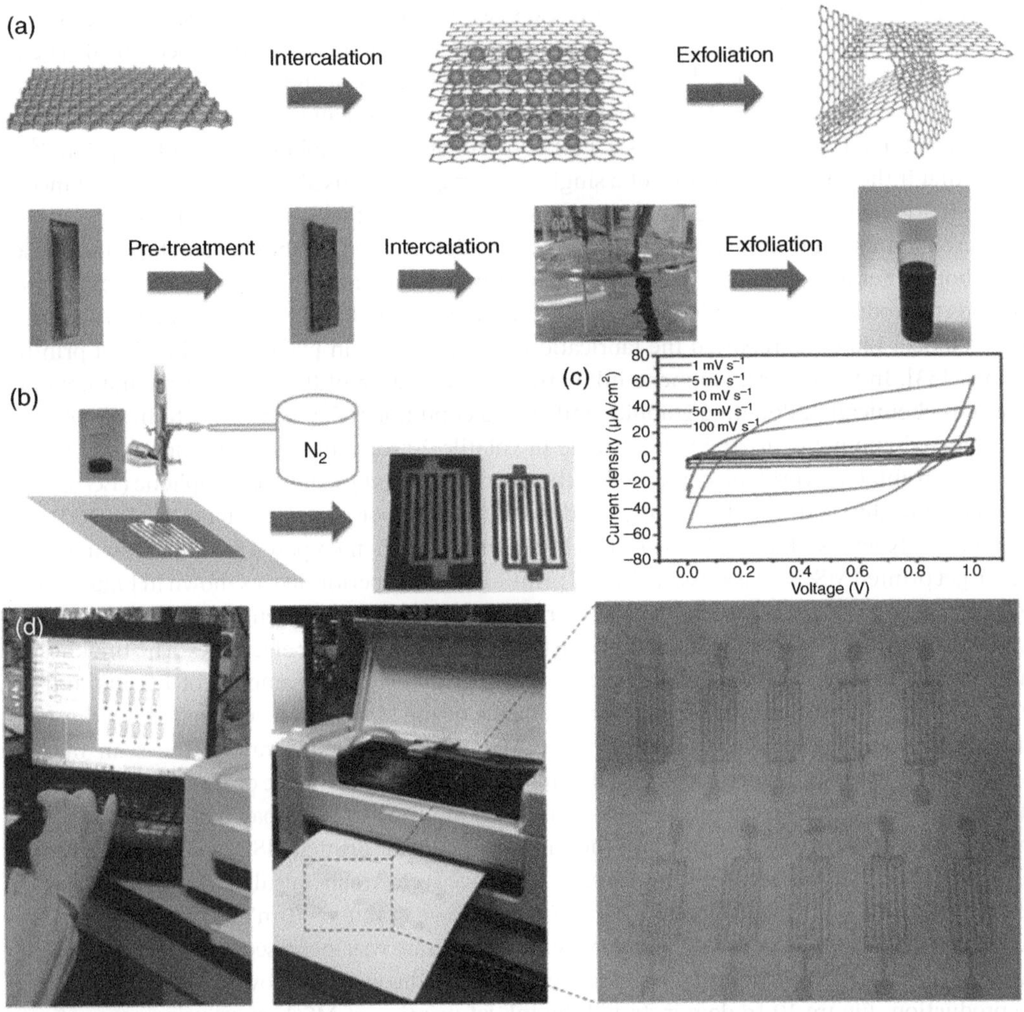

Figure 10.1 (a) Schematic illustration of the electrochemical exfoliation of graphene and photographs of the graphite foil volume change during exfoliation. (b) Schematic illustration of the mask-assisted spray deposition method for the direct printing of single and arrayed MSC devices. (c) Cyclic voltammetry curves of a pristine EG ink (total amount 3 ml) MSC on a paper substrate at scan rates of 1–100 $mV s^{-1}$. (d) Inkjet printing of custom-designed MSC arrays from a "home computer and printer" using pristine EG ink. (*Source:* Reproduced with permission from Ref. [53], © 2016, Wiley-VCH Verlag GmbH & Co. KGaA, Weinheim.)

designed cylindrical or pad stamps. Figure 10.2c showed the photograph of a 3D printed pad with 12 interdigitated stamps. After brushing the printable ink, the pad was covered with a piece of A4 paper and 12 MXene-based MSCs can be produced in minutes (Figure 10.2d). Further shortening the production time is possible by design interdigitated cylindrical stamps (Figure 10.2e). With this stamp, dozens of MXene-based MSCs can be rapidly produced in second.

In a recent work, Zhang et al. developed additive-free MXene inks and printed MSCs via extrusion printing and inkjet printing techniques [24]. Organic inks were prepared for the inkjet printing and aqueous inks for extrusion printing. Figure 10.3a showed the schematic illustration of direct MXene printing using the additive-free inks. Inkjet-printed patterns with the MXene ink

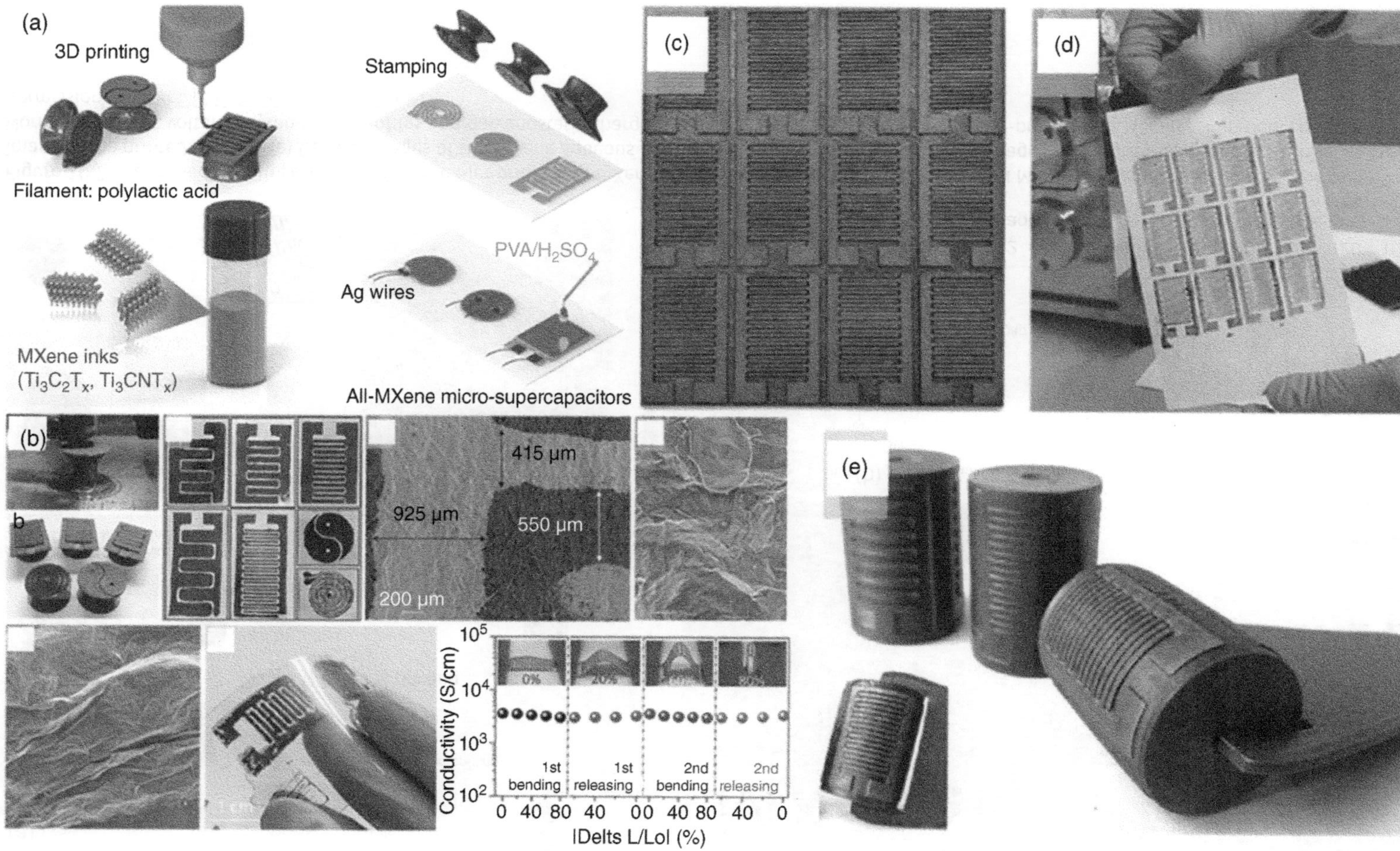

Figure 10.2 (a) Schematic illustration of the fabrication of MSCs using the stamping printing method. (b) As-stamped MSCs with different structures. (c) Scalable production of a pad with 12 interdigitated stamps. (d) As-stamped MSCs by rolling. (e) 3D printed cylindrical stamps. (*Source:* Reproduced with permission from Ref. [62], © 2018, Wiley-VCH Verlag GmbH & Co. KGaA, Weinheim.)

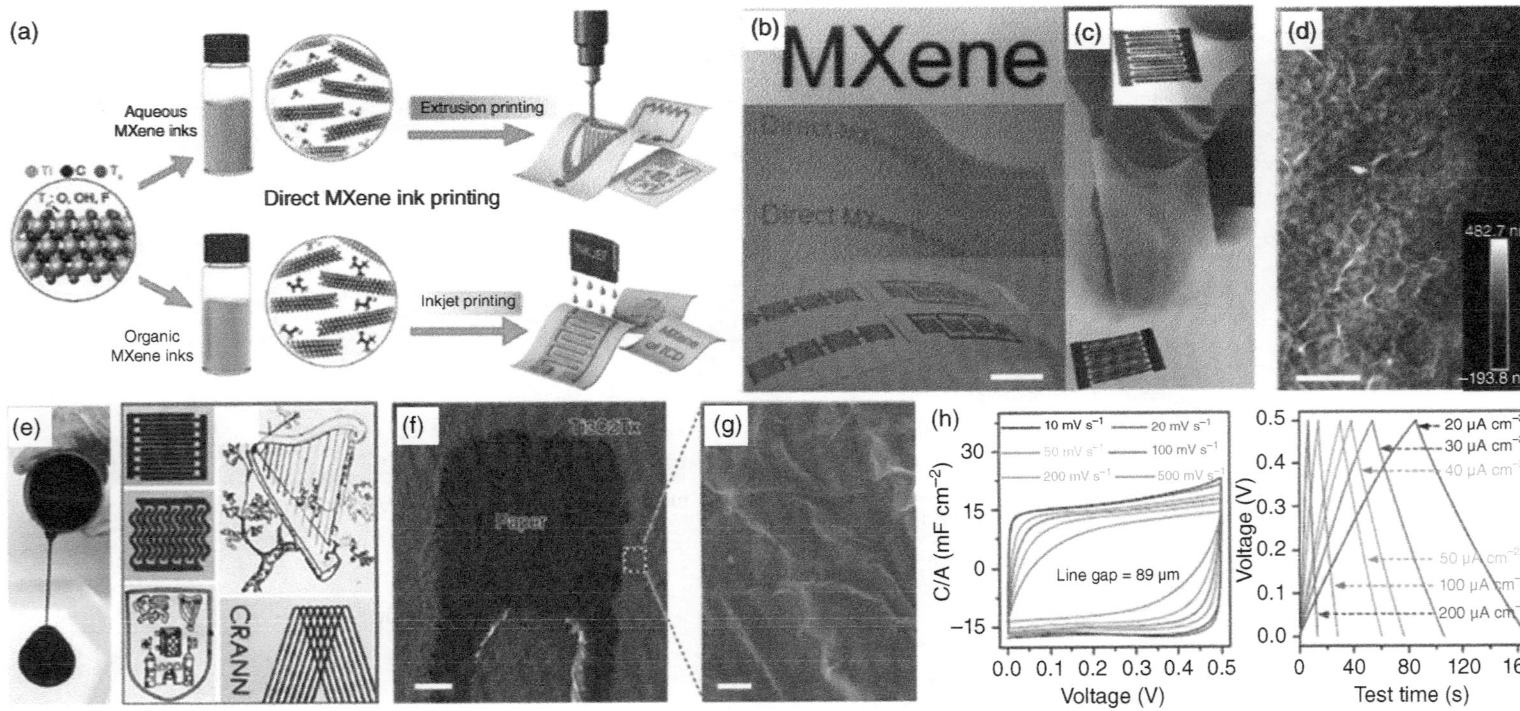

Figure 10.3 (a) Schematic illustration of direct ink-jet printing of MXene ink. (b–d) Optical images of the printed MSCs and atomic force microscope (AFM) image of the printed MXenes. (e) Optical images of the MXene aqueous ink and the printed patterns. (f, g) SEM images of the printed MXene MSC. (h) Normalized cyclic voltammograms (CV) profiles and galvanostatic charge-discharge (GCD) curves of the extrusion-printed MSC. (*Source:* Reproduced with permission from Ref. [24], © 2019, the Authors.)

were shown in Figure 10.3b–d, including “MXene” word and MSCs in series and parallel. As-printed MSC adhered strongly to the substrate, as confirmed by the clean scotch tape observed after peeling for ten times. Uniform MXene films with interconnected nanosheet network can be printed on various substrates. Figure 10.3e gave the optical images of the MXene patterns using the extrusion printing. The aqueous MXene ink showed suitable fluidic properties with good viscosity of ~0.71 Pa·s. Different patterns can be printed using the MXene ink with excellent resolution. Using this method, printing two paths gave on-chip MSCs with resolved line gap of 120 μm, width of 438 μm, and spatial uniformity within 5.6% (Figure 10.3f, g). The normalized cyclic voltammograms (CV) profiles and galvanostatic charge-discharge (GCD) curves of the extrusion-printed MSCs were depicted in Figure 10.3h. From the electrochemical measurement results, the printed MSCs exhibited high volumetric capacitance up to 562 F cm^{-3} and high energy density of 0.32 μW h cm^{-2}.

Pseudocapacitors are based on the fast and reversible redox reactions or Faradic charge transfer reactions at the surface of the electrode materials. Typical pseudocapacitive electrodes include transition metal oxides (e.g. Co_3O_4, RuO_2, or multiple metal oxides), transition metal sulfides (e.g. CoS_2, MoS_2, $NiCo_2S_4$), transition metal selenides (NiSe, CoSe, etc.), and conducting polymers (e.g. polyaniline, polypyrrole, polythiophene, and derivatives) [63–71]. Pseudocapacitors can also be fabricated using the printing techniques [72–75]. For example, Wang et al. fabricated SnSe nanoparticles based supercapacitors using spray printing method [76]. SnSe nanoparticles with the sizes of tens of nanometers were first prepared via a sol-gel process, which were dispersed in distilled water to form the SnSe based ink. The ink was then spray printed on various substrates including porous carbon cloth, PET substrate, etc. (Figure 10.4a, b). Assisted with mask, patterned SnSe electrode could be obtained. Figure 10.4c, d showed the CV curves of the device in the potential range of −0.3 to 0.3 V at scan rates of 20–200 mV s^{-1}, and the specific capacitances at various current densities, respectively. As-printed supercapacitor exhibited high capacitance of 1800 μF cm^{-2} at the current density of 20 μA cm^{-2}, excellent flexibility, and long cycle life and good rate capability.

Printing techniques were also applied to MSCs with composite electrodes. For instance, Wang et al. reported the screen printing of flexible MSCs with MnO_2/onion-like carbon (MnO_2/OLC) as the active material [77]. As-printed MSCs exhibited capacitance of 7.04 mF cm^{-2} at a current density of 20 μA cm^{-2}.

10.2.2 Printable Current Collector

Current collector is one of the most important parts of supercapacitor. Its primary function is to collect the charge generated by the active electrode material to produce a large current output. To facilitate charge transport, current collectors should be fully attached with the active electrode materials. For printed MSCs, the current collectors should be printed into interdigitated configuration; as a result, the current collectors should be capable to endow good mechanical compliance to withstand deformation. Typical current collectors for printed MSCs can be divided into three categories: metals, carbon-based materials, and conductive polymers.

Metal nanoparticles are the most commonly used current collector for printed MSCs. Printable inks with metal nanoparticles were first printed on device substrate. Then it was annealed at desired temperature to get current collector suitable for following MSC fabrication. As the metal nanoparticles have high surface to volume ratio, they usually possess very low melting point compare to their bulk counterparts. Typical 150–200 °C for 1–2 h is sufficient to recover the majority of the electrical conductivity of the metal nanoparticles. For example, Liu et al. fabricated on-chip

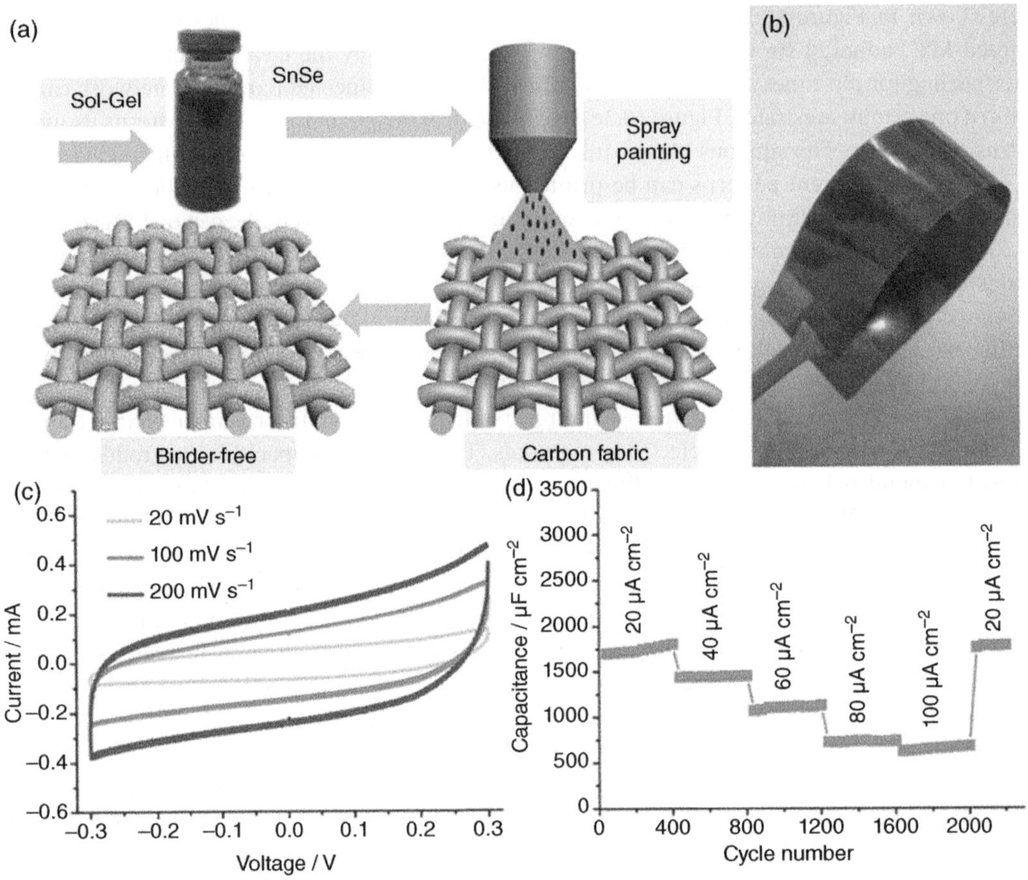

Figure 10.4 (a) Schematic illustration of the fabrication of SnSe based supercapacitors using the spray printing process. (b) Optical image of printed electrode on the flexible PET substrate. (c) CV curves measured at different scan rates. (d) Specific capacitance at various current densities. (*Source:* Reproduced with permission from Ref. [76], © 2014, Wiley-VCH Verlag GmbH & Co. KGaA, Weinheim.)

MSCs using screen printing technique [78]. Figure 10.5a showed the optical image of the used screen-printing facility, including a brush, screen, and screen fixer. To print the MSC, Ag-based current collector was first printed on PET substrate with conductive silver paste (SPI Supplies, USA), which was then heated in vacuum at 200 °C for one hour to remove the polymer in the paste. N-doped rGO was then screen printed onto the Ag current collector and dried at 110 °C under vacuum overnight. After dropped gel electrolyte (PVA-H_3PO_4) on the substrate, flexible MSCs were obtained, as shown in Figure 10.5b. As-fabricated flexible MSC showed excellent flexibility. The electrochemical performance of the device was studied, as depicted in Figure 10.5c,d. An areal capacitance of 3.4 mF cm^{-2} was obtained for the device, which is among the high values of graphene-based MSCs.

Although metal nanoparticles contained inks are easy to be printed, cracks usually occurred even under small bending strains, which limited their practical applications in flexible electronics. To overcome this problem, metallic nanowires are used instead. For example, Liang et al. developed a water-based Ag nanowire ink for screen printing Ag nanowire films [79]. The Ag nanowire ink contains a low solid content of 6.6 wt%, a mixture of (hydroxypropyl)methyl cellulose,

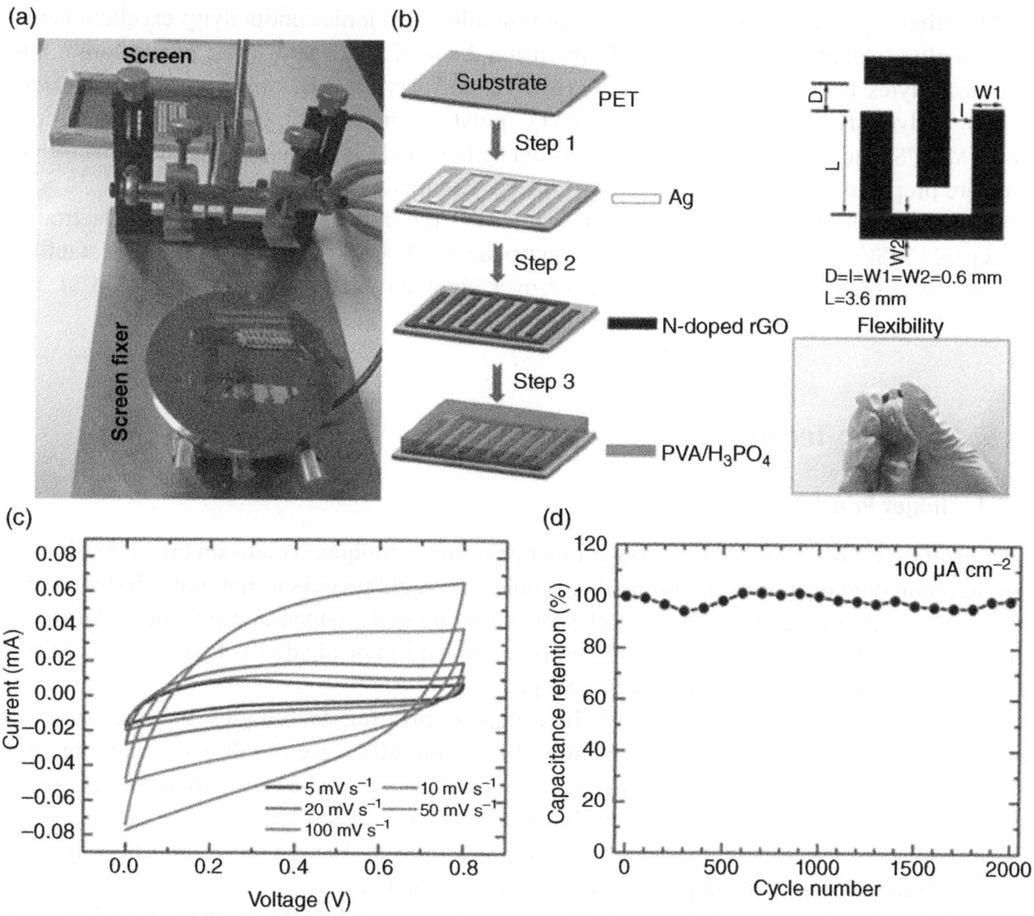

Figure 10.5 (a) Optical image of the screen printing facility. (b) Schematic illustration of the fabrication process of the flexible MSCs. (c) CV plots of the device at various scan rates of 5–100 $mV\,s^{-1}$. (d) Cycling stability of the device at 100 $\mu A\,cm^{-2}$. (*Source:* Reproduced with permission from Ref. [78], © 2014, The Royal Society of Chemistry.)

fluorosurfactant, and an antifoaming agent as organic additives. The interactions between the organic additives and the Ag nanowire networks enabled the screen printing of Ag nanowire based lines with 50 μm resolution, high electrical conductivity of $4.67 \times 10^4\,S\,cm^{-1}$, and very excellent stretchability.

10.2.3 Printable Electrolyte

Electrolyte is an important component of supercapacitors. It mainly affects the electrochemical properties of the devices including voltage window, rate capability, and cycling stability. Basically, electrolytes can be classified into two types, the liquid electrolytes (aqueous or organic based solution) and the solid-state electrolytes (or gel electrolytes) [33, 80–84]. However, for printing MSCs aiming at wearable electronics, liquid electrolytes often faced with the leakage problem, thus leading to safety issues. As a result, gel-like or solid-state electrolytes are the most commonly used electrolytes for printable MSCs. Ideal gel-like or solid-state electrolytes for printable MSCs should exhibit high electrochemical and thermal stability, high ionic conductivity, excellent bending/stretching tolerance, good safety and dimensional stability. Till now, most of the commonly used electrolytes for printable MSCs include PVA, polyacrylonitrile, poly(vinylidene fluoride) (PVDF), poly(4-styrenesulfonic acid) (PSSH), poly(styrene-b-methylmethacrylateb-styrene) (PS-PMMA-PS), etc. One good example is reported by Li et al. [85] In their work, PSSH electrolyte inks were prepared by mixing PSSH with deionized water, ethylene glycol, and phosphoric acid. The as-prepared electrolyte inks were then printed on the surface of interdigitated electrodes. As-prepared MSCs exhibited very attractive electrochemical performance with excellent stability. They can be stored at ambient condition for over eight months without the addition of any encapsulation.

10.3 Printing Techniques

10.3.1 Inkjet Printing

Inkjet printing is an additive-based, non-contact printing technique, which can create mask-free, high precision digital patterns. By propelling droplets of liquid precursor materials, designed patterns can be deposited on various substrates including metallic, paper, and polymers. An inkjet printer consists of a print head and an ink reservoir, which can be divided as two modes: continuous inkjet printing and drop-on-demand inkjet printing.

Inkjet printing is one of the mostly investigated printing techniques to print on-chip MSCs [22, 85–88]. Key issues of inkjet printing to on-chip MSCs are the droplet formation and morphology optimization of printed patterns. Generally, the quality and resolution of the printed patterns are dependent on the capability of the apparatus, the viscosity surface tension, dispersibility of the inks and the wettability of the substrate. As a result, it is the most important thing to develop printable inks to print high performance on-chip MSCs.

Li et al. developed a simple full-inkjet-printing technique to fabricate graphene-based on-chip MSCs [85]. To get the full printed MSCs, two kinds of printable inks were developed, including the electrochemically exfoliated graphene (EEG) inks (electrodes) and the PSSH inks (electrolytes), as shown in Figure 10.6a. The prepared inks exhibited a long stable period over six months with excellent jetting performance. They can be printed on various substrates including glass, Kapton, and silicon substrates, etc. (Figure 10.6b–d). The thickness and device configurations greatly affected the electrochemical performance of the printed MSCs. Optimized MSC exhibited a highest areal capacitance of about $0.7\,mF\,cm^{-2}$. Such printing process is easily scaled up and MSC arrays could be printed in short time. Figure 10.6e showed the photograph of a MSC arrays printed on Kapton substrate, which contained 9 parallel rows of 12 series-connected MSCs (denoted as 12S×9P). Compared with single MSC, large area MSC arrays delivered higher power density (Figure 10.6f–h).

Another interesting work on inkjet-printed MSCs was done by Choi et al., which was printed on a conventional A4 paper [86]. As shown in Figure 10.7a,b, to improve the quality of the printed patterns, a layer of cellulose nanofibril (CNF) nanomats was first printed on the A4 paper. The ink containing single-walled carbon nanotubes (SWCNTs) and AC were then inkjet-printed on top of the CNF-coated paper, followed by the printing of Ag nanowire inks. To improve the electrical conductivity of the printed electrodes, SWCNTs assisted photonic sintering of Ag nanowires were performed. Afterwards, the ionic liquid/ultraviolet-cured triacrylate polymer-based solid-state electrolyte ink was inkjet-printed on the top of the treated electrodes. The fully inkjet-printed

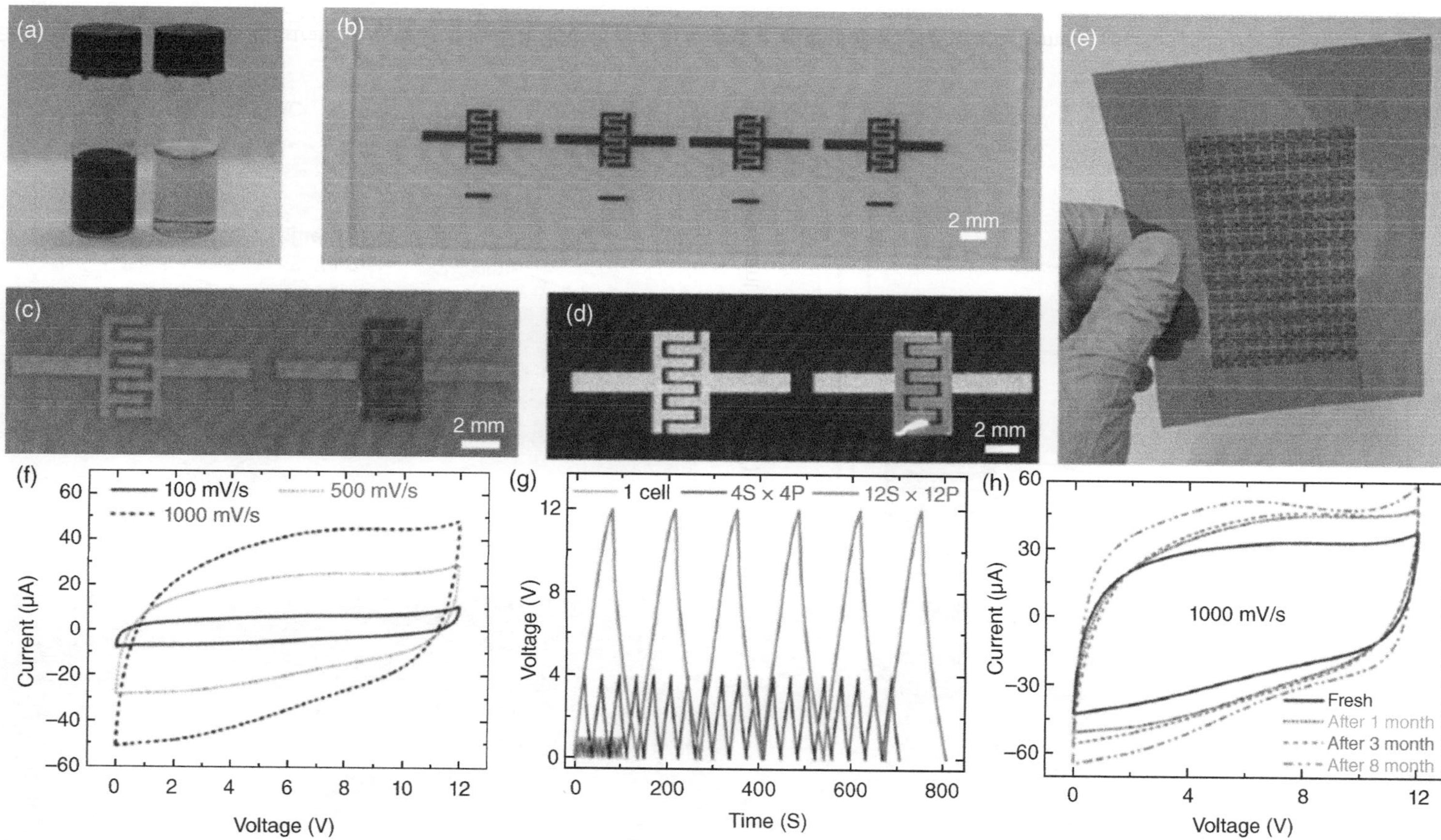

Figure 10.6 (a) Photograph of the EEG ink and PSSH ink. (b) Photograph of the printed MSCs on glass substrate. (c, d) Photographs of the printed EEG electrodes (left) and the MSCs with printed PSSH electrolytes (right) on Kapton and silicon substrate, respectively. (e) Photograph of the fully inkjet printed MSC arrays. (f–h) Electrochemical measurements of the MSC arrays. (*Source:* Reproduced with permission from Ref. [85], © 2017, American Chemical Society.)

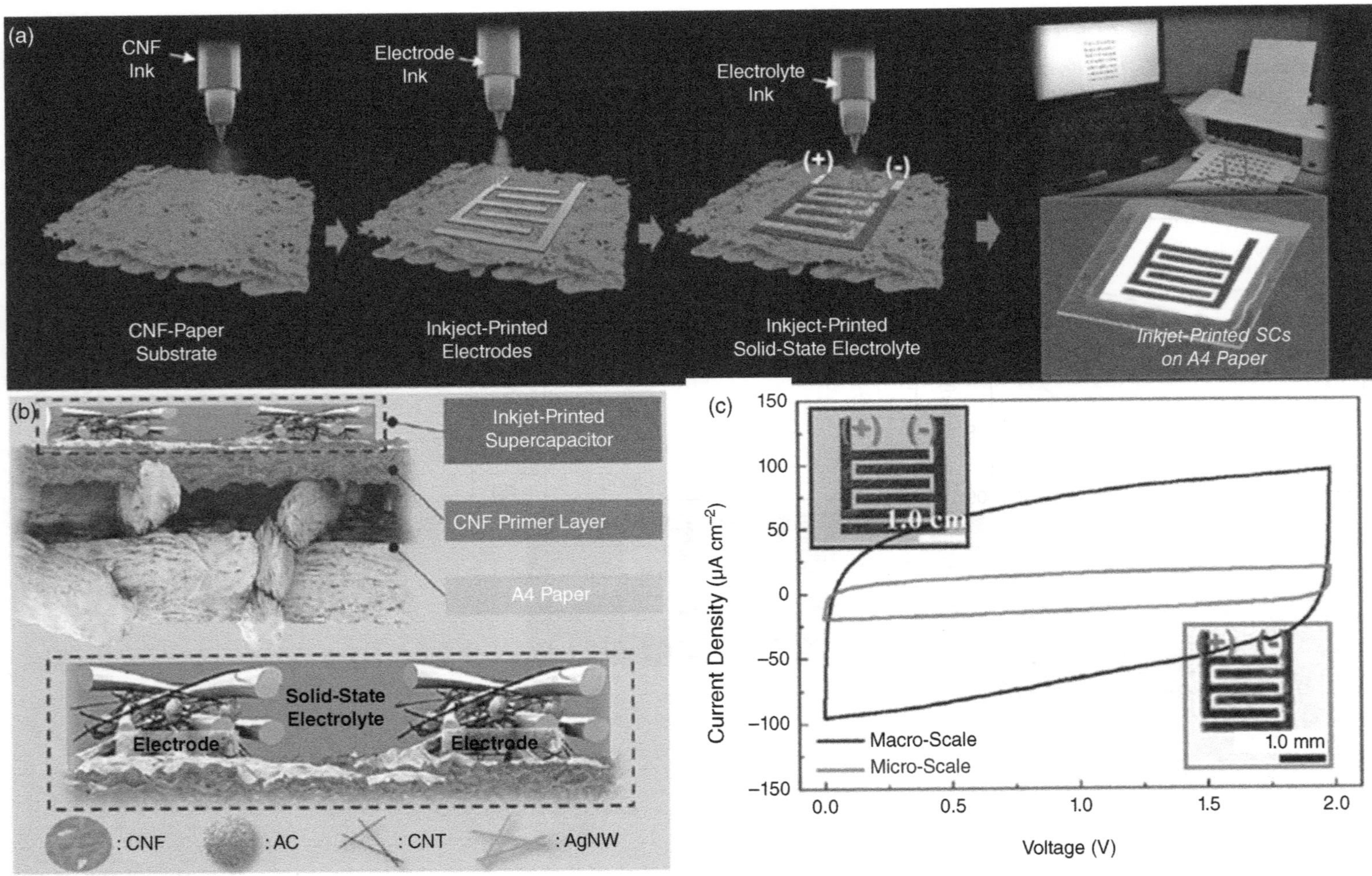

Figure 10.7 (a) Schematic illustration of the stepwise fabrication procedure of the inkjet-printed MSCs. (b) Schematic illustration of the architectural uniqueness of the inkjet-printed MSCs. (c) CV profiles of the inkjet-printed MSCs with different dimensions. (*Source:* Reproduced with permission from Ref. [86], © 2016, The Royal Society of Chemistry.)

MSCs demonstrated dimensional integrity, mechanical flexibility, and good electrochemical performance (Figure 10.7c). These MSCs can be readily connected in series or parallel without the aid of metallic interconnects, leading to user-customized control of cell voltage and capacitance.

Besides on-chip MSCs governed by the EDLC mechanism, on-chip MSCs with pseudocapacitive mechanism were also prepared using the inkjet printing process. For instance, Giannakou et al. printed NiO nanoparticles based MSCs [88]. The process includes: inkjet-printing of silver current collector, inkjet-printing NiO active electrode on the top of the silver layer, annealing the electrode and electrolyte drop casting. As-printed NiO nanoparticles based MSCs showed up to 14 orders of magnitude higher electrical conductivity than single crystal NiO. Remarkable areal and volumetric specific capacitances of 155 $mF\,cm^{-2}$ and 705 $F\,cm^{-3}$ were obtained. Besides, Li et al. combined laser induced graphene (LIG) as the current collector and inkjet-printed bis-terpyridyl based molecular cobalt complexes (TPy-Co) as the active electrodes [73]. The pseudocapacitive TPy-Co based MSC exhibited stable electrochemical performance with a capacitance two orders of magnitude higher than pristine LIG.

10.3.2 Spray Printing

Spray printing has been considered as a low-cost, easy operation, and industrial up scaling method and has been widely used in traditional coatings [89–91]. It is a noncontact printing method where the ink droplets (containing nanoparticles or microparticles) pass through a nozzle to deposit onto the target substrates. As spray printing is a noncontact process, the particles can be coated onto uneven or curved surfaces, allowing for easy selection and modification of printed patterns.

In recent years, spray printing was also widely used to fabricate on-chip MSCs. Xiong et al. recently reported the produce of on-chip MSCs using the spray printing method. Graphene oxide (GO) with the lateral area of 14.6 μm^2 and thickness of 0.93 nm was first prepared by the oxidation of natural graphite powder. GO was then mixed with PANI nanofibers in water to form into GO/PANI colloidal gels for further spray printing. Well controlled GO/PANI colloidal gels were able to be sprayed out through 350 μm nozzle under a shear rate of 10^5–$10^6\,s^{-1}$. Figure 10.8a showed the optical microscopy image of the spray-printed patterns of GO/PANI (GP-3.0-3.0) gels on PET substrate, which showed complementary structure. GO/PANI gels were also sprayed on gold current collectors to fabricate MSC arrays, as demonstrated in Figure 10.8b,c. The optimized MSCs exhibited high areal capacitance of 35–57 $mF\,cm^{-2}$ at the current density of 0.1 $mA\,cm^{-2}$, about four to six times higher than the other reported spray-printed MSCs (Figure 10.8d,e). Spray printability of GO/PANI colloidal gels make it possible to be printed on nonplanar flexible surfaces. For example, Figure 10.8f showed the photograph of the MSCs printed on the curved columnar fingers of gloves. The patterned shapes were well retained. Compared with the MSCs printed on planar substrates, the MSCs on nonplanar flexible surfaces did not show capacitance degradation even under bending states. To demonstrate the possible applications of the printed MSCs, the authors integrated the MSCs on gloves with flexible resistive-type strain sensor. The strain sensors were successfully powered by the MSCs to detect the finger motion (Figure 10.8g).

Shi et al. fabricated an interesting graphene-based linear tandem MSC via the spray printing method [91]. Linear arrays of patterned film electrodes were first prepared by spray printing of graphene-based ink through an inconsecutively parallel linear customized mask. Subsequently, a polymer gel electrolyte of H_2SO_4/PVA was drop-casted onto the linear electrodes to finish the fabrication of MSC. Studies found that the printed linear tandem MSC array consisting of 10 MSCs presented high voltage output of 8.0 V with remarkable flexibility. Both symmetric and asymmetric MSCs could be fabricated using the spray printing technique.

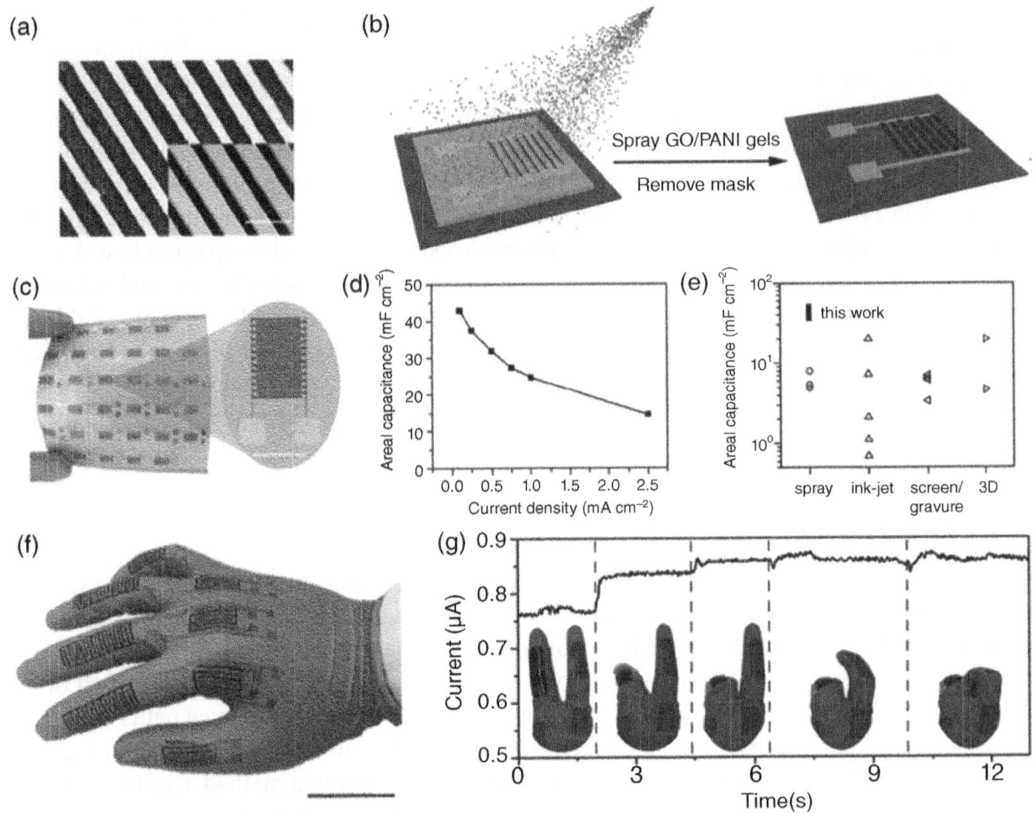

Figure 10.8 (a) Optical microscopy image of spray-printed patterns of GP-3.0-3.0 gels on PET substrate. (b) Schematic illustration of spray printing of GO/PANI gels for MSCs. (c) Photograph of GP-3.6-2.4 gel-based MSC array on Au-coated polyimide substrate. (d) Areal capacitance of GP-3.6-2.4 gel-based MSCs. (e) Comparison of the spray-printed MSCs with other printed devices in literature. (f) Photograph of GP-0.52-0.34 gel-based MSC array on the columnar surface of gloves prepared from the spray printing process. (g) Current-time curve of the integrated devices during finger motion. (*Source:* Reproduced with permission from Ref. [89], © 2019, Wiley-VCH Verlag GmbH & Co. KGaA, Weinheim.)

10.3.3 Screen Printing

Screen printing is a widely used mass-printing process. It is a contact process that presses the ink down with a squeegee with sufficient force to penetrate through the prepatterned stencil screen onto a desired substrate. A wide variety of inks can be used to screen printing on-chip MSCs on various planar substrates due to its simplicity and versatility [77, 92–96]. Wang et al. fabricated all-solid on-chip MSCs by utilizing screen printing technique. The corresponding fabrication process was depicted in Figure 10.9a. It started from the printing of Ag current collectors on PET substrate with Ag conductive paint (SPI Supplies, USA). After drying and annealing in vacuum at 200 °C for 1 h, the active electrode ink was screen-printed on top of the Ag current collector, followed by drying at 110 °C. Finally, the PVA/H_3PO_4 electrolyte was coated on top of the device to finish the fabrication process. Using the screen-printing method, on-chip MSCs can be printed on various substrates (Figure 10.9b), including PET, glass, and printing paper. MSCs printed on PET

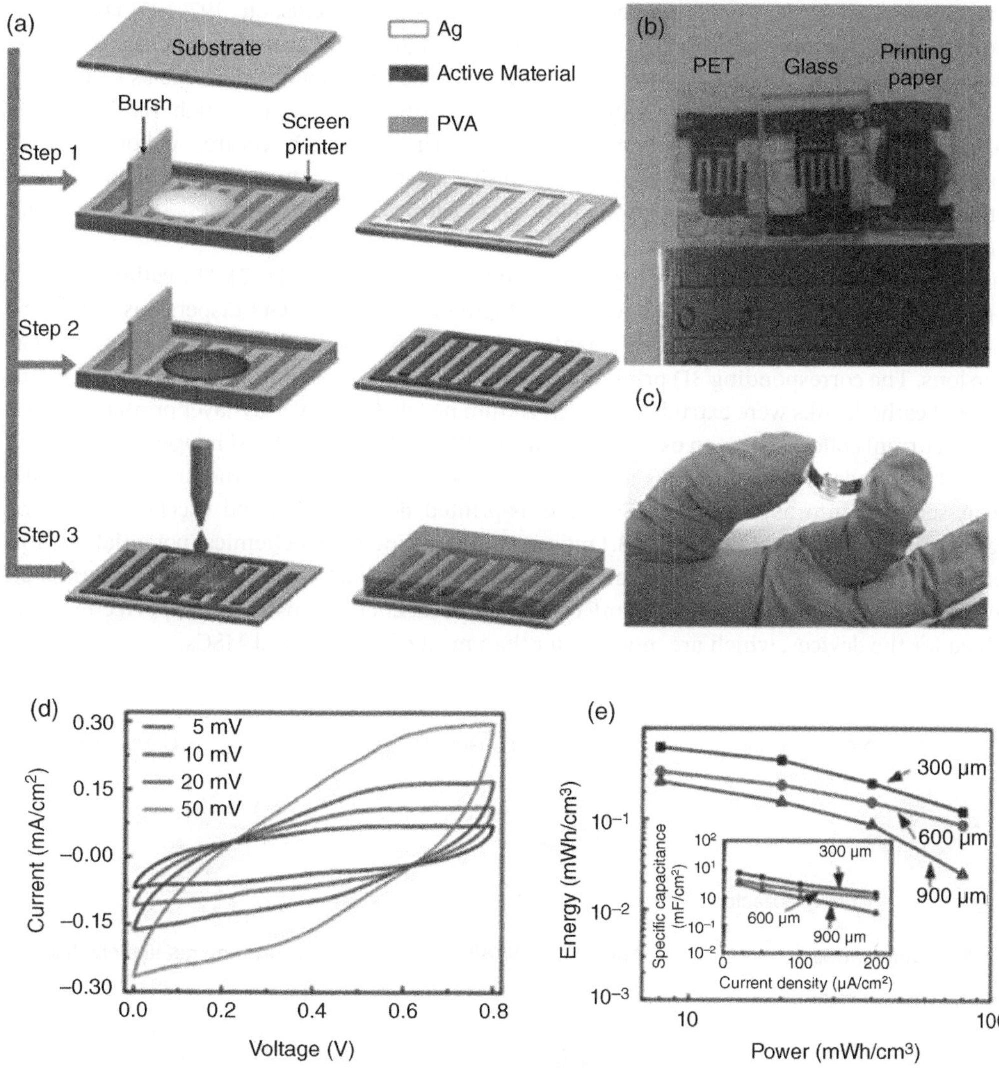

Figure 10.9 (a) Schematic diagram of the screen printing process to on-chip MSCs. (b) Photographs of the MSCs printed on different substrates. (c) Photograph of the flexible MSCs. (d) Cycling performance of a single printed MSC. (e) Specific energy with the power density of the printed MSCs with various channel lengths. (*Source:* Reproduced with permission from Ref. [77], © 2014, IOP Publishing Ltd.)

substrate exhibited excellent flexibility, as shown in Figure 10.9c. Considering the electrochemical measurements shown in Figure 10.9d,e, the printed MSCs exhibited capacitance of 7.04 mF cm^{-2} at a current density of 20 μA cm^{-2} with excellent cycling stability.

10.3.4 3D Printing

Three-dimensional (3D) printing or additive manufacturing is an extrusion-based technology that has been successfully demonstrated to construct complex structures with a wide application in various areas, such as electronic, biomedical, and energy storage fields, etc. [97–105] During the printing process, the functional inks were extruded through nozzle and directly printed on substrates layer by layer in the vertical dimension. Each of these layers can be regarded as a thin horizontal cross-section of the eventual object. To develop suitable inks with high viscosities and shear-thinning rheological properties, it is essential to print ideal architectures without collapse. Compared with other printing techniques, 3D printing offers great advantages, such as excellent printing flexibility and geometry controllability, good control over the thickness of the patterns, and cost effectiveness and environment friendliness.

Shen et al. printed asymmetric on-chip MSCs using the 3D technique [102]. The cathode ink was composed of vanadium pentoxide (V2O5) with highly concentrated GO dispersions, while the anode was graphene-vanadium nitride quantum dots (G-VNQDs) with highly concentrated GO dispersions. The corresponding 3D printing procedures were illustrated in Figure 10.10a. Both the anode and cathode inks were extruded through 200 μm nozzle to be layer-by-layer printed on interdigitated current collectors. As an example, Figure 10.10b–d showed the SEM images of the printed VOx/rGO electrode. The continuous and stacked closely layer-by-layer structure can be clearly seen. Assembled into asymmetric MSCs, the as-printed devices exhibited excellent structural integrity, a large areal mass loading of 3.1 mg cm^{-2}, and a wide electrochemical potential window of 1.6 V. Besides, the 3D printed asymmetric MSCs showed excellent electrochemical performance. An ultrahigh areal capacitance of 207.9 mF cm^{-2}, high areal energy density of 73.9 μWh cm^{-2} were obtained for the devices, which are much better than most of the reported MSCs.

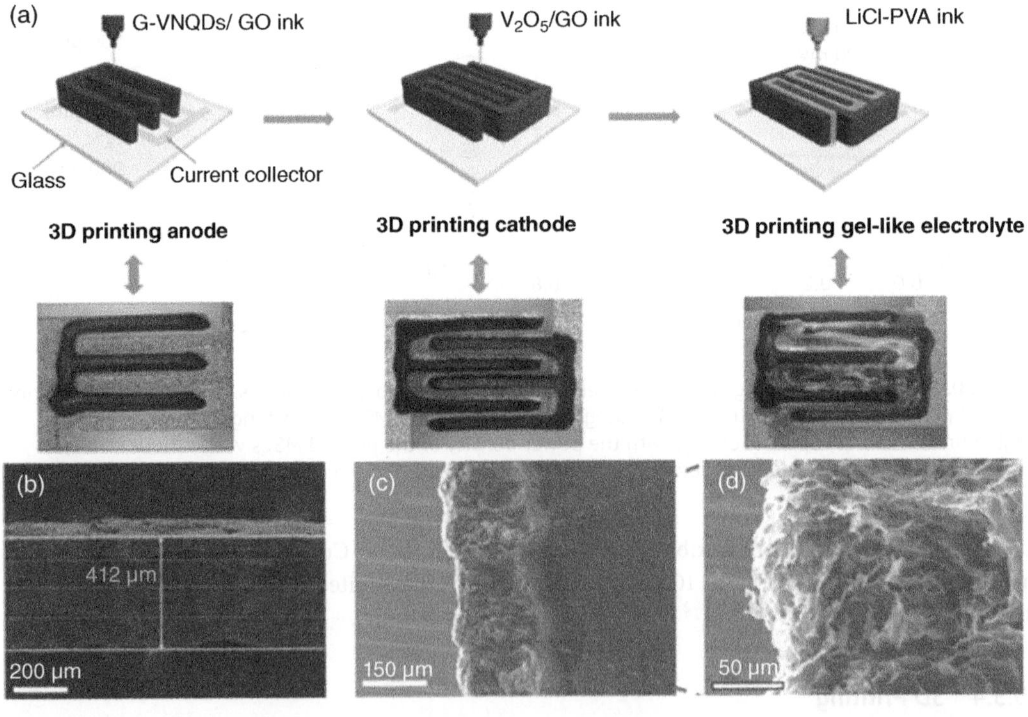

Figure 10.10 (a) Schematic diagram of the 3D printing process for MSCs. (b–d) SEM images of the 3D printed VO_x/rGO electrode. (*Source:* Reproduced with permission from Ref. [102], © 2018, Wiley-VCH Verlag GmbH & Co. KGaA, Weinheim.)

10.4 Summary

In summary, we summarized in this review the recent progress of printing on-chip MSCs. Printable materials including active electrodes, current collectors, and electrolytes, were highlighted in detail. Following this part, the recent progress of fabricating different types of on-chip MSCs using different printing techniques was also discussed, mainly focused on inkjet printing, screen printing, spray printing, and 3D printing techniques. In fact, besides the introduced printing techniques, many other printing techniques were also applied to fabricate on-chip MSCs, including laser-printing, stamp-assisted printing, transfer printing and their combinations.

Although great progress has been obtained on printing on-chip MSCs, there are still a lot of issues should be considered. First, printable inks suitable for different printing techniques should be developed. Many parameters including solid loading, particle dispersion, viscosity, rheology behavior, etc. should be considered to form printable inks. Second, the uniformity in both device structures and electrochemical performance should be improved. Although on-chip MSCs could be mass produced via the printing method, the uniformity was still far from practical application requests. With the great succeed in producing on-chip MSCs, how to use these devices as power units and integrate them with other functional devices should be considered.

References

1 Zamarayeva, A.M., Ostfeld, A.E., Wang, M. et al. (2017). Flexible and stretchable power sources for wearable electronics. *Sci. Adv.* 3 (6): e1602051.

2 Beidaghi, M. and Gogotsi, Y. (2014). Capacitive energy storage in micro-scale devices: recent advances in design and fabrication of micro-supercapacitors. *Energy Environ. Sci.* 7 (3): 867–884.

3 Wang, C.J., Sim, K., Chen, J. et al. (2018). Soft ultrathin electronics innervated adaptive fully soft robots. *Adv. Mater.* 30 (13): 1706695.

4 Kim, J., Shim, H.J., Yang, J. et al. (2017). Ultrathin quantum dot display integrated with wearable electronics. *Adv. Mater.* 29 (38): 1700217.

5 Wang, X.F., Liu, B., Wang, Q.F. et al. (2013). Three-dimensional hierarchical GeSe2 nanostructures for high performance flexible all-solid-state supercapacitors. *Adv. Mater.* 25 (10): 1479–1486.

6 Xu, J., Wang, Q.F., Wang, X.W. et al. (2013). Flexible asymmetric supercapacitors based upon Co_9S_8 Nanorod//Co_3O_4@RuO_2 nanosheet arrays on carbon cloth. *ACS Nano* 7 (6): 5453–5462.

7 Liu, B., Zhang, J., Wang, X.F. et al. (2012). Hierarchical three-dimensional $ZnCo_2O_4$ nanowire arrays/carbon cloth anodes for a novel class of high-performance flexible lithium-ion batteries. *Nano Lett.* 12 (6): 3005–3011.

8 Wang, X.F., Liu, B., Liu, R. et al. (2014). Fiber-based flexible all-solid-state asymmetric supercapacitors for integrated photodetecting system. *Angew. Chem. Int. Ed.* 53 (7): 1849–1853.

9 Wang, X.F., Lu, X.H., Liu, B. et al. (2014). Flexible energy-storage devices: design consideration and recent progress. *Adv. Mater.* 26 (28): 4763–4782.

10 Wang, X.F., Jiang, K., and Shen, G.Z. (2015). Flexible fiber energy storage and integrated devices: recent progress and perspectives. *Mater. Today* 18 (5): 265–272.

11 Wu, Z.S., Parvez, K., Feng, X.L. et al. (2013). Graphene-based in-plane micro-supercapacitors with high power and energy densities. *Nat. Commun.* 4: 2487.

12 Xiao, H., Wu, Z.S., Chen, L. et al. (2017). One-step device fabrication of phosphorene and graphene interdigital micro-supercapacitors with high energy density. *ACS Nano* 11 (7): 7284–7292.

13 Chmiola, J., Largeot, C., Taberna, P.L. et al. (2010). Monolithic carbide-derived carbon films for micro-supercapacitors. *Science* 328 (5977): 480–483.

14 Peng, L.L., Peng, X., Liu, B.R. et al. (2013). Ultrathin two-dimensional MnO_2/graphene hybrid nanostructures for high-performance, flexible planar supercapacitors. *Nano Lett.* 13 (5): 2151–2157.

15 Lu, Y., Jiang, K., Chen, D. et al. (2019). Wearable sweat monitoring system with integrated micro-supercapacitors. *Nano Energy* 58: 624–632.

16 Lu, Y., Lou, Z., Jiang, K. et al. (2019). Recent progress of self-powered wearable monitoring systems integrated with microsupercapacitors. *Mater. Today Nano* 8: 100050.

17 Huang, T.T., Jiang, K., Li, L. et al. (2018). Large-scale fabrication of flexible on-chip micro-supercapacitors by a mechanical scribing process. *ChemElectroChem* 5 (13): 1652–1657.

18 Huang, T.T., Jiang, K., Chen, D. et al. (2018). Recent progress and perspectives of metal oxides based on-chip microsupercapacitors. *Chin. Chem. Lett.* 29 (4): 553–563.

19 Li, L., Fu, C.W., Lou, Z. et al. (2017). Flexible planar concentric circular micro-supercapacitor arrays for wearable gas sensing application. *Nano Energy* 41: 261–268.

20 Gu, S.S., Lou, Z., Li, L.D. et al. (2016). Fabrication of flexible reduced graphene oxide/Fe_2O_3 hollow nanospheres based on-chip micro-supercapacitors for integrated photodetecting applications. *Nano Res.* 9 (2): 424–434.

21 Xu, J. and Shen, G.Z. (2015). A flexible integrated photodetector system driven by on-chip microsupercapacitors. *Nano Energy* 13: 131–139.

22 Sundriyal, P. and Bhattacharya, S. (2019). Scalable micro-fabrication of flexible, solid-state, inexpensive, and high-performance planar micro-supercapacitors through inkjet printing. *ACS Appl. Energy Mater.* 2 (3): 1876–1890.

23 He, W., Ma, R.J., and Kang, D.J. (2020). High-performance, flexible planar microsupercapacitors based on crosslinked polyaniline using laser printing lithography. *Carbon* 161: 117–122.

24 Zhang, C.F., McKeon, L., Kremer, M.P. et al. (2019). Additive-free MXene inks and direct printing of micro-supercapacitors. *Nat. Commun.* 10: 1795.

25 Li, H.P. and Liang, J.J. (2020). Recent development of printed micro-supercapacitors: printable materials, printing technologies, and perspectives. *Adv. Mater.* 32 (3): 1805864.

26 Cui, Z. (2016). *Printed Electronics: Materials, Techniques and Applications*, 316–338. Fusionopolis, Solaris South Tower, Singapore: Wiley.

27 Xu, J., Wang, X.F., Wang, X.W. et al. (2014). Three-dimensional structural engineering for energy-storage devices: from microscope to macroscope. *ChemElectroChem* 1 (6): 975–1002.

28 Chen, P.C., Shen, G.Z., Shi, Y. et al. (2010). Preparation and characterization of flexible asymmetric supercapacitors based on transition-metal-oxide nanowire/single-walled carbon nanotube hybrid thin-film electrodes. *ACS Nano* 4 (8): 4403–4411.

29 Conway, B.E. (2013). *Electrochemical Supercapacitors: Scientific Fundamentals and Technological Applications*. Springer.

30 Pandolfo, A.G. and Hollenkamp, A.F. (2006). Carbon properties and their role in supercapacitors. *J. Power Sources* 157 (1): 11–27.

31 Chen, S., Zhu, J.W., Wu, X.D. et al. (2010). Graphene oxide-MnO_2 nanocomposites for supercapacitors. *ACS Nano* 4 (5): 2822–2830.

32 Laforgue, A., Simon, P., Sarrazin, C. et al. (1999). Polythiophene-based supercapacitors. *J. Power Sources* 80 (1–2): 142–148.

33 Wang, G.P., Zhang, L., and Zhang, J.J. (2012). A review of electrode materials for electrochemical supercapacitors. *Chem. Soc. Rev.* 41 (2): 797–828.

34 Frackowiak, E., Jurewicz, K., Delpeux, S. et al. (2001). Nanotubular materials for supercapacitors. *J. Power Sources* 97–98: 822–825.

35 Jiang, H., Lee, P.S., and Li, C.Z. (2013). 3D carbon based nanostructures for advanced supercapacitors. *Energy Environ. Sci.* 6 (1): 41–53.

36 Gonzalez, A., Goikolea, E., Barrena, J.A. et al. (2016). Review on supercapacitors: technologies and materials. *Renew. Sustain. Energy Rev.* 58: 1189–1206.

37 Wu, N.L. (2002). Nanocrystalline oxide supercapacitors. *Mater. Chem. Phys.* 75 (1–3): 6–11.

38 Beguin, F. and Frackowiak, E. (2013). *Supercapacitors: Materials, Systems, and Applications*. KGaA: Wiley-VCH Verlag GmbH & Co.

39 Lokhande, C.D., Dubal, D.P., and Joo, O.S. (2011). Metal oxide thin film based supercapacitors. *Curr. Appl. Phys.* 11 (3): 255–270.

40 Peng, X., Peng, L.L., Wu, C.Z. et al. (2014). Two dimensional nanomaterials for flexible supercapacitors. *Chem. Soc. Rev.* 43 (10): 3303–3323.

41 Zhu, Y.W., Murali, S., Stoller, M.D. et al. (2011). Carbon-based supercapacitors produced by activation of graphene. *Science* 332 (6037): 1537–1541.

42 Zhang, L.L. and Zhao, X.S. (2009). Carbon-based materials as supercapacitor electrodes. *Chem. Soc. Rev.* 38 (9): 2520–2531.

43 Pech, D., Brunet, M., Durou, H. et al. (2010). Ultrahigh-power micrometre-sized supercapacitors based on onion-like carbon. *Nat. Nanotechnol.* 5 (9): 651–654.

44 Pan, H., Li, J., and Feng, Y.P. (2010). Carbon nanotubes for supercapacitor. *Nanoscale Res. Lett.* 5 (3): 654–668.

45 An, K.H., Kim, W.S., Park, Y.S. et al. (2001). Supercapacitors using single-walled carbon nanotube electrodes. *Adv. Mater.* 13 (7): 497.

46 Kaempgen, M., Chan, C.K., Ma, J. et al. (2009). Printable thin film supercapacitors using single-walled carbon nanotubes. *Nano Lett.* 9 (5): 1872–1876.

47 Xie, K., Qin, X.T., Wang, X.Z. et al. (2012). Carbon nanocages as supercapacitor electrode materials. *Adv. Mater.* 24 (3): 347.

48 Chen, T. and Dai, L.M. (2013). Carbon nanomaterials for high-performance supercapacitors. *Mater. Today* 16 (7–8): 272–280.

49 Li, X. and Wei, B.Q. (2013). Supercapacitors based on nanostructured carbon. *Nano Energy* 2 (2): 159–173.

50 Novoselov, K.S., Geim, A.K., Morozov, S.V. et al. (2004). Electric field effect in atomically thin carbon films. *Science* 306 (5696): 666–669.

51 Allen, M.J., Tung, V.C., and Kaner, R.B. (2010). Honeycomb carbon: a review of graphene. *Chem. Rev.* 110 (1): 132–145.

52 Dai, L.M. (2013). Functionalization of graphene for efficient energy conversion and storage. *Acc. Chem. Res.* 46 (1): 31–42.

53 Liu, Z.Y., Wu, Z.S., Yang, S. et al. (2016). Ultraflexible in-plane micro-supercapacitors by direct printing of solution-processable electrochemically exfoliated graphene. *Adv. Mater.* 28 (11): 2217–2222.

54 Lukatskaya, M.R., Mashtalir, O., Ren, C.E. et al. (2013). Cation intercalation and high volumetric capacitance of two-dimensional titanium carbide. *Science* 341 (6153): 1502–1505.

55 Naguib, M., Mashtalir, O., Carle, J. et al. (2012). Two-dimensional transition metal carbides. *ACS Nano* 6 (2): 1322–1331.

56 Naguib, M., Kurtoglu, M., Presser, V. et al. (2011). Two-dimensional nanocrystals produced by exfoliation of Ti3AlC2. *Adv. Mater.* 23 (37): 4248–4253.

57 Anasori, B., Lukatskaya, M.R., and Gogotsi, Y. (2017). 2D metal carbides and nitrides (MXenes) for energy storage. *Nat. Rev. Mater.* 2 (2): 16098.

58 Xia, Y., Mathis, T.S., Zhao, M.Q. et al. (2018). Thickness-independent capacitance of vertically aligned liquid-crystalline MXenes. *Nature* 557 (7705): 409–412.

59 Wang, H., Wu, Y., Yuan, X.Z. et al. (2018). Clay-inspired MXene-based electrochemical devices and photo-electrocatalyst: state-of-the-art progresses and challenges. *Adv. Mater.* 30 (12): 1704561.

60 Ghidiu, M., Lukatskaya, M.R., Zhao, M.Q. et al. (2014). Conductive two-dimensional titanium carbide "clay" with high volumetric capacitance. *Nature* 516 (7529): 78–U171.

61 Lukatskaya, M.R., Kota, S., Lin, Z.F. et al. (2017). Ultra-high-rate pseudocapacitive energy storage in two-dimensional transition metal carbides. *Nat. Energy* 2 (8): 17105.

62 Zhang, C.F., Kremer, M.P., Seral-Ascaso, A. et al. (2018). Stamping of flexible, coplanar micro-supercapacitors using MXene inks. *Adv. Funct. Mater.* 28 (9): 1705506.

63 Lu, Q., Chen, J.G.G., and Xiao, J.Q. (2013). Nanostructured electrodes for high-performance pseudocapacitors. *Angew. Chem. Int. Ed.* 52 (7): 1882–1889.

64 Brezesinski, T., Wang, J., Tolbert, S.H. et al. (2010). Ordered mesoporous alpha-MoO_3 with iso-oriented nanocrystalline walls for thin-film pseudocapacitors. *Nat. Mater.* 9 (2): 146–151.

65 Li, H.B., Yu, M.H., Wang, F.X. et al. (2013). Amorphous nickel hydroxide nanospheres with ultrahigh capacitance and energy density as electrochemical pseudocapacitor materials. *Nat. Commun.* 4: 1894.

66 Liu, T.Y., Finn, L., Yu, M.H. et al. (2014). Polyaniline and polypyrrole pseudocapacitor electrodes with excellent cycling stability. *Nano Lett.* 14 (5): 2522–2527.

67 Wang, G.M., Lu, X.H., Ling, Y.C. et al. (2012). LiCl/PVA gel electrolyte stabilizes vanadium oxide nanowire electrodes for pseudocapacitors. *ACS Nano* 6 (11): 10296–10302.

68 Wu, C.Z., Lu, X.L., Peng, L.L. et al. (2013). Two-dimensional vanadyl phosphate ultrathin nanosheets for high energy density and flexible pseudocapacitors. *Nat. Commun.* 4: 2431.

69 Muller, G.A., Cook, J.B., Kim, H.S. et al. (2015). High performance pseudocapacitor based on 2D layered metal chalcogenide nanocrystals. *Nano Lett.* 15 (3): 1911–1917.

70 Xiao, J.W., Wan, L., Yang, S.H. et al. (2014). Design hierarchical electrodes with highly conductive $NiCo_2S_4$ nanotube arrays grown on carbon fiber paper for high-performance pseudocapacitors. *Nano Lett.* 14 (2): 831–838.

71 Li, L., Raji, A.R.O., Fei, H.L. et al. (2013). Nanocomposite of polyaniline nanorods grown on graphene nanoribbons for highly capacitive pseudocapacitors. *ACS Appl. Mater. Interfaces* 5 (14): 6622–6627.

72 Liu, X.H., Jervis, R., Maher, R.C. et al. (2016). 3D-printed structural pseudocapacitors. *Adv. Mater. Technol.* 1 (9): 1600167.

73 Li, G.J., Meng, Z.G., Qian, J.S. et al. (2019). Inkjet printed pseudocapacitive electrodes on laser-induced graphene for electrochemical energy storage. *Mater. Today Energy* 12: 155–160.

74 ten Elshof, J.E. and Wang, Y. (2019). Advances in ink-jet printing of MnO_2-nanosheet based pseudocapacitors. *Small Methods* 3 (8): 1800318.

75 Wang, Y., Zhang, Y.Z., Dubbink, D. et al. (2018). Inkjet printing of delta-MnO_2 nanosheets for flexible solid-state micro-supercapacitor. *Nano Energy* 49: 481–488.

76 Wang, X.F., Liu, B., Xiang, Q.Y. et al. (2014). Spray-painted binder-free SnSe electrodes for high-performance energy-storage devices. *Chemsuschem* 7 (1): 308–313.

77 Wang, Y., Shi, Y.M., Zhao, C.X. et al. (2014). Printed all-solid flexible microsupercapacitors: towards the general route for high energy storage devices. *Nanotechnology* 25 (9): 094010.

78 Liu, S.Y., Xie, J., Li, H.B. et al. (2014). Nitrogen-doped reduced graphene oxide for high-performance flexible all-solid-state microsupercapacitors. *J. Mater. Chem. A* 2 (42): 18125–18131.

79 Liang, J.J., Tong, K., and Pei, Q.B. (2016). A water-based silver-nanowire screen-print ink for the fabrication of stretchable conductors and wearable thin-film transistors. *Adv. Mater.* 28 (28): 5986.

80 Zhang, Y.Z., Wang, Y., Cheng, T. et al. (2019). Printed supercapacitors: materials, printing and applications. *Chem. Soc. Rev.* 48 (12): 3229–3264.

81 Dubal, D.P., Chodankar, N.R., Kim, D.H. et al. (2018). Towards flexible solid-state supercapacitors for smart and wearable electronics. *Chem. Soc. Rev.* 47 (6): 2065–2129.

82 Hyun, W.J., Secor, E.B., Kim, C.H. et al. (2017). Scalable, self-aligned printing of flexible graphene micro-supercapacitors. *Adv. Energy Mater.* 7 (17): 1700285.

83 Zhong, C., Deng, Y.D., Hu, W.B. et al. (2015). A review of electrolyte materials and compositions for electrochemical supercapacitors. *Chem. Soc. Rev.* 44 (21): 7484–7539.

84 Cheng, X.L., Pan, J., Zhao, Y. et al. (2018). Gel polymer electrolytes for electrochemical energy storage. *Adv. Energy Mater.* 8 (7): 1702184.

85 Li, J.T., Delekta, S.S., Zhang, P.P. et al. (2017). Scalable fabrication and integration of graphene microsupercapacitors through full inkjet printing. *ACS Nano* 11 (8): 8249–8256.

86 Choi, K.H., Yoo, J., Lee, C.K. et al. (2016). All-inkjet-printed, solid-state flexible supercapacitors on paper. *Energy Environ. Sci.* 9 (9): 2812–2821.

87 Cheng, T., Wu, Y.W., Chen, Y.L. et al. (2019). Inkjet-printed high-performance flexible micro-supercapacitors with porous nanofiber-like electrode structures. *Small* 15 (34): 1910830.

88 Giannakou, P., Masteghin, M.G., Slade, R.C.T. et al. (2019). Energy storage on demand: ultra-high-rate and high-energy-density inkjet-printed NiO micro-supercapacitors. *J. Mater. Chem. A* 7 (37): 21496–21506.

89 Xiong, Z.Y., Yun, X.W., Qiu, L. et al. (2019). A dynamic graphene oxide network enables spray printing of colloidal gels for high-performance micro-supercapacitors. *Adv. Mater.* 31 (16): 1804434.

90 Garlapati, S.K., Divya, M., Breitung, B. et al. (2018). Printed electronics based on inorganic semiconductors: from processes and materials to devices. *Adv. Mater.* 30 (40): 1707600.

91 Shi, X.Y., Wu, Z.S., Qin, J.Q. et al. (2017). Graphene-based linear tandem micro-supercapacitors with metal-free current collectors and high-voltage output. *Adv. Mater.* 29 (44): 1703034.

92 Guo, R.S., Chen, J.T., Yang, B.J. et al. (2017). In-plane micro-supercapacitors for an integrated device on one piece of paper. *Adv. Funct. Mater.* 27 (43): 1702394.

93 Xu, Y.F., Schwab, M.G., Strudwick, A.J. et al. (2013). Screen-printable thin film supercapacitor device utilizing graphene/polyaniline inks. *Adv. Energy Mater.* 3 (8): 1035–1040.

94 Zhang, X.Y., Zhao, W., Wei, L. et al. (2019). In-plane flexible solid-state microsupercapacitors for on-chip electronics. *Energy* 170: 338–348.

95 Li, H.P., Liu, S.R., Li, X.R. et al. (2019). Screen-printing fabrication of high volumetric energy density micro-supercapacitors based on high-resolution thixotropic-ternary hybrid interdigital micro-electrodes. *Mater. Chem. Front.* 3 (4): 626–635.

96 Suikkola, J., Bjorninen, T., Mosallaei, M. et al. (2016). Screen-printing fabrication and characterization of stretchable electronics. *Sci. Rep.* 6: 25784.

97 Ning, L.Q. and Chen, X.B. (2017). A brief review of extrusion-based tissue scaffold bio-printing. *Biotechnol. J.* 12 (8): 1600671.

98 Zarek, M., Layani, M., Cooperstein, I. et al. (2016). 3D printing of shape memory polymers for flexible electronic devices. *Adv. Mater.* 28 (22): 4449.

99 Wang, J.F., Liu, Y.Y., Fan, Z.M. et al. (2019). Ink-based 3D printing technologies for graphene-based materials: a review. *Adv. Comp. Hybrid Mater.* 2 (1): 1–33.

100 Tian, X.C., Jin, J., Yuan, S.Q. et al. (2017). Emerging 3D-printed electrochemical energy storage devices: a critical review. *Adv. Energy Mater.* 7 (17): 1700127.

101 Chang, P., Mei, H., Zhou, S.X. et al. (2019). 3D printed electrochemical energy storage devices. *J. Mater. Chem. A* 7 (9): 4230–4258.

102 Shen, K., Ding, J.W., and Yang, S.B. (2018). 3D printing quasi-solid-state asymmetric micro-supercapacitors with ultrahigh areal energy density. *Adv. Energy Mater.* 8 (20): 1800408.

103 Yu, W., Li, B.Q., Ding, S.J. et al. (2018). 3D printing of interdigitated electrode for all-solid-state microsupercapacitors. *J. Micromech. Microeng.* 28 (10): 105014.

104 Hu, H.B., Pei, Z.B., Fan, H.J. et al. (2016). 3D interdigital Au/MnO_2/Au stacked hybrid electrodes for on-chip microsupercapacitors. *Small* 12 (22): 3059–3069.

105 Yu, W., Zhou, H., Li, B.Q. et al. (2017). 3D printing of carbon nanotubes-based microsupercapacitors. *ACS Appl. Mater. Interfaces* 9 (5): 4597–4604.

11

Recent Advances of Flexible Micro-Supercapacitors

*Songshan Bi, Hongmei Cao, Rui Wang and Zhiqiang Niu**

Key Laboratory of Advanced Energy Materials Chemistry (Ministry of Education), College of Chemistry, Nankai University, Tianjin, 300071, P. R. China

11.1 Introduction

With the development of portable and wearable micro-electronic devices, there has been an increasing demand for flexible and miniaturized energy storage systems, which can be integrated on a chip and compatible with other micro-electronics [1]. Miniaturizing power sources onto a chip would potentially increase the density of various functional devices [2]. In addition, they can reduce the complexity of overall chip design through removing complex interconnections with bulky power sources to build self-powered micro-device systems [3].

Among various energy storage devices, supercapacitors (SCs), which exhibit high power density, long cycling performance, and fast charge/discharge process, have attracted particular attention in recent years [3]. They can bridge the gap between dielectric capacitors and batteries. Therefore, they can be applied in some fields, such as electric vehicles and portable electronics. Conventional SCs are usually assembled into buckle or spiral-wound configuration with sandwiched architecture. Their general characteristic features of inflexibility, heavy weight, and bulky size, limit their integration in flexible or portable micro-electronic devices. In contrast, flexible micro-supercapacitors (MSCs), which are based on micro-scale electrodes with interdigitated architecture on a flexible substrate, could meet the demand for the miniaturization and integration of flexible electronics [4]. Therefore, flexible MSCs are highly desired for the integration of self-powered micro-electronics.

The flexible MSCs with in-plane interdigitated architecture have some primary advantages over flexible SCs with sandwiched architecture. Firstly, the narrow gaps between electrodes could shorten the diffusion pathways of ions in electrolytes, leading to a higher power density [5]. Secondly, the interdigitated architecture of flexible MSCs benefits their integration with other micro-electronics on a chip [6]. Furthermore, MSCs could be integrated together as close as possible on a chip to enhance the output potential or current, increasing the density of MSC units by reducing complex connection [7]. When the MSCs bear strains, their unique interdigitated architecture could reduce the risk of short circuit of electrodes. Compared with the sandwiched architecture, the in-plane structure could also avoid sliding or dislocation between electrodes and separator. Therefore, flexible MSCs are the promising candidate of energy storage devices for

*e-mail: zqniu@nankai.edu.cn

Flexible Supercapacitors: Materials and Applications, First Edition. Edited by Guozhen Shen, Zheng Lou and Di Chen.

flexible miniaturized on-chip electronics. Recently, various MSCs with excellent electrochemical performance and high flexibility have been achieved by the design of electrode materials and device configurations [6, 8, 9].

To extend the application of MSCs in portable electronic devices, flexible MSCs are often integrated with other micro-electronics (e.g. sensors [10], solar cells [11], and nanogenerators [12]) with internal connections on a chip to build a flexible self-charging or self-powered system [13]. The space utilization and compatibility of the integrated devices are important for evaluating the integration degree. In addition, functional materials could be added into micro-electrodes, electrolytes or packing materials, endowing the MSCs with other functions (e.g. self-healing [14], electrochromism [15], thermal self-protection [16], and sensing [17]) while maintaining the ability of storing energy, which could be called smart MSCs [6]. The intelligence level of them can be reflected by how much the performance of the smart MSCs declines compared to original MSCs.

In this chapter, we will discuss recent advances of flexible MSCs. It begins with a brief introduction of the general features of flexible planar MSCs, including electrodes, electrolytes and the interdigitated architecture, followed by a review of the electrode materials and fabrication strategies for flexible MSCs. Then, we outline the achievements of their integration with other electronics and intelligent behaviors. Finally, we try to summary and view the future development of flexible MSCs.

11.2 General Features of Flexible MSCs

In general, a conventional SC includes four components: current collectors, electrode materials, a separator, and electrolyte, which are stacked in a sandwiched configuration (Figure 11.1a) [18]. Different from conventional SCs, MSCs consist of interdigitated micro-electrode arrays on a substrate, and the electrolyte is coated on them (Figure 11.1b). They often employed flexible films, such as polyethylene terephthalate (PET) [19], polydimethylsiloxane (PDMS) [17], and cellulose fiber papers [20], as the substrates to build the flexible devices. Furthermore, compared with conventional SCs, there is no separator in MSCs, and the electrolyte filled in the gap could act as separator (Figure 11.1c). The open side edges of active materials are exposed to electrolyte, contributing to the diffusion of electrolyte into electrodes (Figure 11.1d).

According to their energy storage mechanisms, the electrodes of flexible MSCs could be generally classified into two types: (i) capacitors-type electrodes. They store energy by the reversible ion adsorption/desorption on the surface of active materials without chemical reaction, resulting in excellent cycle stability and high-power density. The materials with high specific surface area and good electrical conductivity, such as activated carbon (AC), carbon nanotubes (CNTs), and graphene, have been employed as the active materials of capacitors-type electrodes [21–23]. (ii) Pseudocapacitors-type electrodes. They are based on the fast and reversible redox or Faradic reactions to achieve electrochemical storage of charges, leading to a higher energy density, but lower power density and limited cycling performance than capacitors-type electrodes. Metal oxides/hydroxides (e.g. manganese dioxide (MnO_2) and hydroxyl oxidize iron (FeOOH)), conducting polymers (e.g. polypyrrole (PPy), and polyaniline (PANI)) and the emerging new materials (e.g. metal organic frameworks (MOFs), covalent organic frameworks (COFs), and MXenes) are the main materials of pseudocapacitors-type electrodes [24–27]. Most pseudocapacitor materials have low electrical conductivity and limited mechanical properties than nanocarbon materials. Therefore, they are often combined with CNTs or graphene to achieve micro-electrodes with high conductivity and high flexibility simultaneously [25, 27].

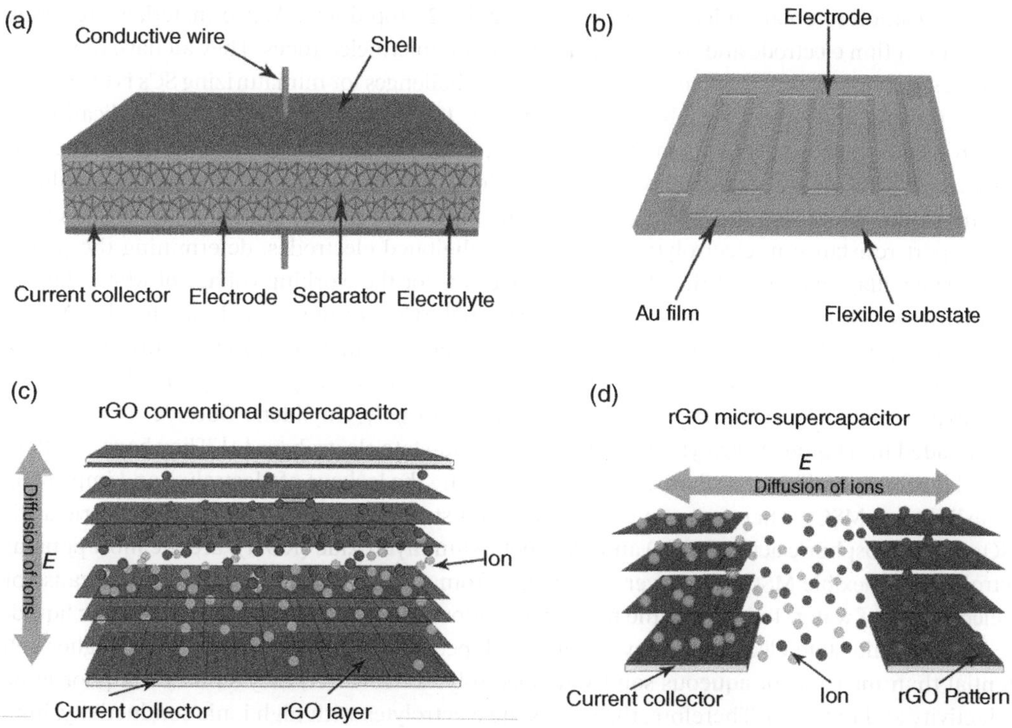

Figure 11.1 Schematics of (a) conventional SCs and (b) MSCs; *Source:* [4]. Reproduced with permission of Springer Nature. Schematics of the ion-diffusion pathway in (c) conventional film electrodes and (d) micro-electrodes. *Source:* [5]. Reproduced with permission of John Wiley and Sons.

To date, various flexible symmetrical MSCs have been designed because of their simple fabrication process [23]. Compared with symmetric MSCs, asymmetrical MSCs, which consist of two different electrodes with a combination of capacitors-type and pseudocapacitor-type electrodes, suffer from more complicated fabrication steps [28]. However, they can reach higher operating voltage by combing different potential windows of two electrodes. For MSCs, the calculation equations of the energy density (E) and power density (P) could be written as below [9]:

$$E = \frac{1}{2}CV^2 \tag{11.1}$$

$$P = \frac{E}{t} \tag{11.2}$$

in which C is the specific capacitance, V is the voltage range in charge/discharge process, and t is the charge/discharge time. Since the voltage (V) of asymmetrical MSCs is higher in comparison with symmetrical MSCs, asymmetrical MSCs could achieve a higher energy density.

The fabrication process of flexible MSCs in general involves additional steps of patterning to achieve interdigitated architectures. Various micro-fabrication techniques have been developed to pattern electrode finger arrays, such as photolithography [5], laser scribe [29, 30], spray printing [19] and screen printing [31]. Moreover, active materials deposition strategies should be matched with the micro-fabrication techniques to fabricate micro-electrodes. In general, there are two strategies to fabricate micro-electrodes [3]: 1) bottom-up. Active materials are controllably deposited on the patterned position to obtain micro-electrodes on a chip. 2) Top down. Active materials are firstly fabricated to a film electrode and then patterned to obtain micro-electrodes. They all have their own advantages and limitations. Until now, one of the main challenges for miniaturizing SCs is to develop a simple, reliable, and large-scale fabrication technology to achieve the industry level applications.

Electrolytes, which provide ions in the charge/discharge process, are also important for the performance of MSCs. Ionic conductivity, voltage window, and ionic radius of electrolytes would be taken into consideration in the design of MSCs. The ionic conductivity and ionic radius affect the ion-transport resistance in electrolytes between interdigitated electrodes, determining the power density. The voltage window of the electrolyte is decisive for the working voltage of MSCs. Liquid electrolytes, which are often employed in conventional SCs, are not suitable for flexible MSCs, because they are difficult to be packaged onto the surface of micro-electrodes, causing the risk of leakage. To overcome this issue, polymer additives, such as poly(vinyl alcohol) (PVA), poly(vinylpyrrolidone) (PVP), polyacrylonitrile (PAN) and poly(vinylidene fluoride) (PVDF), are usually added into liquid electrolytes to obtain gel or solid-state electrolytes [4]. They have excellent physical flexibility and high mechanical integrity, avoiding the leakage of electrolyte and improving the flexibility of MSCs. Therefore, aqueous gel or solid-state electrolytes (e.g. PVA/ (sulfuric acid) H_2SO_4, PVA/phosphoric acid (H_3PO_4) and PVA/potassium hydroxide (KOH)) are the most popular electrolytes for flexible MSCs. However, they suffer from a low operated voltage (~1 V) because of the electrolysis of water. To improve the potential window of flexible MSCs, organic or ionic liquid-based gel or solid-state electrolytes have been developed [32, 33]. They can deliver a higher cell potential than the cases of aqueous solid-state electrolytes. However, they suffer from poor ionic conductivity and high cost. Therefore, the solid-state electrolytes with high ionic conductivity, high safety, low cost, and wide voltage window have to be further developed for flexible MSCs.

In addition to the intrinsic properties of the electrode materials and electrolytes, the performance of flexible MSCs is also influenced by the architecture of the electrodes, including the thickness and width of interdigitated electrodes as well as the space between them [5]. By combining electrode-patterning steps and thin-film deposition methods, the electrode thickness of MSCs could be well-controlled and reach to nanoscale. The thin thickness of micro-electrodes could not only greatly improve the gravimetric capacitance of MSCs, but also endow micro-electrodes with excellent flexibility. However, ultrathin thickness is not desirable for MSCs, which would lead to a low areal energy density because of the low mass loading of active materials. As a result, a suitable thickness of electrodes should be evaluated. In addition, with the development of electrode-patterning techniques, the width of electrodes and the space between them could be precisely designed. Owing to the unique interdigitated electrodes, the side edges of micro-electrodes are exposed to the electrolyte. The decrease of the electrode widths is beneficial for the infiltration of electrolyte into electrodes, increasing the accessibility of electrolyte into the active electrode materials, especially for materials with 2D layered structures such as graphene, transition metal dichalcogenides, and MXenes [26, 32]. Compared with conventional SCs, the narrow spaces between interdigitated electrodes could replace the separator. It could shorten ionic diffusion path and decrease the electrolyte ion-transport resistance, resulting in an improved frequency response and ultrahigh power density.

11.3 Active Materials of Flexible MSCs

Electrode materials play an important role in flexible MSCs, which determine the electrochemical performances of the devices, including capacitance, voltage, cycling performance and rate capability. To meet the demand of MSCs with high flexibility and excellent electrochemical performance, electrode materials are expected to possess high surface area, excellent conductivity, and good mechanical durability [23]. Up to now, various materials have been used to fabricate flexible MSCs, such as carbon-based materials, metal oxides/hydroxides and conductive polymers [34]. Moreover, different micro-fabrication techniques such as photolithography technique, laser engraving process, and printing techniques, have been employed to cooperate with electrode material deposition strategies for the fabrication of flexible MSCs [2].

11.3.1 Graphene-based Materials

Graphene has attracted much attention in the fabrication of flexible MSCs due to its good conductivity, high specific surface area ($2630\,m^2\,g^{-1}$), and remarkable Young's modulus (~1.0 TPa) [4]. Moreover, graphene sheets with unique 2D conjugated structures are identified to be ideal building blocks to construct micro-electrodes of flexible MSCs [22]. Based on their excellent properties, pure graphene and graphene-based composite materials are considered as the promising active materials of flexible MSCs.

Graphene can be directly prepared through many approaches such as chemical vapor deposition (CVD) method [35], laser direct writing process [36], and direct exfoliation of graphite [37], which can be classified as pristine graphene. For example, by CVD method, monolayer graphene with good conductivity, high transparency, and low structural defects can be obtained, which is a desired active material of transparent devices [37]. Moreover, the CVD-grown graphene can be easily transferred onto many flexible substrates [38]. Therefore, it can be used to fabricate transparent and flexible MSCs with excellent electrochemical properties (Figure 11.2a) [39]. However, the synthetic conditions of CVD method are usually restrained and difficult to handle, which limits its extensive use. A low cost and one-step process can be successfully applied to produce pristine graphene with three-dimensional (3D) porous network by laser inducing on polyimide substrate [41]. After transferring the graphene onto elastic rubber, the MSCs could exhibit an extremely stretchability up to 400% of the original state. However, the capacitances of the devices decreased obviously with the increasing of the tensile strains since the electrodes exhibited high resistance at strain states. After the devices recovered to the original state, the resistance of the electrodes could almost recover. Moreover, the capacitance retention of the devices maintained over 90% after being stretched or twisted for over 1000 times, indicating their good stretchability and flexibility.

Graphene can also be obtained by exfoliating graphite in liquid phase, which is a more effective technique for scalable production of graphene [37]. Impressively, those exfoliated graphene-based inks are well compatible with printing techniques [42]. For example, using low cost and environmentally friendly ethanol and terpineol as organic solvent, and ethyl cellulose as stabilizer, the obtained graphene/ethyl cellulose ink could be utilized to prepare micro-electrodes via inkjet printing method (Figure 11.2b) [40]. The graphene/ethyl cellulose with high conductivity could be used as both electrode material and current collector, contributing to their durability under the deformation conditions. Moreover, flexible MSCs with smaller size and higher production throughput can be fabricated by combining the pristine graphene-based ink with a self-aligned inkjet printing process [21]. Those devices achieved high resolution and 100% yield, and the footprint was smaller than $1\,mm^2$. In addition, the resultant devices exhibited favorable electrochemical performances and excellent stability against mechanical deformation.

Graphene oxide (GO), which is obtained from the chemical exfoliation of graphite, is more widely used to fabricate flexible MSCs [43]. Since the exfoliated GO sheets exhibit amphiphilic properties, they are highly soluble in aqueous solution and some organic solvents with controllable concentrations, which can be well compatible with various micro-fabrication techniques [44]. For

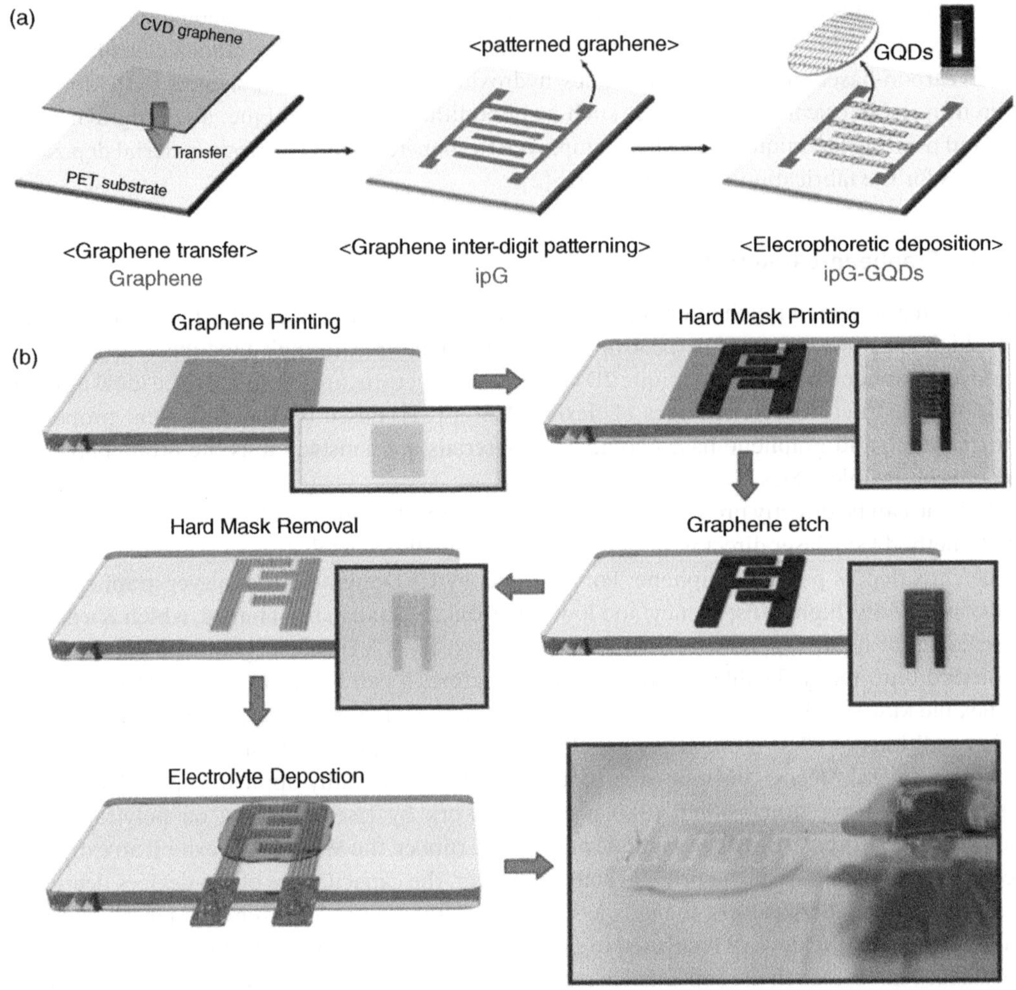

Figure 11.2 (a) The fabrication process of a highly transparent and flexible MSC based on pristine graphene. ipG: interdigitated pattern of graphene. GQDs: graphene quantum dots. *Source:* [39]. Reproduced with permission of Elsevier. (b) Schematic diagram and optical image of the MSCs fabricated by inkjet printing method. *Source:* [40]. Reproduced with permission of Royal Society of Chemistry.

example, the GO dispersion with high concentration (20 mg ml^{-1}) could be directly used as viscous ink to combine with micro-extrusion printing and 3D printing techniques for the fabrication of reduced graphene oxide (RGO)-based micro-electrodes [34, 36]. Besides, in GO solution, the GO sheets with many negatively charged functional groups can be deposited on the designed substrates by an electrophoretic process, and then ultrathin RGO electrodes were achieved after further reducing above GO electrodes (Figure 11.3a) [5]. The mass loading and thickness of the GO layers can be easily controlled by the electrophoretic deposition process. In addition, since the reduction potential of some metals is lower than that of RGO/GO, the RGO electrodes can be obtained by reducing GO on those metal surfaces [46, 47]. As shown in Figure 11.3b, in the GO solution, GO sheets were in-situ self-assembled and selective reduced on the interdigitated metal surfaces at room temperature, and then the flexible MSCs were fabricated based on the obtained RGO electrodes and solid electrolyte [45]. Impressively, by metal reduction self-assembly, the area

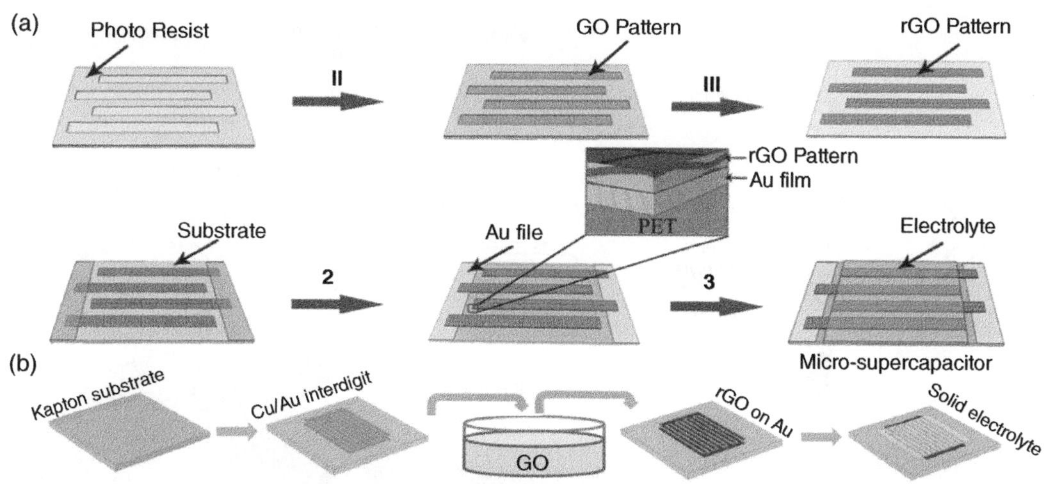

Figure 11.3 Schematic illustration of preparing RGO-based MSCs. (a) Fabricating MSCs directly on a substrate by combining photolithography technique with an electrophoretic method. *Source:* [5]. Reproduced with permission of John Wiley and Sons. (b) Fabricating MSCs by in situ reducing GO on interdigitated metal surfaces. *Source:* [45]. Reproduced with permission of Elsevier.

and shapes of RGO micro-electrodes could be well controlled through tuning the size and pattern of metal foils.

Owing to unique 2D structure, the GO sheets can be used as building blocks to construct 2D films by various methods [23, 48]. As shown in Figure 11.4a, the GO films can be obtained by a spin coating method [22]. However, the GO films are intrinsically insulating, they cannot directly serve as the electrodes of MSCs, Therefore, the films were then converted into interdigitated RGO films through the CH_4 plasma reduction and patterning processes. This method can be conducted on arbitrary substrates. However, the reduction and patterning processes cannot be performed at the same time. In comparison, laser direct writing method can achieve the reduction and patterning of GO films simultaneously without using any additive reductants and photomasks (Figure 11.4b), which displays high efficiency to fabricate MSCs on a large scale [29, 30]. Furthermore, more than one hundred flexible MSCs were prepared on a single disk in 30 minutes using a LightScribe digital video disk (DVD) burner (Figure 11.4c), and the devices could exhibit a high-power density of $200\,W\,cm^{-3}$ [29]. Moreover, these micro-electrodes fabricated by this method exhibited all-in-one RGO-GO-RGO configurations, which contributed to the superior mechanical flexibility of those MSCs.

Graphene with high surface area and excellent conductivity is a desirable conductive scaffold to support other functional materials [49]. The introduction of pseudocapacitive materials into graphene materials is a promising method to boost their performances by the synergistic effect of the electrical double layer and Faradaic capacitance mechanisms [50–52]. PANI, FeOOH, and MnO_2, representing characteristic conductive polymers, transition metal hydroxides, and metal oxides, were electrodeposited onto the surface of laser induced graphene to form composite electrodes of MSCs [25]. Both the symmetric and asymmetric MSCs were fabricated, and they exhibited enhanced specific capacitances. In addition, RGO-based composite film electrodes could be fabricated through synchronously selective reducing/assembling GO sheets with Mn_3O_4 nanoparticles on Zn foils in alkaline solution (Figure 11.5) [28]. The voltage window of the resultant asymmetric MSCs based on the interdigitated RGO and RGO/Mn_3O_4 electrodes could be enhanced to 1.6 V, contributing to an improved energy density.

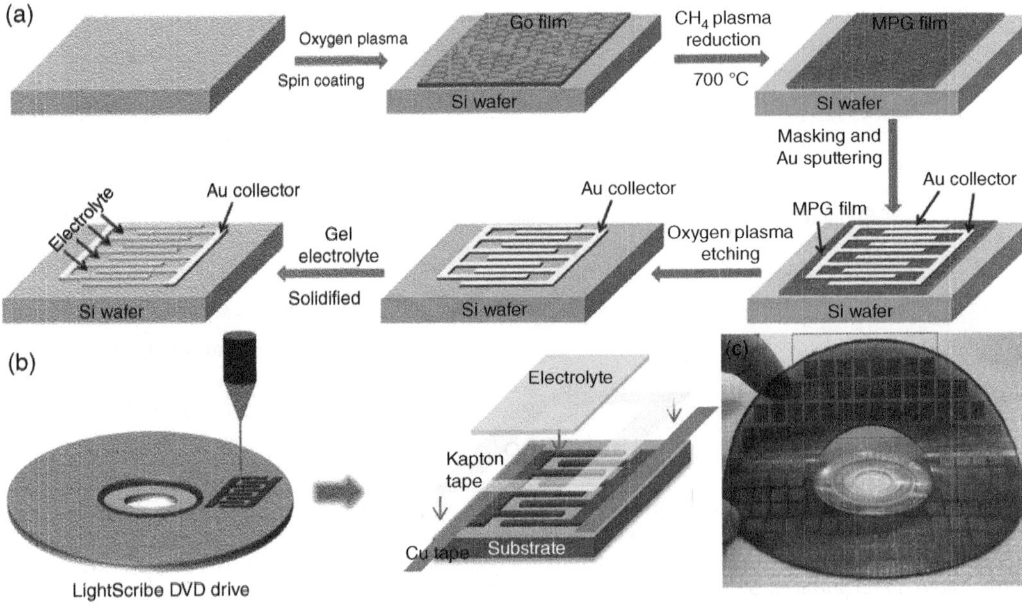

Figure 11.4 (a) Schematic illustration of fabricating flexible RGO-based MSCs on a silicon wafer. MPG: reduced graphene. *Source:* [22]. Reproduced with permission of Springer Nature (b) Schematic diagram showing the fabrication process of RGO-based MSCs using a standard LightScribe DVD burner; (c) The digital photograph of more than 100 MSCs produced on a single disc. *Source:* [29]. Reproduced with permission of Springer Nature.

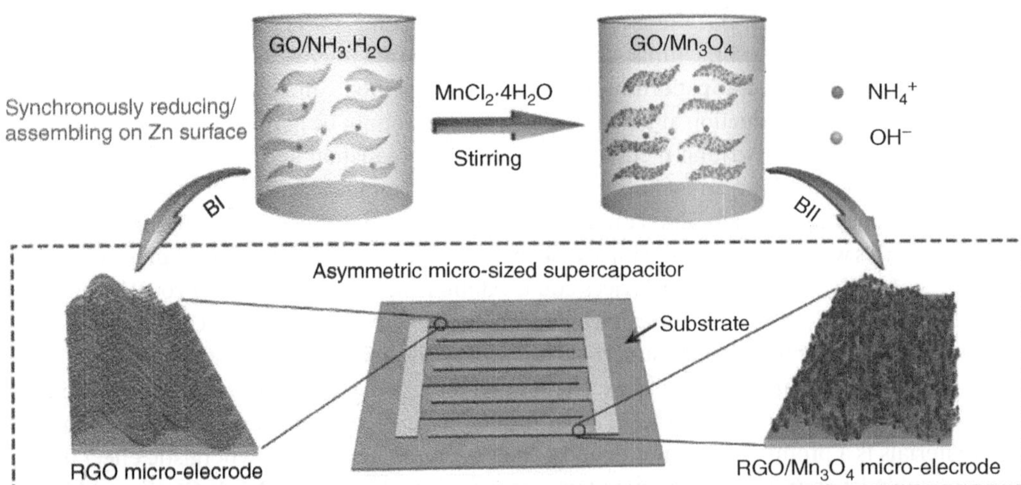

Figure 11.5 Schematic illustration of fabricating the RGO and RGO/Mn_3O_4 electrodes of MSCs on Zn foils in alkaline solution. *Source:* [28]. Reproduced with permission of John Wiley and Sons.

11.3.2 CNT-based Materials

CNTs are also prospective electrode materials of flexible MSCs due to their high electrical conductivity ($2\times10^7\,S\,m^{-1}$), high length/diameter ratio, large surface area and extraordinary mechanical properties with a tensile strength of up to 100 GPa and Young's modulus higher than 1 TPa [4]. Benefiting from their high length/diameter ratio, CNTs can be easily fabricated into flexible thin films through many strategies, and the film electrodes generally offer superior robustness under different deformation conditions [53].

The single-walled carbon nanotubes (SWCNTs) film micro-electrodes can be fabricated by a spray coating method. The thickness of the SWCNT layers could be controlled by adjusting the times of spray coating process. The spray-coated SWCNT electrodes could work well even at the scan rate of $1000\,V\,s^{-1}$, and deliver a high power density of $1125\,W\,cm^{-3}$ [54]. The spray coating method is considered to be a simple approach to fabricate micro-electrodes. Nevertheless, this method usually needs the assist of photomasks during the spraying process, and the active materials will be deposited on the surfaces of both interdigitated substrates and photomasks, leading to the waste of materials. A facile selective wetting-induced patterning fabrication approach can avoid this problem and be applied to prepare interdigitated CNT electrodes under mild condition [55]. In the fabrication process, the injected flow of CNTs dispersion was patterned by the hydrophilous channels on PDMS with high utilization, resulting in well-defined electrodes with closely spaced micro-architecture after drying. Subsequently, PVA/H_3PO_4 solid-state electrolyte was well integrated with the interdigitated CNTs architecture. After peeling off the PDMS, the all-solid-state MSCs exhibited superior flexibility and reliable electrochemical performances.

CNT-based electrodes have been widely used in flexible MSCs. However, the CNTs are generally randomly distributed in the electrodes, which will display lower accessible surface area to electrolyte ions in comparison with the vertically aligned carbon nanotube (VACNT) films [24]. Therefore, VACNT carpets were prepared on the graphene film by CVD method and applied in the fabrication of MSCs (Figure 11.6a and b) [56]. The VACNT arrays can efficiently reduce the bundling and overlap of the CNTs (Figure 11.6c–e), facilitating the ion diffusion into VACNT films. The entire axial portion of CNTs in VACNT arrays is accessible to electrolyte, enlarging their surface area to

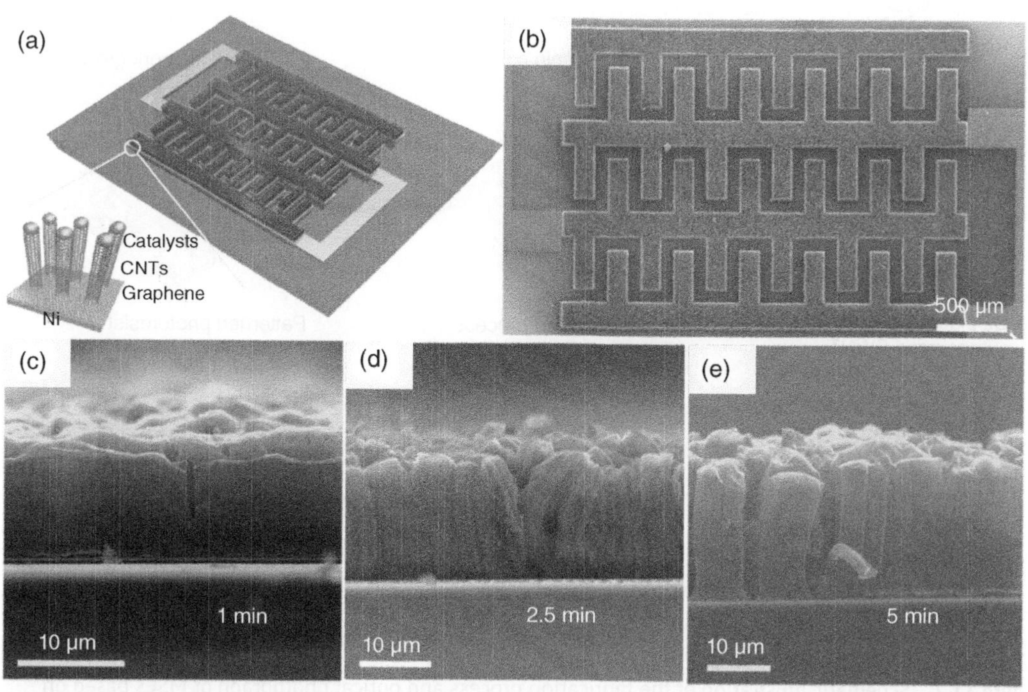

Figure 11.6 (a) Schematic illustration and (b) SEM image of the MSCs based on the VACNT carpet electrodes; (c–e) Cross-section SEM images of VACNT carpets grown for different time. *Source:* [56]. Reproduced with permission of American Chemical Society.

contact with the electrolyte ions. Therefore, the flexible MSCs based on VACNT film electrodes can displayed enhanced energy and power performances [57].

Compared to graphene, CNTs exhibit smaller specific surface area but higher electrical conductivity. Combining the CNTs with graphene can take full advantage of the synergistic effect between graphene and CNTs [58], where the CNTs can provide superior conductivity and efficiently prevent the aggregation of the graphene sheets while the graphene with large surface area can offer high capacitance [59]. Therefore, flexible all-solid-state MSCs based on graphene/CNTs composite electrodes could display an enhanced energy density of 10.7 mWh cm^{-3} and power density of 3.17 W cm^{-3} [60]. Moreover, the MSCs displayed outstanding mechanical flexibility and electrochemical stability even suffering bending deformation.

To further optimize the electrochemical properties of CNT-based flexible MSCs, pseudocapacitive materials (e.g. MnO_2 [27, 61, 62], Mn_3O_4 [63], vanadium nitride (VN) [64], polythiophene (PEDOT) [33, 65], PANI [66]) were often introduced into the CNT-based electrodes. For example, MnO_2/CNTs composite micro-electrodes have been applied in the flexible MSCs (Figure 11.7) [27]. The specific capacitance and energy density can be significantly enhanced through adding MnO_2 nanoflowers into the CNT-based micro-electrodes. The areal capacitance of the composite electrodes containing CNTs and MnO_2 was almost 43.6 times larger than that of the electrodes without MnO_2. Similarly, the hybrid PANI/multi-walled carbon nanotubes (MWCNTs)-RGO/thermoplastic

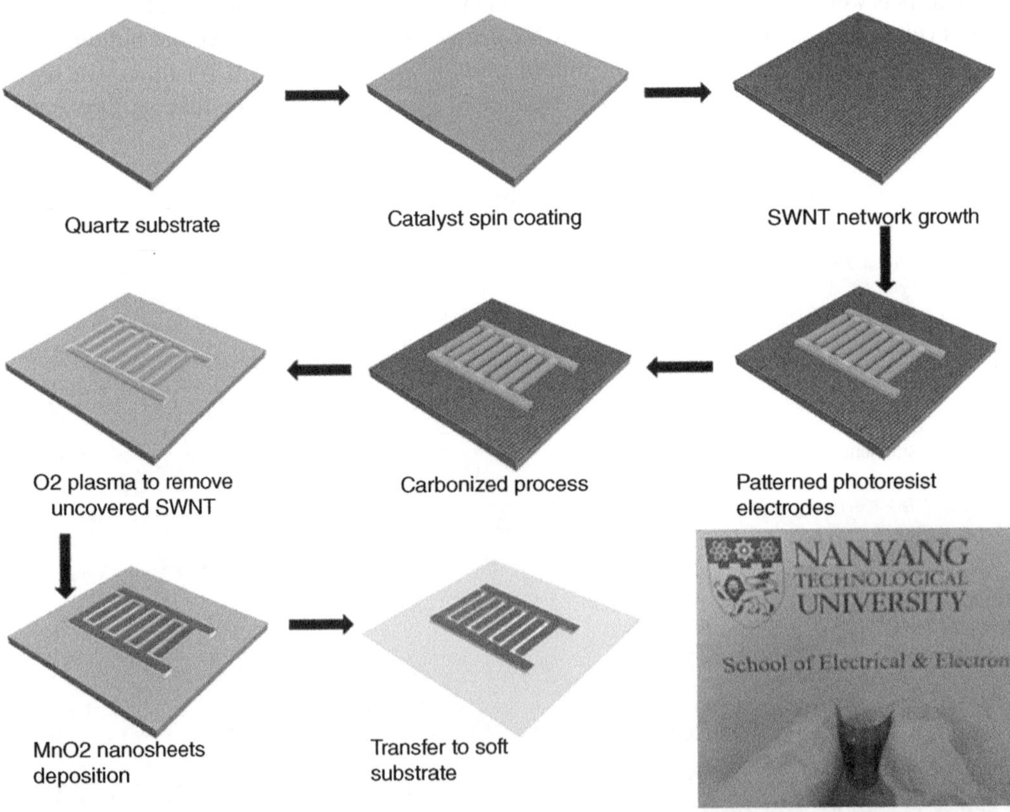

Figure 11.7 Schematic illustration of the fabrication process and optical photograph of MSCs based on the SWNT/carbon/MnO_2 electrodes. SWNT: single-wall carbon nanotube. *Source:* [27]. Reproduced with permission of Elsevier.

polyurethane (TPU) electrodes were also prepared by microfluidic-spinning method combined with in situ polymerization process [66]. The introduction of PANI dramatically enhanced their specific capacitance from 42.1 to 155.5 $mF\,cm^{-2}$.

11.3.3 Other Carbon-based Materials

Other carbon materials such as AC, onion-like carbon, carbon fiber, and laser induced carbon materials have also been used in the fabrication of MSCs due to their high surface area, good conductivity, and porous structure [67]. AC with high surface area, good conductivity and tunable pore size is a representative electrode material of electronic double layer capacitor (EDLC). AC nanospheres with small particle size are facile to achieve a homogeneous viscous ink for printing techniques [68]. As shown in Figure 11.8a, the interdigitated micro-electrodes based on AC nanospheres were produced via a screen-printing method [69]. The AC nanospheres possessed a high surface area of 2534 $m^2\,g^{-1}$ and good conductivity, benefiting to the excellent electrochemical performances of MSCs even under the bending state of 90° (Figure 11.8b). Moreover, the connected MSCs could be integrated with a wristband to power an electronic watch (Figure 11.8c), illustrating the potential of such devices in integrating with flexible and wearable electronics. Furthermore, N-doped carbon nanospheres with high surface area and optimized mesoporous structure were produced via a soft-template assisted strategy [71]. Heteroatom doping can efficiently enhance their electrochemical performance. Based on the N-doped carbon nanospheres and PVA/H_2SO_4 gel electrolyte, the resultant solid-state MSCs reached a high energy density of 3.06 $\mu Wh\,cm^{-2}$.

In addition, porous carbon films with uniform and continuous 3D large porous structures could be produced by laser direct writing technique, as displayed in Figure 11.8d and e [70]. The polymide (PI) sheet could be directly used as a precursor of carbonization and flexible substrate. The resultant MSCs exhibited good flexibility (Figure 11.8f). Moreover, the interdigitated carbon films can be also prepared by pyrolyzing the patterned photoresist [67]. After transferring the carbon films to polymer substrates, the flexible and robust micro-electrodes could be achieved.

11.3.4 Transition Metal Oxides and Hydroxides

Transition metal oxides/hydroxides are widely used as the pseudocapacitive materials of MSCs [6]. However, it is well known that these materials usually have low electrical conductivity and poor structural stability, always leading to unsatisfying rate capability and cycle life. Recently, some strategies have been developed to overcome these issues [52]. As discussed above, carbon materials with porous architecture have been used as conductive skeletons to support pseudocapacitive materials, and excellent energy and power densities can be simultaneously achieved in these hybrid electrodes [50]. In addition, the active material layers in the MSCs are generally thin, which can enable the electrons transport between active materials and current collector. Therefore, the metal compounds can be directly deposited on the metal current collector to serve as micro-electrodes. For instance, MnO_2-based electrodes were fabricated by electrochemically depositing nanostructured MnO_2 on Ni current collector [72]. The thickness of MnO_2 layer will greatly affect the electrochemical performance of the MSCs. Large MnO_2 loading will cause the crack and exfoliation of the active materials from the current collector under deformation conditions. Moreover, thick metal oxide layers can result in lower conductivity of the electrodes, leading to the degraded rate performance of the MSCs.

To improve the conductivity of MnO_2-based micro-electrodes, nanoporous gold (NPG) with high surface area and brilliant conductivity was employed to serve as current collector [73]. NPG/MnO_2

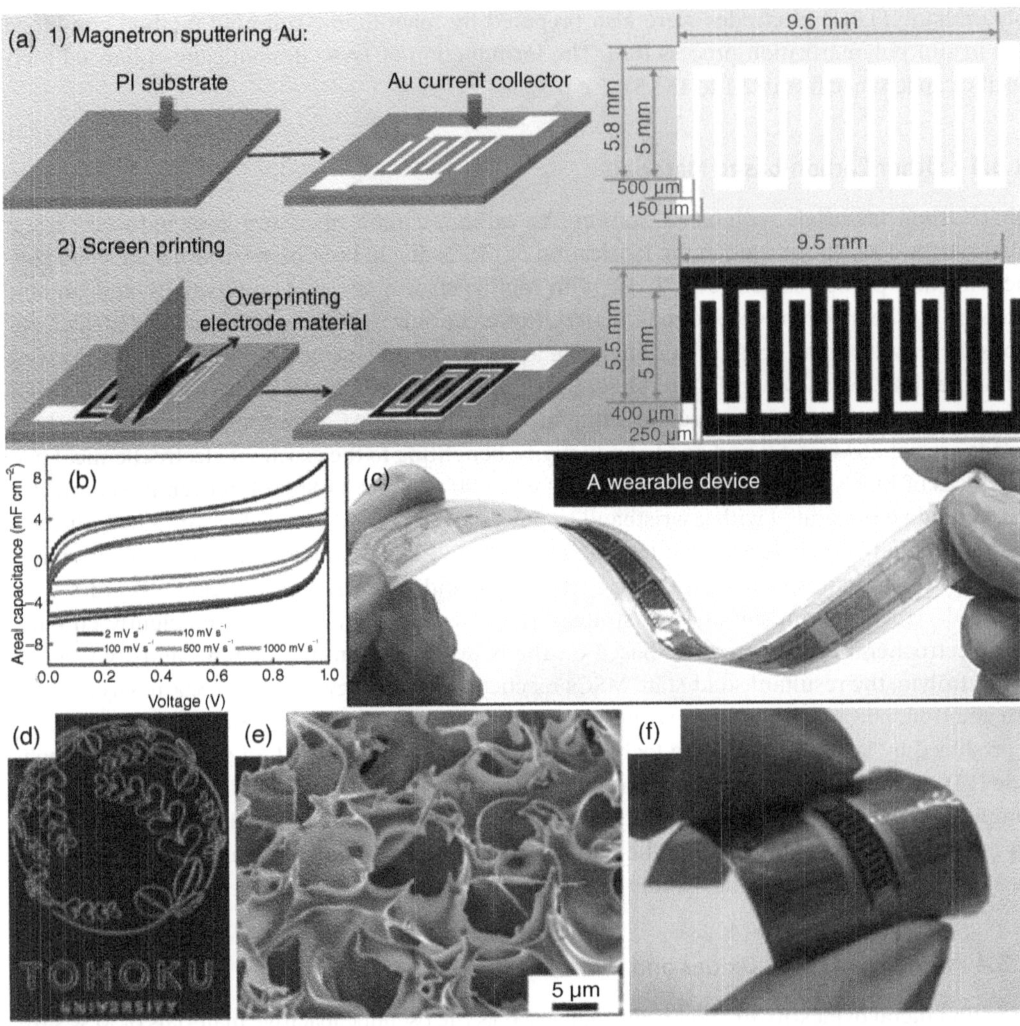

Figure 11.8 (a) Schematic illustration of fabricating interdigitated electrodes via screen-printing technique; (b) Cyclic voltammetry curves of the MSC based on AC nanosphere electrodes under bending state of 90°; (c) Photograph of a wristband constructed by the connected MSCs. *Source:* [69]. Reproduced with permission of Elsevier. (d) Optical image of a badge based on carbon structures; (e) SEM image of carbon structures; (f) Optical photograph of a flexible MSC obtained by laser writing technique. *Source:* [70]. Reproduced with permission of Royal Society of Chemistry.

interdigitated electrodes were fabricated via combining template plasma etching technique with electrochemical deposition process (Figure 11.9a) [73]. The morphologies of the NPG and NPG/MnO_2 composites with different MnO_2 mass loading were shown in the Figure 11.9b–f. The resultant flexible micro-devices delivered high energy density of 0.693–1.03 μWh cm^{-2}, and power density of 46.3–312 μW cm^{-2}. The loading and crystalline degree of the MnO_2 could significantly affect the energy and power densities of those electrodes. MnO_2 with proper mass loading and low crystallization degree exhibited enhanced capacitance.

The thin film electrodes can exhibit high power density, and remarkable cycling performances, but the ultrathin thickness of active materials determines their limited capacitance. To increase

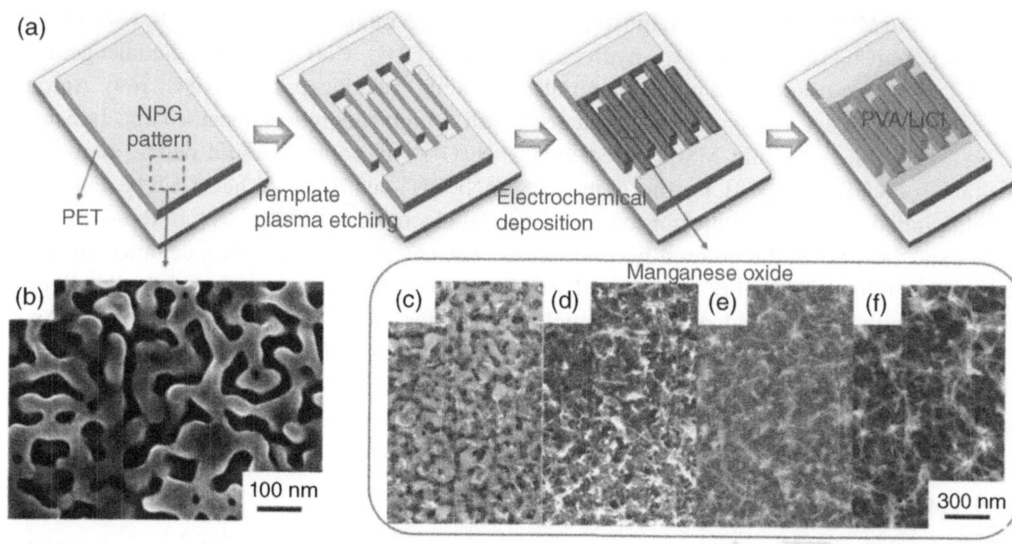

Figure 11.9 (a) The fabrication process of the flexible MSCs based on NPG/MnO_2 electrodes; (b) SEM image of NPG current collector; (c–f) SEM images of NPG/MnO_2 electrode materials with different MnO_2 mass loading. *Source:* [73]. Reproduced with permission of Elsevier.

the capacitance of MSCs, MnO_x/Au multilayer structure and 3D interdigitated Au/MnO_2/Au electrodes were designed [74, 75]. The novel alternated configuration of active materials and current collector significantly benefited to the electron transport in such thick film electrode, and further enhanced the area capacitance of the MSCs [75]. In addition, pseudocapacitive materials can be designed into the asymmetric MSC systems, which can achieve the high energy density by extending the potential window [20]. For instance, the NPG/MnO_2//NPG/PPy based asymmetric MSCs were fabricated (Figure 11.10), and the devices could maintain good capacitive behavior even at a voltage window of up to 1.6 V [76]. Benefiting to the good conductivity of NPG current collector and the suitable configuration, the MSCs could deliver a high energy density of 45.3 $mWh\,cm^{-3}$ and power density of 440.4 $W\,cm^{-3}$.

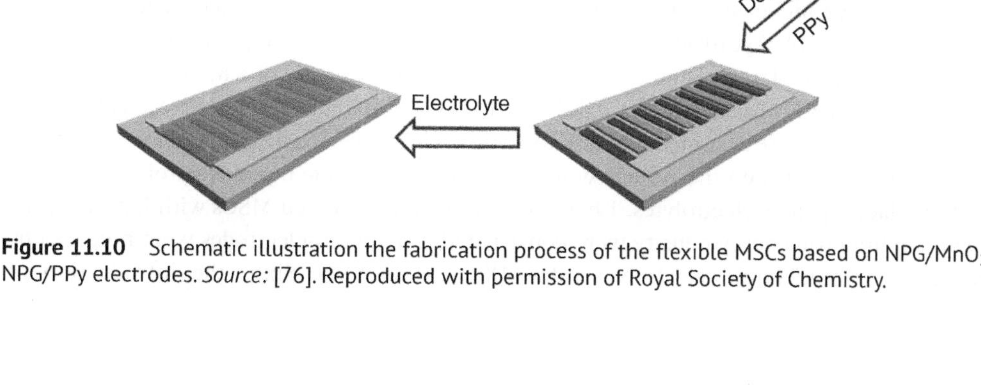

Figure 11.10 Schematic illustration the fabrication process of the flexible MSCs based on NPG/MnO_2// NPG/PPy electrodes. *Source:* [76]. Reproduced with permission of Royal Society of Chemistry.

In addition to MnO_2, many other nanostructured transition metal oxides (e.g. RuO_2, NiO, Co_3O_4, V_2O_5, Bi_2O_3, $ZnCo_2O_4$, and $NiFeO_4$), hydroxides (e.g. $Ni(OH)_2$, $Co(OH)_2$), and dichalcogenides (e.g. VS_2) were also used as the active materials of flexible pseudocapacitive MSCs [69]. Among them, $Ni(OH)_2$ can act as a battery-type material with a high theoretical specific capacitance of $2082\,F\,g^{-1}$ [77]. The one-chip MSCs based on the $Ni(OH)_2$ nanoplates could deliver a specific capacitance of $8.80\,F\,cm^{-3}$ [78]. Furthermore, CuSe@$Ni(OH)_2$ hybrid nanosheets can be deposited on interdigitated Au current collector to serve as micro-electrodes of MSCs [77]. In the vertically-aligned nanosheet films, the inherently conducting CuSe can provide partial capacitance and also serve as 3D conductive network to support $Ni(OH)_2$ nanosheets. Moreover, the vertically aligned architecture with hierarchical porous channels could offer larger accessible surface area. As a result, the MSCs exhibited an enhanced volumetric capacitance of $38.9\,F\,cm^{-3}$. In addition, the devices displayed maximum energy and power densities of $5.4\,mWh\,cm^{-3}$ and $833.2\,mW\,cm^{-3}$, respectively.

11.3.5 MXenes

As a novel class of 2D transition metal nitrides and carbides, MXenes have been proved to be attractive pseudocapacitive materials for SCs [43]. This material can be designed and synthesized by removing "A" layers from their precursors MAX (e.g. MAlX and MGaX). Their general formula can expressed as $M_{n+1}X_nT_x$, where M mains an early transition metal, X is an element of carbon or nitrogen, and T refers to the surface functional group (e.g. OH, O, and F) [42]. MXenes with 2D layered structure are promising to directly act as the active materials of flexible MSCs due to their excellent ion intercalation behaviors, metallic conductivity, high packing densities, and good hydrophilicity [23].

Typically, Gogotsi and co-workers designed all-MXene-based MSCs by using two layers of $Ti_3C_2T_x$ as electrode materials [19]. The bottom layer with large-sized (3–6 μm) stacked $Ti_3C_2T_x$ MXene flakes was mainly used as current collector, and the top layer consisted of small-sized (~1 μm) stacked $Ti_3C_2T_x$ MXene flakes was mainly used as active materials. The all-MXene-based devices displayed much lower contact resistance, high capacitance, and excellent cyclic stability with a high capacitance retention of nearly 100% after 10 000 cycles. Since the MXenes possess superior electrical conductivity, the MXene films can be directly used as current collector and active materials of MSCs. Therefore, thick electrodes can be fabricated by simply superposing the number of thin MXene films, avoiding the use of binder and conductive additives, which is beneficial to improve the energy density of the whole devices [79]. Similarly, the MXene dispersions can be directly used as additive-free inks to combining with printing techniques, such as extrusion printing, inkjet printing (Figure 11.11) [32], and 3D printing techniques [26]. Impressively, the areal capacitance of the 3D-printed MSCs based on $Ti_3C_2T_x$ MXene ink could reach $1035\,mF\,cm^{-2}$, showing great promise to achieve flexible MSCs with high areal capacitance [26].

To further achieve higher energy density, asymmetric MSCs were fabricated by using $Ti_3C_2T_x$ and Co-Al layered double hydroxide nanosheets as electrode materials [31]. The devices achieved a high working voltage of 1.45 V and an improved areal energy density of $8.8\,\mu Wh\,cm^{-2}$. In addition, Qin's group synthesized conjugated PEDOT-MXene films, and the energy density of the asymmetric MSCs with those composite film and MnO_2 electrodes reached to $250.1\,mWh\,cm^{-3}$ at 1.6 V [80]. To further broaden the voltage window of the MSCs, ionic liquid electrolytes could be applied to replace aqueous electrolytes. Typically, flexible ionogel-based MSCs with high working voltage and volumetric energy density were constructed. The micro-electrodes were prepared by pre-intercalating ionic liquid electrolyte into MXene films [81]. Since the introduction of

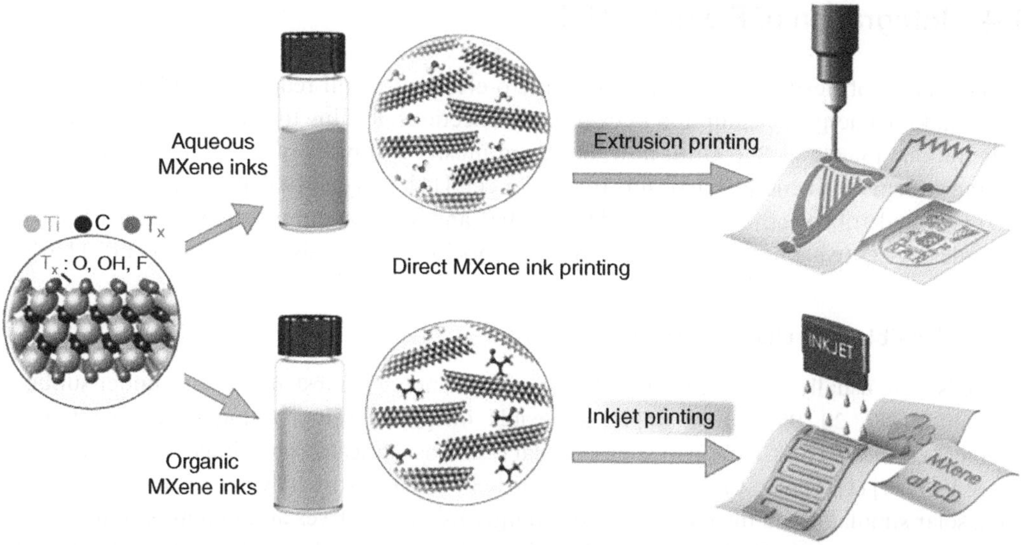

Figure 11.11 The schematic diagram of fabricating all-MXene printed MSCs. *Source:* [32]. Reproduced with permission of Springer Nature.

1-butyl-3-methylimidazolium tetrafluoroborate ionic liquid electrolyte, the resultant devices could work stably at a high voltage of 3 V, and their volumetric energy density could be enhanced to 41.8 mWh cm^{-3}. Moreover, the MSCs could maintain stable performance under repeated deformation conditions, displaying outstanding mechanical durability.

11.3.6 Conductive Polymer

Similar with metal oxides/hydroxides, besides cooperating with carbon-based materials, conductive polymers can be directly deposited on the patterned current collectors to act as micro-electrodes of flexible MSCs [3]. For instance, micro-electrodes based on PANI nanowires were obtained by in-situ electropolymerization of aniline on the surface of interdigitated Au arrays [82]. The resultant MSCs possessed high mechanical flexibility and achieved an energy density of 5.83 mWh cm^{-3}. In addition to PANI, PPy nanofibers can also be directly written on the flexible substrate via a near-field electrospinning technique [83]. Based on those flexible micro-electrodes, the MSCs could remain their excellent electrochemical performances under different bending states. However, the PPy usually has the weakness of moderate conductivity and unstable cycling performance. To overcome these issues, Ag nanoparticles were selected to be incorporated with PPy, achieving Ag@PPy nanocomposites [84]. The core structure of Ag nanoparticles acted as an efficient conductive scaffold to support PPy, resulting in good conductivity and structural integrity of the composite materials during the charge/discharge process.

Compared to PANI and PPy, PEDOT often displays lower specific capacitance since the higher molar mass of its monomer unit and lower doping level [43]. But, it has the capability to exhibit high electrochemical activity in a wide potential window in different electrolytes. Porous PEDOT electrodes could be produced by electrochemical polymerizing EDOT on Au film [85]. Assisted with the mediated process of an anionic surfactant, optimized electrode architecture was achieved. The energy density of the MSCs based on PEDOT micro-electrodes was 7.7 mWh cm^{-3}, which could be comparable to some lithium-based batteries with thin film electrodes.

11.4 Integration of Flexible MSCs

The integration of flexible MSCs with other micro-electronics will reduce the volume of whole devices and the energy consumption in external connections, leading to compact and lightweight configurations [3, 34, 86–89]. Until now, there are mostly two types of integrated systems: self-charging systems with energy harvesters (e.g. solar cell [11], fuel cell [90], piezoelectric generator (PEG) [91] and triboelectric generator (TENG) [92] and self-powering systems with functional electronic devices (e.g. photodetectors [93], gas sensors [94], piezoresistance sensors (PRS) [10]).

11.4.1 Flexible Self-charging MSCs

Solar cells were usually integrated with MSCs to realize energy harvest and storage under sunlight illumination [11, 20, 95]. For example, a flexible integrated self-charging system containing 12 parallel-connected MSCs and two Si-based solar cells was fabricated on a single Ecoflex substrate (Figure 11.12a) [11]. The MSCs could be charged to 0.8 V for approximately two seconds under light from a solar simulator and the corresponding charge/discharge curves at different current density were shown in Figure 11.12b. Apart from solar cells, fuel cells can be also applied in the self-charging integrated system. A flexible energy chip composed of four photocatalytic fuel cells (PFCs) in series and two asymmetric MSCs was fabricated [90]. The PFCs with TiO_2 photoanodes and Ag counter electrodes were used to deliver stable electrical energy. The MSC could be charged to 1.2 V after 288 seconds with ultraviolet (UV) intensity of 100 mW cm^{-2} and then discharged for 703 seconds (0.1 mA cm^{-2}) (Figure 11.12c). Moreover, this process could be cycled for several times, exhibiting great potential of supplying energy for wearable electronics constantly (Figure 11.12d).

In addition to solar and chemical energy, the energy from ambient environment or regular human motions in low frequency can be harvested and converted into electricity in a safe, efficient, and environmentally friendly way. Therefore, integrated self-charging systems driven by PEG or TENG were also developed [12, 91, 96, 97]. PEG supplies energy through the flow of electrons in piezoelectric materials, which are driven by compressive stress [13]. For instance, a flexible self-charging integrated system containing PEG and MSC was reported [91]. The voltage of MSC could reach 0.1 V through continuous pressing the PEG for 400 seconds, and then the stored electricity was discharged at 0.2 μA cm^{-2} for about 300 seconds, demonstrating the application potential for energy harvest and storage system.

Besides, TENG is also an attractive mechanical energy harvester that can be integrated with MSCs [12, 97, 98]. The working mechanism of TENG relies on the coupling effect of contact electrification and electrostatic induction. Compared with PEG, it possesses advantages of high output voltage and low cost [92]. A TENG and MSC could be integrated through the double-faced polyimide substrate and a rectifier [97]. After applying stress to the device for 117 minutes, the MSC could be charged up to 3 V. Furthermore, the electrochemical performance of the integrated device was almost unchanged under 90° bending state. To achieve higher energy harvesting efficiency, a self-charging system that integrated hybrid piezo/TENG with flexible MSC was further demonstrated [12]. The piezo/TENG could deliver a high output voltage of 150 V as well as an output current of 20 μA. Moreover, the MSCs could be charged up to 3 V to light an light-emitting diode (LED) under continuous human palm impacts in 900 seconds.

11.4.2 Flexible Self-powering MSCs

Portable sensors or detectors have attracted wide attention due to their capability of monitoring personal health and environment through various signals (e.g. light, gas, pressure, strain and

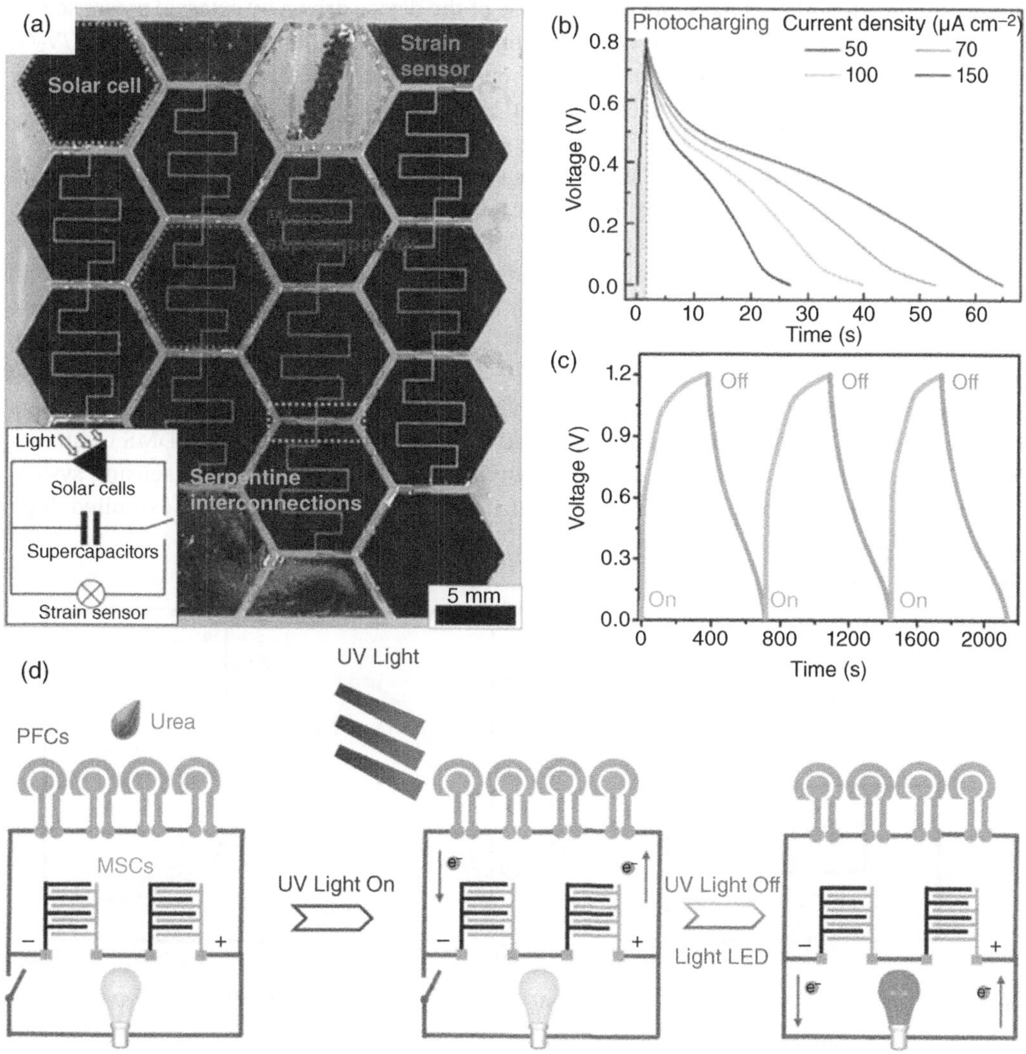

Figure 11.12 (a) Optical image of the self-charging system containing MSCs and solar cells (inset shows the circuit diagram); (b) Photo-charging curves of the integrated system under light, and discharging curves at different current densities in the dark states. *Source:* [11]. Reproduced with permission of Elsevier. (c) Photo-charging and discharging curves and (d) schematic illustration of the self-charging system containing MSCs and PFCs. *Source:* [90]. Reproduced with permission of American Chemical Society.

temperature) [2]. Miniaturizing and integrating them with MSCs can not only avoid the bulky external power supply, but also decrease the overall volume and weight of integrated system. For this motivation, recently, different sensor or detectors, such as photodetectors, gas sensors, and PRS has been integrated with MSCs.

Photodetectors are able to convert light into electrical signal, which can be applied in many fields such as environmental monitoring and intelligent robots [49, 93, 99, 100]. A flexible photodetecting system was fabricated by integrating a rGO-based MSC with a CdS-based photodetector [93]. The electrodes in MSC were utilized as the drain electrodes of the CdS-based photodetector simultaneously. With illuminating light on/off, the photocurrent was regularly increased and disappeared. The photoelectric stimulate time and recover time were 0.83 and 2.90 seconds, respectively, which is almost the same with the case of the device driven by external power source (0.73 and 2.90 seconds). In addition, a flexible photodetectable MSC based on MWCNT/V_2O_5 electrodes and SnO_2-based UV detector was also reported (Figure 11.13a) [100]. The discharging rate of MSCs was significantly improved when the UV intensity increased from 0 to 1.4 mW cm^{-2} due to the photocurrent generation of SnO_2 (Figure 11.13b).

Besides photodetectors, gas sensors can also be integrated with MSCs [94, 101, 102]. For instance, a stretchable NO_2 gas sensor driven by a flexible MSC on planar Ecoflex substrate was developed (Figure 11.13c and d) [94]. The sensor could detect NO_2 gas for 50 minutes even under stretching by 50%. In addition to NO_2 sensor, an ethanol sensor with high detection sensitivity of less than 1 ppm was also integrated with MSC on the same flexible substrate, displaying a quick response time (13 seconds) and recovery time (4.5 seconds) [102]. Based on this integrated device, a drunken driving monitoring system was realized.

PRS possesses wide strain ranges, high sensitivity and stability. An all-in-one sensing system integrated with a PRS and a MSC was fabricated based on the porous CNT/PDMS elastomer (Figure 11.14a) [10]. In virtue of the piezoresistivity of porous structure, the sensing system not only showed outstanding areal capacitance (249.89 μF cm^{-2}), but also exhibited high

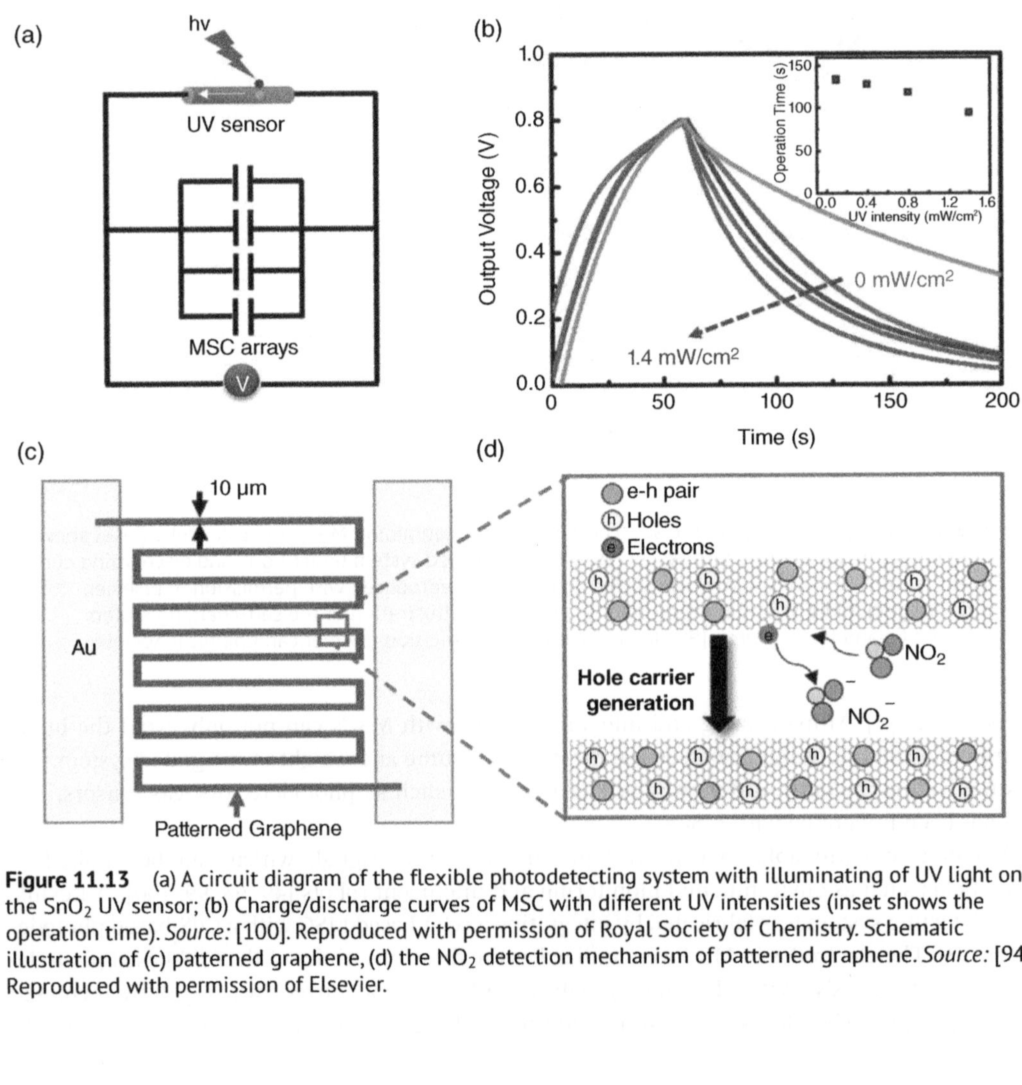

Figure 11.13 (a) A circuit diagram of the flexible photodetecting system with illuminating of UV light onto the SnO_2 UV sensor; (b) Charge/discharge curves of MSC with different UV intensities (inset shows the operation time). *Source:* [100]. Reproduced with permission of Royal Society of Chemistry. Schematic illustration of (c) patterned graphene, (d) the NO_2 detection mechanism of patterned graphene. *Source:* [94]. Reproduced with permission of Elsevier.

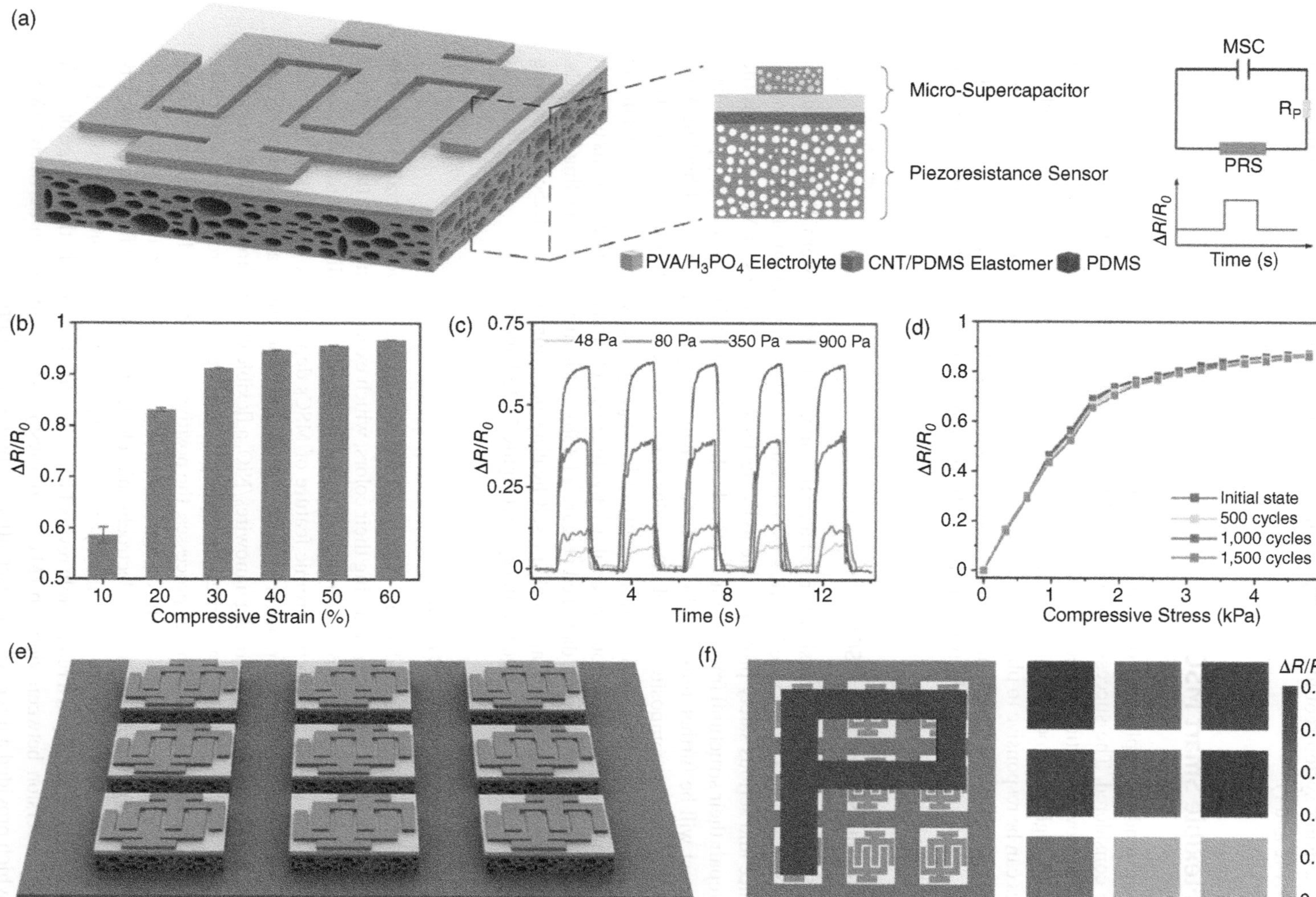

Figure 11.14 (a) Schematic illustration of an all-in-one sensing system composed of a PRS and a MSC; (b) Resistance performance with different compressive strains; (c) Resistance change after compressing/releasing cycles with different stresses; (d) Resistance change after repeated compressing/releasing cycles; (e) Schematic illustration of the 3D touch device; (f) The 3D touch device loaded by a "P" shape plate and the corresponding resistance signals of the device. *Source:* [10]. Reproduced with permission of Elsevier.

sensitivity (0.51 kPa^{-1}) and wide detection range (0–2 kPa^{-1}) (Figure 11.14b–d). Moreover, the all-in-one sensing system could be designed into a 3D touch device for user identification (Figure 11.14e and f).

11.5 Flexible Smart MSCs

In order to extend the application of flexible MSCs into some specific fields, flexible smart MSCs should be considered. The successful assembly of flexible smart MSCs depends mainly on the fabrication of novel functional electrode and electrolyte materials [4, 23, 103]. They should not only provide high electrochemical performance but also display smart behaviors. Therefore, the devices can be responsive to physical or chemical stimuli originating from themselves or external environment, such as mechanical deformation, voltage, light and temperature, which are also called self-healing [14], electrochromism [12], photodetection [17] and thermoreversible self-protecting [104].

11.5.1 Flexible Self-healing MSCs

Flexible MSCs may undergo undesirable mechanical damage caused by external deformation in practical applications. The fracture in MSCs would result in the degradation of electrochemical performance and serious safety problems due to the leakage of electrolytes. If MSCs possess the ability to repair their structural fracture automatically, their lifetime will be significantly prolonged and their cost will be further reduced [14, 105]. As shown in Figure 11.15a, a self-healing MSC based on MXene/rGO composite aerogel electrodes was packaged by self-healing carboxylated polyurethane (PU) shell [14]. As there were plenty of hydrogen bond acceptors and donors in the polymer network of PU, the shell could re-establish hydrogen bonds and dynamical chains between the fractured surfaces after being damaged (Figure 11.15b), achieving the self-healing of MSCs. The self-healing MSC presented an excellent self-healing behavior (Figure 11.15c), remaining 81.7% specific capacitance after five healing cycles (Figure 11.15d and e).

11.5.2 Flexible Electrochromic MSCs

Owing to the electrochemical reduction/oxidation processes, some active materials in flexible MSCs have the ability of reversibly changing their colors, which exhibit great potential to monitor the real-time working status through intuitive visual inspection, avoiding unexpected overuse [18, 106]. In general, the electrochromic feature of MSCs depends on their active electrode materials or electrolytes. Transition metal oxides and conducting polymers are the common electrochromic materials [103]. Based on Ag nanowires/NiO, a flexible electrochromic MSC was fabricated [12]. In such device, Ag nanowires/NiO worked as both electrochromic and capacitive materials. During the charge/discharge processes, the positive electrodes in MSC changed to dark color and then recovered to transparent on account of the reversible Faradaic redox process of Ni^{2+}/Ni^{3+} couple. Apart from employing electrochromic electrode materials, a flexible electrochromic MSC was also fabricated by using methyl viologen based electrochromic electrolyte [15]. The resulting device displayed a remarkable reversible electrochromic effect in the voltage of 0–1 V due to the conversion between viologen (MV^{2+}, colorless) and its radical ion derivative (MV^{+}, purple), which provided a direct visual observation without extra equipment.

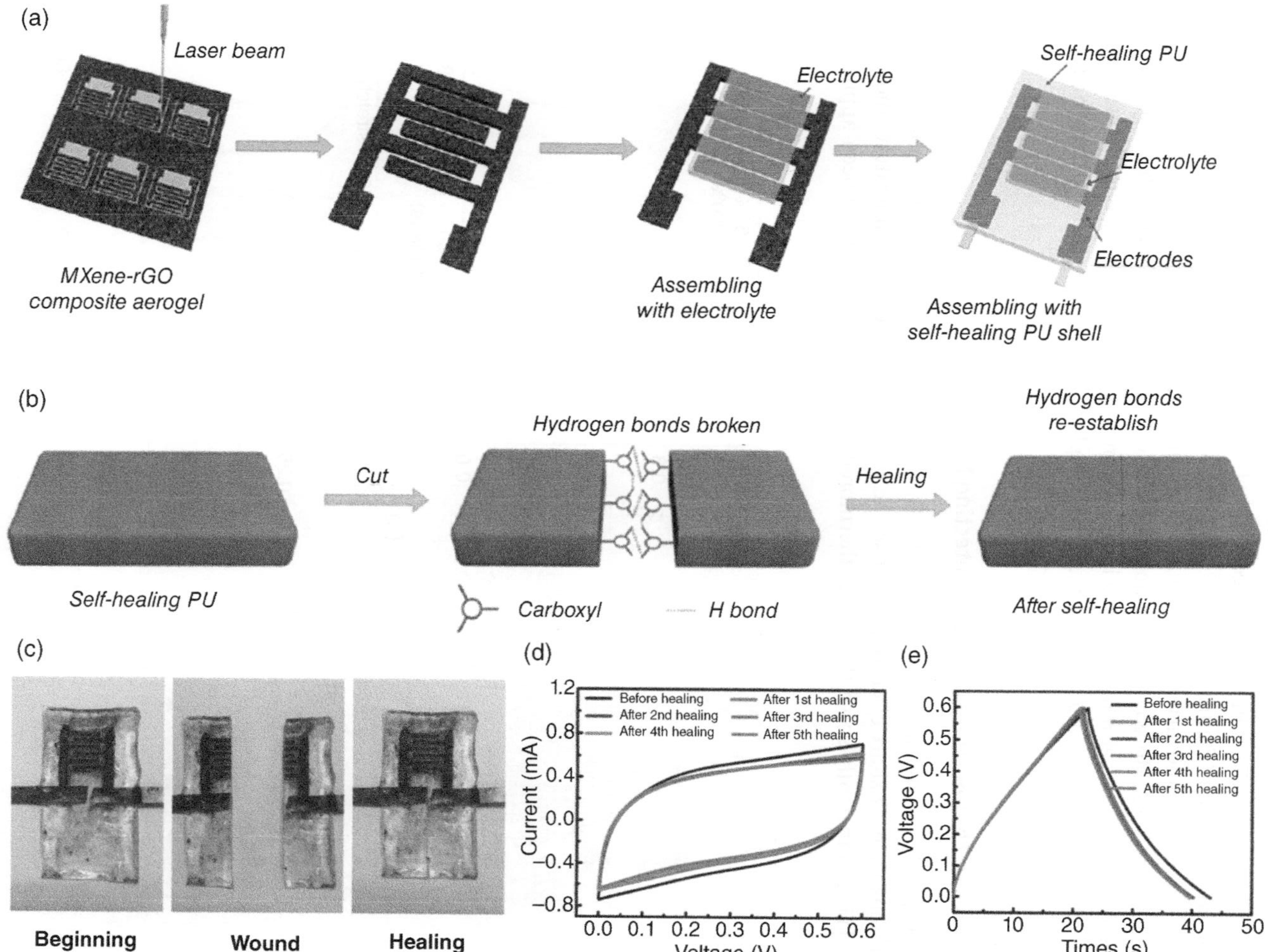

Figure 11.15 (a) The fabricating process and (b) self-healable mechanism of the flexible self-healing MSCs; (c) Optical images of the healing process of a flexible self-healing MSC; (d) Cyclic voltammogram curves and (e) galvonostatic charge/discharge curves of the self-healing MSCs. *Source:* [14]. Reproduced with permission of American Chemical Society.

11.5.3 Flexible Photodetectable MSCs

Different from integrating photodetector with flexible MSCs, introducing active materials with photodetecting ability into electrodes can endow flexible MSCs with the photodetecting behavior. It will simplify the configuration of whole devices [107, 108]. For instance, TiO_2 nanoparticles were coated onto the SWCNTs film micro-electrodes of flexible MSC to serve as the working electrode of the photodetector (Figure 11.16a) [17]. When it was exposed to UV light with an intensity of $5\,mW\,cm^{-2}$, a stable photocurrent of 0.32 mA was obtained and the photodetection sensitivity of integrated device was 6.2. Moreover, the flexible photodetectable MSC exhibited stable capacitance photocurrent and response at repeated stretching 100 times (Figure 11.16b–d). Such strategy will pave the way for fabricating integrated devices with high flexibility and stable electrochemical performance.

11.5.4 Flexible Thermoreversible Self-protecting MSCs

MSCs are generally worked at room temperature. However, the fast energy consumption or abusive conditions such as short circuits will cause the increase of internal temperature and even explosion [1]. To efficiently solve this issue, thermoreversible electrolytes have been applied in

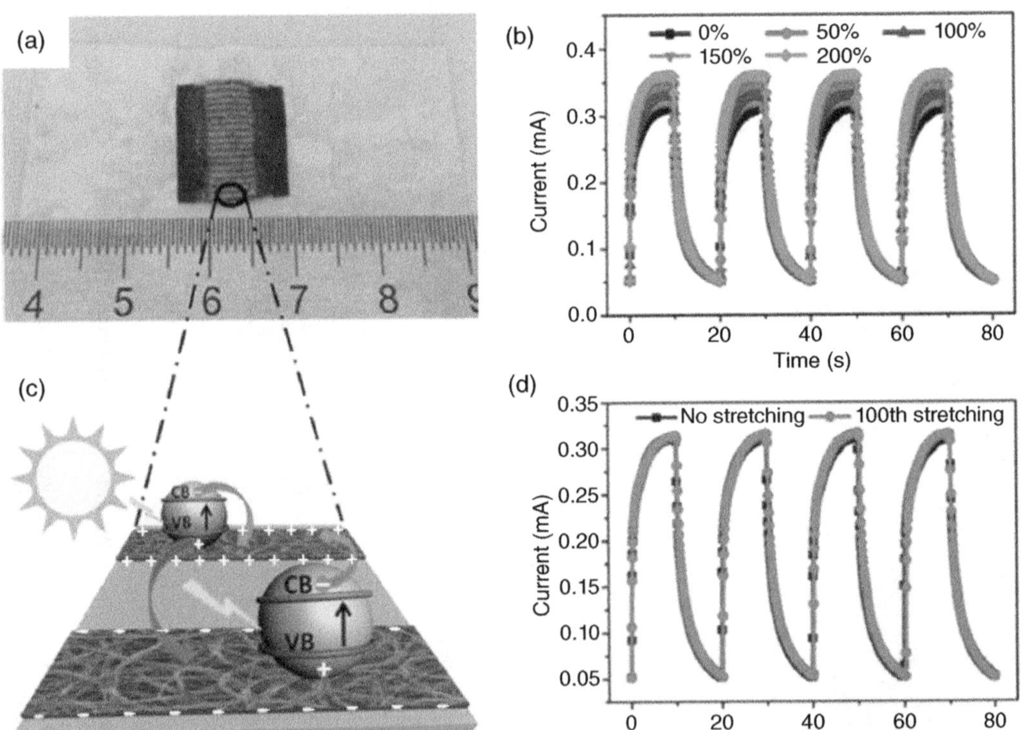

Figure 11.16 (a) Optical image of a flexible photodetectable MSC; (b) Photocurrent response of the photodetectable MSC at different strain states; (c) Schematic illustration of the photodetecting mechanism: the electron-hole pairs were separated under UV light irradiates, resulting in photocurrent. CB: conduction band. VB: valence band; (d) Photocurrent response of the photodetectable MSC before and after repeated stretching cycles. *Source:* [17]. Reproduced with permission of Elsevier.

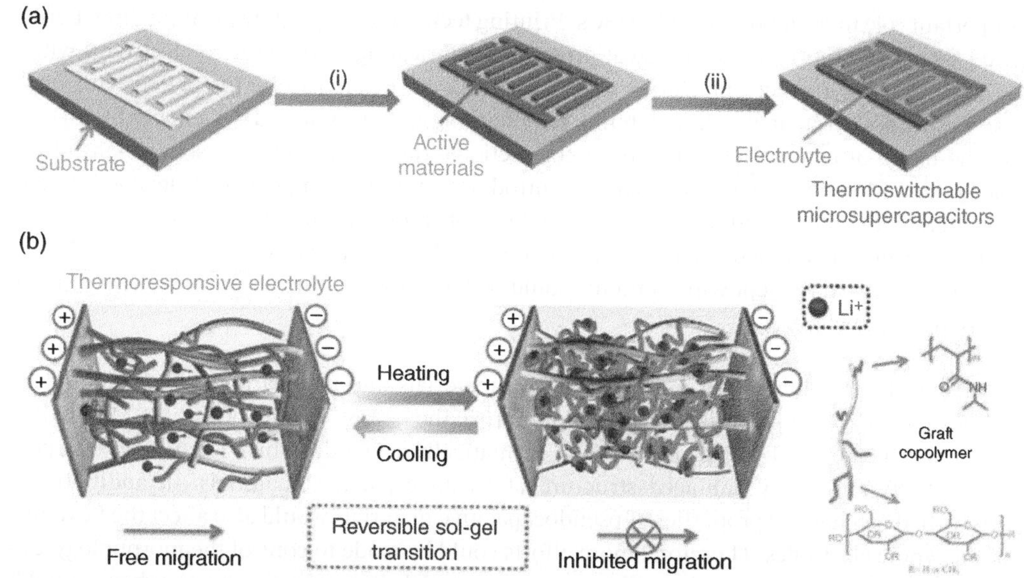

Figure 11.17 (a) Schematic illustrations of the fabrication process of flexible thermoreversible self-protecting MSCs based on PNIPAAm/MC electrolyte; (b) Schematic illustrations of the reversible sol-gel transition and ion transport between electrodes under different temperatures. *Source:* [104]. Reproduced with permission of Royal Society of Chemistry.

MSCs on account of their abilities of reversibly controlling the operation of MSCs. A flexible thermoreversible self-protecting MSC based on the poly(N-isopropylacrylamide)-g-methylcellulose (PNIPAAm/MC) electrolyte with sol-gel transition was developed (Figure 11.17a) [104]. The graft copolymer in electrolyte could control the migration of Li^+ ions according to the temperature change (Figure 11.17b). When the environment temperature increased from 30 to 80 °C, the graft copolymer formed hydrogels through hydrophobic association and restrained the migration of Li^+ ions, thus the MSC achieved a switch-off behavior. It illustrates the promising application for protecting integrated electronics. In addition, poly(ethylene oxide)-block-poly(propylene oxide)-blockpoly(ethylene oxide) (PEO-PPO-PEO) electrolyte with similar sol-gel transition at different temperature was also applied to realize self-protecting of MSCs [16]. The transition temperature and the degree of capacitance suppression could be controlled by optimizing molecular weight of the polymers and solution concentration.

11.6 Summary and Prospects

In conclusion, we summarized the recent advances of flexible MSCs in the aspects of electrode materials, fabrication strategies, the integration with other electronics, and the design of smart behaviors. Various active materials, including graphene, CNTs, transition-metal oxides/hydroxides, conductive polymers, and their composites, have been employed as micro-electrodes for flexible MSCs. In addition to electrode materials, the electrochemical performance of MSCs is also affected by the architecture of micro-electrodes (e.g. thickness, width and spacing gaps between them). Narrowing the width and spacing gaps between micro-electrodes could improve the energy and power densities of flexible MSCs. Therefore, the micro-fabrication techniques play an important role in designing flexible MSCs. Printing techniques (e.g. spray printing, inkjet printing and screen printing) possess the great prospect in fabricating flexible micro-electrodes with a large scale. For the further applications, flexible MSCs are desired to be integrated with other micro-electronics on a single chip to build self-charging or self-powered systems, which could avoid the usage of external electric circuit connections, thus decreasing the size/weight of the whole system. Moreover, smart materials are introduced into the composites of flexible MSCs to endow them with other functions for extending the application in some fields.

Despite the aforementioned encouraging progress, the development of flexible MSCs is still at its early stage. Efforts in the following contents could be taken toward further improvement of high-performance flexible MSCs.

1) Although the introduction of pseudocapacitive materials into nanocarbon-based materials to fabricate composite micro-electrodes could improve the energy density of flexible MSCs, their power density and cycling performance are dramatically degraded, which is caused by the poor conductivity and easily damaged structure of pseudocapacitive materials. In addition, the unsatisfied mechanical properties of pseudocapacitive materials would also affect the flexibility of the whole electrodes. Therefore, more efforts could be made to control the morphology, distribution and mass loading of pseudocapacitive materials in conductive nanocarbon scaffolds to optimize the synergistic effects between them.
2) Many strategies have been developed to prepare micro-electrodes, however, most of them have their own limitations. For example, laser scribing is an effective and scalable technique to pattern micro-electrodes in a short time, whereas it has some disadvantages such as high cost and the waste of electrodes materials. Besides, although printing strategies have shown great potential in fabricating micro-electrodes with a large-scale, most of their precisions are limited on hundreds of micrometers. Therefore, novel micro-fabrication techniques with higher precision, low cost and scalable production are highly desired.
3) Self-powered systems based on MSCs are still at the stage of proof-of-concept. The current fabrication techniques of MSCs are not fully compatible with various micro-electronics, especially for the devices with complex composition, leading to a great challenge for the wide application. In addition, the total energy conversion efficiency between them is also of great significance, which could be achieved by optimizing the design of overall structures and the match degree between energy storage and conversion. For smart flexible MSCs, the current research focuses on the exploration of novel smart designs. However, their smart performance is often limited. Therefore, it is critical to achieve high performance of energy storage and other functions, simultaneously. Moreover, more attention could be paid on some special macroscopic functional MSCs with other promising properties.
4) Several evaluation methods based on weight, area and volume have been employed to evaluate the electrochemical performance of MSCs. In MSCs, since the active materials loaded on interdigitated micro-electrodes are ultra-thin, the weight of the active materials is nearly negligible compared with the whole device. Therefore, the gravimetric specific capacitance (C, $F\,g^{-1}$), power density, and energy density of active materials are not inappropriate for evaluating the performance of MSCs. The specific capacitance, energy and power density based on the area of MSCs would be the key performance metrics. Besides, some MSCs are evaluated with the area of active materials, ignoring the footprint of the gaps between electrodes. Therefore, the total area of MSCs instead of just the size of active materials has to be considered to evaluate the performance of MSCs.
5) There has not been a standardized measurement method of the actual flexibility of flexible MSCs, leading to much difficulty to compare them with that of previous reports. In general, the

flexibility of MSCs is tested by measuring the electrochemical performance under different deformation conditions. However, the size or thickness of these flexible MSCs are varied, resulting in different strains even at the same deformation degree. Thus, it is desired to develop standardized measurement methods for the flexibility of MSCs.

6) Although great experimental efforts have been made for the development of flexible MSCs, the theoretical simulations of flexible MSCs under bending conditions are absent. They could help us to better understand the influence of strains on the device structure. In addition, in-situ characterization techniques could witness the structural evolution of electrode materials in different bending states, which are also not well-established. Therefore, theoretical simulations and in-situ characterization techniques should be improved to guild us for the design of flexible MSCs.

References

1 Kyeremateng, N.A., Brousse, T., and Pech, D. (2017). Microsupercapacitors as miniaturized energy-storage components for on-chip electronics. *Nat. Nanotechnol.* 12 (1): 7–15.

2 Zheng, S., Shi, X., Das, P. et al. (2019). The road towards planar microbatteries and micro-supercapacitors: from 2D to 3D device geometries. *Adv. Mater.* 31 (50): 1900583.

3 Liu, L., Niu, Z., and Chen, J. (2016). Unconventional supercapacitors from nanocarbon-based electrode materials to device configurations. *Chem. Soc. Rev.* 45 (15): 4340–4363.

4 Liu, L., Niu, Z., and Chen, J. (2017). Design and integration of flexible planar micro-supercapacitors. *Nano Res.* 10 (5): 1524–1544.

5 Niu, Z., Zhang, L., Liu, L. et al. (2013). All-solid-state flexible ultrathin micro-supercapacitors based on graphene. *Adv. Mater.* 25 (29): 4035–4042.

6 Zhang, P., Wang, F., Yu, M. et al. (2018). Two-dimensional materials for miniaturized energy storage devices: from individual devices to smart integrated systems. *Chem. Soc. Rev.* 47 (19): 7426–7451.

7 Lethien, C., Bideau, J.L., and Brousse, T. (2019). Challenges and prospects of 3D micro-supercapacitors for powering the internet of things. *Energy Environ. Sci.* 12 (1): 96–115.

8 Da, Y., Liu, J., Zhou, L. et al. (2019). Engineering 2D architectures toward high-performance micro-supercapacitors. *Adv. Mater.* 31 (1): 1802793.

9 Liu, N. and Gao, Y. (2017). Recent progress in micro-supercapacitors with in-plane interdigital electrode architecture. *Small* 13 (45): 1701989.

10 Song, Y., Chen, H., Chen, X. et al. (2018). All-in-one piezoresistive-sensing patch integrated with micro-supercapacitor. *Nano Energy* 53: 189–197.

11 Yun, J., Song, C., Lee, H. et al. (2018). Stretchable array of high-performance micro-supercapacitors charged with solar cells for wireless powering of an integrated strain sensor. *Nano Energy* 49: 644–654.

12 Qin, S., Zhang, Q., Yang, X. et al. (2018). Hybrid piezo/triboelectric-driven self-charging electrochromic supercapacitor power package. *Adv. Energy Mater.* 8 (23): 1800069.

13 Wang, J., Li, F., Zhu, F. et al. (2019). Recent progress in micro-supercapacitor design, integration, and functionalization. *Small Methods* 3 (8): 1800367.

14 Yue, Y., Liu, N., Ma, Y. et al. (2018). Highly self-healable 3D microsupercapacitor with mxene-graphene composite aerogel. *ACS Nano* 12 (5): 4224–4232.

15 Zhang, P., Zhu, F., Wang, F. et al. (2017). Stimulus-responsive micro-supercapacitors with ultrahigh energy density and reversible electrochromic window. *Adv. Mater.* 29 (7): 1604491.

16 Ma, S., Shi, Y., Zhang, Y. et al. (2019). All-printed substrate-versatile microsupercapacitors with thermoreversible self-protection behavior based on safe sol-gel transition electrolytes. *ACS Appl. Mater. Interfaces* 11 (33): 29960–29969.

17 Chen, C., Cao, J., Wang, X. et al. (2017). Highly stretchable integrated system for micro-supercapacitor with AC line filtering and UV detector. *Nano Energy* 42: 187–194.

18 Jia, R., Shen, G., Qu, F. et al. (2020). Flexible on-chip micro-supercapacitors: efficient power units for wearable electronics. *Energy Storage Mater.* 27: 169–186.

19 Peng, Y.-Y., Akuzum, B., Kurra, N. et al. (2016). All-MXene (2D titanium carbide) solid-state microsupercapacitors for on-chip energy storage. *Energy Environ. Sci.* 9 (9): 2847–2854.

20 Guo, R., Chen, J., Yang, B. et al. (2017). In-plane micro-supercapacitors for an integrated device on one piece of paper. *Adv. Funct. Mater.* 27 (43): 1702394.

21 Hyun, W.J., Secor, E.B., Kim, C.-H. et al. (2017). Scalable, self-aligned printing of flexible graphene micro-supercapacitors. *Adv. Energy Mater.* 7 (17): 1700285.

22 Wu, Z.S., Parvez, K., Feng, X. et al. (2013). Graphene-based in-plane micro-supercapacitors with high power and energy densities. *Nat. Commun.* 4: 2487.

23 Zhang, H., Cao, Y., Chee, M.O.L. et al. (2019). Recent advances in micro-supercapacitors. *Nanoscale* 11 (13): 5807–5821.

24 Beidaghi, M. and Gogotsi, Y. (2014). Capacitive energy storage in micro-scale devices: recent advances in design and fabrication of micro-supercapacitors. *Energy Environ. Sci.* 7 (3): 867–884.

25 Li, L., Zhang, J., Peng, Z. et al. (2016). High-performance pseudocapacitive microsupercapacitors from laser-induced graphene. *Adv. Mater.* 28 (5): 838–845.

26 Orangi, J., Hamade, F., Davis, V.A. et al. (2020). 3D printing of additive-free 2D $Ti_3C_2T_x$ (MXene) ink for fabrication of micro-supercapacitors with ultra-high energy densities. *ACS Nano* 14 (1): 640–650.

27 Sun, L., Wang, X., Zhang, K. et al. (2016). Metal-free SWNT/carbon/MnO_2 hybrid electrode for high performance coplanar micro-supercapacitors. *Nano Energy* 22: 11–18.

28 Wang, X., Wan, F., Zhang, L. et al. (2018). Large-area reduced graphene oxide composite films for flexible asymmetric sandwich and microsized supercapacitors. *Adv. Funct. Mater.* 28 (18): 1707247.

29 El-Kady, M.F. and Kaner, R.B. (2013). Scalable fabrication of high-power graphene micro-supercapacitors for flexible and on-chip energy storage. *Nat. Commun.* 4: 1475.

30 Gao, W., Singh, N., Song, L. et al. (2011). Direct laser writing of micro-supercapacitors on hydrated graphite oxide films. *Nat. Nanotechnol.* 6 (8): 496–500.

31 Xu, S., Dall'Agnese, Y., Wei, G. et al. (2018). Screen-printable microscale hybrid device based on MXene and layered double hydroxide electrodes for powering force sensors. *Nano Energy* 50: 479–488.

32 Zhang, C.J., McKeon, L., Kremer, M.P. et al. (2019). Additive-free MXene inks and direct printing of micro-supercapacitors. *Nat. Commun.* 10 (1): 1795.

33 Song, D., Zare Bidoky, F., Secor, E.B. et al. (2019). Freestanding ion gels for flexible, printed, multifunctional microsupercapacitors. *ACS Appl. Mater. Interfaces* 11 (10): 9947–9954.

34 Zhao, C., Liu, Y., Beirne, S. et al. (2018). Recent development of fabricating flexible micro-supercapacitors for wearable devices. *Adv. Mater. Technol.* 3 (9): 1800028.

35 Lim, B.Y., Yoon, J., Yun, J. et al. (2014). Biaxially stretchable, integrated array of high performance microsupercapacitors. *ACS Nano* 8 (11): 11639–11650.

36 Li, W., Li, Y., Su, M. et al. (2017). Printing assembly and structural regulation of graphene towards three-dimensional flexible micro-supercapacitors. *J. Mater. Chem. A* 5 (31): 16281–16288.

37 Sang Tran, T., Dutta, N.K., and Roy Choudhury, N. (2019). Graphene-based inks for printing of planar micro-supercapacitors: a review. *Materials* 12 (6): 978.

38 Suk, J.W., Kitt, A., Magnuson, C.W. et al. (2011). Transfer of CVD-grown monolayer graphene onto arbitrary substrates. *ACS Nano* 5 (9): 6916–6924.

39 Lee, K., Lee, H., Shin, Y. et al. (2016). Highly transparent and flexible supercapacitors using graphene-graphene quantum dots chelate. *Nano Energy* 26: 746–754.

40 Sollami Delekta, S., Smith, A.D., Li, J. et al. (2017). Inkjet printed highly transparent and flexible graphene micro-supercapacitors. *Nanoscale* 9 (21): 6998–7005.

41 Song, W., Zhu, J., Gan, B. et al. (2018). Flexible, stretchable, and transparent planar microsupercapacitors based on 3D porous laser-induced graphene. *Small* 14 (1): 1702249.

42 Li, H. and Liang, J. (2020). Recent development of printed micro-supercapacitors: printable materials, printing technologies, and perspectives. *Adv. Mater.* 32 (3): 1805864.

43 Wang, X. and Zhang, Q. (2019). On-chip microsupercapacitors: from material to fabrication. *Energ. Technol.* 7 (11): 1900820.

44 Liu, L.-J., Liu, C.-F., Li, X. et al. (2019). Graphene-based planar on-chip micro-supercapacitors with whole series/parallel configuration for integration. *Integr. Ferroelectr.* 199 (1): 95–104.

45 Wu, Z.K., Lin, Z.Y., Li, L.Y. et al. (2014). Flexible micro-supercapacitor based on in-situ assembled graphene on metal template at room temperature. *Nano Energy* 10: 222–228.

46 Qi, D., Liu, Z., Liu, Y. et al. (2015). Suspended wavy graphene microribbons for highly stretchable microsupercapacitors. *Adv. Mater.* 27 (37): 5559–5566.

47 Zhong, M., Zhang, F., Yu, Y. et al. (2018). Flexible micro-supercapacitors assembled via chemically reduced graphene oxide films assisted by a laser printer. *Nanotechnology* 29 (43): 43LT01.

48 Wang, X., Wang, R., Zhao, Z. et al. (2019). Controllable spatial engineering of flexible all-in-one graphene-based supercapacitors with various architectures. *Energy Storage Mater.* 23: 269–276.

49 Gu, S., Lou, Z., Li, L. et al. (2015). Fabrication of flexible reduced graphene oxide/Fe_2O_3 hollow nanospheres based on-chip micro-supercapacitors for integrated photodetecting applications. *Nano Res.* 9 (2): 424–434.

50 Amiri, M.H., Namdar, N., Mashayekhi, A. et al. (2016). Flexible micro supercapacitors based on laser-scribed graphene/ZnO nanocomposite. *J. Nanopart. Res.* 18 (8): 237.

51 Song, B., Li, L., Lin, Z. et al. (2015). Water-dispersible graphene/polyaniline composites for flexible micro-supercapacitors with high energy densities. *Nano Energy* 16: 470–478.

52 Kurra, N., Jiang, Q., and Alshareef, H.N. (2015). A general strategy for the fabrication of high performance microsupercapacitors. *Nano Energy* 16: 1–9.

53 Yu, Y., Zhang, J., Wu, X. et al. (2015). Facile ion-exchange synthesis of silver films as flexible current collectors for micro-supercapacitors. *J. Mater. Chem. A* 3 (42): 21009–21015.

54 Pu, J., Wang, X., Xu, R. et al. (2018). Highly flexible, foldable, and rollable microsupercapacitors on an ultrathin polyimide substrate with high power density. *Microsyst. Nanoeng.* 4: 16.

55 Kim, S.K., Koo, H.J., Lee, A. et al. (2014). Selective wetting-induced micro-electrode patterning for flexible micro-supercapacitors. *Adv. Mater.* 26 (30): 5108–5112.

56 Lin, J., Zhang, C., Yan, Z. et al. (2013). 3-dimensional graphene carbon nanotube carpet-based microsupercapacitors with high electrochemical performance. *Nano Lett.* 13 (1): 72–78.

57 Hsia, B., Marschewski, J., Wang, S. et al. (2014). Highly flexible, all solid-state micro-supercapacitors from vertically aligned carbon nanotubes. *Nanotechnology* 25 (5): 055401.

58 Xiao, H., Wu, Z.-S., Zhou, F. et al. (2018). Stretchable tandem micro-supercapacitors with high voltage output and exceptional mechanical robustness. *Energy Storage Mater.* 13: 233–240.

59 Beidaghi, M. and Wang, C. (2012). Micro-supercapacitors based on interdigital electrodes of reduced graphene oxide and carbon nanotube composites with ultrahigh power handling performance. *Adv. Funct. Mater.* 22 (21): 4501–4510.

60 Chih, J.-K., Jamaluddin, A., Chen, F. et al. (2019). High energy density of all-screen-printable solid-state microsupercapacitors integrated by graphene/CNTs as hierarchical electrodes. *J. Mater. Chem. A* 7 (20): 12779–12789.

61 Shi, L., Wang, Y., Zou, P. et al. (2018). Laser processed micro-supercapacitors based on carbon nanotubes/manganese dioxide nanosheets composite with excellent electrochemical performance and aesthetic property. *Chin. Chem. Lett.* 29 (4): 592–595.

62 Sun, L., Wang, X., Liu, W. et al. (2016). Optimization of coplanar high rate supercapacitors. *J. Power Sources* 315: 1–8.

63 Lee, G., Kim, D., Kim, D. et al. (2015). Fabrication of a stretchable and patchable array of high performance micro-supercapacitors using a non-aqueous solvent based gel electrolyte. *Energy Environ. Sci.* 8 (6): 1764–1774.

64 Ouldhamadouche, N., Achour, A., Lucio-Porto, R. et al. (2018). Electrodes based on nano-tree-like vanadium nitride and carbon nanotubes for micro-supercapacitors. *J. Mater. Sci. Technol.* 34 (6): 976–982.

65 Liu, W., Lu, C., Li, H. et al. (2016). Paper-based all-solid-state flexible micro-supercapacitors with ultra-high rate and rapid frequency response capabilities. *J. Mater. Chem. A* 4 (10): 3754–3764.

66 Tong, Y.-L., Xu, B., Du, X.-F. et al. (2018). Microfluidic-spinning-directed conductive fibers toward flexible micro-supercapacitors. *Macromol. Mater. Eng.* 303 (6): 1700664.

67 Kim, M.S., Hsia, B., Carraro, C. et al. (2014). Flexible micro-supercapacitors with high energy density from simple transfer of photoresist-derived porous carbon electrodes. *Carbon* 74: 163–169.

68 Zhao, W., Wei, L., Fu, Q. et al. (2019). High-performance, flexible, solid-state micro-supercapacitors based on printed asymmetric interdigital electrodes and bio-hydrogel for on-chip electronics. *J. Power Sources* 422: 73–83.

69 Zhang, X., Zhao, W., Wei, L. et al. (2019). In-plane flexible solid-state microsupercapacitors for on-chip electronics. *Energy* 170: 338–348.

70 Cai, J., Lv, C., and Watanabe, A. (2016). Cost-effective fabrication of high-performance flexible all-solid-state carbon micro-supercapacitors by blue-violet laser direct writing and further surface treatment. *J. Mater. Chem. A* 4 (5): 1671–1679.

71 Qiu, Y., Hou, M., Gao, J. et al. (2019). One-step synthesis of monodispersed mesoporous carbon nanospheres for high-performance flexible quasi-solid-state micro-supercapacitors. *Small* 15 (45): 1903836.

72 Guo, K., Wan, Y., Yu, N. et al. (2018). Hand-drawing patterned ultra-thin integrated electrodes for flexible micro supercapacitors. *Energy Storage Mater.* 11: 144–151.

73 Shi, X., Zeng, Z., Liao, C. et al. (2018). Flexible, planar integratable and all-solid-state micro-supercapacitors based on nanoporous gold/manganese oxide hybrid electrodes via template plasma etching method. *J. Alloys Compd.* 739: 979–986.

74 Si, W., Yan, C., Chen, Y. et al. (2013). On chip, all solid-state and flexible micro-supercapacitors with high performance based on MnO_x/au multilayers. *Energy Environ. Sci.* 6 (11): 3218–3223.

75 Hu, H., Pei, Z., Fan, H. et al. (2016). 3D interdigital au/MnO_2/au stacked hybrid electrodes for on-chip microsupercapacitors. *Small* 12 (22): 3059–3069.

76 Zhang, C., Xiao, J., Qian, L. et al. (2016). Planar integration of flexible micro-supercapacitors with ultrafast charge and discharge based on interdigital nanoporous gold electrodes on a chip. *J. Mater. Chem. A* 4 (24): 9502–9510.

77 Gong, J., Li, J.C., Yang, J. et al. (2018). High-performance flexible in-plane micro-supercapacitors based on vertically aligned CuSe@Ni$(OH)_2$ hybrid nanosheet films. *ACS Appl. Mater. Interfaces* 10 (44): 38341–38349.

78 Wu, H., Jiang, K., Gu, S. et al. (2015). Two-dimensional $Ni(OH)_2$ nanoplates for flexible on-chip microsupercapacitors. *Nano Res.* 8 (11): 3544–3552.

79 Hu, H., Bai, Z., Niu, B. et al. (2018). Binder-free bonding of modularized mxene thin films into thick film electrodes for on-chip micro-supercapacitors with enhanced areal performance metrics. *J. Mater. Chem. A* 6 (30): 14876–14884.

80 Qin, L., Tao, Q., Liu, X. et al. (2019). Polymer-mxene composite films formed by mxene-facilitated electrochemical polymerization for flexible solid-state microsupercapacitors. *Nano Energy* 60: 734–742.

81 Zheng, S., Zhang, C., Zhou, F. et al. (2019). Ionic liquid pre-intercalated mxene films for ionogel-based flexible micro-supercapacitors with high volumetric energy density. *J. Mater. Chem. A* 7 (16): 9478–9485.

82 Hu, H., Zhang, K., Li, S. et al. (2014). Flexible, in-plane, and all-solid-state micro-supercapacitors based on printed interdigital Au/polyaniline network hybrid electrodes on a chip. *J. Mater. Chem. A* 2 (48): 20916–20922.

83 Shen, C., Wang, C.-P., Sanghadasa, M. et al. (2017). Flexible micro-supercapacitors prepared using direct-write nanofibers. *RSC Adv.* 7 (19): 11724–11731.

84 Liu, L., Lu, Q., Yang, S. et al. (2018). All-printed solid-state microsupercapacitors derived from self-template synthesis of Ag@ppy nanocomposites. *Adv. Mater. Technol.* 3 (1): 1700206.

85 Kurra, N., Hota, M.K., and Alshareef, H.N. (2015). Conducting polymer micro-supercapacitors for flexible energy storage and ac line-filtering. *Nano Energy* 13: 500–508.

86 Shen, C., Xu, S., Xie, Y. et al. (2017). A review of on-chip micro supercapacitors for integrated self-powering systems. *J. Microelectromech. Syst.* 26 (5): 949–965.

87 Xu, S., Liu, W., Hu, B. et al. (2019). Circuit-integratable high-frequency micro supercapacitors with filter/oscillator demonstrations. *Nano Energy* 58: 803–810.

88 Lu, Y., Jiang, K., Chen, D. et al. (2019). Wearable sweat monitoring system with integrated micro-supercapacitors. *Nano Energy* 58: 624–632.

89 Ai, Y., Lou, Z., Chen, S. et al. (2017). All rGO-on-PVDF-nanofibers based self-powered electronic skins. *Nano Energy* 35: 121–127.

90 Qiu, M., Sun, P., Cui, G. et al. (2019). A flexible microsupercapacitor with integral photocatalytic fuel cell for self-charging. *ACS Nano* 13 (7): 8246–8255.

91 Zhu, M., Huang, Y., Huang, Y. et al. (2017). A highly durable, transferable, and substrate-versatile high-performance all-polymer micro-supercapacitor with plug-and-play function. *Adv. Mater.* 29 (16): 1605137.

92 Song, Y., Wang, H., Cheng, X. et al. (2019). High-efficiency self-charging smart bracelet for portable electronics. *Nano Energy* 55: 29–36.

93 Xu, J. and Shen, G. (2015). A flexible integrated photodetector system driven by on-chip microsupercapacitors. *Nano Energy* 13: 131–139.

94 Yun, J., Lim, Y., Jang, G.N. et al. (2016). Stretchable patterned graphene gas sensor driven by integrated micro-supercapacitor array. *Nano Energy* 19: 401–414.

95 Lin, Y., Chen, J., Tavakoli, M.M. et al. (2019). Printable fabrication of a fully integrated and self-powered sensor system on plastic substrates. *Adv. Mater.* 31 (5): 1804285.

96 Jiang, Q., Wu, C., Wang, Z. et al. (2018). MXene electrochemical microsupercapacitor integrated with triboelectric nanogenerator as a wearable self-charging power unit. *Nano Energy* 45: 266–272.

97 Luo, J., Fan, F.R., Jiang, T. et al. (2015). Integration of micro-supercapacitors with triboelectric nanogenerators for a flexible self-charging power unit. *Nano Res.* 8 (12): 3934–3943.

98 Ye, J., Tan, H., Wu, S. et al. (2018). Direct laser writing of graphene made from chemical vapor deposition for flexible, integratable micro-supercapacitors with ultrahigh power output. *Adv. Mater.* 30 (27): 1801384.

99 Yue, Y., Yang, Z., Liu, N. et al. (2016). A flexible integrated system containing a microsupercapacitor, a photodetector, and a wireless charging coil. *ACS Nano* 10 (12): 11249–11257.

100 Kim, D., Yun, J., Lee, G. et al. (2014). Fabrication of high performance flexible micro-supercapacitor arrays with hybrid electrodes of MWNT/V_2O_5 nanowires integrated with a SnO_2 nanowire UV sensor. *Nanoscale* 6 (20): 12034–12041.

101 Kim, D., Kim, D., Lee, H. et al. (2016). Body-attachable and stretchable multisensors integrated with wirelessly rechargeable energy storage devices. *Adv. Mater.* 28 (4): 748–756.

102 Li, L., Fu, C., Lou, Z. et al. (2017). Flexible planar concentric circular micro-supercapacitor arrays for wearable gas sensing application. *Nano Energy* 41: 261–268.

103 Wang, R., Yao, M., and Niu, Z. (2020). Smart supercapacitors from materials to devices. *InfoMat* 2 (1): 113–125.

104 Zhang, P., Wang, J., Sheng, W. et al. (2018). Thermoswitchable on-chip microsupercapacitors: one potential self-protection solution for electronic devices. *Energy Environ. Sci.* 11 (7): 1717–1722.

105 Huang, S., Wan, F., Bi, S. et al. (2019). A self-healing integrated all-in-one zinc-ion battery. *Angew. Chem. Int. Ed.* 58 (13): 4313–4317.

106 Patil, S. and Lee, D.-W. (2019). Status review on the MEMS-based flexible supercapacitors. *J. Micromech. Microeng.* 29 (9): 093001.

107 Zhang, P., Wang, F., Yang, S. et al. (2020). Flexible in-plane micro-supercapacitors: progresses and challenges in fabrication and applications. *Energy Storage Mater.* 28: 160–187.

108 Yu, C., An, J., Chen, Q. et al. (2020). Recent advances in design of flexible electrodes for miniaturized supercapacitors. *Small Methods* 4 (6): 1900824.

Index

Flexible Supercapacitors: Materials and Applications, First Edition. Edited by Guozhen Shen, Zheng Lou and Di Chen.

d

e

n

o

V

W

Z

Printed and bound by CPI Group (UK) Ltd, Croydon, CR0 4YY
17/09/2026
14964322-0004